Composite Materials,

Volume II:

Processing, Fabrication,

and Applications

Composite Materials, Volume II:

Processing, Fabrication, and Applications

MEL M. SCHWARTZ

To join a Prentice Hall PTR internet mailing list,
point to: http://www.prenhall.com/register

Prentice Hall PTR, Upper Saddle River, New Jersey 07458
http://www.prenhall.com

Library of Congress Cataloging-in-Publication Data

Schwartz, Mel M.
    Composite materials / Mel Schwartz.
        p.   cm.
    Includes bibliographical references and index.
    Contents: v.   1. Properties, nondestructive testing, and repair.
    ISBN 0-13-300039-7
    1. Composite materials.   I. Title.
TA418.9.C6S37   1996
620.1′18—dc20   96-28559
                                                               CIP

Acquisitions editor: *Bernard M. Goodwin*
Editorial/production supervision
    and interior design: *bookworks*
Manufacturing manager: *Alexis Heydt*
Cover design: *Scott Weiss*
Cover design director: *Jerry Votta*

The background material on the cover is a reproduction of a photograph of Kevlar, a fiber produced by DuPont which typifies raw materials used to fabricate composite parts. The Kevlar displayed here is used by Sikorsky Aircraft in the manufacturing of helicopter parts. Photograph courtesy of Sikorsky Aircarft.

 © 1997 by Prentice Hall PTR
Prentice-Hall, Inc.
A Simon & Schuster Company
Upper Saddle River, New Jersey 07458

The publisher offers discounts on this book when ordered
in bulk quantities. For more information, contact:

    Corporate Sales Department
    PTR Prentice Hall
    One Lake St.
    Upper Saddle River, NJ 07458

    Phone: 800-382-3419
    FAX: 201-236-7141
    E-mail: corpsales@prenhall.com

Printed in the United States of America
10 9 8 7 6 5 4 3 2 1

ISBN 0-13-300039-7

Prentice-Hall International (UK) Limited, *London*
Prentice-Hall of Australia Pty. Limited, *Sydney*
Prentice-Hall Canada Inc., *Toronto*
Prentice-Hall Hispanoamericana, S.A., *Mexico*
Prentice-Hall of India Private Limited, *New Delhi*
Prentice-Hall of Japan, Inc., *Tokyo*
Simon & Schuster, Asia Pte. Ltd., *Singapore*
Editora Prentice-Hall do Brasil, Ltda., *Rio de Janeiro*

To Mother and Dad, whose emotional strength,
free-flowing love, and caring concern
have helped shape my maturation.

Contents

Preface

This second volume on composite material fabrication, processing, and future reinforced composite material systems seeks to cover this vast field of materials and engineering. It starts by considering the various reinforcements, the matrices, and the processes involved in carrying out the novel methods of incorporating the matrix into fiber systems. For example, vapor phase reaction techniques are being used to permeate carbon fiber structures with silicon carbide in order to avoid a high-temperature sintering step that may degrade the system. Reinforcements, matrices, and other processing techniques are also covered for polymer, metal, and ceramic systems.

Postprocessing and manufacturing are one of the most significant areas of composite fabrication. The various joining, machining, forming, drilling, cutting and finishing methods are described as well as the different types of composite families.

Additionally, assembly techniques are discussed, such as cocuring an integral tank as part of cocuring a whole aircraft wing, which excites manufacturers. In fact, maximum use of this approach could halve both part count and assembly time and reduce fastener use 67 percent in future fighter aircraft.

Finally, what's ahead for the composite material systems—functionally gradient materials, intermetallic composites, interpenetrating polymer networks, liquid crystal polymers, nanocomposites—and applications of composite materials and structures is considered.

New synthesis methods may be required for these new composite materials and may force the development of previously uneconomical process routes if they can offer the path to a technical solution for advanced system capability. For example, intermetallic materials present opportunities for reducing the number of stages in turbine engines and in so doing may be economically beneficial even at a higher materials cost because the smaller number of stages leads to greater economy of use.

To fabricate these future composite systems of extreme conceptual size, nanomachines or "molecular assemblers" will be used with tiny robotic arms that can seize molecules in solution, placing them in the desired positions. Some critics believe that nanotechnology is science fiction and "too far out." Others think the concept not to be feasible because thermal vibrations, quantum mechanical effects, and radiation will prevent accurate positioning. However, after reading and digesting Chapters 1–3, you can be the judge.

Mel M. Schwartz

Composite Materials, Volume II: Processing, Fabrication, and Applications

1

Processing of Composite Materials

1.0 INTRODUCTION

Volume 1 covered the various types of families of composite materials:

1. Polymer (organic) matrix composites (PMCs) and glass-ceramics
2. Metal matrix composites (MMCs)
3. Carbon-carbon composites (CCCs)
4. Ceramic matrix composites (CMCs)
5. Intermetallic matrix composites (IMCs)

and the reinforcing agents:

1. Short and chopped fibers [glass, (Gl), graphite (Gr)]
2. Whiskers (SiC, Si_3N_4)
3. Particulates (SiC, Al_2O_3)
4. Plastic, metallic, and ceramic fibers [polyacrylonitrile (PAN), aramid, Al_2O_3, SiC, glass, pitch, rayon, etc.].

Also described were the plastic, metallic, and ceramic materials used as binding agents [resins, epoxies, polyether ether ketone (PEEK), aluminum and titanium alloys, ZrO_2, Al_2O_3, Si_3N_4, etc.] (Table 1.1).

The main concern of designers, engineers, and composite manufacturers and suppliers is not the cost of the materials listed above but the processing and manufacturing

TABLE 1.1.   Matrix Materials

Polymeric
   Thermoset polymers (resins)
      Epoxies: principally used in aerospace and aircraft applications
      Polyester, vinyl esters: commonly used in automotive, marine, chemical, and electrical applications
      Phenolics: used in bulk molding compounds
      Polyimides, polybenzimidazoles (PBI), polyphenylquinoxaline (PPQ): for high-temperature aerospace
         applications (temperature range 250–400°C)
   Thermoplastic polymers
      Nylons (such as nylon 6, nylon 6,6), thermoplastic polyesters (such as PET, PBT), polycarbonate (PC),
         polyacetals: used with discontinuous fibers in injection-molded articles
      Polyamide-imide (PAI), polyether ether ketone (PEEK), polysulfone (PSUL), polyphenylene sulfide
         (PPS), polyether imide (PEI): suitable for moderately high-temperature applications with continuous
         fibers
Metallic
   Aluminum and its alloys, titanium alloys, magnesium alloys, copper-based alloys, nickel-based superalloys,
      stainless steel: suitable for high-temperature applications (temperature range 300–500°C)
Ceramic
   Aluminum oxide (Al_2O_3), carbon, silicon carbide (SiC), silicon nitride (Si_3N_4): suitable for high-temperature
      applications

costs, the fabrication techniques and methods being economical (Chapter 2), and as a result being able to put a competitive product into the application marketplace.

1.1 FIBERS

Before cost-effective composites can be realized, the low-cost manufacture of reinforcements must be accomplished. This can be done by identifying the critical variables controlling the process economics for fabrication of the various types of reinforcements.

1.1.1 Glass Fiber

Glass fiber is an inorganic, synthetic, multifilament material. Glass fibers are the most common of all reinforcing fibers for polymeric (plastic) matrix composites. Glass fiber composites are strong, low in cost, nonflammable, nonconductive (electrically), and corrosion-resistant. The disadvantages are low tensile modulus, relatively high specific gravity (among the commercial fibers), sensitivity to abrasion with handling (which frequently decreases tensile strength), relatively low fatigue resistance, and high hardness (which causes excessive wear on molding dies and cutting tools).

    Glass fiber is made by melting the raw materials in a high-temperature furnace and then drawing the molten material into filaments. It is a capital-intensive process. There are two main categories of glass fibers: E-glass and high-strength glass. High-strength glass can be further subdivided into the following categories: S-glass, S-2 glass, and S-2 hollow glass fiber.

    Another type, known as C-glass, is used in chemical applications requiring greater corrosion resistance to acids than is provided by E-glass. E-glass has the lowest cost of all commercially available reinforcing fibers, which is the reason for its widespread use in the fiber-reinforced plastics (FRP) industry. S-glass, originally developed for aircraft components and missile casings, has the highest tensile strength among all fibers in use. However, the compositional difference and higher manufacturing cost make it more expensive than E-glass. The lower-cost version of S-glass, S-2 glass, has been made available in recent years. Although S-2 glass is manufactured with less stringent nonmilitary specifications, its tensile strength and modulus are similar to those of S-glass. High-strength S-glass fibers have a specially formulated glass composition that yields physical properties superior of those of E-glass and competes with aramid fibers in markets such as aerospace and ballistic armor. Impregnated strand tensile strengths of high-strength glass are approximately twice that of E-glass and equal to that of aramid fibers. The modulus of high-strength glass is 25% greater than that of E-glass, and high-strength glass performs at temperatures of about 149°C, or about 15% higher than E-glass.[1]

Glass fibers are amorphous (noncrystalline) and isotropic (equal properties in all directions) and are a long, three-dimensional network of silicon, oxygen, and other atoms arranged in a random fashion.

The manufacturing process for glass fibers is depicted in the flow diagram in Figure 1.1. Various ingredients in the glass formulation are first dry-mixed and melted in a refractory furnace at about 1370°C. The molten glass is extruded through a number of orifices contained in a platinum bushing and rapidly drawn into filaments of approximately 10-μm diameter. A protective coating (*size*) is then applied on individual filaments before they are gathered together into a strand and wound on a drum. The size is a mixture of lubricants, antistatic agents, and a binder. The binder packs the filaments together into a strand. The size may also contain small percentages of a coupling agent that promotes adhesion between fibers and the specific matrix for which it is formulated.

The basic commercial form of continuous glass fibers is a strand, which is a collection of parallel filaments numbering 204 or more. A *roving* is a group of untwisted parallel strands (*end*) wound in a cylindrical *forming package*. Rovings are used in continuous molding operations such as filament winding and pultrusion. They can also be preimpregnated with a thin layer of polymeric resin matrix to form *prepregs*. Prepregs are subsequently cut into required dimensions, stacked, and cured into the final shape in batch molding operations such as compression molding and hand layup molding.

Chopped strands are produced by cutting continuous strands into short lengths. The ability of the individual filaments to hold together during or after the chopping process depends largely on the type and amount of size applied during the fiber manufacturing operation.

1.1.2 Aramid (Para-aramid) Fiber

Para-aramid is the precise generic term for material known as aramid. These fibers possess a tensile strength seven to eight times that of steel wire and have an excellent stability against temperature change. Therefore, para-aramid fibers are widely used in the automotive industry for tire cords, timing belts, and brake friction materials. In addition, these fibers have been used as concrete reinforcing materials and in high-performance ropes.

Kevlar 49 belongs to a group of highly crystalline aramid (aromatic polyamide) fibers that have the lowest specific gravity and the highest tensile strength-to-weight ratio among current reinforcing fibers. As a reinforcement, aramid fibers have been used in many marine and aerospace applications where light weight, high tensile strength, and resistance to impact damage (e.g., caused by accidentally dropping a hand tool) are impor-

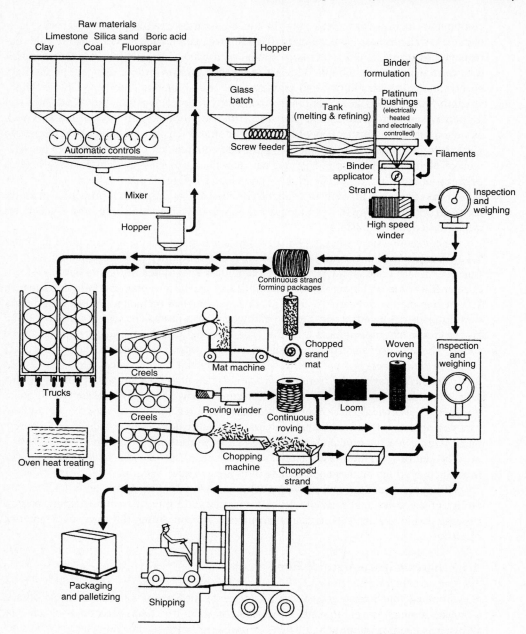

Figure 1.1 Flow diagram for glass fiber manufacturing. (Courtesy of PPG Industries.)

tant. Like carbon fibers, they also have a negative coefficient of thermal expansion (CTE) in the longitudinal direction, which is utilized in designing low-thermal-expansion composite printed circuit boards. The major disadvantages of aramid fiber-reinforced composites are their low compressive strengths and the difficulty in cutting and machining them.

Kevlar 49 filaments are manufactured by extruding an acidic solution of a proprietary precursor (a polycondensation product of terephthaloyol chloride and *p*-phenylene-diamine) from a spinneret. During the filament-drawing process, Kevlar 49 molecules become highly oriented in the direction of the filament axis. Weak hydrogen bonds between hydrogen and oxygen atoms in adjacent molecules hold them together in the transverse direction. The resulting filament is highly anisotropic, with much better physical and mechanical properties in the longitudinal direction than in the radial direction.

Kevlar 49 fibers are commercially available as untwisted yarns (with 134, 267, 768, and 1000 filaments per yarn), roving (3072 and 5000 filaments per roving), and fabrics.

A second-generation Kevlar fiber is Kevlar 149, which has the highest tensile modulus of all commercially available aramid fibers. The tensile modulus of Kevlar 149 is 40% higher than that of Kevlar 49; however, its strain to failure is lower. Kevlar 149 has an equilibrium moisture content of 1.2% at 65% relative humidity and 22°C, which is nearly 70% lower that of Kevlar 49 under similar conditions. Kevlar 149 also has a lower creep rate than Kevlar 49.

1.1.3 Graphite (Carbon) Fiber

Carbon fibers are commercially available with a variety of tensile moduli ranging from 207 GPa on the low side to 1035 GPa on the high side. In general, the low-modulus fibers have lower specific gravities, lower cost, higher tensile and compressive strengths, and higher tensile strains to failure than the high-modulus fibers. Among the advantages of carbon fibers are their exceptionally high tensile strength-to-weight ratios as well as their tensile modulus-to-weight ratios, very low CTE (which provides dimensional stability in such applications as space antennas), and high fatigue strengths. The disadvantages are their low impact resistance and high electric conductivity, which may cause "shorting" in unprotected electrical machinery. Their high cost has so far excluded them from widespread commercial applications. They are used mostly in the aerospace industry, where weight savings is considered more critical than cost.

Structurally, carbon fibers contain a blend of amorphous carbon and graphitic carbon. Their high tensile modulus results from the graphitic form in which carbon atoms are arranged in crystallographically parallel planes of regular hexagons.

Carbon fibers are manufactured from two types of precursors (starting materials), namely, textile precursors and pitch precursors. The manufacturing process from both precursors is outlined in Figure 1.2. The most common textile precursor is polyacrylonitrile. Filaments are wet-spun from a solution of PAN and stretched at an elevated temperature during which the polymer chains are aligned in the filament direction. The stretched filaments are then heated in air at 220°C for a few hours. In the next step, PAN filaments are carbonized by heating at a controlled rate at 1000°C in an inert atmosphere. With the elimination of oxygen and nitrogen atoms, the filaments now contain mostly carbon atoms arranged in aromatic ring patterns in parallel planes.

As the carbonized filaments are subsequently heat-treated at or above 2000°C, their structure becomes more ordered and turns toward a true graphitic form with increasing heat treatment temperature. The graphitized filaments have attained a high tensile modulus, but their tensile strength may be relatively low. Their tensile strength can be increased by hot stretching them above 2000°C, during which the graphitic planes align in the filament direction.

Pitch, a by-product of petroleum refining or coal coking, is a lower-cost precursor

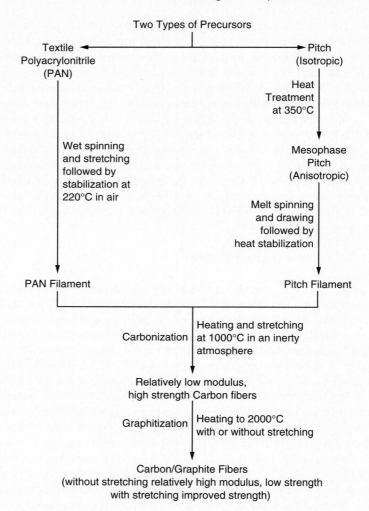

Figure 1.2 Flow diagram for carbon manufacturing.

than PAN. The carbon atoms in pitch are arranged in low-molecular-weight aromatic ring patterns. Heating to temperatures above 300°C polymerizes (joins) these molecules into long, two-dimensional, sheetlike structures. The highly viscous state of pitch at this stage is referred to as *mesophase*. Pitch filaments are produced by melt spinning mesophase through a spinneret (Figure 1.3). While passing through the spinneret die, the mesophase pitch molecules become aligned in the filament direction. The filaments are cooled to freeze the molecular orientation and subsequently heated between 250 and 400°C in an oxygen-containing atmosphere to stabilize them as well as to make them infusible (to avoid fusing the filaments together). In the next step, the filaments are carburized at temperatures around 2000°C. The rest of the process of transforming the structure to graphitic form is similar to that followed for PAN precursors.

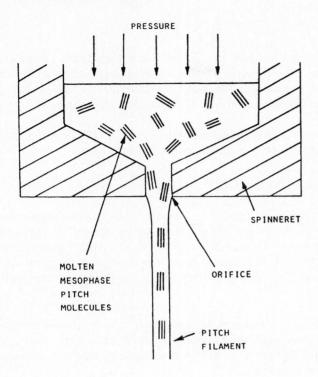

PRESSURE

SPINNERET

MOLTEN
MESOPHASE
PITCH
MOLECULES

ORIFICE

PITCH
FILAMENT

Figure 1.3   Alignment of mesophase pitch in a pitch filament.

PAN carbon fibers are generally categorized into high-strength, high-modulus, and ultrahigh-modulus types. The high-strength PAN carbon fibers, such as T-300 and AS-4, have the lowest modulus, while the ultrahigh-modulus PAN carbon fibers, such as GY-70, have the lowest tensile strength as well as the lowest tensile strain to failure. Recently, a number of intermediate-modulus, high-strength PAN carbon fibers, such as T-40 and IM-7, have been developed that also possess the highest strain to failure. Another point to note is that the pitch carbon fibers have very high modulus values, but their tensile strength and strain to failure are lower than those of the PAN carbon fibers.

Carbon fibers are commercially available in three basic forms, namely, long and continuous tow, chopped (6–50 mm long), and milled (30–3000 μm long). The long and continuous tow, which is simply a bundle of 1000–160,000 parallel filaments, is used for high-performance applications. The price of carbon fiber tow decreases with increasing filament count. Although high filament counts are desirable for improving productivity in continuous molding operations such as filament winding and pultrusion, it becomes increasingly difficult to wet them with the matrix. Dry filaments are not conducive to good mechanical properties.

Carbon fiber tows can also be woven into two-dimensional (2D) fabrics of various styles. Hybrid fabrics containing commingled or cowoven carbon and other fibers, such as E-glass, Kevlar, PEEK, and polyphenylene sulfide (PPS), are also available. Techniques of forming three-dimensional (3D) weaves with fibers running in the thickness direction have also been developed.

1.1.4 Boron Fiber

The most prominent feature of boron fibers is their extremely high tensile modulus, which is in the range 379–414 GPa. Because of their relatively large diameter, boron fibers offer excellent resistance to buckling, which in turn contributes to high compressive strength for boron fiber-reinforced composites. The principal disadvantage of boron is its high cost.

Boron fibers are manufactured by chemical vapor deposition (CVD) of boron onto a heated substrate (either a tungsten wire or a carbon monofilament).

1.1.5 Ceramic Fibers—SiC and Al_2O_3

SiC and Al_2O_3 fibers are examples of ceramic fibers notable for their high-temperature applications in metal and ceramic matrix composites. Their melting points are 2830 and 2045°C, respectively. SiC retains its strength well above 650°C, and Al_2O_3 has excellent strength retention up to about 1370°C. Both fibers are suitable for reinforcing metal matrices in which carbon and boron fibers exhibit adverse reactivities. SiC fibers are available in three different forms:

1. Monofilaments are produced by CVD of β-SiC on a 10- to 25-μm-diameter carbon monofilament substrate. The carbon monofilament is coated with ~1-μm-thick pyrolytic graphite to smoothe its surface as well as to enhance thermal conductivity. The average fiber diameter is 140 μm.

2. Multifilament yarns are produced by melt spinning of a polymeric precursor, such as polycarbosilane at 350°C in nitrogen gas. The average fiber diameter in the yarn is 14.5 μm, and a commercial yarn contains 500 fibers. Yarn fibers have a considerably lower strength than monofilaments.

3. Whiskers, which are 0.1–1 μm in diameter and about 50 μm in length, are produced from rice hulls containing 10–20 wt% SiO_2. The rice hulls are first heated in an oxygen-free atmosphere to 700–900°C to remove the volatiles and then to 1500–1600°C for 1 h to produce SiC whiskers (SiC_w). The resulting SiC_w contain 10 wt% of SiO_2 and up to 10 wt% Si_3N_4. The tensile strength and modulus of these whiskers are reported as 13 and 700 GPa, respectively.

Continuous multifilament Al_2O_3 yarn has been available from DuPont under the trade name Fiber FP.* It is a high-purity (>99%) polycrystalline α-Al_2O_3 fiber dry spun from a slurry mix of alumina and proprietary spinning additives. The fired filaments may be coated with a thin layer of silica to improve their strength (by healing the surface flaws) as well as their wettability with the matrix. The filament diameter is 20 μm, and there are 210 filaments in the yarn. Experiments have shown that Fiber FP retains almost 100% of its room-temperature tensile strength after 300 h of exposure in air at 1000°C.

*Production of Fiber FP has recently been discontinued.

1.2 MATRICES

The role of the matrix in a fiber-reinforced composite is (1) to transfer stresses between the fibers, (2) to provide a barrier against an adverse environment, and (3) to protect the surface of the fibers from mechanical abrasion. The matrix plays a minor role in the tensile load-carrying capacity of a composite structure. However, selection of a matrix has a major influence on the interlaminar shear as well as in-plane shear properties of the composite material. The interlaminar shear strength is an important design consideration for structures under bending loads, whereas the in-plane shear strength is important for structures under torsional loads. The matrix provides lateral support against the possibility of fiber buckling under compression loading, thus influencing to some extent the compressive strength of the composite material. The interaction between fibers and matrix is also important in designing damage-tolerant structures.

1.2.1 Polymeric Matrices

A polymer is defined as a long-chain molecule containing one or more repeating units of atoms joined together by strong covalent bonds. A polymeric material (commonly called a *plastic*) is a collection of a large number of polymer molecules of similar chemical structure (but not of equal length). In the solid state, these molecules are frozen in space either in a random fashion (for amorphous polymers) or in a mixture of random and orderly (folded) fashions.

1.2.1.1 Thermoplastic (TP) and thermoset (TS) polymers.   Polymers are divided into two broad categories: thermoplastics and thermosets. In a thermoplastic polymer, individual molecules are linear in structure with no chemical linking between them. They are held in place by weak secondary bonds (intermolecular forces) such as van der Waals forces and hydrogen bonds. With the application of heat and pressure, these intermolecular bonds in a solid thermoplastic polymer can be temporarily broken and the molecules can be moved relative to each other to flow into new positions. Upon cooling, the molecules freeze in their new positions, restoring the secondary bonds between them and resulting in a new solid shape. Thus a thermoplastic polymer can be heat-softened, melted, and reshaped (postformed) as many times as desired.

In a thermoset polymer, on the other hand, the molecules are chemically joined together by cross-links, forming a rigid, three-dimensional network structure. Once these cross-links are formed during the polymerization reaction (also called the curing reaction), the thermoset polymer cannot be melted and reshaped (postformed) by the application of heat and pressure. However, if the number (frequency) of cross-links is low, it may be possible to soften it at an elevated temperature.

1.2.1.2 Thermoplastic versus thermoset.   The primary consideration in the selection of a matrix is its basic mechanical properties which include tensile modulus, tensile strength, and fracture toughness.

Traditionally, thermoset polymers (also called *resins*) have been used as a matrix material for fiber-reinforced composites. Starting materials used in the polymerization of thermoset polymers are usually low-molecular-weight liquid chemicals with very low viscosities. Fibers are either pulled through or immersed in these chemicals before the polymerization reaction begins. Since the viscosity of the polymer at the time of fiber incorporation is very low, it is possible to achieve a good wet-out between the fibers and the matrix without the aid of high temperature or pressure.

The most important advantage of thermoplastic polymers over thermoset polymers is their high impact strength and fracture resistance, which in turn impart excellent damage tolerance characteristics to the composite material. In general, thermoplastic polymers have higher strains to failure than thermoset polymers, which may provide a better resistance to matrix microcracking in the composite laminate.

Before carbon fiber (or any other fiber) is used to manufacture a composite, it has to be made into an item known as a *prepreg*. A prepreg consists of bundles of fibers (typically containing 1200, 2400, or a greater number of individual strands of carbon fiber), which are carefully arranged by tensioning rollers and fed into tanks containing resin which coats the fiber to form the first step in composite assembly. Resins comprise 40% of the volume of a prepreg, and fiber comprises the other 60%. Carbon fiber can also be used in the dry form with wet resin in the filament winding and resin transfer molding (RTM) processes. These processes are discussed further later in this chapter.

There are 10 broad types of base resins for advanced composite matrices: (1) bismaleimide (BMI), (2) epoxy (Ep), (3) phenolic, (4) polyester, (5) polyether sulfone (PES), (6) polyether ether ketone, (7) polyamide (PA), (8) polyphenylene sulfide, (9) polyamide-imide (PAI), and (10) vinylester phenol.

1.2.2 Metal Matrix Composites

The MMCs are materials consisting of metal alloys reinforced with continuous fibers, particulates, or whiskers. The addition of these reinforcements gives MMCs superior mechanical properties and unique physical characteristics which can be tailored to fit a variety of applications. Because of their ability to provide the needed strength at the lowest weight and least volume, they are attractive for structural and nonstructural applications.

The two most commonly used metal matrices are based on aluminum and titanium. Both of these metals have comparatively low specific gravities and are available in a variety of alloy forms. Although magnesium is even lighter, its great affinity for oxygen promotes atmospheric corrosion and makes it less suitable for many applications. Beryllium is the lightest of all structural metals and has a tensile modulus higher than that of steel. However, it suffers from extreme brittleness, which is the reason for its exclusion as a potential matrix material. Nickel- and cobalt-based superalloys have also been used as matrices, but the alloying elements in these materials tend to accentuate the oxidation of fibers at elevated temperatures.

Aluminum and its alloys have attracted the most attention as matrix material for MMCs. Commercially, pure aluminum has been used because of its good corrosion resistance. Aluminum alloys, such as 6061, 1100, and 201, have been used because of their higher tensile strength-to-weight ratios. Carbon fiber is used with aluminum alloys; however, at typical fabrication temperatures of 500°C or higher, carbon reacts with aluminum to form Al_4C_3, which severely degrades the mechanical properties of the composite. Protective coatings of TiB_2 are added to carbon fibers to reduce the problem of fiber degradation as well as to improve their wetting with the aluminum alloy matrix. Carbon fiber-reinforced aluminum composites are inherently prone to galvanic corrosion, and therefore a more common reinforcement for aluminum alloys is SiC.

The titanium alloys that are the most useful in MMCs are α, β alloys (e.g., Ti-6Al-4V) and metastable β alloys (e.g., Ti-10V-2Fe-3Al). These titanium alloys have higher tensile strength-to-weight ratios as well as better strength retentions at 400–500°C than those of aluminum alloys. The CTE of titanium alloys is closer to those for reinforcing fibers, which reduces the thermal mismatch between them. One of the problems with titanium alloys is their high reactivity with boron and Al_2O_3 fibers at normal fabrication temperatures. Borsic (boron fibers coated with SiC) and SiC fibers show less reactivity with titanium. Improved tensile strength retention is obtained by coating boron and SiC fibers with carbon-rich layers.

MMC engine applications are produced and used for automobile engine cylinders die-cast from a carbon fiber-aluminum-Al_2O_3 material.

Titanium MMCs are used in applications where performance is demanded without regard to cost-effectiveness. This is where one obtains high-temperature performance [national aerospace plane (NASP) engine components, etc.] unattainable with conventional materials.

1.2.3 Ceramic Matrix Composites

Ceramic fibers such as SiC and Si_3N_4 use polysilane as the base material. CMCs, in which ceramic or glass matrices are reinforced with continuous fibers, chopped fibers, whiskers, platelets, or particulates, are emerging as a class of advanced engineering structural materials. They currently have limited high-temperature applications but a large potential for much wider use in military, aerospace, and commercial applications such as energy-efficient systems and transportation.

There are also other specialty CMCs such as nanocomposites (made from reactive powders) and electroceramics. CMCs are unique in that they combine low density with high modulus, strength, and toughness (contrasted with monolithic ceramics) and strength retention at high temperatures. Many have good corrosion and erosion characteristics for high-temperature applications. CMCs have been used in jet fighters (the Mirage series). Industrial uses of CMCs include furnace materials, energy conversion systems, gas turbines, and heat engines.

1.3 FABRICATION PROCESSES FOR POLYMERIC MATRICES

Processes for incorporating reinforcements into a polymeric matrix can be divided into two categories. In one category, the continuous and discontinuous fibers and matrix are processed directly into the finished structure. Examples of such processes are filament winding and pultrusion. In the second category, the reinforcements are incorporated into the matrix to prepare ready-to-mold sheets that can be stored and later processed to form laminated structures by autoclave molding and compression molding (Table 1.2).

Ready-to-mold fiber-reinforced polymer sheets are available in two basic forms, prepregs and sheet molding compounds (SMCs).

TABLE 1.2. Plastics Fabrication Technologies[1]

Process	Predominate Polymer Processed	Material Type or Form	Shape Characteristics and Common Products	Process Description
Bag molding	Reinforced thermoset	Prepregs (partly cured), epoxy or polyester-soaked sheets	Simple, contoured, large	A number of layers of reinforced material are placed (hand layup) between two inexpensive mold halves; the entire assembly is placed in a plastic bag and low pressure applied (via vacuum, external air, or autoclave methods); the assembly is heated and cured.
Centrifugal casting	Thermoset	Liquid resin, fiber reinforcement	Uniform wall thickness	Resin is applied to the inside of a rotating cylindrical mold and uniformly forced against the reinforcing material. Heating aids flowing and curing.
Continuous laminating	Polyester	Resin, reinforcement	Thin, flat, or curved profiles	Reinforcement is passed through a resin bath, faced with cellophane, passed through squeeze rolls, and heated.
Filament winding	Epoxy, polyester	Resin	Round, rigid	Continuous filaments (usually glass) in the form of roving are saturated with resin and machine-wound onto mandrels having the shape of the desired finished part. Once winding is completed, part and mandrel are placed in an oven for curing. Mandrel is then removed through porthole at the end of the wound part.
Laminates high pressure—industrial and decorative	Thermoset	Reinforcement fiber sheet and prepregs, liquid resin	Thick-walled, simple shapes	Reinforcing material (layers) is pre-impregnated or soaked with resin (sometimes a thermoplastic) and compressed under pressure over 1000 psi into a flat sheet. Tube and rod shapes are also made from rolled or closed mold methods. Prepregs are semicured reinforced materials used to make the above forms. Decorative laminates may have a separate patterned face sheet laminated to the structure.
Match die molding (low pressure)	Thermoset-epoxy, polyester, others	Reinforced plastic compounds (SMC, BMC, and variations) or resin and reinforcing material	Contoured, medium-wall thickness	Either preforms (chopped-glass fibers and resin shape) or a premix molding compound (resin and short glass fiber strands) are placed in a mold, heated, and pressed to shape.

Process	Predominate Polymer Processed	Material Type or Form	Shape Characteristics and Common Products	Process Description
Open mold (contact molding)	Thermoset (primarily reinforced plastics) forming process 40% of total.	Liquid resin (spray, hand-applied) or thermoplastic sheet; reinforcing material	Large, complex	Reinforced shapes are formed by using open molds and room-temperature curing resins. Hand layup process: resin is brushed, poured, or sprayed onto reinforcing material laid in mold. Spray-up method: resin and chopped fibers sprayed onto mold. Reinforced vacuum-formed sheet method: mainly acrylic sheet is vacuum-formed and reinforced resin applied to back surface. The process also includes encapsulating or potting.
Pultrusion	Thermoset polyesters, some epoxies, and others	Liquid resins, fiber reinforcement	Uniform cross section	Resin-swelled fibers are pulled through a heated shaping die, which initiates cure in the thermosetting resin and forms the shape of the continuous length of the pultruded product.
Reaction injection molding	Thermoset (urethanes)	Liquid	Large, intricate, high-performance, solid, or cellular	Reactive components are generally mixed by impingement in a chamber and then injected into a closed mold.
Rotational molding	Thermoplastic	Powder, liquid, precatalyzed	Hollow bodies, complex	Premeasured material is poured into a mold. The mold is closed, heated, and rotated in the axis of two planes until the contents have fused to the inner walls of the mold. The mold is opened and the part removed.
Thermoforming	Thermoplastic	Film sheet	Simple, thin-walled	Film is continuously roll-fed or sheet is cut and placed over either a male or a female mold. The part is formed by vacuum forming, draping and heating, pressing, or many other variations.
Transfer molding	Thermoset	Pellet (compound)	Simple configurations	The material is placed and heated in a transfer chamber. It is then fed by means of a plunger into a closed mold, heated, and ejected.

1.3.1 Prepregs

Prepregs are thin sheets of fiber impregnated with predetermined amounts of uniformly distributed polymeric matrix. Fibers may be in the form of continuous rovings, mat, or woven fabric. Epoxy is the primary matrix matrial in prepreg sheets, although other thermoset and thermoplastic polymers have also been used. The width of prepreg sheets may vary from < 25 mm to >457 mm. Sheets wider than 457 mm are called broad goods. The thickness of a ply cured from prepregs is normally in the range 0.13–0.25 mm. Resin con-

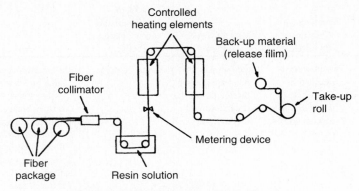

Figure 1.4 Schematic of prepreg manufacturing.

tent in commercially available prepregs is between 30 and 45% by weight. A typical set of steps in prepreg manufacturing is shown in Figure 1.4.

1.3.2 Reactive Processing

Reactive processing, whereby the final shape and molecular structure are achieved simultaneously, is among the oldest polymer processing technologies. The scope of reactive processing (RP) has evolved steadily to encompass a broad range of forming methods: compression molding of prepreg composites and sheet and bulk molding compounds, injection molding of bulk molding compounds (BMCs), pultrusion and filament winding of fibers wet with liquid resins, and various liquid injection molding techniques. The processing of reactive liquids to generate composites has seen a decade of feverish research and development and includes the following methods.

1. *Reinforced reaction injection molding* (RRIM). The extremely rapid impingement mixing of reactive liquid streams, containing reinforcements and fillers, which are then injected into a closed mold. Polymerization, cross-linking and part formation occur simultaneously.

2. *Resin transfer molding.* A method characterized by placement of dry reinforcement in the mold before the mold is closed and resin admitted. In most cases, a low-viscosity resin must be used and either low pressure or low vacuum employed to assist resin flow and wet-out of the reinforcement.

3. *Structural reaction injection molding* (SRIM). A fast RTM method that exploits the productivity advantages of standard reaction injection molding (RIM) machines to achieve rapid impingement mixing and injection into the mold chamber containing the preplaced reinforcement.

1.3.2.1 Sheet molding compounds[2]   Sheet molding compounds are thin sheets of fiber precompounded with a thermoset resin and are employed primarily in compression molding processes. Common thermoset resins used for SMC sheets are polyesters and vinyl esters.

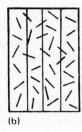

(a)   (b)   (c)

Figure 1.5   Various types of sheet molding compounds. (*a*) SMC-R. (*b*) SMC-CR. (*c*) XMC.

There are three primary types of SMCs:

1. SMC-R, containing randomly oriented discontinuous fibers. Nominal fiber content (by weight percent) is usually indicated by a two-digit number after the letter *R*. For example, the fiber common in SMC-R30 is 30% by weight.

2. SMC-CR, containing a layer of unidirectional continuous fibers on top of a layer of randomly oriented discontinuous fibers. Nominal fiber contents are usually indicated by two-digit numbers after the letters *C* and *R*. For example, the nominal fiber contents of SMC-C40 R30 are 40% by weight unidirectional continuous fibers and 30% by weight random discontinuous fibers.

3. XMC (a trademark of PPG Industries), containing continuous fibers arranged in an X pattern, where the angle between the interlaced fibers is between 5 and 7°. Additionally, it may also contain randomly oriented discontinuous fibers interspersed with the continuous fibers (Figure 1.5).

SMC-R and SMC-CR sheets are manufactured on a SMC machine (Figure 1.6), whereas the XMC sheets are manufactured by a filament winding process in which continuous strand rovings are pulled through a tank of resin paste and wound under tension around a rotating cylindrical drum.

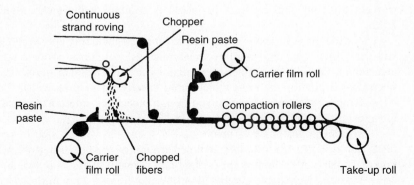

Figure 1.6   Schematic of a sheet molding compounding operation.

While SMC has been extremely successful in a remarkably broad range of applications, as in automobile hoods, the process possesses certain inherent limitations. Compression molding of these compounds involves relatively high molding pressures, which result in high machine costs and place a practical limit on part size. Then too, as normally practiced, SMC requires considerable flow to achieve optimum mechanical and surface properties. This flow in turn results in complex and difficult-to-control fiber orientations and concentrations with a considerable distribution of properties at different locations. It is not surprising that a broad range of issues have been under intense investigation in the past few years, including control of composition uniformity, control of flow and shrinkage, the origin of surface imperfections, the influence of pressure distribution on the curing process, the rheology of reactive systems, and so on.[3] With increased knowledge and faster curing chemistries, the composite manufacturing system should continue to grow throughout the 1990s to compete with the other evolving methods.

1.3.2.2 Reaction injection molding. RIM is generally considered an energy-efficient process since it involves the metering of relatively low-viscosity reactants and a moderate mold temperature of roughly 65°C for most exothermic systems. Polyurethane-based polymers have been highly successful and widely used. The need for higher moduli and reduced thermal expansion has dictated the use of various mineral fillers (e.g., milled glass fibers, mica, flake glass, glass beads, wollastonite, and clay). However, the use of fibrous, flake, and particulate reinforcements in RRIM increases the difficulty of pumping the slurries, leading as well to more rapid machine wear and to fiber orientation, with resultant part anisotropy.

1.3.2.3 Resin transfer molding and structural reaction injection molding. RTM, with its ability in principle to achieve precise control in the placement of fibers in high concentrations combined with rapid processability, should satisfy many of the fundamental requirements for meeting the economic needs of a mass production industry. Also, the capability of producing large, integrated sections and complex part geometries appears to make this an ideal composites fabricating method provided the part production time can be shortened to a time scale of minutes or less and variable costs can be kept at acceptable levels. Various techniques are potentially available to meet this end, not the least of which is combining the technologies of RTM and RIM.

A wide range of resins are available for RTM including epoxy, modified polyesters, and vinyl ester, as well as reinforcements such as continuous-filament random glass mats and woven fabrics. Most fiber constructions are available in glass, aramid, or carbon fibers; but E-glass is the most widely used because of its price-versus-performance balance. To meet various application and design objectives, combinations of reinforcements are generally needed.

The development issues involving RTM apply as well to SRIM. Additionally, high-reactivity systems, fast fill, complete penetration of the preform, good wet-out, elimination of voids, lack of movement of the reinforcement fibers, and avoidance of resin-rich areas present a formidable combination of requirements. Low initial viscosity and a precisely controlled gel time followed by a rapid cure are essential features of these systems. RTM is discussed further later in this chapter.

1.3.3 Molding

Although, as we shall see shortly, there are many ingenious variations on the theme of molding, the basic steps are simply as follows.

1. The entire surface of the reinforcement must be coated with matrix. A number of procedures are available, including soaking, brushing, or spraying, either with neat resin or a solution of polymer. Sometimes one can coat the polymer onto a film of release agent and transfer it to a reinforcement by passing both between heated rollers.
2. The fibers must be placed at their correct ratios in appropriate directions to resist the worst stresses applied to the component. The orientation of the fibers to withstand stress must be taken seriously.
3. The completed layup is shaped by forcing it against the surface of a suitable mold. When an accurate external contour is needed, a female mold is used; where accurate internal dimensions are required, the mold is male.
4. After a period of flow, the matrix is caused to harden or *cure* and the composite is brought to its final shape. Cure is brought about by heat and pressure or by appropriate chemical agents which cause chemical cross-linking. In the case of thermoplastics, once heat and pressure have caused sufficient flow, cooling alone is enough to harden the matrix.

1.3.3.1 Forms of molding.   In any molding process, it is an advantage to the fabricator if the reinforcing fiber can be presented in a form that is convenient to handle. When a reinforcing fiber is available in continuous lengths, as in glass carbon or aramid, it is possible to carry out many of the classical textile processes such as weaving, knitting, and braiding.[4]

Machinery now exists for measuring the length of continuous tows very precisely so that a number of reels can be matched with respect to length. This makes it easy to produce an array of parallel-laid impregnated fibers in the form of a narrow tape, which is useful in tape laying (Figure 1.7), or as a wide sheet of prepreg.

Matching lengths of fiber are also an advantage in the processes of filament winding and pultrusion. Continuous lengths can also be put through a chopper (25–50 mm) and allowed to fall onto a screen in a random, overlapping manner to form a wide, resilient mat known as a *chopped strand mat.*

1.3.4 Fabrication Techniques

There are over a dozen molding methods that have been absorbed into the skills of the composites industry. Even though some may not be recognized as molding methods, they still fall into the category of molding. It is instructive to start with the simplest of all and then proceed to others that are more complex.

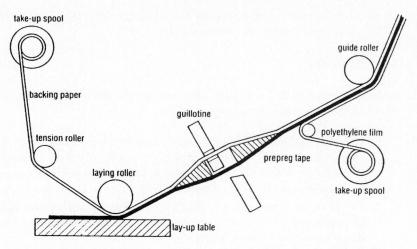

take-up spool

guide roller

backing paper

guillotine

tension roller

polyethylene film

laying roller

prepreg tape

take-up spool

lay-up table

Figure 1.7   Tape-laying machine. The machine peels back the protective film, then lays down the tape and cuts it to the exact length before moving across a distance equal to the tape width. The sequence is automatically controlled.[4]

1.3.4.1 "Leaky mold" technique utilizing carbon fiber and a cold-setting resin.

The apparatus consists of an open-ended metal trough with a loose-fitting top force that is T-shaped in cross section.

Also required are a few grams of carbon fiber in the form of straight tow and about three times the quantity of low-viscosity liquid cold-setting resin. Both epoxy and polyester resins work well.

After the mold is lightly coated with a stearate grease parting agent, a quantity of freshly catalyzed resin with a pot life of about 20 min (longer for an epoxy resin, ~50 min) is poured onto the bottom of the mold, and a combed and weighed amount of fiber is dropped into it. The resin gradually wets the fiber bundle, and bubbles of air can be seen coming away from the top. After 10 min the top force of the mold is placed over the array and a heavy weight balanced on top. The apparatus then acts as a filter press, the fiber being trapped but the excess resin escaping from the ends of the trough and into the clearance between the top and bottom forces (Figure 1.8).[4] After the resin has hardened, the mold may be opened and the smooth parallel-sided rectangular specimen removed. It is trimmed to size and is suitable for evaluation testing or use.

1.3.4.2 High-pressure compression molding.[2,4,5]

Compression molding is one of the oldest manufacturing techniques in the plastics industry. Traditionally, it has been used for molding thermosetting (such as phenolic and alkyd) powders and rubber compounds. However, compared with the injection molding process, it has found only limited use with either thermoplastic or thermosetting resins. With the recent development of high-strength SMC and a greater emphasis on the mass production of composite materials, the compression molding process is receiving a great deal of attention from both industry and the research community.

Compression molding has a number of advantages over the injection molding proc-

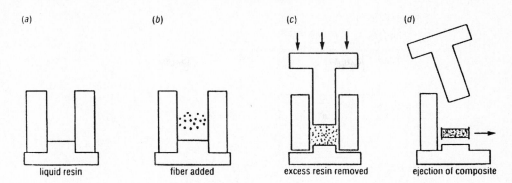

Figure 1.8 The leaky mold technique. After the fiber bundle has been saturated with liquid resin, the mold is closed to act as a simple filter press and remove excess matrix.[4]

ess. It is performed using relatively simple tools with no sprues, runners, or gates. Consequently, very little material is wasted in a compression molding operation. High fiber volume fractions and long fiber lengths can be easily accommodated in compression molding, whereas injection molding is limited to low fiber volume fractions and fiber lengths of 3 mm or less. Thus, in general, better physical and mechanical properties can be achieved in compression-molded parts. The molding pressure in the compression molding process is lower than that in the injection molding process. This means that for comparable part surface areas, a lower-capacity press is required for compression molding than for injection molding. For this reason, compression molding is more suitable than injection molding for producing parts with large surface areas (Table 1.3).

The process of compression molding can be divided into three basic steps (Figure 1.9).

1. *Charge preparation and placement.* The stack of SMC plies placed in the preheated mold is referred to as the charge. The plies are die-cut in the desired shape and size from a properly matured SMC roll, polyethylene carrier films are peeled from each ply, and the plies are stacked into a charge outside the mold (Figure 1.10).

TABLE 1.3. Principal Parameters in a Compression Molding Process[2]

Molding parameters	Tool parameters
Mold temperature	Mold design (radius, ribs, etc.)
Molding pressure	Shear edges
Mold closing speed	Vents
Charge specification (geometry, placement, and size)	Ejection system
	Parting line
Material parameters	Draft
Resin paste formulation	Tool material
Resin-catalyst-inhibitor reactivity	Surface finish
Maturation time	
Sheet thickness	
Sheet temperature	

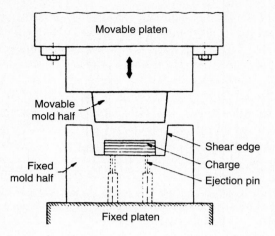

Figure 1.9 Schematic of a composite molding press.[2]

Rectangular ply patterns are commonly used in the charge, however, circular, elliptical, or any other ply patterns can also be used. The ply dimensions are selected to cover 60–70% of the mold surface area with the charge; the charge weight is measured just before placing it in the mold. The location of the charge placement in the mold is one of the key factors in determining the quality of the molded part since it influences fiber orientation, void content, knit line formation, and so on.

2. *Mold closing.* After the charge has been placed in the bottom mold half, the top mold is quickly moved down to contact the top surface of the charge. Then the top mold is closed at a slower rate, usually at 5–10 mm/s. With the rising temperature in the charge, the viscosity of the SMC is reduced. As the molding pressure increases with continued mold closure, the SMC flows toward the extremities, forcing the air in the cavity to escape through the shear edges or other vents. Mold closing speed is another key molding parameter that influences the quality of a compression-molded part.

   The molding pressure based on the projected part area ranges from 1 to 40 MPa depending on the part complexity, flow length, and SMC viscosity at the mold temperature used. The common mold surface temperature is approximately 149°C. Both top and bottom molds are externally heated to maintain the mold surface tem-

Compression molding

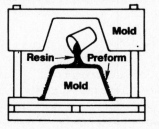

Figure 1.10 Compression molding.

perature within ±5°C of the desired mold temperature. The amount of flow in a compression molding process is relatively small compared with that in an injection molding process. However, it is still a critical parameter that determines the quality of the part. It controls the void content due to air entrapment and fiber orientation in the part and thus affects the mechanical properties of the part.

3. *Curing.* After the cavity has been filled, the mold remains closed for a predetermined period of time to ensure a reasonable level of curing and ply consolidation throughout the part. The curing time, which may vary from one to several minutes, depends upon several factors, including the resin-catalyst-inhibitor reactivity, part thickness, and mold temperature. At the end of the curing time, the top mold is opened and the part is removed from the bottom mold with the aid of ejector pins. The part is then allowed to cool outside the mold, while the mold surfaces are cleaned of any remaining debris and sprayed with an external mold release agent in preparation for the molding of the next part.

   As the part cools outside the mold, it continues to cure and shrink. The temperature distribution during cooling is important in determining the residual stresses in the final part; however, it is very much influenced by part design.

Short fiber-reinforced thermoplastics may be manufactured in heated compression molds, although this process at present represents a small volume of the total consumption of these materials. With this method the material in the form of pellets or chopped rod is placed in a mold located in a hydraulic press. Pressure is applied to the composite material which flows under the influence of heat and pressure to fill the mold cavity. The cost of a compression mold is low compared to that of an injection mold, and for the fabrication of a few large parts with simple geometry, compresion molding may be an attractive alternative.

Good mechanical properties can be achieved because fiber length is not degraded by compression molding; but fiber orientation does not occur, so the advantage of preferred orientation in the plane of the part is lost. Cycle times in compression molding are long because the material must be cooled well below the matrix melt temperature before it can be removed from the mold.[6,7]

Advances have been made in carbon fiber-reinforced polyether ketone ketone (PEKK) compression molding technology, which is well suited for aero panel applications. Potential applications for this technology within the aircraft market include access panels, component housings, selected interior trim panels, and other semistructural panels—both flat and shaped, including rimmed or ribbed panels.

The composite materials formed by compression flow molding provide several potential advantages over aluminum, including the capability of forming complex curvatures, the ability to vary part thickness without secondary machining, and the potential for parts integration.

For most panel applications, the final part is somewhat thicker than aluminum, but part weight is 10–25% lighter with equivalent stiffness.

DuPont XTC, a new experimental thermoplastic moldable sheet product, is based on a long glass fiber reinforced thermoplastic polyester and is 30% by volume carbon fiber-reinforced PEKK. This flow compression-molded panel technology is a valid cost-driven alternative to aluminum for a variety of semistructural aircraft applications. The fuel resistance, static dissipation, and flame resistance of PEKK-carbon compression-molded panels are well suited to meet aircraft requirements.

Applications. Compression-molded SMC composites are used in many automotive, business machine, appliance, construction, and industrial applications. Four such applications are the following.

1. *Computer enclosures.* Computer enclosures are commonly fabricated from sheet steel panels, which are expensive to finish and paint to an exact color, gloss, and texture. They are now being compression-molded using integrally pigmented SMCs. The surface is textured in the mold to produce the desired finish.

   The enclosures are molded as thin as 1.5 mm. The flatness of the panels is ensured by designing proper ribs on the nonappearance side of the panels. For example, it is observed that X-patterned ribs are better in maintaining flatness than circular patterned ribs.

2. *Dishwasher inner doors.* Compression-molded SMC composites are being used in many appliances, such as dishwashers, refrigerators, laundry equipment, and air conditioners, to replace porcelainized and painted steel. An example is the inner door of a dishwashing machine,[6] which is compression molded using a thick molding compound (TMC). The material is pigmented to produce a white color, which eliminates the need for painting. The TMC door is molded in one piece with mounting bosses and reinforcing ribs and has fewer seams and fasteners than a steel door. The molding time is reported to be 47 s.

   The principal advantage of compression molding the inner door is parts consolidation, which reduces the production cost significantly over that of a steel door. The CTE of TMC is close to that of steel. As a result, the assembly of the inner TMC door panel and the outer steel door panel keeps a tight seal at varying temperatures.

3. *Light truck tailgate.* The SMC tailgate used in Ford light trucks is a two-panel design replacing a steel production tailgate that required seven spot welds. The materials used are a vinyl ester (SMC-R50) for the inner panel and a polyester (SMC-R50) for the outer panel. The inner panel is ribbed to provide stiffness for the tailgate.

   The two panels are compression-molded separately and bonded together with a urethane adhesive. Two steel hinge cups and two latch tapping plates are also included in the assembly. Excluding the cups and plates, the SMC tailgate is nearly 27% lighter than a similar steel tailgate.[7]

4. *Automotive road wheels.* Standard automotive road wheels are made of low-carbon steel, such as SAE 1010 or 1020 alloy. Recently, high-strength SMCs have been used in prototype automotive road wheels. The SMC wheels are compression molded in one step in a four-piece mold consisting of a top mold, a bottom mold, and two sides.

   The material used in SMC wheels is a combination of SMC-R50 and XMC. The SMC wheels provide 40–50% weight savings over standard steel wheels. Other advantages of SMC wheels are their uniform weight distribution (therefore, less difficulty in balancing) and their corrosion resistance. It has been reported in the literature that in accelerated fatigue tests SMC wheels perform as well as simulated road tests used for standard steel wheels.

1.3.4.3 Preform molding.

Preform technology has been used extensively in the automotive industry for many years. Although it has been replaced by the use of SMCs for many applications, items fabricated from preforms generally have higher strength and stiffness and less variability in these properties.[8] This is due in part to the fact that fiber placement is less well distributed in preform molding than in the molding of SMCs. A significant amount of flow of resin and fibers occurs during cure of SMCs, and this can lead to variations in fiber content and changes in local fiber orientation.

The use of low-profile liquid resins for preform molding can improve the surface of the moldings, and the higher mechanical properties that can be achieved with the use of preforms make this technology an attractive alternative to the use of SMCs for applications where a high degree of complexity is not required. In fact, preformed short-fiber composite parts made a reappearance in the 1990 Corvette.[8] The most common method of resin application is still pouring of liquid resin by hand followed by closed mold compression molding of the part.

1.3.4.4 Incremental forming.

Incremental forming is a variation of compression molding that may be used in the processing of thermoplastic composites. In incremental forming a flat packet of thermoplastic composite layers is clamped in a transfer mechanism, and a large part is formed a section at a time. The sections of material to be formed are heated locally. This method has several advantages, one of which is that a large part may be formed using a relatively small oven and press.

1.3.4.5 Stamp molding.

Stamp molding is a form of rapid compression molding used with fiber mat-reinforced composite materials. In this process precut sheets or *blanks* of mat-reinforced material are first heated in an oven to a temperature of about 10–37.8°C above the melting or softening point of the matrix.

The heated blanks are then stacked and transferred to a mold located in a rapidly closing press. The mold is maintained at temperatures well below the solidification point of the resin, so when the press is closed, the resin and mat flow to fill the mold and solidify rapidly. This process does not degrade fiber length, and cycle times are often competitive with those of injection molding. New semitoggle link systems on hot stamping machines are able to provide fast, smooth action which is adjustable for variations in material thickness. Microprocessor controls are available that provide accurate temperature control and self-diagnostics plus total count and batch count facilities.

1.3.4.6 Autoclave molding.

For large components the autoclave method is almost universal where double curvature and the highest-quality molding are specified.

The autoclave is a cylindrical pressure vessel that can generate a pressure of several atmospheres. It is also equipped with a means of producing a vacuum within any airtight membranes placed within the vessel so that volatile matter such as solvents or water vapor can be removed. Heating is closely controlled by electric heaters that warm the atmosphere (usually nitrogen), and this transfers heat to the composite layup by convection and conduction. A set of rail tracks transports molds in and out of the apparatus.[9]

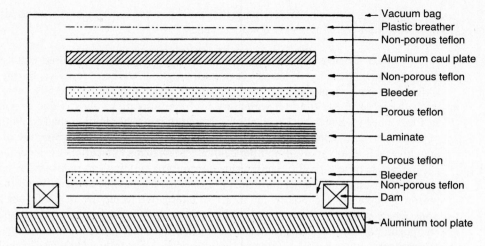

Vacuum bag
Plastic breather
Non-porous teflon
Aluminum caul plate
Non-porous teflon
Bleeder
Porous teflon
Laminate
Porous teflon
Bleeder
Non-porous teflon
Dam
Aluminum tool plate

Figure 1.11   Schematic of the prepreg layup used in autoclave cure and schematic of a bag molding process.

A number of auxiliary materials are required when molding in the autoclave, including (1) mold, (2) nonporous glass, (3) reinforcement prepreg, (4) porous glass, (5) glass paper bleed, (6) glass cloth bleed, (7) undersized vacuum membrane, (8) glass cloth venting layer, (9) vacuum membrane, (10) cork framing, and (11) rope (Figure 1.11).

Autoclave molding is a modification of pressure bag and vacuum bag molding. The method produces denser, void-free moldings because higher heat and pressure are used in the cure. Curing pressures are generally in the range 3.4–6.9×10^5 Pa. Cures are faster (but still involve many hours), and the method accommodates higher-temperature matrix resins having higher properties than conventional resins. Autoclave size limits part size (Figure 1.11).

A new line of thermoplastic composite prepreg tapes for autoclaving, compression molding, filament winding, and thermoforming was recently introduced by Quadrax Advanced Material Systems.[10] The tapes are based on technology and production equipment employing nylon 6 and polymethyl methacrylate (PMMA), as well as PPS. Quadrax currently utilizes polyethermide (PEI) and PEEK as matrix resins. The tapes can be unidirectional (widths of 15.24 or 30.48 cm) or biaxial, with glass, carbon, or aramid fibers at levels of 50–70% by weight. The proprietary tape manufacturing process appears to enhance the wet-out of fiber and resin, yielding minimal voids and microcracking.

Pressure Bag Molding.   Pressure bag molding is similar to the vacuum bag method except that air pressure, usually 2.04–3.4×10^5 Pa, is applied to a rubber bag or sheet that covers the laid-up composite to force out entrapped air and excess resin. Pressurized steam may be employed instead to accelerate the cure. Cores and inserts can be used with the process, and undercuts are practical, but only female molds can be employed (Figure 1.12). Both prepreg and wet layup materials are amenable to this process.

Vacuum Bag Molding.   Vacuum bag molding, a refinement of hand layup, uses a vacuum to eliminate entrapped air and excess resin. After the layup is made (on either a male or female mold) from precut plies of glass mat or fabric and resin, a nonadhering

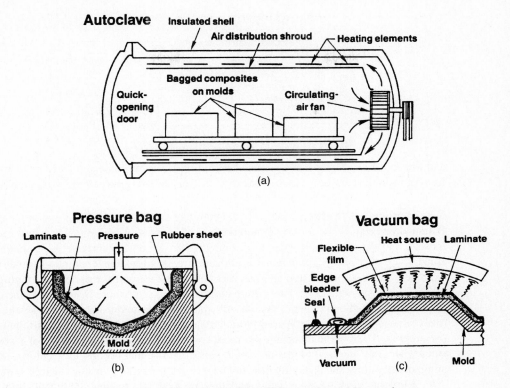

Figure 1.12   (*a*) Autoclave molding. (*b*) Pressure bag. (*c*) Vacuum bag.

film [usually polyvinyl alcohol (PVA) or nylon] is placed over the layup and sealed at the edges. A vacuum is drawn on the bag formed by the film, and the composite is cured at room temperature. Compared to hand layup, the vacuum method provides higher reinforcement concentration and better adhesion between layers.

1.3.4.7 Hand layup.   Hand layup, a simple, old composite fabrication process, is a low-volume, labor-intensive method suited especially for large components such as boat hulls. Glass or other reinforcing mat or woven roving is positioned manually in the open mold, and resin is poured, brushed, or sprayed over and into the glass. Entrapped air is removed manually with squeegees or rollers. Room-temperature curing polyesters and epoxies are the most commonly used matrix resins. Curing is initiated by a catalyst or accelerator in the resin system, which hardens the fiber-resin composite without external heat. For a high-quality part surface, a pigmented gel is first applied to the mold by spraying (Figure 1.13). After curing, the item is removed from the mold and finished on the exterior side.

Hand Layup to Automation.   Most continuous-fiber composites are cured either by compression in a heated press or tool or by autoclave cure. Normally autoclave cure is performed in a vacuum bag in order to provide for the escape of volatilized reaction by-products

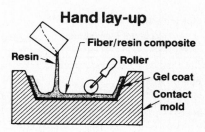

Figure 1.13 Hand layup.

and entrapped air. These methods may be used on wet layup prepared by hand or by some form of resin transfer, on continuous-fiber SMCs, or on B-staged (partially cured) prepregs.

Automated Prepreg Layup.   Vacuum bag autoclave and compression molding of B-staged prepregs is the most common method of processing continuously reinforced thermosets, and significant advances in this process have come from increasing automation. Hand layup of prepregs into laminates of the desired reinforcement and part geometry is an extremely labor-intensive step and has accounted in large part for the cost of high-performance composite structures. Fabrication currently represents more than 70% of the cost of producing finished components. As a result, automated techniques have been developed for many aspects of laminate preparation including storage, retrieval, cutting, and delivery of the prepreg. These techniques are the most cost-effective in the fabrication of large parts since when they are used to produce small parts, the machine is constantly cutting, accelerating, decelerating, and changing direction and is in the air more than on the tool surface.

Modern automatic tape layup machines are available from a number of commercial sources.[11-13] These machines are able to cut, layup, and compact a variety of tape widths much more rapidly and effectively than first-generation machines. For a one-stage process, the use of "natural path" programming minimizes wrinkles or buckling on a contoured surface.

In addition, more advanced two-stage machines are now available that are able to precut various widths and patterns from the prepreg in the first stage. These are laid down in a second stage. During the second stage the tape is held under tension and then applied with the appropriate compacting pressure. The two-stage machines avoid the natural path limitations of the one-stage machines because the use of separately cut pieces allows the same adjustments and relative slipping across the tape that occur in hand layup.

Combination machines that are able to deliver wet roving, prepreg roving, tape, stitched or braided reinforcement, or even commingled fibers, are also available. These newer machines combine the capabilities of tape layup with those of a filament-winding machine and can be used to cofabricate complicated composite structures.[14]

The use of low-tack thermosets such as BMI and thermoplastics in these machines requires controlled heating of the prepreg tape as it is delivered to the part. A tape temperature control system is now available, as is an automatic tape layup machine specially designed for thermoplastic tape. Parts prepared using these systems do not show any decrease in structural properties compared to those prepared using conventional techniques. (Figure 1.14).

Figure 1.14 Thermoplastic tape-laying machine.[14]

Automated Prelayup Techniques.   If hand layup is used for preparation of the laminate, several different automated prelayup steps may be employed to improve the efficiency of the process. These include automated nesting, cutting and identification of prepregs, and the use of laser light or ink jet printers to project assembly instructions and layup patterns directly onto tooling or work in progress, thereby eliminating the need for templates.[13,15,16]

Nesting is the process by which full-scale composite plies are placed within a broad goods material boundary as tightly as possible to increase material utilization. Software is available that computes and carries out the most efficient nesting, tracks raw materials being used, and identifies the cut plies to facilitate layup. This software can be used with many different types of cutters and has been shown to markedly increase prepreg utilization to as high as 85–95%. New-generation high-speed automatic cutters can also provide significant savings in labor hours per part.

The use of templates in the hand layup of complex composite parts is expensive and is a source of error and inaccuracy in the finished part. The cost of producing the templates is considerable, and it is sometimes difficult to install the template accurately. Laser or ink jet systems can perform the same function simply and accurately. Laser systems can be used within safe limits, and observation of the projections over long periods of time does not cause eye fatigue. Ink jet printers use fast-drying dye to print marks on

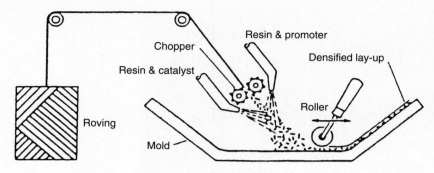

Figure 1.15   Schematic of the spray-up process.

the master tool and on each subsequent ply as it is laid down. White-pigmented ink is used for marking graphite composite prepreg.[17,18]

1.3.4.8 Spray layup.   Spray-up is a processing technique that can be used to produce complex shapes. It is applicable to continuous glass and carbon but is of little use for aramids, owing to the difficulty of chopping this tough fiber cleanly. It involves simultaneous spraying of resin, catalyst, and chopped reinforcement onto an open mold shape, usually after a gel coat has been applied to the surface. Figure 1.15 is a schematic of the spray-up process.

Rollers or squeegees are then used to remove entrapped air and work the resin into the reinforcement. General-purpose resins are usually used because they cure at room temperature, although external heat is sometimes applied to accelerate curing. As in hand layup, gel coats can be used to produce a smooth, pigmented surface. The structure may be selectively stiffened by the local addition of an extra thickness of material, and stiffeners prefabricated from other materials may be easily incorporated into the structure by hand placement and overspraying. Reinforced material may be sprayed onto the back of a previously thermoformed thermoplastic shell to produce a strong, stiff composite item with a high-quality surface. If spray-up is used directly to produce a finished end item, a room-temperature curing resin system can be employed or heat may be applied to speed the cure and increase production.

Variations of spray-up techniques may also be used to prepare chopped fiber preforms if a binder rather than a room-temperature curing resin is used, and the materials are sprayed onto a screen rather than a mold. The end item is prepared from such a preform by addition of the desired resin and molding under pressure. Spray-up lends itself to automation, and numerically controlled multiaxis machines are available that can be used to produce complex parts utilizing this technique.

1.3.4.9 Injection molding.   At high shear rates (e.g., 10^4–10^5/s) the difference between the viscosities of reinforced and unreinforced materials can decrease significantly, allowing the use of conventional injection molding with reinforced thermoplastics.[19]

Injection molding is the most widely used process for high-volume production of thermoplastic resin parts, reinforced or otherwise. With modifications, the process is also

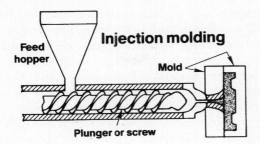

Figure 1.16   Injection molding.

used for molding thermoset resins. Pellets of resin containing fiber reinforcement are fed into a hopper and then into a heated barrel containing a rotating screw that mixes and heats the material. The heated resin is then forced at high pressure through sprues and runners into a matched-metal mold. Molding is rapid, and parts can be very precise and complex (Figure 1.16).

There are several configurations of injection units in use today. The simplest, first-generation plunger and torpedo machines are still being made, but the most widely used type is the reciprocating single-screw injection unit. Descriptions of some of the more complex machines, which may combine a screw with a plunger or two independently operated screws, have been published in books on injection molding.[20–23]

The most important process parameters controlled by the injection unit are the following.

- *Melt temperature.* The temperature of the melt when it penetrates into the mold is controlled by the temperature control sytem of the injection unit but may also be affected by the injection speed and by the level of back pressure.[20]
- *Injection speed.* This is the speed at which the screw advances during the mold filling step. Modern machines are equipped with variable injection speed control—a profile of speeds rather than a single constant value is used to fill the mold. Typical mold filling starts at a slow speed to prevent jetting; speed is increased during the middle part of filling and reduced again toward the end to allow smooth and accurate transition to pressure control, which takes over when the mold is full.[20,24]
- *Injection pressure.* The pressure exerted by the screw on the melt is not constant during the mold filling stage. Injection pressure builds up as the mold is filled and as the resistance to flow increases. It is only when the mold is full that a transfer from speed control to pressure control takes place. Injection pressure is the principal variable during this holding stage.

Thermoplastic Materials.   Injection molding of thermoplastic materials is a cyclic process in which a molten polymer is injected into a closed, cold mold where it solidifies, taking the shape of the mold cavity. The mold is then opened, the molding is removed, and the cycle is repeated. Since the first hand-operated injection machines were introduced more than 60 years ago, injection molding has evolved into a complex, sophisticated process particularly suitable for large production runs—many thousands or millions of nominally identical parts.

Virtually every thermoplastic resin is injection-molded, and most resins are also available in several filled and/or reinforced forms. The terms *filled* and *reinforced* indicate that a second, discontinuous, usually more rigid, phase has been blended into the polymer. When the aspect ratio (i.e., the ratio of the largest to the smallest dimension) is close to 1, the second phase is referred to as *filler*. If the aspect ratio is much larger than 1, as is the case with fibers, the term *reinforcement* is used to describe the second phase. Filled and reinforced thermoplastics are used because they allow a significant and easy modification of the base resin properties. For example, glass fibers provide a much higher room- and high-temperature rigidity than unfilled polypropylene.[19] Talc, a platelike material, provides a reinforcement intermediate between those of calcium carbonate and glass fibers. Similar changes in properties have been observed with other commodity and engineering resins. In fact, fillers and reinforcements extend the range of properties available from a given base polymer toward values that would otherwise require the use of a different, more expensive, resin rather than reduce its cost. The glass fiber reinforced polypropylene can in certain applications replace unfilled polyamide or polycarbonate. While it is possible to reinforce efficiently any thermoplastic resin, only certain fiber-resin combinations have found widespread use.

Reinforcing material can be of either fibrous or planar shape. In practice fibrous reinforcements are almost exclusively used, with glass fibers dominating this market segment. Carbon or aramid fibers provide higher stiffness at lower weight, but their use in large-volume applications has been limited by their high cost. Planar reinforcements such as talc, mica, or glass flake are also available and can be used where stiffness and isotropy are required.

The strengths and limitations of reinforced thermoplastics have the same origin; they can be processed on the same equipment as unreinforced resins. Fibers or fillers can be easily blended into the molten resin using a single- or twin-screw extruder; the resulting compound is processed on an ordinary injection molding machine at a high production rate. Unfortunately, the necessity for making the matrix-fiber mix flow during processing imposes serious limitations on the product. In particular, fiber length and maximum fiber loading are far less than optimal. Moreover, fiber orientation in the molding is determined by flow. Both these phenomena affect and control the ultimate properties available from this class of reinforced plastics.

Orientation and redistribution of the reinforcing fibers occurs during injection molding and can exert a strong influence on the mechanical properties of the composite part. It has been shown that filler particles may form a "surface of accumulation" toward which the filler migrates. This occurs because near a wall normal stress effects cause the filler to move inward, whereas far from the wall shear thinning causes filler to move outward. The surface of accumulation forms in between these two regions. Filler migration may also result in variation in filler contact depending on distance from the gate.

All these effects can occur with short-fiber reinforcement, but the analysis has been performed only for spherical filler. It is nearly impossible to predict what the fiber distribution and orientation will be a priori, but many factors are known to influence fiber orientation including gate and mold geometry, fill rate, viscosity, fiber loading, channel diameter-to-fiber-length ratio, fiber aggregation state, pressure changes, and flow instabilities. Such knowledge can be used to build a desired orientation into a reinforced molded or extruded part.

The conventional injection molding process limits the fiber length that can be achieved since the action of the screw in the barrel and the passage of fibers through narrow gates and openings in the mold cause significant fiber breakage. Under normal operating conditions the maximum fiber length in an injection-molded part are in the range 50–500 μm range regardless of starting length. This normally means that less than one-quarter of the fibers present in the end item will be longer than the so-called critical length and that most of the fibers are too short to contribute their full strength to the composite.

Among the new thermoplastic materials, PEEK, a semicrystalline thermoplastic that exhibits both amorphous and crystalline characteristics depending on its thermal history during the forming operation, was evaluated by researchers. Two grades of material, neat (unfilled) and 30% glass-filled PEEK, were molded and evaluated.

They concluded that

- PEEK can be processed on current molding equipment if the geometry of the part is compatible.
- Voids and shrinkage are difficult to eliminate in thick-walled parts.
- A 0.09- and a 0.18-kg injection molding machine can fill a charge housing mold (which has a 0.508-mm wall thickness) with the mold temperature at 204°C, the injection pressure at 196 MPa, and the band heaters set at their maximum output (388°C).
- Annealing can change the physical characteristics of parts by providing a higher percentage of crystallinity.

Long Fiber-Reinforced Injectable Thermoplastics.   The need for longer fibers in injectable compounds has been recognized for a long time. Initially it led to the development of a compounding process derived from extrusion wire coating. Multifilament rovings are pulled through an extrusion die, coated with polymer, cooled, and cut into pellets. Fibers in the pellets are not dispersed in the matrix—they stay bundled together. In injection molding, good dispersion can be achieved by sufficiently shearing the material (e.g., by using high back pressure), however, this leads to severe fiber length degradation and machine wear.[25-30]

Not only are these long fiber-reinforced thermoplastics injection-molded successfully, but they are also being compounded on-line in a special extruder that maintains the integrity and distribution of fibers and are being compression-molded in bulk form into a variety of automotive applications. The bulk product can yield the same structural components as high-modulus, reinforced thermoplastic sheet but at a much lower materials cost. The process is called *long-fiber thermoplastic direct melt-phase molding* and can be adapted to run several resins, including polypropylene, polyester, and nylon, as well as alloys. In fact, almost any thermoplastic can work with this process. Fiber lengths run to 50 mm.

A long-fiber polypropylene cab back panel made using the process was recently specified for a new pickup truck. Interior, underhood, and structural automotive parts are under development, and appliance, furniture, and recreational equipment markets also are being examined for applications.

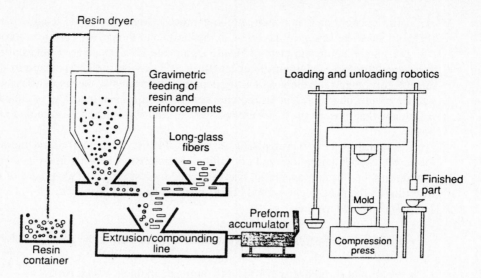

Figure 1.17 Extrusion-compounding line processes glass fibers almost 50.8 cm long without damage, thereby optimizing their structural properties in compression-molded parts.[31]

The key to the technology is the ability to extrude compound long-fiber composite materials without sacrificing the length integrity of the glass fibers used in the process. The thermoplastic composites are compounded in an extruder adjacent to the molding press, and the bulk materials or billets (round cylinders or logs) are transferred into the mold after accumulating to a desired weight at a preform station between the extruder and the press (Figure 1.17).

As resin and additives are gravimetrically fed by a blender into the extruder, long glass fibers are also gravimetrically fed into the resin stream by a separate feeder. Fillers can be added to the fiber stream. The feeders are computer-controlled, with composite formulation and system output rate centrally monitored.[31]

Truckenmüller and Fritz[32] developed an injection molding process whereby roving strands are directly incorporated into the polymer melt by using a reciprocating-screw plasticating unit. The direct incorporation of continuous fibers (DIFs) offers the possibility of substituting the relatively expensive and limited pultrusion process currently used to produce long fiber pellets. In their preliminary study experimental investigations on glass fiber-reinforced polyamide 6.6 were carried out, starting with short- and long-fiber pellets in comparison to the direct incorporation of roving strands into the polymer melt.

They found that compared to their pultrusion-compounded counterparts, all DIF materials examined showed exactly the same high mechanical and/or physical properties. Compared to a conventional extrusion compounded short fiber-reinforced material, the long fiber-reinforced composites showed a significant increase of at least 100% in impact toughness, even at low temperatures, as well as a significantly increased shear modulus above the glass-transition temperature. On the contrary, the tensile and flexural properties of the short and long fiber-reinforced materials were found to be similar, except that the short-fiber parts exhibited higher values of ultimate tensile strength and strain.

The incorporation of roving strands with comparably low tex (fineness of strands) values, higher filament diameters, and reduced sizing contents leads to an optimal combination of mechanical, physical, and optical properties. The flow behavior of the long-fiber filler melt was found to be dependent on the fiber length distribution, leading to an irregularly shaped melt front.

Applications.   Structural analysis is only one of the several important aspects of product design of injection-molded products. Processing considerations usually related to mold design and dimensional stability of molded parts (shrinkage and warpage) may impose solutions other than those based on the results of structural analysis alone. Aesthetic factors or a need for additional processing (painting, welding) may tip the scale toward one or another otherwise equivalent choices.

Housings for Electrical Tools.   One of the early applications in which fiber-reinforced plastics have replaced die-cast metals is housings for drills, jig and circular saws, and similar electrical tools. Smooth and aesthetically appealing on the outside, the internal faces are quite complex, with structural ribs, areas provided for tight insertion of motors and switches, metal inserts for screws, and so on. Compared with die-cast metals, reinforced plastics provide the advantage of safety (double insulation) and a surface finish that does not require painting. Glass-reinforced polyamides (66 or 6) offer the best cost/performance combination. An acceptable surface finish is achieved by formulation of the resin.

Automotive Under-the-Hood Applications.   Functional requirements include a combination of high-temperature, chemical, and mechanical resistance. An excellent example is provided by the radiator tanks (parts attached to the radiator top and bottom). Formerly these parts were made from copper alloys. Coolant pipe fittings had to be brazed or soldered to the radiator tank body. The plastic has to withstand a temperature of 130°C and the effect of many chemicals (hydrocarbons, antifreeze, cleaning products, calcium chloride, etc.). Today glass-reinforced polyamide 66 is almost universally accepted. Plastic radiator tanks can be used with aluminum radiators (no electrolytic corrosion). To date tens of millions of polyamide tanks have been successfully molded.

Other under-the-hood parts made from glass-reinforced polyamide include fans, fan brackets, and gearbox covers. Some recently developed passenger car models contain as much as 6 kg of glass-reinforced polyamides and polybutylene terephthalate, the latter being mainly used in electrical applications. For noncritical applications (fan shrouds, horn shells, heater housing), glass-reinforced polypropylene is a material of choice because of its cost.

Plastic Drawers.   Plastic drawers for kitchen furniture used to be made of high-impact polystyrene, which is susceptible to stress cracking when in contact with oils, fats, and household cleaners. Thus this material was replaced with another low-cost plastic—polypropylene. By adding talc to the polypropylene researchers were able to maintain a cycle time identical to that experienced with polystyrene and thus be cost-competitive.

Additionally, talc-filled polypropylene (40 wt%) has found a major market in injection-molded lawn furniture.

Metal Inserts and Attachments. Composite parts rarely form the entire structure or system. Hence, integration of the composite members with the rest of the system is a critical part of the design process. The designer has the option of choosing among a variety of generic approaches to joining that include adhesive bonding, mechanical fastening, advanced bonding processes (fusion bonding, etc.), and entrapment. The last mentioned technique, in particular the encapsulation of metal inserts within a composite part, is depicted by two generic schemes for attachment in Figure 1.18. In Figure 1.18a the attachment protrudes out of the composite assembly, whereas in Figure 1.18b the insert provides for the bolting in of another part.

It is well established that relatively thin-walled structures are most efficiently joined through adhesives, whereas thick-walled structures lend themselves to mechanical fastening. The technique of incorporating metal inserts and attachments into composites is expected to provide significant improvements in both part performance and fabrication efficiency (and thereby a reduction in cost simultaneous with an increase in reliability) over adhesive bonding and direct mechanical fastening (i.e., fastening through holes, molded in or drilled, in the composite).[33]

Injection molding is probably the best developed method for producing fiber-reinforced plastic composite parts with integrally molded metal parts. A multitude of

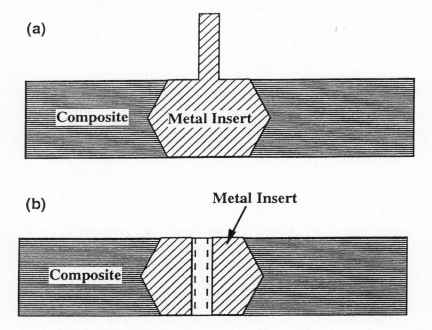

Figure 1.18 Generic schemes for attachment.[33] (*a*) Insert protrudes out from composite structure, providing an external attachment point. (*b*) Insert is completely encapsulated by the composite.

electrical and electronic devices (e.g., capacitors, relays) are produced by overmolding the conductive metal components with either reinforced or unreinforced polymers. In these devices, the molded-in metal components are the primary functioning parts, while the overmolded polymer is for appearance, electrical insulation, and mechanical and environmental protection.[34]

In spite of its frequent use, the molding in of metal inserts using the injection molding process can be difficult. In fact, texts on injection molding typically recommend avoiding the practice altogether.[35,36] The difficulties arise in properly positioning the insert within the mold and then maintaining the position within the dimensional tolerances required for attachment purposes during injection and solidification.

Injection molding of composites with molded-in inserts offers several advantages over multiple-operation composite-metal fabrication methods (e.g., molding followed by hole stamping and then ultrasonic staking of the inserts). The process allows for a good degree of part integration, reduction of labor and manufacturing operations, and improvements in thermal, electrical, and mechanical performance. However, there are several disadvantages. First, compared to other composite manufacturing processes, injection molding requires relatively high clamping forces (2.22×10^7 to 3.11×10^7 N for a flat 5-kg part),[37] which drives up tooling costs. Second, maintaining insert positioning during molding can present the major problems discussed previously. Furthermore, molded-in inserts cause the formation of *weld* or *knit* lines (i.e., lines along which two separate polymer streams meet and then consolidate) along the centerline of the insert, downstream of the insert. This causes the preferential alignment of fibers along the insert edge as a result of surface effects, and in the vicinity of the weld line, the fibers become preferentially aligned along it. These effects lead to significantly reduced and locally varying mechanical properties in the part. Finally, the process allows for only very short (0.2- to 10-mm) discontinuous reinforcing fibers, resulting in mechanical properties too low to justify their use in structural components, and of course a network of continuous fibers encapsulating the insert is not possible.

Thermoset Materials.   Thermosetting molding compounds typically consist of a base resin and either a fibrous reinforcement or a filler. In addition, the molding compound may contain a variety of other components such as catalysts, colorants, and processing aids. The thermosetting resin is capable of undergoing an irreversible crosslinking reaction, which results in an infusible network structure. The process and extent of this cross-linking reaction depend on temperature, and therefore care must be exercised in controlling the thermal history of the material during processing. As a consequence of their three-dimensional network structure, thermosets inherently possess excellent mechanical, heat resistance, and chemical resistance properties. Their ability to retain their performance characteristics and dimensional stability at elevated temperatures can usually be attributed to the nature of the filler or reinforcement.

Various types of thermosetting molding compounds are available commercially including allyls, aminos, epoxies, phenolics, thermoset polyesters, thermoset polyimides, and silicones.

The injection molding of thermosets is similar in many respects to the injection molding of thermoplastics, with the notable difference relating to solidification of the product. In the case of thermoplastics, solidification is accomplished by means of cooling, whereas for thermosets solidification occurs by means of the cure reaction, which is achieved in a heated mold cavity.

The reactivity and changing rheological characteristics of thermosetting systems made processing difficult initially and hampered the widespread development and application of the screw injection method for thermosets. Nevertheless, the injection molding process permits great flexibility for complex part geometry and design, facilitates good control of the process and product, obviates the need for special preheating equipment or procedures, and is more suitable for short cycles. Recent developments in material handling, process control, robots, and automatic mold change equipment have resulted in improved production efficiency, scrap reduction, part consistency, and conformity to tight tolerances. Reliable data acquisition and automated part removal, sorting, and finishing are becoming more commonplace in the industry, with concomitant improvements in productivity.

Recently, more productive molding techniques have been developed including hot cone molding, runnerless injection-compression (RIC), and lost core hollow part injection. These innovations have resulted in dramatically shorter molding cycles, reduced scrap, and improved finished part properties.

Hot cone or hollow sprue molding simply involves placing a mandrel or core into the sprue. This results in a hollow sprue and a consequent reduction in sprue weight. Since the sprue is often one of the thickest sections, the overall cure time may be substantially reduced. Durez Resins has developed and licensed a patented runnerless injection-compression molding process. A cold manifold plate and cooled bushing are employed with modified injection-compression tooling to minimize the production of scrap material in the form of cured sprues and runners.

A variety of fillers have been employed with thermosetting composites.

Glass Fillers.   Both hollow and solid glass microspheres are widely used in conjunction with thermosetting resins. The spherical shape enhances the dispersion uniformity and provides for a more uniform distribution of stress in the molded part. Hollow spheres may be used as extenders to lower costs and to provide for a reduction in density without substantial loss of mechanical performance. Solid spheres are used to improve modulus. Glass spheres are easily wetted and may be treated with special surface coatings to improve the bonding characteristics to the matrix. Compared with other fillers, spheres possess a minimum surface-to-volume ratio, and thus high filler loadings can be achieved without unacceptable increases in viscosity. For the same loading level, glass spheres yield improved mold shrinkage, warpage, and cycle time. Glass microspheres are inert, nontoxic, and stable, and they yield improved abrasion and corrosion resistance. They are easily mixed by low shear processes into epoxies, polyesters, and other thermosetting systems.

Mineral Fillers.   Although commodity mineral fillers are used as extenders to reduce resin cost, the compounding operation is expensive. Thus the selection of an appropriate mineral filler is generally justified by the ability of the filler to provide specific performance advantages or characteristics such as increased modulus, increased heat distortion temperature, or a lower thermal expansion coefficient. Clay, mica, talc, and wollastonite have high ratios of length to thickness or length to width and thereby perform a reinforcing function.

Alumina trihydrate is used in conjunction with unsaturated polyesters, epoxies, and phenolics because of its flame-retardant and smoke-suppressant properties. Calcium carbonate (limestone) is another commonly used filler in polyester systems. Other mineral fillers include barium sulfate, calcium sulfate, silica, and feldspar. In general, mineral fillers are readily available and comparatively inexpensive.

Metallic Powders.   Metallic powders (e.g., aluminum flake, aluminum-coated glass, bronze, zinc, and nickel) are used principally to enhance thermal and electric conductivity of the material and to provide shielding from electromagnetic and radiofrequency interference (EMI/RFI). Metallic powders are a generally more expensive class of filler material.

Organic Fillers.   Organic fillers include wood flour, ground nutshells, starches, carbohydrate by-products, and various synthetic organic materials. They are frequently used as fillers and reinforcing agents with ureas, melamines, and phenolics. Their major attribute is low cost, and their major disadvantage is water absorption.

Thermoset molding compounds may contain fibrous reinforcements that are strong and inert and form an effective bond with the matrix resin. Fibers yield dramatic improvements in both tensile and flexural strength in the fiber direction. Some of the common fibrous reinforcements are aramid fibers, carbon fibers, cellulosic fibers, ceramic fibers, glass fibers, metallic fibers, and thermoplastic fibers.

Aramid Fibers.   Aromatic polyamide fibers, such as Kevlar, have low density, high tensile strength, excellent toughness, outstanding heat resistance, low dielectric properties, and fairly good chemical rersistance. They also possess excellent vibration damping, low thermal expansion coefficients, good frictional properties at elevated temperatures, and exceptional wear resistance. Chopped aramid fibers ranging in length from 6.35 to 25.4 mm are commonly used to reinforce thermosets. Aramid-carbon hybrid composites are used with epoxies, polyimides, phenolics, and polyesters.

Carbon Fibers.   Carbon fibers possess a number of inherent advantages including low density, outstanding mechanical properties, excellent electric conductivity, good chemical inertness and corrosion resistance, excellent high-temperature properties, and good frictional and wear characteristics. The highest composite mechanical properties are obtained by employing 60% by volume or more of a unidirectional continuous fiber in a high-strength epoxy or similar matrix. Carbon-glass hybrids are encountered with SMC systems.

Cellulosic Fibers.   Cellulosic fibers are being used as low cost extenders and reinforcements with a variety of thermosetting resins including ureas, melamines, polyesters, phenolics, and elastomers. The tensile strength of cellulosic fibers is approximately 545 MPa compared to 3.5 GPa for E-glass fibers and 4.5 GPa for S- and R-glass fibers.

Glass Fibers.   Glass fibers are the most commonly encountered reinforcements for injection-molded composite parts because of their outstanding cost-performance characteristics. Glass fibers are commercially available in a myriad of different forms and possess good dimensional stability, a high strength-to-weight ratio, good corrosion resistance, and ease of fabrication. Glass fibers are treated with coupling agents or other surface modifiers, or they may be coated with metals. Continuous-strand roving is fed to a chopper assembly, and the chopped fibers are randomly arranged in typical SMC systems.

Metallic Fibers.   A wide variety of metallic and metal alloy fibers are commercially available, including aluminum, nickel, and stainless steel fibers. At low loading levels they essentially maintain the mechanical properties and processing characteristics of the base resin yet impart EMI shielding or electrostatic discharge (ESD) protection to the molded part. Unsized chopped fibers are typically used with thermosetting compounds.

Thermoplastic Fibers.   Thermoplastic fibers, such as polyester and nylon, are used to impart better fatigue life and impact strength and to reduce the brittleness or propensity for microcracking of the thermoset matrix. Thermoplastic fibers are inherently nonabrasive and are readily used in conjunction with BMC, SMC, and TMC applications. The main disadvantages relate to compatibility with the resin for proper wet-out and adhesion, surface finish, and adequate dimensional stability over the temperature range for processing the thermoset composite.

1.3.4.10 Filament winding.   Filament winding is usually thought of as a process in which a filamentary yarn or tow is first wetted by a resin and then uniformly and regularly wound about a rotating mandrel. The finished pattern is cured, and the mandrel removed. The result can be something as simple as a piece of pipe or as complex as an aircraft fuselage or an automobile frame.

The principal advantages of filament winding over other composite material fabrication methods are its low material and labor costs and its reproducibility due to the robotic motions. The greatest disadvantages are the tooling limitations for removable mandrels and the inability to wind on negatively curved (concave) surfaces.

Materials.   There are a number of material options available for filament winding. Typical fibers are fiberglass, carbon, and aramid. The most commonly used resins include thermoset polyesters, vinyl esters, epoxies, and phenolics. The material combination can be classified as either a wet system (i.e., the fiber is wetted with resin just before winding on the mandrel) or a prepreg system (i.e., the resin has been applied to the fiber in an earlier operation and a "staged" yarn is delivered at the winding station).

Fibers.   Fiberglass for filament winding is available as either single-end or multistrand roving. A single-end roving is one strand of glass filament collected into a discrete bundle during the spinning operation. Most glass fiber used for filament winding is either E-glass or S-2 glass.

Fiberglass is packaged on either internal or external payoff spools. For the majority of filament-winding operations, it is desirable to have the spool fed from the outside to maintain tension through the machine.

Aramid fiber, most notably DuPont's Kevlar, was widely accepted as a filament-winding fiber beginning with the strategic missile motor case industry. Its principal asset is its high strength-to-weight ratio, which is considerably improved over that of the fiberglass previously used for motor cases.

The fibrilative nature of aramid fiber gives it the property of being able to withstand abrasive wear. The property suggests aramid's use as an external layer for structures that receive considerable wear and abrasion. As a companion property to its ability to take abuse, aramid itself abrades materials that come into contact with it. This is a major consideration for the pulleys and the various components of the payoff system in a filament winder.

Carbon fiber for filament winding is generally available as 3000, 6000, 12,000, and 50,000 filament tows. Unlike glass and aramid, carbon is not made as multistrand rovings. Carbon fiber is very different from fiberglass and aramid. Being brittle, it has a tendency to abrade and break; thus more care must be taken in handling it. To minimize breakage of the filaments, the number of turns and twists in the process must be kept to an absolute minimum.

From the standpoint of initial material damage, wet winding is better than prepreg tow winding because the additional unrolling and rerolling that occur during prepregging are eliminated. Whether a wet winding process or a prepreg fiber is used, the key factor is still to minimize the number of turns to which the carbon fiber is subjected.

Also, unlike the other fibers, carbon is electrically conductive. This means that dry fiber fly must be controlled to avoid electrical shorts in nearby equipment. This control is normally achieved by enclosing the creel with negative pressure, vacuuming the eyelet boards, and impregnating the fiber at the earliest possible point in the delivery stream.

Resins.   Filament winding can utilize resin in three distinct forms. The predominant one is as a liquid, where the fiber is wetted as it passes through a resin bath. Another form is as a prepreg tow, where the fiber is impregnated in an early step, staged to a tack-free consistency, and rewound on a bobbin. A third form utilizes thermoplastic resins, which may be in the form of a dry bobbin, a powdered coating, or a commingled fiber.

Wet thermoset filament winding requires a resin to have a viscosity in the range 1–3 Pa's. Resin components can be chosen so that the combination of pot life, winding temperature, viscosity, gel time, and cure time can be optimized. Wet resins include epoxy, vinyl ester, polyester, phenolic, polyimide, and bismaleimide. With each resin there are a number of recognized curative and accelerator combinations that give acceptable end use performance. Epoxies have the widest range of properties of the resins used in filament winding. They are the predominant resins used in the aerospace market. However, as end use temperature requirements have increased, more interest has developed in systems that are more difficult to process such as phenolics and polyimides.

Preimpregnated tow, typically carbon with epoxy resin, has the advantages of accurate resin content control and a higher level of quality control and traceability as compared with wet systems. Also, the resin systems may be identical to conventional prepreg tape forms, and this can be particularly attractive when aerospace material qualifications are needed. The disadvantage is higher cost (i.e., higher than that of the wet systems).

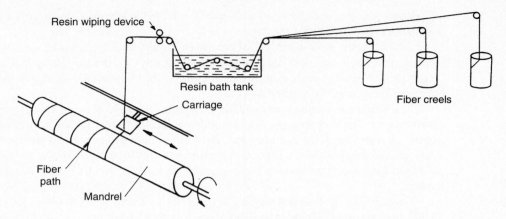

Figure 1.19   Schematic of a filament-winding process.

The use of thermoplastics in filament winding has recently begun and is discussed further later in chapter.

Filament winders which can wind at elevated temperatures are now available for use with polyimides and other resin systems requiring high-temperature processing. The composite is normally cured at elevated temperature without pressurization, and removal of the mandrel completes the process, although some finish machining may be performed.

Basics of Process.   Figure 1.19 is a schematic of a basic filament-winding process. A large number of fiber rovings are pulled from a series of creels into a liquid resin bath containing liquid resin, catalyst, and other ingredients such as pigments and ultraviolet (UV) absorbers. Fiber tension is controlled by fiber guides or scissor bars located between each creel and the resin bath. Just before entering the resin bath, the rovings are usually gathered into a band by passing them through a textile thread board or a stainless steel comb.

At the end of the resin tank, the resin-impregnated rovings are pulled through a wiping device that removes excess resin from the rovings and controls the resin coating thickness around each roving. The most commonly used wiping device is a set of squeeze rollers in which the position of the top roller is adjusted to control the resin content as well as the tension in fiber rovings. Another technique for wiping resin-impregnated rovings is to pull each roving separately through an orifice, very much like the procedure in a wire-drawing process. This latter technique provides better control of resin content. However, in the case of fiber breakage during a filament winding operation, it becomes difficult to rethread the broken roving line through its orifice.

Once the rovings have been thoroughly impregnated and wiped, they are gathered together in a flat band and positioned on the mandrel. The traversing speed of the carriage and the winding speed of the mandrel are controlled to create the desired winding angle patterns. Typical winding speeds range from 90 to 110 linear m/min, although for more precise winding slower speeds are recommended.

The filament-winding process can generally be classified as helical or polar winding (Figures 1.20*a* and 1.20*b*). In the first case, fiber is fed from a horizontally translating

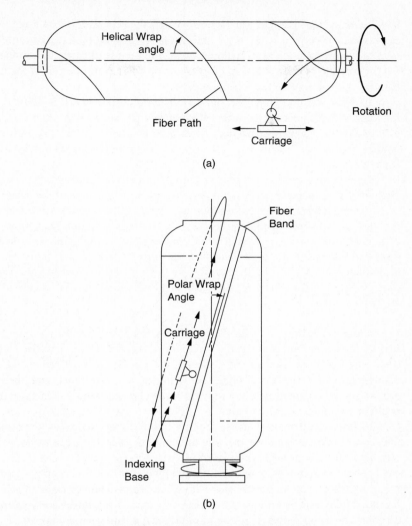

Figure 1.20   (*a*) Helical winding. (*b*) Polar winding.

delivery head to a rotating mandrel, while in the second case, a delivery unit races around a slowly indexing mandrel. All filament-winding processes have a number of required subsystems. These include fiber delivery (resin bath and band alignment), mandrel, mechanical and/or electronic control, curing, and mandrel preparation and removal.[38]

Helical Winding.   The angle of the roving band with respect to the mandrel axis is called the *wind angle*. By adjusting the carriage feed and the mandrel rotational speed, any wind angle between near 0° (longitudinal winding) and near 90° (hoop winding) can be obtained. Since the feed carriage moves backward and forward, fiber bands crisscross at plus and minus the wind angle and create a weaving or interlocking effect. It is also possible to produce a helical winding by keeping the feed carriage stationary and traversing the rotating mandrel back and forth. The mechanical properties of the helically wound part depend strongly on the wind angle.

Helical winding is the predominant method used today. It is particularly well suited for long, slender geometries such as pressure pipe and launch tubes, where winding angles of 20–90° (hoop) are needed. Most pipe is wound at 54.7°, which is derived from netting theory and assumes a 2:1 (hoop-to-longitudinal) stress field in a cylindrical, capped pressure vessel.

There are a number of practical limitations to helical winding. These include machine bed size, mandrel and wound part weight, and clearance of the turning diameter.

In helically winding large structures, special consideration must be given to mandrel design. Mandrel weight causes wear on bearings. Mandrel deflection is a critical and often controlling issue, particularly for sand and plaster units. Inertia considerations often limit winding speed. In large diameter structures such as rocket motor cases with length-to-diameter ratios of 2:1 or greater, low angle helicals are essential to absorb the longitudinal stresses in the structure. Low winding speeds and precise geodesic winding paths are dictated here.

Polar Winding.   In another type of filament winding process, called *polar winding,* the carriage rotates about the longitudinal axis of a stationary (but indexable) mandrel. After each rotation of the carriage, the mandrel is indexed to advance one fiber bandwidth. Thus the fiber bands lie adjacent to each other and there are no fiber crossovers. A complete wrap consists of two plies oriented at plus and minus the wind angle on two sides of the mandrel.

Contrary to helical winding, polar winding favors very low wind angles. Since the fiber is wound in a plane intersecting the mandrel ends, the angle must be less than about 20°. Typically, polar angles are 5–15°.

The principal advantage of polar winding is that it is a simple, rapid winding technique for short, stubby geometries with a length-to-diameter ratio of less than 2, where balanced fiber placement is required. In polar-wound structures that require combinations of polar and hoop wraps, it is common practice to intersperse the patterns. This tends to load the hoop fiber more evenly, as well as to provide a compaction for each set of polar wraps.

Examples of polar-wound structures include virtually all third-stage missile motors, apogee kick motors, inertial upper-stage motors of space shuttle payloads, and even certain spherical petroleum storage tanks.

The potential of filament-wound thermoplastics has been studied by researchers and engineers at Thiokol.[39] Their studies focused on launch vehicles. Compared with those for filament-wound epoxies, costs using present technology put thermoplastics at a disadvantage. But if material costs can be lowered and the winding process speeded up, thermoplastics like PEEK, PPS, and Radel-X could become attractive alternatives to metals and epoxy-based composites. Thiokol has developed several in situ methods for filament-winding pressure vessels up to 457.2 mm in diameter. The systems used infrared (IR) heat and in-process compaction to process the thermoplastic composites in one step. Fila-

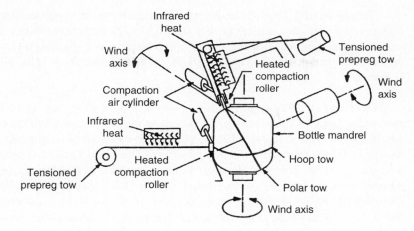

Figure 1.21   Improved polar winding delivery system.[39]

ment-wound epoxy vessels can be wound fast but require an additional step for curing, usually at elevated temperatures. Test methods developed at Thiokol delivered acceptable composite properties as revealed in short-beam shear and Naval Ordnance Laboratory (NOL) ring tests. Though not comparable to epoxy systems in short-beam shear, a filament-wound Radel-X composite exhibited 38.5 MPa; epoxies exhibited a 60.1 MPa short-beam shear strength. The Radel-X filament-wound NOL specimen had the best tensile strength of the thermoplastics tested according to the Thiokol researchers.[39]

Thiokol engineers also designed a polar winder delivery system that doesn't require a heated mandrel. In scaled-up versions that allow standard mandrels to be used, this could mean saving costs and making the process safer. An improved infrared heating system and heated compaction rollers were the key elements in consolidating the thermoplastic tows. Scaling up the features and overall design concepts would of course entail considerable refinement. This work demonstrates a strong potential for thermoplastics, however, faster methods (40 m/min) are required to make thermoplastics more attractive than epoxy filament winding. Thermoplastics offer the attractions of reprocessability, repairability, weldability of attachments, and good toughness. In situ winding would eliminate the usual bagging, autoclave or oven curing, and refrigerated storage (Figure 1.21).

Hoop Winding.   Hoop winding is a special case of helical winding in which the angle approaches 90°. This type of winding provides reinforcement only in the circumferential direction. Hoop winds are often used in conjunction with other angles not only to provide hoop reinforcement but also to compact wet winding at intermediate fabrication points. Winding with constant tension, hoop winding provides more normally directed force to the previously wound plies than does any other angle. Conversely, a 0° wind has no normal force component.

NOL ring   A parallel filament wound test specimen used for measuring strength properties of the material by testing the entire ring, or segments of it.

Examples of helically wound structure include pressure pipe, osmotic tubes, cylindrical pressure vessels, tactical missile launch tubes, auxiliary fuel tanks for aircraft and helicopters, engine nacelles and cowls, main rotor shafts, strategic and tactical missile motor cases, torque tubes, truck driveshafts, electric motor driveshafts, and core sample holders.[40]

Wet and Dry Filament Winding. As mentioned above, epoxy resins are used in two major types of filament-winding processes. The first, and more common, is called *wet filament winding.* In wet filament winding, epoxy resin, curing agent, and additives are mixed by the fabricator to a liquid mixture. The liquid mixture is poured into a bath through which the fibers are drawn. The fibers are then passed through an orifice to remove excess resin and then wound onto a mandrel while wet with resin. The desired article is formed by progressive winding of layers of resin-soaked fibers onto the mandrel. The article is then cured in an oven (or, alternatively, by heating the mandrel with steam or another heat source) followed by removal of the mandrel.

The second method, called *dry* or *prepreg winding,* is a blend of filament winding and prepreg technologies. A fiber roving is passed through a heated resin-curing agent mixture bath by a prepreg manufacturer. The resin-wet fibers are then passed through an orifice to remove excess resin, as in wet filament winding, but are not immediately applied to the mandrel. Rather, the resin-impregnated fibers are cooled to room temperature, causing the liquid resin mixture on the fibers to become a tacky semisolid. The resultant prepreg roving is then shipped to the fabricator of the final article, who winds the roving onto the mandrel and cures the article as in wet filament winding.

Wet and dry winding each have a number of advantages leading to a respective niche in the marketplace. Wet winding is much more commonly used primarily because of its lower cost, with resin impregnation and part fabrication performed by the same machine. In dry winding, these steps are separate and the intermediate prepreg roving must be stored and shipped, with attendant extra costs. Since only a few minutes or hours need elapse between resin-curing agent mixing and part fabrication in wet winding, wet winding allows the use of resin systems with a much shorter pot life. Resin-curing agent mixtures used in dry winding must be stable for at least several days at room temperature in order that the mixture not cure during shipment.

Dry winding is preferred by some fabricators, however, because it relieves the fabricator of the responsibility for maintaining a proper resin-to-curing agent-to-additive ratio and allows closer control of the resin-to-fiber ratio in the final product. Dry winding also reduces the likelihood of exposure of fabricator personnel to potentially toxic curing agents or additives. The result is that parts with a relatively low price per unit weight (such as filament-wound pipe and most pressure vessels) are produced by wet winding, while some higher-value parts in the aerospace and sports equipment industries are produced by dry winding.

In filament winding, as in fiber impregnation processes generally, a mixed resin system viscosity of 1–2 Pa's is generally preferred. At this level, the viscosity is low enough for the system to impregnate the fiber easily in the resin bath but not so low that "resin runout" becomes a problem in fabrication of the final part. Many epoxy systems used for filament winding have a room temperature viscosity far above this level. Hence the resin bath must be heated in order to reduce the viscosity. Heating reduces pot life and hence has limited usability when the resin system has a short pot life even at room temperature, as is the case when aliphatic or cycloaliphatic amines are used as curing agents.

Equipment

Winders (Gear-and-Sprocket and Computer-Controlled).   The principal advantage of filament winding is the ability to accurately place a fiber band on a rotating mandrel. In a basic gear-driven machine, this implies direct coupling between the machine headstock (angle of rotation) and the chain drive on the traversing eyelet. Control of this system is by mechanical or numerical means, and mechanical control is the predominant method. Here a single motor drives both the winding mandrel and a sprocket gear linked to the drive chain and delivery head. The principal disadvantage of this system is the inability to rapidly change winding angles.

Numerically controlled winders with punched tape operation also have been developed. The tapes control hydraulic servo drives, and each axis has its own hydraulic motor. Another control method has replaced the punched tape with an optical tracking device. Here the servo is controlled by an electric eye following a black-white interface on a rotating disk or cylinder.

Within the past 10–15 years, numerical control with microprocessor-controlled servo motors has been developed and refined. The two principal advantages of a micro-computer-controlled machine are the ability to more accurately place the fiber on the mandrel using eyelet manipulations. Some microprocessor-controlled machines have as many as six degrees of freedom coupled with path-smoothing options, acceleration controls, and independent yarn tension control, (Figure 1.22).

In numerically controlled machines, independent drives are used for the mandrel as well as for the carriage. In addition, a transverse feed mechanism and a rotating payout eye (Figure 1.23) allow an unequal fiber placement on the mandrel. The transverse feed mechanism is mounted on the carriage and can move in and out radially; the payout eye can be controlled to rotate about a horizontal axis. The combination of these two motions prevents fiber slippage as well as fiber bunching on mandrels of irregular shape. Although each mechanism is driven by its own hydraulic motor, their movements are related to the

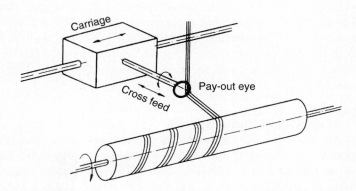

Figure 1.22   Layout of a computer-controlled filament winding machine.[41]

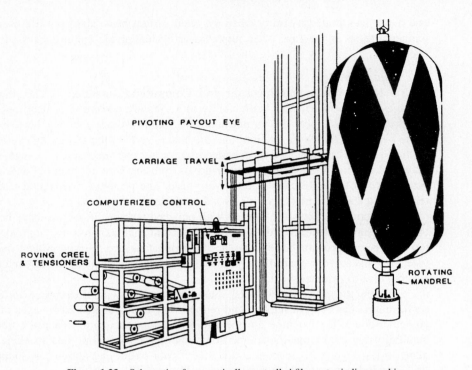

PIVOTING PAYOUT EYE

CARRIAGE TRAVEL

COMPUTERIZED CONTROL

ROVING CREEL
& TENSIONERS

ROTATING
MANDREL

Figure 1.23   Schematic of a numerically controlled filament-winding machine.

mandrel rotation by numerical controls. Since no mechanical connections are involved, wind angles can be varied without much manual operation. With conventional filament-winding machines, the shapes that can be created are limited to surfaces of revolution, such as cylinders of various cross sections, cones, box beams, or spheroids. The computer-controlled multiaxis machines can wind irregular and complex shapes with no axis of symmetry, such as the aerodynamic shape of a helicopter blade.

Recent advances in microprocessor control have included off-line program development, where most of the patterns and parameters can be determined in advance of actual prototype winding. Off-line programming is typically used with a graphics terminal that can simulate the mandrel shape, the fiber band path, and the path of the delivery point in space. The same program can graphically display the closing error and, given the coefficient of friction between the band and the existing layers, define the permissible offset from the optimum geodesic path.

Mandrels and Curing Systems

Mandrel. The mandrel is the geometric basis for the final part. As such, it must support the uncured composite during winding and through cure without deforming beyond acceptable limits. Mandrels can be classified as permanent, removable, and reusable.

A permanent mandrel becomes an integral part of the final structure. In a scuba tank, for example, fiber is wound over a thin metal wall that essentially acts as a gas barrier. Removable mandrels must be separable from the cured part without damage to the part. Reusable mandrels must be removable in a way that maintains the integrity of the part and the mandrel.

When the filament-wound part is open-ended and the opening is the largest diameter of the part, mandrel design and material considerations are straightforward. For instance, pipe and driveshaft mandrels are straight steel cylinders, often with chromed surfaces and very slight tapers (0.17 mm/m) to facilitate extraction. Aluminum has also been used, but it is more prone to damage.

Metal mandrels: Two of the primary types of mandrel structure require the use of metal. One is a permanent shell, while the other is reusable. The permanent use of a metal mandrel is normally for high-pressure (i.e., gas pressure) vessels. For this application, the metal mandrel acts as a leakage barrier for gas inside the vessel. Most composite stuctures are not impervious to the passage of gas, particularly for helium applications. Hence, the internal metal structure is a necessity. Aluminum, stainless steel, Inconel, and titanium have all been used for this application.

The other type of metal mandrel, known as net metal mandrel, is removed from the composite structure after cure. It is reusable without any practical limits on the number of times of reuse. These net metal mandrels provide the precise geometry for the internal dimensions of the solid rocket motor case and its insulation in the industries where they have been used. After cure, the skeletal support structure and the aluminum sheet metal shapes are disassembled and removed from the motor case through the largest boss opening.

Expandable mandrels: All mandrels expand by normal CTE action during the cure process. However, certain mandrels, notably those made out of rubber, can be artificially inflated from the inside to provide either shape or pressure to the curing composite part. Configurations for petroleum storage tanks have used thick-walled, rubber, circular balloons successfully as mandrels. In a second configuration, a thin rubber bladder is formed over a hard tool. When winding is complete, the unit is placed in a clamshell mold and the tool-rubber interface is pressurized to force the windings outward against the clamshell during cure. The result is a dimensionally controlled, smooth outer surface.

Single-use mandrels: The oldest of single-use devices is a plaster mandrel. This structure is created by assembling a skeleton iron core with iron cover plates, wrapping the core with burlap to increase adhesion of the plaster, and then troweling in place on the external surface. The whole structure is then cured in an oven to harden the plaster.

A major disadvantage of this type of mandrel is the presence of water in the plaster in its cured condition. This moisture boils off at resin-curing temperature, degrading the composite matrix properties.

A second type of single-use mandrel is a soluble structure. Originally this type of mandrel was

- Salt paste, which is heated, leaving a cast salt block of the required shape. After cure of the composite the mandrel is dissolved in water.
- Ultrafine sand with polyvinyl alcohol as a binder. Sand is cast and then cured at 93°C, and a hard surface is left that can be machined to meet tight tolerances. When

the composite is cured, the PVA binder is softened, water circulates through the sand, and because PVA is soluble in water, sand and PVA become a slurry which flows out of the motor case into a receiving tank. This type of mandrel has the disadvantage of being the heaviest of all material mandrels, and thus the lifting capabilities of plant equipment may not be available. The second disadvantage is that it cannot be used at temperatures above 150°C; consequently, alternate methods must be found for high-temperature curing resins. The use of sodium silicate or "water glass" as a binder at temperatures up to 343°C was successfully employed recently with this type of mandrel for polyimide (PI) resins.

Curing System. In filament winding, the requirements for a long pot life of the resin and a fast cure of the part are contradictory. In general, resins that can maintain acceptable winding viscosities for long periods of time—from a few hours to days—require either longer cure times at lower temperatures or much higher initiation temperatures, depending on the accelerator or catalyst.

With the exception of microwave energy, the curing process for filament-wound parts is initiated at either the inner or outer surface, depending on the location of the heat source. Both methods have advantages. When curing initiates at the inner surface, the parts tend toward higher fiber content because the resin can be bled out. Also, the void content is reduced since there is less tendency to trap air pockets. With external surface curing first, higher resin contents are possible and drippage can be eliminated.

Ovens: Gas-fired or electrical air-circulating ovens are the predominant mode. They are inexpensive, and they can be very large. Any supplemental curing pressure must be applied with shrink tape or a vacuum bag. Energy costs associated with oven cure are higher than with many other methods because the heated mass consists of the part, the surrounding air, and all associated hardware, including the mandrel and support stand. They also take up considerable floor space.

Hot oil: A hot oil system is typically used with a very fast-curing resin system, normally with the ability to cure in less than 15 min. The use of hot oil ensures a very rapid heat-up of the mandrel and eliminates the need for a curing oven. Passages throughout the mandrel permit this early heating. This has a distinct advantage in mandrel removal. Composites frequently shrink during the cure process and in doing so attach themselves firmly to the winding mandrel, which heats up and expands and then contracts upon cooldown with the part still firmly attached. In the hot oil system, the mandrel heats up and expands and the composite heats up and cures and then contracts under the expanded condition. When the oil flow ceases, the mandrel has cooled down; it shrinks away from the cured composite and permits easy removal. Hot oil systems operate in the range 150–204°C.

Lamps: Heat lamps, used in conjunction with reflective surfaces and a rotating mandrel, can also provide cure temperatures on the order of 171°C. Heat lamps are portable, and care must be taken to provide curing to all sections of the part. Infrared lamps are another means of supplying heat to curing resin systems in a composite structure, however, they are rarely used in production.

Another lamp curing method utilizes a capacitance discharge, a pulsed xenon lamp. To promote the curing process using this type of lamp, a light-sensitive catalyst is used in the resin system.

Steam: A number of pipe manufacturers use hot steam for resin curing. The metal mandrel ends have adaptors for the passage of water and steam. After the tube is wound, hot steam is circulated through the hollow mandrel. When cure is complete, a cold-water flush cools the mandrel, allowing for extraction.

Autoclave: When aerospace quality laminates are needed or more sophisticated epoxy, BMI, or PI resins are used, it may be necessary to cure the filament-wound parts in an autoclave with vacuum assist. Though not normally associated with mass production techniques, autoclave curing can provide pressures of 1.4–2.1 MPa with vacuum augmentation at temperatures as high as 371°C. The principal disadvantage of autoclaves is the long cycle time, coupled with limited use and availability.

Microwave: It has been shown in several development programs that microwave curing can have significant advantages with fiberglass and aramid fiber composites. Microwave energy is absorbed rapidly by both resin and fiber, and results have shown that cures can be effected in minutes for systems that previously took hours. The energy level required for this type of system is high; hence the process is costly. A major drawback, however, is the inability to use microwave curing methods with a conductive fiber such as carbon. For this reason, the majority of composite fabricators have abandoned consideration of microwave energy as a means of curing composites, especially carbon.

Other curing methods: Electron beam, laser, radio frequency (RF) energy, ultrasonics, and induction curing methods have all been studied with filament-winding systems with varying degrees of success. Ultrasonic energy has proved to be unreliable because the energy is imparted to the resin system in a very rapid, nonuniform manner. This has resulted in uncontrolled exotherms, nonuniform curing, and actual burning and charring of the composite structure. Others have devoted significant research and development efforts to the RF curing method and have demonstrated a preliminary feasibility.

Laser methods and electron beam heating have been studied and abandoned, primarily because they do not work with carbon fiber systems. Both of these forms of energy are unable to penetrate the outer layer of carbon fiber in a composite structure.

Laser-directed energy, however, may be an advantage in the consolidation process for thermoplastics. Some researchers have demonstrated the use of a laser beam for melting the zone surrounding the laydown point of prepreg fibers in a filament-wound PEEK system. This method should be applicable to all forms of thermoplastic matrices in filament winding. Since only the top layer is being melted, the inability of the laser beam to penetrate through the carbon layer is not important.

Because carbon fiber is conductive, induction heating may have the most promising future for use with this fiber system in a variety of matrix resins. Induction heating is a process that has been used for more than 50 years in a variety of production processes in various industries. However, its application in the curing of resin systems has not been studied until recently. Several demonstrations suggest that 2- to 3-h cure times can be reduced to only a few minutes with induction heating.

Process Parameters.   The important process parameters in a filament-winding operation are fiber tension, fiber wet-out, and resin content. Adequate fiber tension is required to maintain fiber alignment on the mandrel as well as to control resin content in the wound part. Excessive fiber tension can cause differences in resin content in the inner and outer layers, undesirable residual stresses in the finished product, and large mandrel deflections.

Fiber tension is created by pulling the rovings through a number of fiber guides placed between the creels and the resin bath. Mechanical action on the fibers in the resin bath, such as looping, generates additional fiber tension.

Good fiber wet-out is essential for reducing voids in a filament-wound part. The following material and process parameters control fiber wet-out.

1. Viscosity of the catalyzed resin in the resin bath, which depends on the resin type and resin bath temperature, as well as cure advancement in the resin bath
2. Number of strands (or ends) in a roving, which determines the accessibility of resin to each strand
3. Fiber tension, which controls the pressure on various layers already wound around the mandrel
4. Speed of winding and duration of the resin bath.

As a rule of thumb, each roving should be under the resin surface level for ⅓–½ s. In a line moving at 60 m/min, this means that the length of roving under the resin surface level should be approximately 30 cm. For good wetting, the minimum roving length under the resin surface level is 15 cm.

Proper resin content and uniform resin distribution are required for good mechanical properties as well as for weight and thickness control. Resin content is controlled by proper wiping action at the squeegee bars or stripper die, fiber tension, and resin viscosity.

Fiber collimation in a multiple-strand roving is also an important consideration to create uniform tension in each strand as well as to coat each strand evenly with the resin. For good fiber collimation, single-strand rovings are often preferred over conventional multiple-strand rovings. Differences in strand lengths in conventional multiple-strand rovings can cause sagging (catenary) in the filament-winding line.

The common defects in filament-wound parts are voids, delaminations, and fiber wrinkles. Voids may appear because of poor fiber wet-out, the presence of air bubbles in the resin bath, an improper band width resulting in gapping or overlapping, or excessive resin squeeze-out from the interior layers caused by high winding tension. In large filament-wound parts, an excessive time lapse between two consecutive layers of windings can result in delaminations, especially if the resin has a limited pot life. Reducing the time lapse and brushing the wound layer with fresh resin just before starting the next winding are recommended for reducing delaminations. Wrinkles result from improper winding tension and misaligned rovings. Unstable fiber paths that cause fibers to slip on the mandrel may cause fibers to bunch, bridge, and improperly orient in the wound part.

Filament Winding and Robotics.   Filament-winding equipment manufacturers have molded the developments of computer control systems with their multiaxis machines.[42–44]

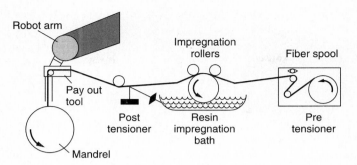

Figure 1.24   Fiber delivery path for thermoset wet winding with a robotic filament winder.[45]

Figure 1.24 is a schematic overview of the fiber path during wet winding with a robotic filament-winding station. Continuous fibers such as glass, carbon, and aramid fibers are dispensed from a tension compensator with variable tension. A resin infiltration bath for up to 12 fiber tows is mounted on the robot's waist. A large cylinder is continuously wetted with resin by rotating through the bath and, with the help of two additional smaller cylinders rotating above the first one, the fibers are impregnated by squeezing the matrix in between the spread fibers.

An important feature of the robotic placement fiber path tensioner is located between the resin bath and the payout tool, which is attached to the robot's end effector. In cases where the distance between the payout tool and the resin bath decreases faster than the fiber delivery speed, the tension controlled by the pretensioner would be lost. One way of keeping the tension is to have a motorized tensioner for rewinding fibers onto the fiber spool; however, this would cause a problem with fibers already wetted in the resin bath.

The fibers are then directed into the payout tool attached to the robot's end effector, and the robot arm control program allows for placement on a stationary or rotating mandrel. Various payout tools have been designed depending on the application and mandrel geometry.

The concept developed by Steiner[45] and his University of Delaware associates incorporates all elements of the winding process into the placement head, thus eliminating the need for delivery path compensation devices. The fiber spool carrying thermoplastic prepreg tape and a couple of friction-tension rollers are mounted in the head. Two nitrogen gas torches with a combined power rating of 1500 W are used to deliver the heating energy, and two air cylinders supply the consolidation pressure (Figure 1.25).

Future developments[46] include complete integration of numerous computer simulation programs and workstations. This research will provide a computer-integrated design, analysis, and manufacturing capability for robotic filament winding. As shown in Figure 1.26, the desired product properties, such as shape and performance, are the input into the design loop. The output, after passing through a number of simulation programs and interaction processes, will be the final filament-wound part.

It should be further noted, as reported by Vaccari,[47] that Haber, Hardtmann, and

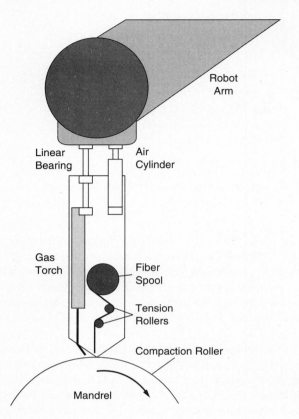

Linear
Bearing

Air
Cylinder

Gas
Torch

Fiber
Spool

Tension
Rollers

Compaction Roller

Mandrel

Robot
Arm

Figure 1.25   Thermoplastic head for robotic fiber placement that can expand current capabilities.[45]

Bubeck have developed a robotic winding system (ROWS) as shown in Figure 1.27. This machine allows automated fiber placement of continuous filaments, tows, and tapes over stationary mandrels. ROWS lends itself to use on parts that could up until now only be made by hand because of their complexity. Nonaxisymmetric parts that cannot be fabricated on a conventional filament winder can now be produced automatically with repeatable results.

The end effector, for example, can wind four tows simultaneously. Four spools, mounted on the rotating drum, rotate with the payout eye to preclude accumulated tow twist.

Since material throughput can also be increased by winding tape instead of tow, the end effecter is also equipped with a mechanism for dispensing tape and rewinding its backup paper on a roll. Because tape is stiffer, however, tape winding, like tape laying, is much more limited in the complexity of shapes that can be produced, tape conformance decreasing with increasing tape width. Simple shapes such as straight cylinders, for example, are ideal for tape winding.

To enhance flexibility, thermoplastic tow must be heated by the dispensing head just prior to winding. Heating then permits the bonding of each winding to the previous one. So-called hot heads for thermoplastics also have been developed for tape layers but so far are limited to the production of flat laminates.

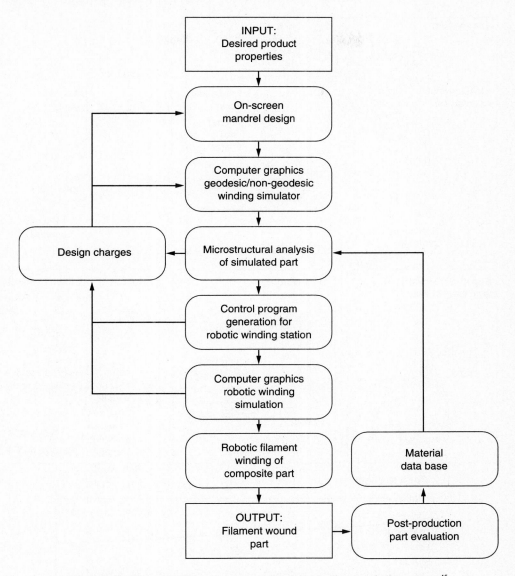

Figure 1.26   Schematic of the computer-integrated robotic filament-winding concept.[45]

Applications

Aerospace and Defense.   The primary aerospace applications for the filament-winding process are motor cases, launch tubes, pressure vessels, and survivable fuel tanks. Filament-wound composite motor cases have been used in the United States since the mid-1960s, beginning with such programs as Minuteman, Polaris, and Poseidon.

In the early 1970s, the U.S. Navy continued its emphasis on composite motor cases by developing a filament-wound C4 missile for use on the submarine *Trident I.* Kevlar 49

The system

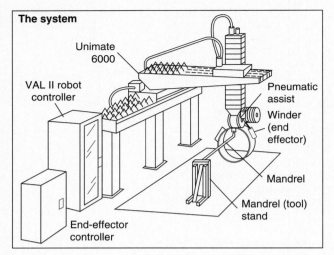

Unimate 6000

VAL II robot controller

Pneumatic assist

Winder (end effector)

Mandrel

Mandrel (tool) stand

End-effector controller

End effector

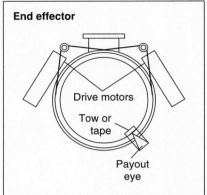

Drive motors

Tow or tape

Payout eye

End effector with long payout arm

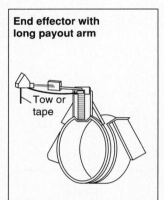

Tow or tape

Tape winding

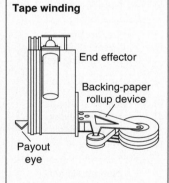

End effector

Backing-paper rollup device

Payout eye

"Hot head" (for thermoplastics)

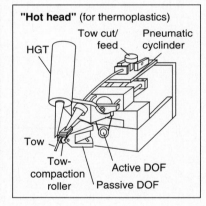

Tow cut/ feed

Pneumatic cyclinder

HGT

Tow

Tow-compaction roller

Active DOF

Passive DOF

Bond strength

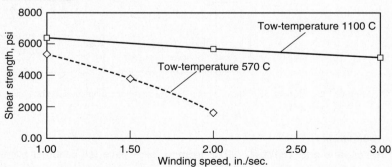

Figure 1.27   ROWS major components.[47]

was the material of choice. The largest intercontinental ballistic missile (ICBM), developed as the Peacekeeper, this 2.4-m-diameter motor case was filament-wound using Kevlar. The Pershing field artillery weapon developed in the late 1970s for the U.S. Army was also filament-wound using Kevlar 49.

The first use of carbon fiber in missile motor cases did not occur until 1980. The Sentry missile was an intercept system developed for the U.S. Army. It used a hybrid of Kevlar and carbon fiber in its high-strength, tapered case. This missile was later canceled and never went into production.

In 1982 the small ICBM or Midgetman system became the first to use an all-carbon-fiber, filament-wound motor case. The change to carbon fiber was made feasible by the development of a high-strength, high-elongation, intermediate-modulus carbon fiber.

U.S. Army and Marine Corps ground-to-ground and ground-to-air missiles have, with very few exceptions, utilized filament-wound E-glass composite launch tubes.

The Viper antitank weapon was developed in the early 1980s and used a small filament-wound motor case for its propulsion system. It also had a filament-wound E-glass two-piece telescoping launch tube. The system became the first man-rated, shoulder-fired, antitank weapon system, and it was to be manufactured at a highly automated production facility. However, the program was canceled in favor of another antitank system that has a filament-wound launch tube with a diameter of 203 mm.

The multiple launch rocket system (MLRS) is a ground-to-ground field artillery missile system in use by the U.S. Army. The motor case for the missile is metallic, but the 279-mm-diameter launch tube is filament-wound using E-glass. After a successful development program, the system went into full production. The system is unique in that it uses helical ribs, cocured to the internal surface of the launch tube, to impart a stabilizing spin to the missile. These ribs increase the complexity of the mandrel and make the tube difficult to remove.

The aerospace pressure vessel industry uses sophisticated filament-winding techniques to produce optimum-design, lightweight pressure vessels for containing gases at pressures of 21–42 MPa. These vessels can be either cylindrical or spherical and are used for flotation devices, pneumatic seat ejection mechanisms for aircraft, pressurization systems for pressure-fed liquid propellant rocket engines, and environmental and breathing systems for space vehicles and spacecraft. All these pressure vessels utilize thin, metallic liners that serve both as leakage barriers to contain the high pressure gases and as winding mandrels for the filament wrapping process. The majority of the liners are aluminum, but titanium and Inconel have also been used. The overwrapping mandrels have been predominantly fiberglass and Kevlar because of their high strength and elongation, damage and abrasion resistance, and fatigue endurance.

The most notable use of filament-overwrapped pressure vessels is in the space shuttle vehicle. Seventeen pressure vessels are used throughout the space shuttle in its environmental, pneumatic actuation, and pressurization systems. These vessels are all spherical in design and use a Kevlar overwrap as the reinforcing fiber.

Franklin[48] reported on several military applications using filament winding for fuselages and other asemblies in the Beech Starship and the CV-22 Osprey tilt-rotor helicopter.

Franklin reported on a material form, towpreg, which is used in the same way as unidirectional prepreg materials. Towpreg was formed into thin tape bands and placed di-

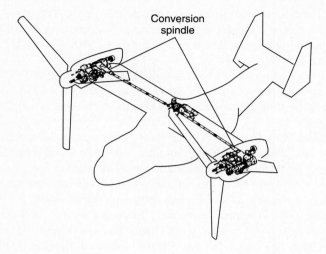

Conversion
spindle

Figure 1.28   Schematic view of the V-22 drive system.[49]

rectly on a mandrel surface, being compacted as it was applied. This in-process compaction required intimate contact between the machine delivery roller and the part at all times. In this case, the mandrel was D-sectioned to provide some variation in the machine capabilities, and the skin structure that was formed was interesting because it used a 8551-7 toughened carbon epoxy system to provide damage resistance.

A second structure fabricated was part of a fuselage developed for the U.S. Air Force in which the skin was fiber placed over a composite male mandrel containing cavities for cocuring I-beam stiffeners and ply reinforcements. Carbon-epoxy was used in the mandrel to control dimensions during thermal expansion in the cure cycle.

Finally, the structure of a composite wing was fabricated. The 609.6-cm wing spar laminate was fiber placed over an aluminum mandrel. Aluminum was chosen for this process to attain maximum expansion during cure and to compensate for the predicted inward springing of the spar flange corners during cure. The spar was optimized with carefully placed ply dropoffs, resulting in a heavy, strong root end and a lighter tip. Fourteen of the 70 plies were dropped off automatically using the fiber placement technology.

Oldroyd[49] described the use of filament-winding technology for the conversion spindle in the V-22 Osprey helicopter (Figures 1.28 and 1.29).

The V-22 conversion spindle is the structural detail that connects the engine and nacelles to the wing. It is the component about which the nacelles rotate to accomplish conversion from helicopter to airplane mode. The spindle contains 339 hand-placed unidirectional tape plies as well as 19 helically wound roving layers. It is a tapered cylinder 1422 mm in length, varying in diameter from 229 to 330 mm and in thickness from 12.7 to 31.8 mm.

Industrial.   The commercial and industrial markets for filament-wound products are varied. Some of the product lines include commercial pressure vessels, high-pressure piping for oil field use, and corrosion-proof underground petroleum storage tanks (Table 1.4).

The majority of commercial pressure vessels fall into the general classification of

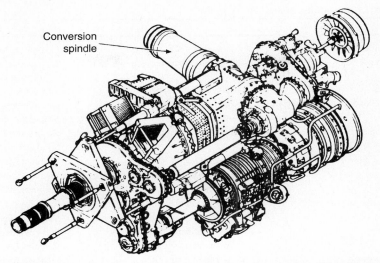

Figure 1.29   Schematic view of the conversion spindle, pylon support, and nacelle components.[49]

breathing apparatus, which includes backpack vessels for firefighters, medical oxygen packs for both hospital and home use, mountain climbing and spelunker backpack units, and scuba diving tanks.

Filament-wound underground fluid storage tanks have been available since the late 1960s. The major advantage of these tanks over their metal counterparts is that they do not corrode and leak the contents into the surrounding environment. Some state regulations have caused a renewal of interest in storing underground fluids in essentially leakproof tankage. Initially, the tanks designed for this market were of single-wall fiberglass filament-wound construction. Now, double-walled tanks are being fabricated. These tanks have a ribbed construction between the two walls to maintain a separation distance, and the between-wall cavity is used to sense any leakage that may take place, either outward or inward.

Filament-wound high-pressure pipe is used extensively in oil field installations, both for surface transport and downhole injection. Though most pipe is rated between 2.7 and 13.8 MPa, there are products available that have continuous-use pressure capabilities at 41 MPa. The joints (elbows, tees, etc.) are also filament-wound. All pipe uses E-glass as the reinforcing fiber, primarily because of cost. There has been some interest, as evidenced by prototype tests, in carbon fiber-reinforced pipe, particularly for containment of extremely corrosive fluids where the only alternatives would be extremely expensive metal pipes such as those constructed of zirconium.

Seifert and Skinner[50] have reported on their recent work in which the key to their approach was software which allows the design of any part shape and enables the operator to calculate the control data. CADWIND software for filament winding gave accurate results. Also, this software was user-friendly, and so machine operators without any computer programming experience were able to use it.

A low-cost multiaxis filament-winding machine (or robot) was required for production. Speed and reliability were vital in the production environment. The machine al-

TABLE 1.4.   Applications of Filament Winding

Examples	Materials	Processes
Corrugated roof slabs	Glass strand mat polyester	Continuous press molding
Cladding panels	Glass strand mat polyester	Laminating (hand layup) vacuum injection molding
Tension rods	Glass roving polyester	Pultrusion
Distribution boxes (telecommunications)	Glass strand mat polyester	Vacuum injection molding SMC press molding
Sanitary cabinets, bath tubs	Glass strand mat polyester	Resin transfer molding Sprayup molding
Wastewater pipes	Glass roving polyester	Filament winding
Driveshafts	Glass roving and carbon roving epoxy	Filament winding
Leaf springs	Glass roving epoxy	Pulforming
Spoilers, bumpers, front ends	Glass strand mat polyester	Resin injection molding SMC press molding
Engine hoods	Glass strand mat thermoplastics	Press molding
Racing cars (monocoque)	Carbon fabric epoxy	Hand layup
Containers, vessels	Glass roving polyester	Filament winding
Pressure vessels	Glass and aramid roving epoxy	Filament winding
Tubes, pipes	Glass roving epoxy	Filament winding
Clearing basin covers	Glass strand mat polyester	Hand layup
Fishing rods	Glass roving epoxy	Pultrusion
Tennis rackets	Glass and carbon fabric epoxy	Press molding
Surfboards	Glass fabric epoxy	Vacuum molding
Masts	Glass and carbon fabric epoxy	Filament winding
Boat hulls	Glass and carbon fabric epoxy	Filament winding
Safety buoys	Glass fabric epoxy	Vacuum molding
Insulating parts	Glass roving epoxy	Filament winding
	Glass chopped phenolic	BMC press molding
Special cases	Glass strand mat polyester	SMC press molding
Slot wedges	Glass roving polyester	Pultrusion
Printed circuit boards	Glass fabric epoxy	Press molding
Skin panels, elevators, rudders, radomes, wings, flaps, covers	Glass, carbon, aramic prepreg, epoxy, Nomex and aluminum honeycomb, foam core	Autoclave molding
Tanks	Glass and aramid roving epoxy	Filament winding
Struts	Carbon prepreg epoxy	Prepreg layup
Rotor blades	Glass, carbon, epoxy prepreg, foam core, honeycomb	Tool molding
Floor panels	Glass, carbon, aramid prepreg, Nomex honeycomb	Prepreg layup Autoclave curing
Cylinders, rolls, shafts, couplings, spindels, robot arms, coverings	Glass and carbon roving, glass and carbon fabric epoxy, polyester thermoplastics	Filament winding Press molding

lowed the manufacture of different types of parts (pipe, elbows, tees, and fittings) as well as different sizes. Filament-winding machines offered maximum flexibility and maximum speed with a PC-based controller, which easily handled the huge control data files and allowed simultaneous running of the winding machine and the CADWIND software.

Engine inlet cowls (for Boeing 747 and 767 commercial aircraft) were fabricated using a seven-axis, 24-tow Viper CNC fiber placement system[51,52] (FPS) which combines the advantages of filament winding, contour tape laying, in-process lamination, and computer control to automate production of complex parts that conventionally require extensive manual labor. The system is also particularly suited to highly contoured structures such as inlet ducts, fan blades, fuselage sections, spars, nozzle cones, tapered casings, and C channels, as well as cowls. Automated fiber placement control allows the FPS to maintain uniform thickness in the process for fabricating parts with tapered geometries such as inlet cowls. Since parts are made to near-net shape (NNS), subsequent machining and material waste are reduced, resulting in lower total costs.

A seven-axis capability permits desired fiber angles to be maintained for maximizing part strength and minimizing part weight. Fiber placement can also be used to fabricate compound-contoured parts with concave and convex surfaces or small radii. A compaction roller mounted on the head laminates the tows onto the layup tool's surface. This removes trapped air and helps to eliminate small gaps between tows for more uniform part quality.

Filament winding[53] has been introduced on thermoplastic tubes for bicycle frames. The process of winding remains nearly the same other than the fact that the fiber is coated with a thermoplastic material prior to winding. The potential benefits include slightly higher damping properties. The task of securely bonding the carbon-thermoplastic tube to an aluminum or a carbon-epoxy lug is the greatest hurdle.

This task is very important and significant for the incorporation of metal inserts in filament-wound structures which is a relatively recent development and appears to be focused on helically wound tubes or shafts with integrally wound metal end attachments. Principal applications include shafts with metal end couplings and pipes with threaded metal ends (Figure 1.18).

Metal attachments can be molded into the composite structure by winding over a metal tube that doubles as the mandrel (often used for pressure vessels) or by putting metal sleeves over the mandrel ends which can be wound over to yield integral end attachments. These end attachments can be tapered to give some degree of encapsulation that will impart greater axial strength to the joint.

There are obvious advantages to the integral filament winding of metal end attachments. The method is relatively quick and economical, requires minimum labor, and is readily automated. Metal end attachments are easy to set up and maintain in the desired position. They are encapsulated (at least circumferentially) by continuous fibers, thus increasing the structural integrity of the part while doing away with the need for drilling holes. Such construction allows the ready use of bolted joints and welds to connect the metal shaft ends to the mating metal universal joints in applications such as drive shafts for automobiles. Hybrid carbon-glass structures with integrally wound metal end attachments have been demonstrated to possess "outstanding reliability" and increased performance levels.[54] Finally the method allows for part integration.

Unfortunately, there are also several disadvantages to this method. Part geometries are limited because filament winding is best suited to the fabrication of axisymmetric parts with positive radii of curvature (e.g., circular cylinders). Although a good deal of work has been done with integrally wound end attachments, more work is required with radially oriented inserts within the body of the composite, which may actually be of more utility such as in connecting rods and high-performance camshafts.[55,56]

Automotive.   In the mid-1970s, interest developed in a one-piece composite driveshaft to replace a two-piece steel shaft used in light trucks. Analysis showed that a composite material, if sufficiently stiff and light, could do the same with a single-piece construction. The challenge was to develop a material combination and fabrication method that would be cost-competitive with the steel assembly. Limited production runs of filament-wound composite driveshafts took place during the mid-1980s, and the shafts were used on Ford Econoline and Astrostar models.

Two distinctly different designs were developed. The first combined a near-axial wrap of carbon fiber with a high-angle wrap of glass in vinyl ester resin. Steel end sleeves were wound and bonded and mechanically pinned to the composite shafts, and yokes were welded to the sleeves. The second shaft combined the two fibers into a single wind, the resin was epoxy, and the steel end sleeves were first welded to the yokes and then attached to the composite shaft. In both cases, filament winding was the manufacturing mode. A third design, currently being used for a van though utilizing hoop filament winding, is basically a pultrusion of carbon fibers with a hoop-wound glass cinching ply over a permanent aluminum core. Aluminum yokes are electron beam-welded to the aluminum core. All three versions are shown schematically in Figure 1.30. These mass-produced shafts were built with specially designed, dedicated winding machines.

1.3.4.11 Other winding techniques

Tape Winding.   Tape winding was formally introduced and patented in 1990 as a proven manufacturing process.

Filament- and tape-winding fabrication processes are quite similar with the exception of the material form being dispensed and the delivery mechanism best suited for each.

Tape winding employs, just as its name implies, a tape form comprised of unidirectional reinforcing fibers bound together with a plastic matrix (resembling fiber-reinforced packaging tape). Figure 1.31 illustrates the tape-winding process.

Tape winding results in circumferential placement only; winding is continuous placement in a true helical as well as circumferential pattern. Tape winding is a relatively economical means of producing a composite structure.[57]

Table 1.5 compares various conventional manufacturing processes with tape winding. In producing an inlet duct skin or a near-cylindrical fuselage skin, tape winding is shown to produce a weight-efficient part and to be a cost-effective process. However, tape winding might not be used if process selection is based solely on cost. Filament winding would probably be employed.

Tape winding is typically selected for products requiring very high quality and the most weight efficient design. Figure 1.32 shows the property retention trends of tape and filament winding compared with hand-laid tape properties. For the properties measured

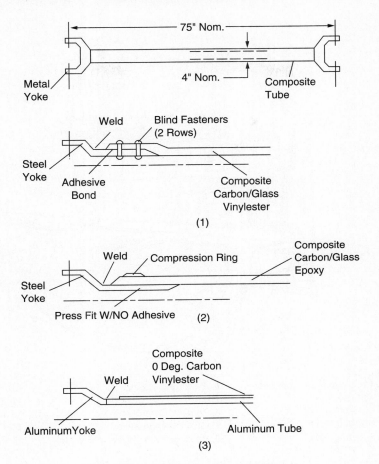

Figure 1.30   Composite driveshaft concepts.[2]

tape-wound properties are as good as those of hand-laid tape, whereas filament-wound properties are as much as 28% lower than tape-wound properties.

The test data used to develop Figure 1.32 was collected over a 5-year period and involved the following material systems: BASF's T300/5250-2 and H46/5250-4, Hercules' IM7/8551-7A and AS4/3502, American Cyanamid's T6300/X3100, and U.S. Polymeric's T6300/V378A. The benefits of designing with the superior properties of unidirectional tape include a significant part-weight reduction and an overall greater structural performance, according to Campbell and Kittelson.[57,58]

Tape winding is a proven process that allows the placement of laterally inflexible tape on irregular mandrel shapes. To fully understand the tape-winding process, a working knowledge of the parts, functions, and attributes of the winding machine is needed. The two major parts of the tape-winding machine are the host filament winder and the tape delivery system (Figure 1.31).

Figure 1.31 illustrates a typical computer-controlled filament winder modified by

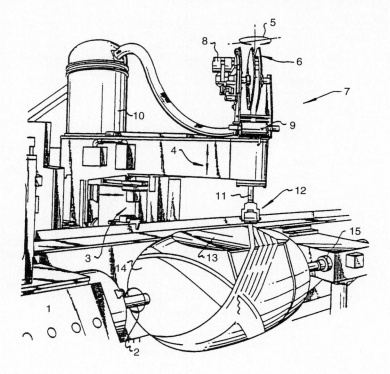

1. Programmable Controller
2. Spindle Axis
3. Horizontal Axis
4. In-Feed Axis
5. Rotating Axis

6. Packaged Tape
7. Creel
8. Tension Controller
9. Guide Components
10. Backing Paper Removal System

11. Roller Turning Axis
12. Delivery Head Roller
13. Mandrel Body
14. End Dome
15. Tail Stock

Figure 1.31   Descriptive overview of a tape-winding machine.[57]

replacing the conventional roving delivery system. The pictured delivery system shows the tape being placed from a vertical direction. However, this particular orientation is one of many possibilities.

The host machine must be understood, especially the function of each controlled axis, accuracy of motion, capabilities, and programming. The machine consists of four basic axes of motion. The spindle axis rotates the mandrel, and the horizontal carriage axis traverses back and forth along the length of the mandrel. The in-feed axis travels perpendicular to the spindle axis of rotation, and it aids in material placement when compound curvatures are present, such as the mandrel end domes. The rotating axis aligns the composite material to match the wind angle being placed.

Programming these machine axes to perform together winds the material onto the mandrel in a predetermined helical pattern.

Tape winding requires a host machine with specific accuracies and capabilities. The delivery system complements the host machine's capabilities by making laterally inflexi-

TABLE 1.5.   Weight and Cost Trade Manufacturing Processes[57]

Material Width [cm (in.)]	Typical Layup Rates[a] [kg/h (lb/h)]	Relative Part Weight	Relative Part Cost
Hand-lay tape			
15 (6)	1.1 (2.5)	1.0	1.6
Hand-lay fabric			
122 (48)	3.4 (7.5)	1.4	1.2
Contoured tape-laying machine			
15 (6)	4.5 (10)	1.0	1.2
Filament winding			
2.5 (1)	9.0 (20)	1.3	0.8
Tape winding			
2.5 (1)	6.8 (15)	1.0[b]	1.0[b]

[a] Comparisons were made considering manufacturing an inlet duct skin or a near-cylindrical fuselage skin.
[b] Baseline.

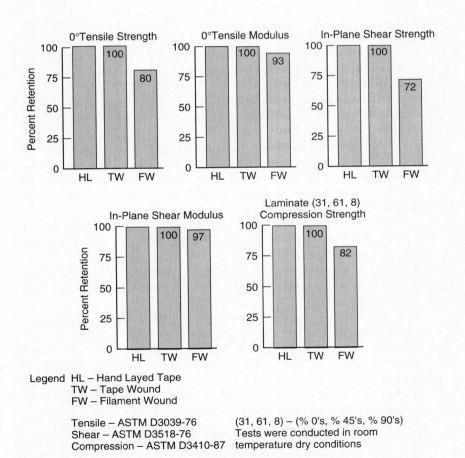

Legend   HL – Hand Layed Tape
          TW – Tape Wound
          FW – Filament Wound

Tensile – ASTM D3039-76   (31, 61, 8) – (% 0's, % 45's, % 90's)
Shear – ASTM D3518-76   Tests were conducted in room
Compression – ASTM D3410-87   temperature dry conditions

Figure 1.32   Mechanical property trends for filament and tape winding.[57]

ble unidirectional tape wind around the mandrel free of wrinkles or other anomalies. This tape delivery apparatus and process consist of two major components, a creel for dispensing the tape and a roller head assembly for placing the tape.

The creel is an apparatus that serves as a dispenser for the packaged tape. The creel uniformly tensions and guides the tape to the roller head assembly and removes and disposes of the backing paper. The creel must rotate to maintain close alignment (in real time) with the wind angle to prevent the tape from twisting or wrinkling. It is recommended that the mechanism for rotating the creel be servo drive-controlled.

The tension controller is a functional part of the creel, and it applies a controlled tensile force uniformly distributed across the tape's width. The applied tension must be sufficient to hold the tape on the mandrel without slipping and must enable the material to be laid down smoothly without wrinkling.

The delivery head roller is the final controlling instrument within the delivery system. The surface of the roller should provide for adequate friction between roller and tape to prevent slippage and material damage. For instance, a roller with a surface made from a polyurethane casting provides a good surface with composite tape. A roller that is convex or has a raised arch in the center (a crown) has the preferred roller geometry for winding composite tape.

Several examples of applications described by Campbell and Kittelson[57,58] include a thin-walled cylindrical fuel tank, a straight C-spar, a scaled aircraft fuselage skin, and an aircraft leading-edge skin.

The fuel tank demonstrated that a more structurally weight-efficient and higher-quality structure could be produced by tape winding rather than by employing filament winding.

Two straight C-spars were fabricated "back to back" by winding tape onto a 15.2 × 15.2-cm cross-sectional mandrel that was 3 m in length. Once again, placement of tape was accomplished without inducing any wrinkling. This project demonstrated that winding over flat contours transitioned by sharp corner radii is well within the capabilities of tape winding.

Aft fuselage skins have been tape-wound back to back using the F-16XL aft upper fuselage skin configuration. Test parts showed that parts with gentle negative complex contours can be produced by tape winding.

One-piece aircraft leading-edge skins lend themselves easily to the winding process, and tape-winding leading-edge skins continue to demonstrate the manufacturing versatility of the process.

In conclusion, there is one primary reason for choosing tape winding instead of filament winding. It produces a more weight-efficient structure. This is achieved by designing with better material properties. Tape-wound properties have been shown to be superior to filament-wound properties and equal to hand-laid tape properties. Tape provides uniform fiber distribution, bandwidth, ply thickness, and resin content, and these factors produce better material properties. Two other positive aspects are that the process is based on proven technology (making it readily available), and that tape winding is a clean manufacturing operation.

Other tubular structural applications come to mind as possibilities, such as the space station's support structure, launch tubes, and torpedo hulls. Tape winding can also be considered for the space station's hull, civilian aircraft fuselages, inlet ducts, and vacuum chambers for nuclear research. With the use of the proper techniques and equipment, any of these structures can be produced by employing the tape-winding process.

Tape-Laying Process.   With a 40% PPS thermoplastic resin and a 60% carbon fiber comprising a 1.27-cm tape, a proprietary thermoplastic fiber placement head and patented hot gas torch technology has been developed by Automated Dynamics Corporation.[59] This firm has created an in situ thermoplastic tape placement technology that has made it possible to replace aluminum in an industrial shaft application.

The shaft has a very high stiffness and load-bearing capability—allowing it to spin at higher speeds than those of the metallic shaft it replaced. A 2.54-cm tape has been demonstrated in an effort to increase throughput, however, in the future a 7.62-cm tape will be evaluated.

The tape-laying machine is capable of laying the fiber in any orientation, from 0 to 90°. The ability to lay the tape at 0° is crucial because of the direct relationship between fiber direction and tube axial stiffness and strength. The process also yields low void contents (on the order of 3–5%) and low residual stresses.

Thermoplastic Filament Winding.   The Institute for Composite Materials [Institut für Verbundwerkstoffe (IVW)] in Kaiserslautern, Germany, claims to have given thermoplastic filament-winding technology an edge in medium- and high-volume production of pressure vessels, stress-resistant pipes, and other structural applications. Laser-, infrared- and direct flame-based techniques for heating the roving and the contact surface were the heating techniques evaluated (Table 1.6).

Consolidation is effected by locally heating the two TP surfaces—one on the inside of the roving and the other on the outside of the object—just prior to their contact at the nip point (i.e., the point where the roving contacts the wound object). High winding speeds of up to 140 m/min and good filament stability are the result of this "dry" process. Unlike thermoset rovings, TP rovings require no postprocess curing—improving cycle time and reducing mandrel cost.

Because TP rovings remain solid during the winding process, precise control of the winding tension allows the filaments to be pretensioned. Higher and more variable winding angles (up to 45°) are said to be achievable.[60]

The IVW says that its laser-based filament-winding system is capable of winding at very high speeds. Polyphenylene (PP) rovings can be wrapped at speeds of up to 137 m/min, and PPS rovings at 90 m/min. Thermoset winding is generally limited to 60 m/min because at high rotational speeds the wet resin tends to sling off. A CO_2 (10.6-μm-

TABLE 1.6.   Comparison of Heating Techniques for Thermoplastic Filament Windings[60]

Technique	System Cost ($1000)	Operating Cost ($/h)	Winding Speed (m/min)	Tape Width (mm)	Heating Control	Crosswinding
Laser	260	12 (NdYag) 7 (CO_2)	60–140	To 6	≤1 ms	Difficult
Infrared	6	0.60	2.5–27	To 6	1 ms–5 s (approx.)	Difficult
Direct flame	4–40	12	30–60	To 30	1–10 s (approx.)	Good

Source: Institut für Verbundwerkstoffe.

wavelength) laser has been employed for its quick response time, and the high radiation density of lasers makes them a good choice for high-speed operation.

Lasers do have limitations, though. Rovings and composite tapes greater than 10 mm in width have not yet proven to be reliably processable with the IVW laser system. Cross-windings of narrow rovings, typically 3–6 mm wide, also tend to exhibit voids at the crossover point. The IVW has therefore developed a direct flame-based system that is capable of winding composite tapes.

Using this technique, the group recently cross-wound structural tubes for a 55.88-cm bicycle frame using winding angles of 17–45°. The resulting C_f/nylon composite structure composed of 3.81- and 4.32-cm-diameter tubes with wall thicknesses ranging from 1 to 2.2 mm, exhibited a rigidity equal to that of conventional steel frames and weighed only 1.01 kg.

Other industrial markets for applications include pumps, compressors, centrifuges, valves, and bushings, where a carbon fiber-reinforced PEEK resin has been developed and used.

In this case, the process is based on hot gas pretreatment of the prepreg, followed by a series of thermal posttreatments that reportedly enhance wear resistance and dimensional stability. XC-2, a prepreg tape form, has a maximum service temperature of 316°C, and its coefficient of friction against hard steel under unlubricated conditions is 0.12.

Fiber Lay Process.   A new fiber lay process is being proposed for the development of complex structures such as pylons, automobile chassis, and aircraft frames. It has already been used in the construction of a demonstration car in which several subframe, tie bar, and track control arms were built.

The fiber lay process is a method of winding or arranging resin-impregnated continuous-fiber rovings under controlled tension in a predetermined pattern and then constraining them in a particular shape. The fibers are held under tension until the resin has cured, similar to the way in which steel is used in prestressed concrete.

The fibers are normally laid to follow the lines of stress orientation, providing maximum strength at minimum weight and cost. An interesting feature of the process is that, whereas the joints are normally the weakest parts of a structure, they are in this case the strongest.

Among the projects recently undertaken was construction of a 75-kg fiber lay bridge. It spanned 4 m, was 800 mm wide, and could carry a 1-metric-ton load. The bridge was designed along conventional lines with straight members, replacing those made from rolled steel sections, in order to conform with customers' expectations.

A recently completed storm tank built using the fiber lay process took a day to complete. A steel tank of the same size (100,000 L) would normally have required a month to build, whereas a similar concrete tank would have taken 3 months.

1.3.4.12 Resin transfer molding

Introduction.   To utilize fully the benefits of advanced materials, the selected manufacturing process must permit the fabrication of large, complex, three-dimensional structures having anisotropic physical properties in selected areas. The ability to integrate multiple parts into single, reliable, reproducible components may also be critical.

RTM has the potential of becoming a dominant low-cost process for the fabrication of large, integrated, high-performance products for the consumer segment of the economy and ultimately for segments now dominated by the higher-precision laminated fabrication techniques.

Process Definition and Variables.   RTM is used today to manufacture a wide variety of articles ranging from small armrests for buses to large water treatment plant components. The process in its most basic form is shown in Figure 1.33. A dry reinforcement material that has been cut and/or shaped into a preformed piece, generally called a preform, is placed in a prepared mold cavity. The preform must not extend beyond the desired seal or pinch-off area in the mold to allow the mold to close and seal properly. Once the mold has been closed and clamped shut, resin is injected into the mold cavity where it flows through the reinforcement preform, expelling the air in the cavity and wetting out or impregnating

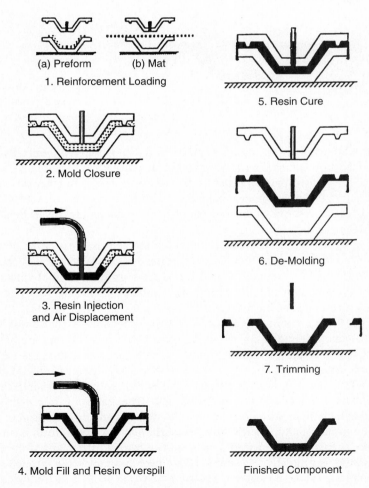

(a) Preform   (b) Mat
1. Reinforcement Loading

2. Mold Closure

3. Resin Injection
and Air Displacement

4. Mold Fill and Resin Overspill

5. Resin Cure

6. De-Molding

7. Trimming

Finished Component

Figure 1.33   RTM process.[61–64]

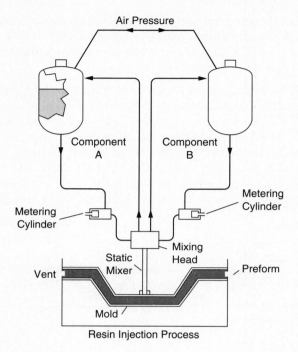

Figure 1.34 SRIM molding process schematic.[2] (Courtesy of Ford Motor Company.)

the reinforcement. The optimum range for the low-viscosity premixed resin is 200–300 cps and the most common resins used to date are polyester, vinyl ester, and epoxy.

The RTM process may also be used with unformed chopped mat or with a variety of unformed or preformed continuous reinforcement geometries. Since this process is carried out at low pressure (1–6 bars), lightweight molds, (even those constructed from PMCs) may be used.

When excess resin begins to flow from the vent areas of the mold, the resin flow is stopped and the molded component begins to cure. When cure is completed, which can take from several minutes to hours, it is removed from the mold and the process can begin again to form additional parts. The molded components may now require a postcure to further complete the resin reaction.

A very similar process is called structural reaction injection molding. The schematic in Figure 1.34 shows the basic SRIM process. Preform and mold preparation are similar to that for the RTM process, with changes in mold release and reinforcement sizings made to optimize their chemical characteristics for the SRIM chemistry. Once the mold has been closed, the SRIM resin is rapidly introduced into the mold and reacts quickly to cure fully within a few seconds. The reaction is in progress as the resin flows through the reinforcement, and therefore wet-out and displacement of the air in the mold must occur rapidly. There is seldom much runoff of excess resin in an SRIM component. Cure is complete shortly after the resin reaches the extremities of the component, and the viscosity is generally too great to allow much resin to escape through vents. When cure is complete, the component is removed from the mold and the process is finished. No postcure is normally done.

Before looking at more sophisticated RTM and SRIM processes, let's examine in more detail the similarities and differences between the two processes.

The greatest difference is in resin chemistry. RTM resins are typically low-viscosity liquids in the range 100–1000 cP. Normally resin systems have two components and require a preinjection mixing ratio in the range 100:1. The liquid reactants, part A and part B, can be mixed at low pressure using a static mixer.

SRIM resins are similarly two-part, low-viscosity liquids in the viscosity range 10–100 cP at room temperature. They are highly reactive in comparison to RTM resins and require very fast, high-pressure impingement mixing to achieve thorough mixing before entering the mold. Mix ratios of typical systems are near 1:1, which is desirable for rapid impingement mixing. A schematic of the SRIM resin delivery system is shown in Figure 1.34.

Preforms for RTM and SRIM systems are similar in most respects. Preforms for SRIM normally are lower in glass content or have additional directional glass which acts as a resin flow channel. Thus resin flow in SRIM is much more sensitive to preform construction than it is in the RTM process.

The planned release of trapped air from within the mold cavity as resin displaces it is achieved differently in SRIM and RTM. The faster cure in the case of the SRIM system minimizes the need to design elaborate seals and overflows into a tool to manage the overflow of resin from the tool cavity.

In numerous other aspects, the two systems are similar. As RTM resin systems are developed to be faster reacting and SRIM systems are slowed down to produce a longer filling time capability, the two alternative systems are approaching a single process. Figure 1.35 shows such an idealized process schematically as it could be applied to the manufacture of an automotive component. In this case, heated, flat sheets of glass reinforcement are stamped into a three-dimensional preform shell and subsequently installed over a molded foam core. After the addition of some small, high-performance attachment point preforms at certain high-stress areas, the assembled preform is transferred to a heated steel tool for resin injection. Up to this point, if the details previously mentioned have been considered in preform design and venting, the choice of RIM or RTM resins would be purely a matter of economics versus performance in the final part. The high-speed resin transfer molding (HSRTM) system envisioned in Figure 1.35 typifies the direction industry is moving in developing a unified approach to liquid molding.

Other differences between the two processes include the following:

1. In RTM there is the ability to position the preform in the mold that provides control of the fiber content and the mechanical properties. But in SRIM there is a tendency for the fiber to move during the filling process as a result of rapid fluid flow. RTM provides minimal movement of the reinforcement during the filling and curing process, which allows optimum performance at a minimum weight.

2. Because the RTM reactions occur inside the mold, RTM offers limited chemical exposure and limits the release of emissions during the process.

Along with RTM's advantages, there are also some disadvantages. Although the RTM process can be automated, considerable difficulty is associated with automating the process. Part of the difficulty involves impregnating high fiber volume fraction mats with

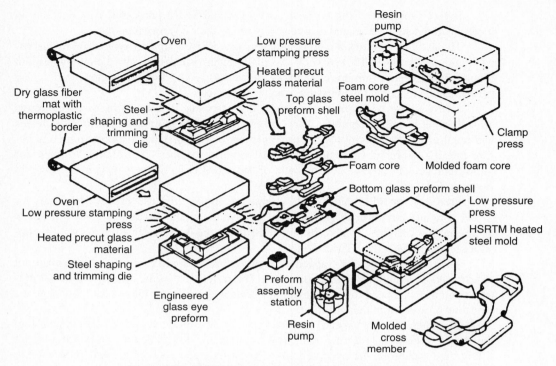

Figure 1.35   HSRTM process schematic.[2] (Courtesy of Ford Motor Company.)

resin-filled additives. The cycle times associated with the process can usually be considered an advantage, but they can also be a disadvantage if the selected resin system causes long cycle times. Another disadvantage of the RTM process is the lack of reinforcement at the edges of the preform inside the mold. Placement of the preform inside the mold is an advantage, but the time and labor associated with preparation of the preform is a disadvantage. Some other problems with the RTM process are associated with filling large parts containing a high glass content at low injection pressures and with the undeveloped nature of higher-speed versions of the process.

Process Variations.   RTM encompasses a range of techniques for fabricating composite structures, all involving the placement of a dry fiber preform in the NNS prior to applying the matrix resin (Figure 1.36). The method previously described utilized resin that was initially placed in a reservoir and flowed by a pressure differential between the reservoir and the mold outlet. The pressure differential could be caused by gravity, vacuum applied to the mold outlet, pressure applied to the reservoir, or a combination of all three. This type of process is generally termed pressure injection (PI) and is shown in Figure 1.36*a*.

Another variation of RTM involves flow through the thickness of the preform as in Figure 1.36*b*. A mold is required on only one side of the preform, although tooling components may also be on the opposite side to support features such as stiffeners. Resin is

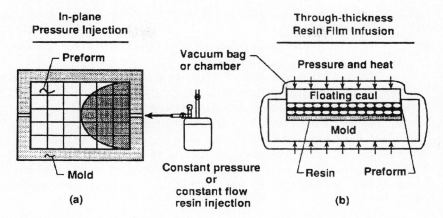

Figure 1.36   RTM variations. (*a*) Pressure injection. (*b*) Resin film infusion.[65]

typically placed on the mold surface in film form so that the preform may be placed and vacuum bagging material applied without uncontrolled flow of resin. To infiltrate the preform, air is evacuated from the vacuum bag and heat is applied. The resin flows through the preform, provided that a good resin seal has been created. The pressure required to compact the preform to the correct volume fraction depends on the type of preform and the structural requirements. If a pressure higher than atmospheric is required, an autoclave is used. This version of RTM is termed resin film infusion (RFI).

Other versions of RTM exist that incorporate features of PI and RFI. One example is the Seeman process.[66] In this procedure, a preform is placed diretly on the mold without a resin film. Within the vacuum bag layers, a series of porous materials is inserted. Resin, in liquid form, is placed in an external reservoir and then flows first into the porous medium. The resin is initially distributed over the area of the preform inside the porous medium, and then through the thickness of the preform.

Girardy[67] described RTM process optimization developmental work using Injectex high-performance textile reinforcements, developing preforming technologies, and perfecting of a defined range of resins for use with this process in France.

Process Variables.   RTM involves a large number of variables linked to the design of the component, the selection and formulation of the constituent materials, and the design of the mold and molding process. Each of these variables may have a significant effect on the events of the molding cycle. The relevant variables are described as follows.

Fiber Fraction.   The fiber fraction of the molding has a very important effect on the mechanical properties of the final part. However, changes in the laminate fiber fraction can also have an influence on the molding process. The effect of the fiber fraction was studied by Rudd, Owen, and Middleton[62] who considered the same constant laminate thickness and varied the mass of fiber in the preform using 18 and 43% fiber by mass [two to six layers of 450 g/m^2 continuous-filament random mat (CFRM), respectively]. Their results indicated that by using a fiber mass fraction of 18% (compared with the standard mass fraction of 43%) the overall cycle time was increased by 100 s or 60%

compared with standard molding conditions. This occurred despite a significant reduction in cavity fill time.

The cavity fill time was reduced by approximately 4 s or 50% owing to the higher permeability of the reduced fiber fraction.

The gel times were increased by approximately 100 s or 60% at all positions in the mold. This resulted from the increased thermal mass, which must be heated to mold temperature, and the reduced thermal conductivity of the laminate.

The peak temperatures recorded in the laminates were approximately 20°C or 12% less than those recorded for the higher fiber fraction. This was considered to result from the longer cure cycle, which released the heat of reaction over a longer period. The predicted temperatures were approximately 25% higher, which was attributed to the shorter cure cycle that was predicted, while in practice the reaction tended further to an isothermal reaction.

Injection Pressure.   Although injection rates can be increased using high-pressure gradients in the mold, increased pressure may also have detrimental effects such as fiber disturbance, losses in part tolerance, and mold damage.

Resin Preheat.   Increasing the temperature at which the resin is introduced into the mold has a twofold effect: the resin viscosity is reduced and the amount of heat that must be supplied by the mold to initiate cure is reduced.

Preform Preheat Temperature.   The preform preheat temperature refers to the temperature of the fiber preform immediately before the start of impregnation. The standard molding conditions involved the impregnation of a preform that was "hot" (i.e., at mold temperature) before the injection of resin. In practice, the temperature of the preform may vary between ambient temperature and the temperature of the mold. Owing to the transient nature of the heating process, it was difficult to produce laminates at intermediate preform temperatures.

Geometric and Design Capabilities.   The characteristics of RTM as a process have been defined, and now we examine the process in terms of part geometries. For an in-depth look at design and process tradeoffs, references 68–70 are recommended.

Size capability is among the major benefits of RTM. Large-area designs are feasible because pressures needed to mold components can be maintained at low levels. In some applications where cycle time is not critical, pressures can be maintained below 0.703 kg/cm². Alternative composite technologies such as compression molding can require pressures up to 140.6 kg/cm² over the part surface to achieve good physical properties and appearance.

In addition to large-area parts, RTM is well suited to the manufacture of parts that have deep draws and minimum draft on the sides of the part. Alternative fabrication technologies such as thermoset compression molding and thermoplastic compression molding are limited in the depth to which they can form by the pressures required for consolidation.

Although RTM is well suited to molding large objects, part complexity may restrict the use of the process. Unlike techniques in which the resin flows into the complex areas of the mold carrying the reinforcement with it, in RTM the reinforcement must be in place before the resin is introduced. While preplacing the reinforcement provides superior physical properties and eliminates flow-induced property variations, it also makes the use of features such as ribs and bosses very difficult. Any small detail or molded-in feature (e.g., holes and grooves) may be better achieved by using another process. There are of course alternatives to the use of some design features. Ribs can conveniently be replaced by closed, cored sections where stiffness is required. RTM allows the use of foam cores for added rigidity in a structure, as well as for providing increasingly complex three-dimensional structures to be molded in one piece. The low pressures used in this process allow foams in the 0.064–0.096 g/cm^3 density range to be molded in place without significant deformation.

One-piece resin transfer-molded components can in some cases provide performances in critical applications that are unattainable using a multipiece assembly of components fabricated by another process. The single-piece Ford Escort front structure integrates 45 individual steel stampings into a single resin transfer-molded component. The resulting component reduces the weight of the structure 30% over that of the normal steel structure while promising improved durability and crash performance. The one-piece nature of the front energy management rail is essential to the structure's performance.

Attachment areas between the structure and the loads it must carry often create a difficult design challenge for composite materials. In designing a resin transfer-molded structure, attachments must be included in the preform construction and they must have sufficient continuity with the bulk of the structure to ensure that stresses are distributed properly. Examples include braided glass preform subassemblies which retain the lower suspension arm of a truck cross-member. Stresses are distributed into the structure through the flange areas of the preformed attachment eye.

Other possibilities for attachments of resin transfer-molded structures include adhesive bonding and molded-in metallic inserts. RTM provides design flexibilty with respect to attachments, a feature that can and should be exploited in a good design. Compression molding, for example, requires the same attention to attachments but presents additional complications involving placement of attachment materials precisely in a hot tool and control of fiber movement during molding. The presence of flow and knit lines in compression-molded materials tends to make attachments less reliable.

Finite element (FE) methods are now available that represent what is likely to be developed in the near future for design with the RTM process. In most computer-assisted component designs produced thus far for RTM manufacturing, a classical NASTRAN* or similar finite element analysis (FEA) is used.[71] Because of the flexibilty available in designing fiber orientations in a preform, the RTM design can be very efficient with respect to material use. Only the reinforcements carrying loads need be included. Presently available laminate analysis can place reinforcements only in the x and y planes, but as more sophisticated preform technology is developed, such as three-dimensional knitting and weaving,[72] work progresses on more sophisticated analytical techniques.[73]

---

* NASTRAN—A general purpose, finite element, computer code used to conduct stress analysis of a laminate.

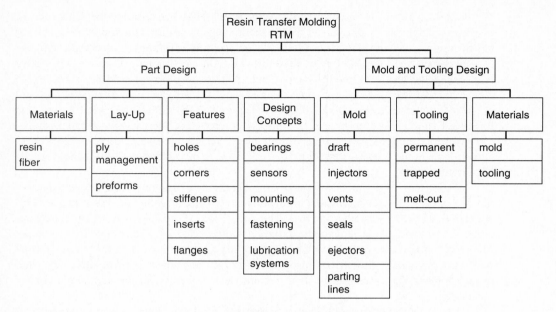

Figure 1.37 Prototype design tool architecture.[14]

Reinforcements and Preform Technology.[74] Design for manufacture (DFM) guidelines have been created (Figure 1.37) to assist engineers in the design of resin transfer-molded components incorporating many factors that affect the performance and producibility of the product, including resin, preform technology, layup, and inserts. These factors lead to new design concepts and result in geometry significantly different from that of the cast metal parts they often replace.

Using preshaped, preoriented reinforcement greatly simplifies RTM layup. Processes for creating preforms include mat forming, weaving, braiding, knitting, and winding. Weaving is used to create a wide range of fabric preforms for the RTM process. Fabric is created by weaving tows of glass, Kevlar, or graphite. The style of the weave affects the resulting properties of the composite. Fabrics that are tightly woven (plain weave) exhibit reduced stiffness and improved damage tolerance. Fabrics that are loosely woven (long-shaft satin or harness weave) exhibit improved stiffness and reduced damage tolerance when compared to tight weaves. A fabric's ability to conform to various geometries is controlled by the weave style. Tight weaves are less capable of conforming to complex geometry, whereas loose weaves exhibit better drape or textile conformity.

One significant aspect of RTM is the ability to add inserts into the mold along with the reinforcement materials. The use of inserts allows many features to be consolidated into one part, resulting in reduced assembly costs. Inserts can be made of any material capable of withstanding the temperature and pressure of the manufacturing process. The insert must also be compatible with the resin system and the design tolerances of the part. Metals, thermoplastics, and syntactic foams have been successfully used as insert materials.

The use of complex preformed shapes was reported by Horn[75] in his use of

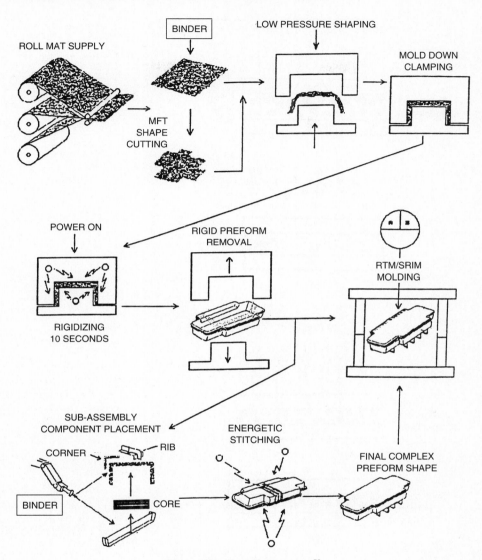

ROLL MAT SUPPLY

BINDER

LOW PRESSURE SHAPING

MOLD DOWN CLAMPING

MFT SHAPE CUTTING

POWER ON

RIGIDIZING 10 SECONDS

RIGID PREFORM REMOVAL

RTM/SRIM MOLDING

SUB-ASSEMBLY COMPONENT PLACEMENT

CORNER   RIB

BINDER   CORE

ENERGETIC STITCHING

FINAL COMPLEX PREFORM SHAPE

Figure 1.38   CompForm process.[75]

CompForm technology. The CompForm process provides a means of forming the fibrous reinforcement material into a shape that will be associated with the final structure of the molded product. The process allows simple shape structures to be formed and molded directly or as "carrier" preforms, allowing component reinforcement sections to be combined progressively into more complex shapes, (Figure 1.38). These preforms are later reconstructed into a final preform consisting of detailed assemblies of different components. This progression of complex shape forming can be called "energetic" stitching and uses the same process methodology as the initial reinforcement forming process.

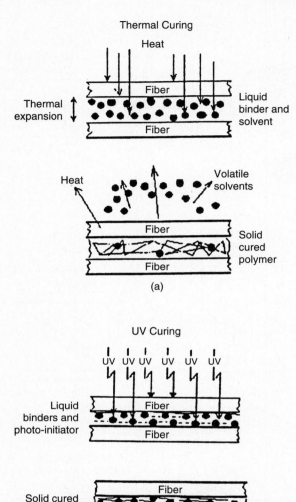

Figure 1.39   (a) Thermal curing.[75] (b) UV curing.[75]

With conventional thermoset preforming, heat and air drying must be incorporated to drive off the volatile solvents and cure the binders. This in turn reduces laydown and causes emission of environmental pollutants (Figure 1.39a). Rather than heating the entire reinforcement mass, UV cure takes place instantly. This rapid curing prevents loss of binder thickness due to thermal expansion, allowing for better compaction of the fibers and rigidity of the preform. The entire amount of binder applied as a liquid remains in the final preform (Figure 1.39b). The heat generated by photopolymerization of the binder immediately dissipates into the reinforcement fibers, and any heat generated by the small mass fraction of the binder to the reinforcement material becomes insignificant.

Therefore, with the high speed of the reaction mechanism of the binder resin and the low energy requirements for achieving rigid reinforcement forms, subassembly of the component sections can be achieved at a high rate. By incorporating subsequent robotics at different stages of an assembly operation, various levels of automation can be set up and integrated into all phases of preform manufacturing. Shapes previously unattainable with conventional technology can now be fabricated through simple, fast, progressive subassembly operations. The ability to assemble component materials in the manner shown in Figure 1.38 allows all the reinforcements to become an integral part of the final molded product, ensuring repeated material placement and structural integrity.

The design freedom allows the high-performance capabilities of the aerospace composites to be merged with the high-volume aspects of automotive manufacturing.

1.3.4.13 Process materials.

The key to choosing a resin transfer-molded component involves the resin chemistry and the forms, shapes, and preforms.

Resin. There are numerous resin chemistry choices for a resin transfer-molded component. Any resin must have the ability to fill the mold completely, wet out the reinforcement, and cure or solidify to a solid state with the desired physical property. The two major criteria in selecting a resin system are processing and performance. The resin rheology must fit the manufacturing requirements of the application so that good-quality parts can be made reproducibly. But just as importantly, the performance of the resin must provide the part with mechanical properties suitable for the environment to which it will be exposed.[76]

Hackett[76] reported on a PR 500 one-part epoxy system that can be used very effectively in the RTM manufacturing procedure and provides a balance of composite properties that are exceptional. The high-glass transition temperature of 205°C Differential Scanning Calorimeter (DSC midpoint) and very low moisture absorption level combine to provide an excellent hot-wet performance. This level of performance may make PR 500 an ideal resin system for aerospace applications that do not require the extremely high-temperature characteristics of BMIs but which require a better performance than can be provided by two-part epoxies. The resin also imparts excellent toughness properties that are not easily obtained in a high-modulus, high-glass-transition-temperature resin.

RTM resins have been developed that combine the low viscosity and long pot life required for good processability with mechanical properties comparable to those of standard 125 and 175°C curing epoxy resins. Several RTM epoxy and polyester[77] systems reported to fulfill these requirements are commercially available. In addition, BMI systems are being developed that are suitable for RTM either at room temperature or above. Successful RTM of demonstration parts has been carried out using both BMI resins and polystyrylpyridines, which are novel heterocyclic aromatic polymers. Other novel resin systems that may be suitable for RTM include acetylene-terminated systems, benzocyclobutenes, allyl nadic imides, BMI styrylpyridines and so-called H-resins which are based on acetylene-terminated polyphenylene polymers.

Bertram and MacPuckett[78] have reported on their work with cross-linkable epoxy thermoplastic (CET) resin technology. They claim that this technology can provide materials that combine excellent compressive strength, damage tolerance, and moisture resistance. XUS 19020.00, an experimental CET resin that is both tough and processable by RTM, is a lower-cost alternative to traditional hand layup of prepregs.

XUS 19020.00 contains a nonamine curing agent that increases the available pot life and molding time while providing elevated-temperature stability for molding flexibility.

The resin can be processed by RTM, hot melt prepregs, or automated tow placement without any modifications in the basic composition. Manufacturers also are saved time and trouble of formulating because the one-part system already contains a curing agent and a latent catalyst.

The temperature window for RTM processing of XUS 19020.00 is 95–120°C, which is easily achieved in standard molding equipment. At 120°C, the resin maintains an ideal RTM viscosity of 0.2 Pa·s (200 cps) for more than 3 h.

Cure temperature has an insignificant effect on the properties of neat (unreinforced) CET resin. This provides flexibility in designing cure schedules for different parts and processing methods.

Its excellent hot-wet performance, compressive strength, toughness, and moisture resistance make RTM-processable XUS 19020.00 CET resin suitable for use in a number of subsonic aircraft primary structures and adhesives. One aircraft company is currently evaluating the resin for primary-structure applications. Other manufacturers are exploring its potential for gas turbine engine airfoils, airplane floor beams, and RTM wing construction.

Salem and his associates[79] have also reported work on fiber glass-reinforced, polycarbonate matrix composites that they successfully fabricated by a RTM process using cyclic (BPA) polycarbonate oligomers. Woven fiberglass composites with glass loadings of up to 72 wt% and polycarbonate molecular weights of up to 50,000 were made. Although parts made by RTM had some matrix voids, they were not intrinsic to either the reaction or the process. Adding a consolidation step after RTM yielded parts with flexural strengths and moduli within 15% of those for comparable compression-molded parts, giving the materials properties suitable for structural applications.

Although this study demonstrated the feasibility of fabricating BPA polycarbonate structural composites by RTM using cyclic oligomers, further developments in reaction chemistry and process engineering are needed to fully exploit the potential of this material system.[79]

Reinforcement.   Reinforcement material selection is likewise seemingly limitless. Glass (both E-glass and S-2 glass), aramid, and carbon fibers are commonly used. Wood fiber and other synthetic fibers such as polyester are possibilities, depending on the requirements. Metal can be an excellent choice for local reinforcement in a structure. Metals should be noncorrosive or properly protected against moisture or other environments prior to inclusion in the preform.

Any reinforcement must be capable of being preformed into the desired configuration. The economics required of the preforming process depends on the application. Aerospace and low-volume applications can utilize cut-and-sew preforming. In a cut-and-sew preform, areas of material, typically fabric, are defined based on the requirements determined in the FEA. The cut-and-sew process lends itself to easy translation from finite element model to component, but it is slow and labor-intensive.

Higher-volume preforming calls for a faster process. Materials used for stamping are typically continuous-strand random glass mats in which the reinforcement is in a

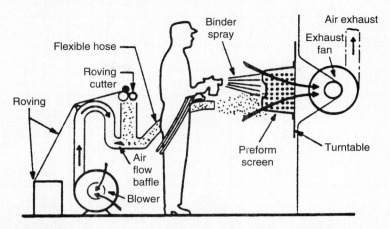

Figure 1.40   Spray-up preform fabrication schematic. (Courtesy of PPG Industries.)

swirled configuration. Both thermoplastic and thermoset binder systems are available to retain the formed shape after stamping. Binder proportion typically ranges from 4% to 8% by weight and can vary throughout the thickness of the preform if desired.

Although rapid and convenient for preforming large structures, stamping of fully random materials is limited in its level of performance. To be more competitive with alternate materials such as steel and aluminum with respect to weight and cost, composite structures normally require the use of either oriented glass or some additional type of reinforcement such as carbon or Kevlar. It is possible to include limited amounts of conformable oriented materials with the random material, subsequently forming the entire sandwich of materials by stamping.

Spray-up of reinforcement presents an additional option for the formation of preforms. Figure 1.40 is a schematic of the random spray-up technique. A perforated screen is made that conforms to the shape of the desired preform. Vacuum is applied to the rear of the screen, and normally the screen is simultaneously rotated about the x axis. A chopper system cuts the reinforcement into short lengths, and a binder spray is also applied.

Once the desired thickness of reinforcement has been achieved, the chopping system is turned off and the preform and screen are transferred to a baking oven where the binder cures and rigidizes the preform. Once stabilized, the preform is cooled, removed from the screen, and trimmed if required. Spray-up systems have demonstrated a capability for manufacturing large preforms at rates as fast as one per minute.[80–83]

Future preform systems will likely use a combination of highly automated technologies. Computerized braiding and filament winding combined with the three-dimensional screen of spray-up are being investigated for use in fully integrated design and preform manufacturing systems.

Forms.   In Figure 1.41 the reinforcement forms are illustrated and textile forms are arranged according to dimension and axis.[84,85] Such forms may be one-, two-, or three-dimensional in nature and have from zero to four or more axes. Examples of two-dimensional forms include woven fabrics and unidirectional prepreg tape. Three-dimensional forms include fibers in the z direction such as integral 3-D weaves and cut-and-sew preforms.

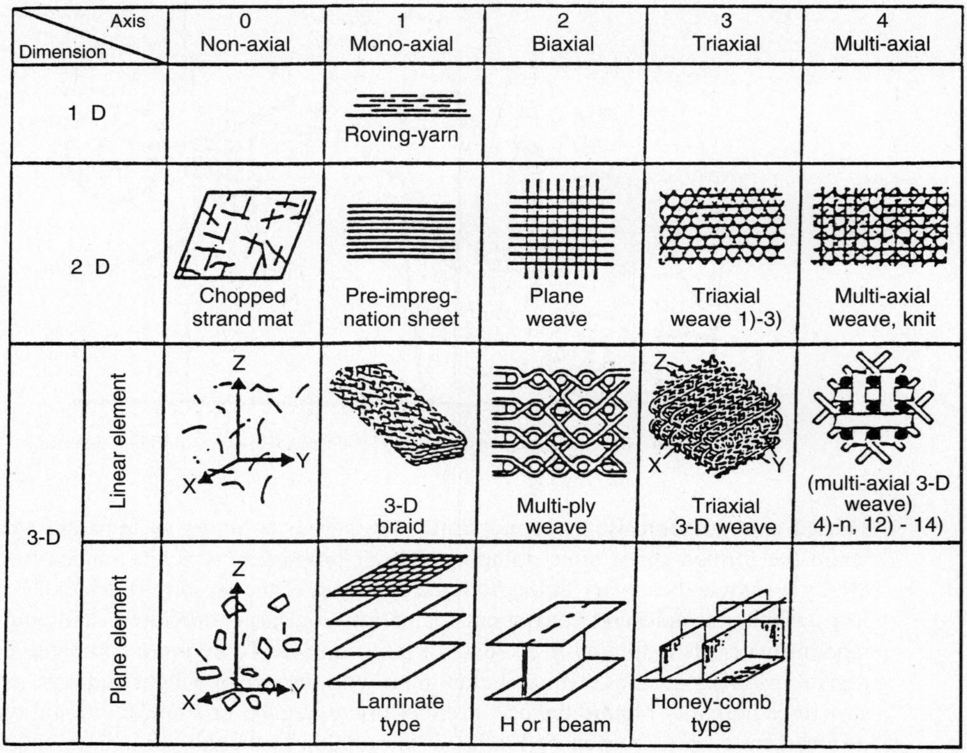

Dimension \ Axis	0 Non-axial	1 Mono-axial	2 Biaxial	3 Triaxial	4 Multi-axial
1 D		Roving-yarn			
2 D	Chopped strand mat	Pre-impregnation sheet	Plane weave	Triaxial weave 1)-3)	Multi-axial weave, knit
3-D / Linear element	(Z, Y, X axes diagram)	3-D braid	Multi-ply weave	Triaxial 3-D weave	(multi-axial 3-D weave) 4)-n, 12) - 14)
3-D / Plane element	(Z, Y, X axes diagram)	Laminate type	H or I beam	Honey-comb type	

Figure 1.41 Classification of textile forms.[84,85]

There are a number of preforming technologies available for RTM. Some of them, however, are not suited for primary load applications. Other technologies are still in the development stage and may yet prove to be viable for advanced composites.[86] The following textile technologies are available for RTM.

1. 2-D weaving
2. 3-D weaving
3. Knitting
4. Cut- and-sew preforming
5. 2-D braiding
6. 3-D braiding.[87–96]

2-D Weaving.   2-D weaving is accomplished by interlacing two sets of yarns perpendicular to each other. In 2-D weaving, the 0° yarns are called the warp and the 90° yarns are called the fill. The 0° direction or the warp is the continuous fabric direction. Variations in standard 0°/90° fabrics are obtained in several ways. Figure 1.42 shows some of the more common styles used in the advanced composites industry.

BASIC WEAVE TYPES

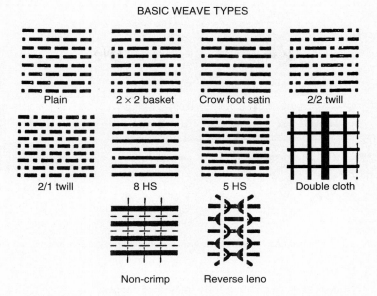

Figure 1.42 Basic weave types used in the advanced composite industry.[97]

For RTM, a series of woven fabrics can be combined to form a dry layup, which is placed in a mold and injected with resin. These fabrics can be preformed via the cut-and-sew technique discussed above.

The advantages of woven fabrics are as follows.

1. They are currently used for prepregs, and used in production processes.
2. Properties in flat laminates are well understood.
3. The advanced composites industry is comfortable with fabric use.
4. They are extremely versatile (0°/90°, 45°, unidirectional, hybrid fabrics).

The disadvantage of woven fabrics is that preforming currently requires extensive manual labor in the layup.

The potential uses for this technology are in manufacturing access doors and curved skins and in downstream cut-and-sew preforming.

3-D Weaving.   Three-dimensional weaving is accomplished using basically the same weaving operations as those described above for 2-D weaving. The differences between the two are in how the warp yarns are set up and how the take-up of the fabric is accomplished. In 2-D weaving, all the warp yarns interlace with the filling yarns. In the case of 3-D-woven fabrics, however, only the z yarns interlace with the filling yarns. In this manner, the warp yarns that do not interlace become the 0° yarns. Figure 1.43 shows different shapes that can be achieved with 3-D weaving.

The advantages of 3-D weaving are as follows.

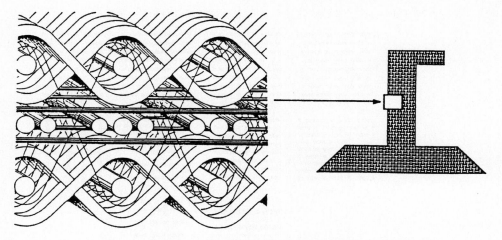

Figure 1.43   Three-dimensional weaving—microstructure and finished preform.[84]

1. It is capable of creating fully integrated shapes.
2. It has high interlaminar strength.

The disadvantages of 3-D weaving are the following.

1. It is a high-cost process.
2. Three-dimensional weaving rates are relatively slow compared to 2-D weaving rates.

Potential uses are for ablative applications and in complex and impact-resistant structures.

Integrally woven blocks have been produced for the fabrication of CCCs for over 25 years. Net-shaped molded composites with a fiber V_f of 35–60%, depending on the type of fibers and fiber architecture selected, and preforms have been produced with ceramic fibers and with carbon fibers having moduli of 230–725 GPa.[96]

Knitting.   There are two kinds of knitting operations: weft and warp. Both of these produce interlooped structures. These methods differ in that weft knits are formed in the weft or horizontal direction, whereas warp knits are formed in the warp or vertical direction.

There are several types of weft-knitted fabrics, including plain, purl, rib, and interlock. They can be made on a pressure foot knitting machine, which can be automated to a high degree from design through production and can produce net shapes, like a T-section and a nose cone. Similar to stockings, they have a high degree of stretch in the weft direction, making them comfortable to a wide variety of shapes as preforms for RTM.

The advantages of weft-warp knits are as follows.

1. They can conform to tool surface by stretching.
2. They can be easily automated for reproducibility.
3. They can form net shape parts.
4. Once in operation, they require little manual labor, hence are low in cost.

Disadvantages of weft-warp knits are the following.

1. Because of the bulk factor, a high fiber V_f cannot be obtained.
2. Isotropy is difficult to achieve.
3. Composite properties are relatively low.

Potentially knitted structures will be used only in non-load-bearing areas such as aircraft ducting, nose cones, and spinners.

Cut-and-Sew Preforming.   The term cut-and-sew is a catch-all phrase for all preforming processes that convert two-dimensional materials into three-dimensional shapes ready to drop into a RTM mold.

The cut-and-sew technique is at least as versatile as the prepreg processes in building a composite part. Materials may be laid up in any orientation, using as many plies as necessary to achieve the desired thickness. Ply dropoffs can be handled with ease, and foam cores or metal inserts can be incorporated.

Once the patterns are cut, they are assembled into the shape of the part being produced, including the positioning of inserts or cores. They are then held together by one of three methods: sewing (by hand or machine), stitching or quilting, or the use of an uncatalyzed solid epoxy resin that is either sprayed between the plies as a molten material or dusted on and the heat set later. Usually only 2–4% epoxy is needed to hold the preform shape.

Several techniques can be used to speed up the production process for cut-and-sew preforms. One is the use of prestitched *blankets* containing the needed plies in the proper orientations.

Advantages of cut-and-sew preforms are as follows.

1. Preforms are produced NNS.
2. Preforms usually have a high degree of structural integrity.
3. All types of woven fabrics and multiaxial, multilayer warp knitting (MMWK) materials can be used.
4. High fiber volumes and properties are obtainable.
5. They are easy to design because flat laminate data can be generated.
6. Parts are infinitely tailorable and reproducible.

The disadvantages of cut-and-sew preforms are the following.

1. In general, the process is still somewhat labor-intensive.
2. Stitching potentially damages the reinforcing fibers.

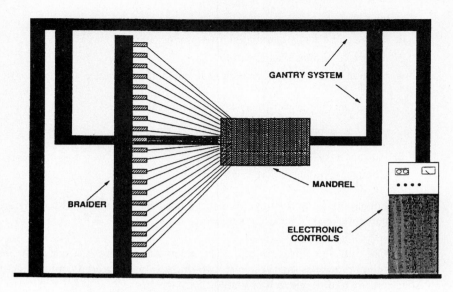

Figure 1.44   Braiding machine with gantry and mandrel.[84]

Potential uses are in integrally stiffened structures, including bulkheads, platforms, housings, and complex 3-D structures with shear walls or foam cores, such as propeller blades.

2-D Braiding.   Two-dimensional braiding is a process that is used universally in the production of items ranging from shoelaces to steel cabling and composite preforms for RTM. Braiding should be selected if the part being made has a high aspect ratio (length divided by diameter), is not large in diameter (typically less than 30.48 cm), and has no areas that are concave (depressions).

The braiding process is fairly simple and is illustrated in Figure 1.44. The braider contains a number of fiber carriers, usually from 72 to 144, which move around it in a maypole fashion, interlocking fibers under each other as shown in the upper right diagram in Figure 1.44. More information on braiding and processing appears later in this chapter.

Braiding is very adaptable to automation because the machine controls all movements and fiber placement. There are some computer routines for designing braided preforms based upon the needed properties and translating them into machine instructions.

The advantages of 2-D braiding are as follows.

1. It has low labor requirements and can be automated.
2. It produces seamless preforms.
3. Parts have high torsional strength and stiffness.
4. Parts have good damage tolerance.

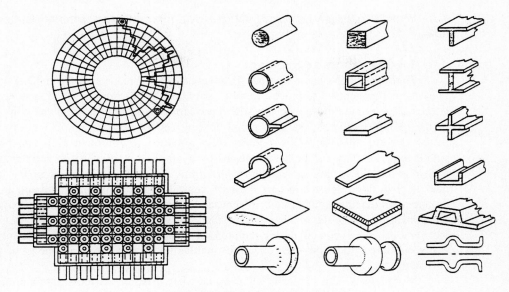

Figure 1.45   3D braiding machines and structures.[84, 98]

The disadvantages of 2-D braiding are the following.

1. It is limited to convex shapes.
2. Quasi-isotropic properties cannot be obtained because of fiber angle limitations.
3. The high ratio of machine diameter to part diameter requirements limit commercially available machines to approximately 30.48-cm-diameter parts.

Potential uses are for long, closed section parts such as missile cases, driveshafts, airfoils, control rods, and other tubular structures.

3-D Braiding.   The 3-D braiding process is an extension of 2-D braiding, similar to the relationship between 2-D and 3-D weaving. In 3-D braiding, the basic motion consists of movement of all carriers one step followed by a compaction step, then carrier motion, and so on. Virtually any three-dimensional shape can be created, some of which are displayed in Figure 1.45.

The advantages of 3-D braiding are as follows.

1. It is able to make complex, fully integrated 3-D shapes.
2. It produces highly damage-tolerant structures.

The disadvantages of 3-D braiding are the following.

1. It is limited to narrow widths.
2. It is a very slow, expensive process.
3. It is not production-viable at the present time.

Potential uses are for small, intricate parts like stiffeners, connecting rods, and tubes.[97,98]

1.3.4.14 Process equipment

Tooling.   RTM has a number of advantages with respect to the tooling used to fabricate components. The process can be carried out at low pressures and in some cases at pressures below atmospheric pressure. Low pressures reduce the cost and complexity of the tooling required. Low-cost epoxy tooling can be used and is in fact employed in the majority of today's low-volume RTM production. Although tools can be very simple and clamping can be accomplished by a simple system of perimeter clamps, the tool must be closed and sealed during the filling process to achieve complete, reproducible resin flow. Epoxy tools are normally used where volumes of production are low; an upper production limit of several thousand parts is typical. As the complexity of a component increases, the number of parts that can be produced decreases because of tool wear and damage occurring during the preform-loading process. Epoxy tools are not normally multipiece tools but can sometimes include removable inserts in one or both tool halves.

Mold releases must be carefully selected when using epoxy tools. Internal mold releases (i.e., releases contained in the resin system) cannot be used if they require a heated tool surface to be effective. Coatings of wax, silicone, or polyvinyl alcohol are normally employed with epoxy tools and may require application after every molding.

Venting of the epoxy tool is required to allow air to escape during injection. Vents may be gaps in the tool seal or tubes inserted through the tool surface. In the case of very low volumes, holes can be drilled through the tool surface at critical points, such as high points where air would naturally be trapped during resin fill. Vents require cleaning between moldings, and so they should be designed for easy maintenance.

The accuracy and surface finish of an epoxy tool depend strongly on the quality of the pattern used to make the initial tool casting. Parts having an acceptable surface finish can be made using these tools, and with hand finishing parts can be made to automotive class A specifications.

A high-quality surface finish can be achieved using nickel shell tools, and because the tool can be readily heated to elevated temperatures (149°C is typical), a wider variety of low-profile resins and mold releases can be used. Cycle times can also be reduced using heat-activated catalyst packages, and there is no danger of damaging tools with excessive exotherms during cure, as in the case of epoxy tools. Cure can even be controlled from a hot surface to a cooler surface to give an improved surface finish. Typical volumes for nickel shell tooling range from several thousand upward to 20,000–40,000 parts annually, which is significantly better than those for epoxy tools.[99]

Venting and sealing of these tools is accomplished in a manner similar to that used for epoxy tools, but vents are more difficult to modify. It is often wise to prove out the filling and venting system for a mold making one epoxy tool before making nickel shell tools.

Cast tooling presents an additional option for medium-volume RTM tooling. Solid zinc alloy castings or thin shells of cast aluminum or steel can be effectively used for RTM molding. These alternatives give advantages similar to those of nickel shell tools but eliminate some of the distortion problems associated with two-sided nickel shell tools. Modification of the sealing and venting are again more difficult than for epoxy tools but easier than for nickel shell tooling. Care must be taken to design tooling that is not too heavy for available molding equipment.

At high volumes or where complex multipiece tools are required, a machined tool steel mold should be used. Steel tools are high in cost but offer lifetimes of 500,000–1,000,000 parts. The complexity of steel tool design dictates that gating, venting, and sealing of the die be determined prior to tool construction. Steel provides the optimum in cycle time, surface finish, part-to-part repeatability, tool durability, and automation; but because of its high cost, volumes must be in the 10,000–50,000 range to justify the added expense.

Presses.   RTM is a low-pressure process that achieves maximum economic viability when used to fabricate large-area, highly integrated composite structures. Parts molded often include deep sections and molded-in foam cores or other inserts. Press equipment used in the process reflects these characteristics of typical molded components—that is, large platen area, low clamping tonnage, and often large daylight openings to facilitate loading of a complex preform.

There are several available options for low-cost, low-pressure presses for RTM molding. An airbag press uses a series of pressurized rubber bags or canvas hoses to apply clamping pressure to a mold. This is a good alternative for low-volume or prototype fabrication but is limited both with respect to daylight and by its inability to lift the upper mold half during demolding. Adding a mold shuttle or using a secondary low-pressure cylinder to manipulate the upper mold half overcomes this limitation but adds to the cost of the press system. Another alternative for press systems is a low-pressure hydraulic press.

Conventional high-pressure presses can of course be used for RTM molding, but the added expense reduces the advantage of RTM as a candidate fabrication process. When using high-pressure presses with low-cost or lightweight tooling, care must be taken to ensure that the press is controllable at the low pressures required to avoid over-compression of the tool.

Preform Equipment.   The development of an economical preforming process for the component to be molded is a key factor in the overall economic viability of an RTM alternative. Selection of the appropriate preform technique, reinforcement material, and preform equipment can make the difference between a successful RTM application and an economic failure.

Cut-and-sew preforming requires a minimum of equipment but is least efficient with respect to productivity. The process can consist of hand-held cutters and a cutting table, or it can be automated by adding steel rule cutting dies or possibly a robotically controlled cutting table. Preform assembly can be done by hand, which is typical for cut and sew, or it can be automated. Automation of the cut-and-sew process is being slowly implemented for prepreg layup for aerospace applications, but this preform technique is likely to be very expensive and primarily suited to low volumes of complex, very high-performance components.

The spray-up preform process yields NNS preforms, with higher productivity levels possible. The equipment is large and consumes considerable floor space but is low in cost. A cycle time of one preform per minute is achievable using this technique.

Stamped preforms can be produced at very fast rates. Although the process requires more trimming equipment than the spray-up method, continuous reinforcement can be used with this technique, increasing the performance level of the components made.

Preforming processes such as braiding and filament winding may be appropriate for preforming simple geodesic shapes. Braiders can provide an interlocked triaxial preform that exhibits high values of tensile strength and stiffness in the finished composite. Braiding applies all three orientations of reinforcement simultaneously and normally produces a preform structure faster and with less capital cost than filament winding. Filament winders normally apply reinforcement along a single well-defined path, resulting in slower preforming rates but increased placement accuracy and an associated increase in performance. Equipment complexity and cost are higher for filament winding relative to braiding. In either technique, care must be exercised to keep the reinforcement content below the upper volume percentage limit required to achieve the cycle time needed for resin impregnation. Both processes are capable of placing reinforcement in a preform at levels that are too dense to impregnate with resin.

Resin Injection Equipment.   Liquid molding resin-handling equipment varies in complexity and cost with cycle time and resin reactivity. The most basic RTM setup uses a one-component resin system and a simple pressure pot for resin injection.

Resins that cure at room temperature can be used with a pressure pot, but the system must be thoroughly cleaned after each shot to prevent resin cure in the system. A note of caution: inexpensive systems often contain zinc or brass components, which can react with components in the resin system, causing acceleration or retardation of the cure; therefore, to be safe, use stainless steel or plastic if there is concern with respect to a reaction.

More rapid cycles, a wider selection of resin systems, and less frequent cleanup can be achieved using a two-component pumping system. In this type of system an A-side resin composition is mixed with a B-side resin composition in a static mixer just before entering the mold. Once combined, the resin reacts rapidly to begin the cure process, with a 1- to 5-min cure time being typical. Resins can easily be made highly reactive, and care must be taken to provide adequate time to flush out the mixing portion of the system before the resin cures in the static mixer.

Injection of the resin into the mold is normally accomplished using either a hand-held gun or a fixed injection nozzle. A hand-held gun with an integral static mixture is normally used with a pressure pot or two-component piston pump system. An automated fixed nozzle has also been used for a resin injection-type molding system.

1.3.4.15 Prototyping with RTM.   RTM is an excellent process choice for making prototype components. Unlike processes such as compression molding and injection molding, which require tools and equipment similar to those used in the actual process to accurately simulate the physical properties achievable in production-level components, RTM allows accurate prototypes to be molded at low costs. It should be noted that in some cases, RTM can also be used to make prototype components designed for these other processes, such as compression molding. When simulating other processes such as SMC, the RTM component typically has properties that exceed those of the production-level product.

When prototypes are made with RTM, generally, less reactive resins are used, allowing long fill times and easy manipulation of the vents. Tooling is normally low-cost epoxy, but it can be made from any impervious material that can contain the resin, such as plaster or wood. Prototype preforms can be made by cut-and-sew methods, and any foam cores used may be machined to shape. Sizes can range from small components to very large, complex three-dimensional structures.

Other processes used to make prototypes, such as hand layup and wet molding, give only a single finished surface; dimensions in the thickness direction are uncontrolled. RTM provides two finished surfaces and controlled thickness, and it requires no auxiliary vacuum or autoclave equipment.[2,100]

1.3.4.16 Productivity.

Cycle times for RTM can be adjusted as required to meet productivity targets. If a long cycle time is acceptable, equipment complexity and cost can be significantly reduced. For example, if several large composite structures per day are to be molded, a low-cost system can be easily configured. A one-part resin system having a cure time of 1 or 2 h can be premixed and a pressure pot system used to inject the resin into the tool. For a large structure, an epoxy tool can provide a moderate number of moldings at low cost.

Tool damage normally determines tool life. The number of parts molded depends on the care with which the preform is loaded into the tool. With low production rates, no special resin inlet or vent provisions are required.

Cycle times for RTM can be reduced from hours to literally seconds by changes in resin chemistry and associated changes in equipment complexity. The distinction between RTM and RIM resin systems is continuing to diminish as the area of liquid molding develops. Cycle times of 1–2 min are achievable for moderate-sized components. Tooling for these rapid cycles must be heated metal, and resin-handling equipment must be two-component, self-cleaning units.

Process automation is rapidly developing for both low-productivity and high-productivity molding.

Preforming automation is an area to be thoroughly studied relative to productivity requirements. The cost associated with operations such as preform assembly can be very high. If volumes are low, hand assembly may be preferable. If high volumes are required, the preform and resulting physical properties must be optimized for automation. Processes such as stamping and trimming of preforms are being automated in systems. It is unlikely that automation of a cut-and-sew preforming approach, similar to aerospace prepreg layup, can be cost effective in the commercial sector.

Automation of the deflashing, trimming, and machining of resin transfer-molded components is no different from that used on components molded using other processes. The good dimensional accuracy associated with resin transfer-molded components aids in ensuring the repeatability of location required in automated systems.

Scrap management can significantly affect the overall economics of the RTM process. Preforms are often large and expensive to produce. Inspection and repair of a preform are necessary when large components are produced. After molding, disposal of a large molding composed of mixed materials is complex and expensive. Increased process control is an effective way to ensure that fewer molded components are rejected because of molding variations.

1.3.4.17 Applications

1. Fiberlite Composites[101] developed a patented modification of the RTM process called network injection molding (NIM) in which the single-point filling action was dynamically controlled, permitting high injection speeds, a low final molding pressure, and excellent wetting.

   The filling time for a 42.8-kg manhole cover (for a 106.68-cm-diameter hole) was approximately 8 min. This is very good considering that the part's complex winglike design with internal ribbing as thin as 4.1–5.1 cm and a glass fiber volume percentage of 50% was not adequate in filling the mold using conventional RTM.

   A second product, a 6-kg fiberglass train seat, was produced at a rate of 1200/week, with cycle times averaging 15 min, a mold close-to-open period of 7–10 min, and a filling time of less than 60 s. Another key asset of the high injection rate was practical elimination of the filtering and binding effect that the glass-fiber preform had on the alumina trihydrate-filled resin.

   Perhaps most importantly, the resulting low molding pressures of 6 bar or less made possible the use of smaller, economical presses, as well as inexpensive P20 steel for the tooling. Epoxy-faced tools were not considered reliable enough for the relatively high production.

2. Molina, Boero, and Smeriglio[102] described their integrated computer-assisted engineering (CAE) approach used in structural analysis and numerical process simulation to design and develop an experimental composite car floor made using the RTM process.

   By substituting variables, the equations that govern the RTM process become practically the same as those used for pure injection molding. Under these conditions, a commercial finite element code was modified to simulate the RTM injection manufacturing process.

   Finally, RTM filling simulation of the floor, in the case of constant injection pressure, provided filling times that were in line with experimental data.

3. Aerospatiale, a French aerospace company,[103] has applied RTM to secondary aircraft items including the production of A321 engine pylon rear cones. The next application of the RTM process was in manufacturing aircraft hoist fairings for an ATR 72 airplane. Future efforts are aimed at the feasibility of producing carbon fiber fishplates using the RTM process. The fishplates are necessary in joining wing boxes. They could be used on an aircraft having an all-carbon-fiber wing box.

4. BP Chemical's Aerospace Composites Division[104] uses the RTM process to build stronger composite thrust reverser door back structures for the Boeing 777. The procedure places dry graphite woven cloth into preforms at a set fiber orientation. The preforms are then placed in the mold tool together with any inserts to be incorporated into the part. Once loaded, the mold is evacuated, heated, and injected. When filled, the tool is elevated to a higher temperature to trigger resin curing. The company produces complex, highly repeatable parts with integral metallic inserts such as fittings and bushings.

5. The application of RTM to small missile production is demonstrated[105] in Figure 1.46.

Application of RTM to small missile production.

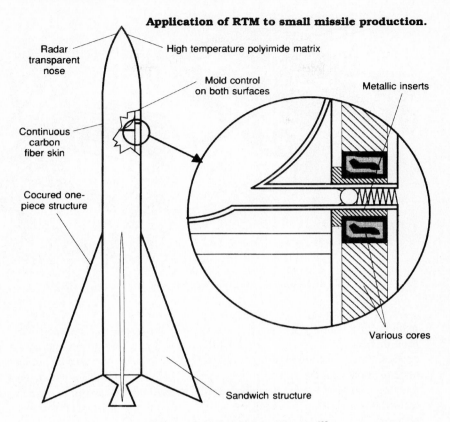

High temperature polyimide matrix

Radar
transparent
nose

Mold control
on both surfaces

Metallic inserts

Continuous
carbon
fiber skin

Cocured one-
piece structure

Various cores

Sandwich structure

Figure 1.46   RTM and missile application.[105]

6. Figure 1.47 illustrates the airbox-plenum in the air intake system of the M1A1 Abrams main battle tank which has been redesigned and prototyped using RTM. Rock and Mosset[106] described the problems of the current aluminum part such as air leaks, weld cracking, and reproducibility, which were eliminated by RTM while maintaining the required structural properties. The composite design and fabrication process allowed RTM of the airbox-plenum into a one-piece structure that is 30% lighter than the current aluminum plenum and economically competitive. The composite airbox-plenum also enhanced airflow by integrating rounded corners and more gradual directional changes into the design.

1.3.4.18 Variations of RTM

High-Speed Resin Transfer Molding.   This process has been under development at Ford[107] and has a number of advantages that may enable cost-effective manufacture of structural composites. In this process the reinforcement materials can be accurately placed and a variety of reinforcements can be used. The method provides the ability to form closed sections as well as to make deep sections that are not achievable and are dif-

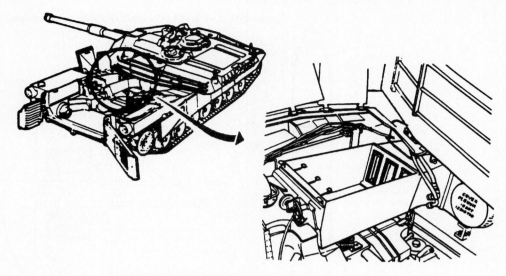

Figure 1.47   Location of airbox-plenum in an M1A1 tank.[106]

ficult to form with high-pressure processes such as compression molding. The low pressures used in HSRTM, typically 0.34–1.4 MPa, allow large parts to be molded on low-tonnage presses. Since there are existing materials for this process, new ones do not have to be developed to implement the process of high-speed RTM. The differences among hand layup, press molding, and HSRTM are described below. In hand layup the reinforcement material can be formed to a high level of complexity as it is placed in the mold and manipulated into its final position. The resin and fiber are normally precombined in this process, and the presence of the resin complicates the forming process. In press molding, the material is not in full coverage of the part. A large volume of material is centrally located in the mold, and during press closure this material flows, filling the complex areas in the mold. In this case the fibers are not in the final desired position, and flow during mold filling causes variability in the physical properties of the part. In HSRTM the dry reinforcement material is located precisely in its final position, and then the resin is rapidly introduced into the mold and wets out the reinforcement, forming the final component. Preplacement of the reinforcement in the mold results in more reproducible physical properties for the part.

The nomenclature of HSRTM is quite confusing at present, and there are other processes very similar to it that are often called SRIM or structural RIM. The SRIM process is different from HSRTM only in the equipment utilized to hold, convey, and mix the resin mixture. A better term for the entire process that incorporates both HSRTM and SRIM is liquid molding (LM).

By far the largest barrier or objective in achieving a viable LM process is cost-effective preforming and preform trimming. Performance and complexity must be blended with cost-effectiveness to develop an optimum process for preforming large, complex, integrated composite structures if LM is going to be the process of choice in future automotive designs.

Figure 1.48 demonstrates a potential goal in the area of low-investment derivative

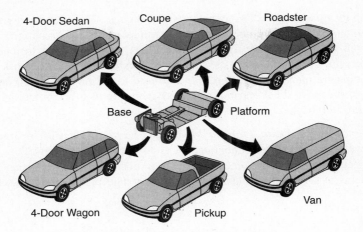

Figure 1.48 Low-investment derivative vehicles.[107]

vehicles. The vehicle platform that is the central portion in the schematic, which in this example is fabricated from steel, is an expensive, complex part of the vehicle to design and engineer. In this scenario that portion of the vehicle would be manufactured in high volumes to realize the economies of high volumes possible for steel, and structural composite body modules of various designs would subsequently be added. The manufacture of such a vehicle using either existing compression molding techniques or LM techniques could be feasible within the near future.

One can hypothesize even more extensive use of composite materials in future vehicle structures such as the body tub structure shown in Figure 1.49 which incorporates the entire underbody of the vehicle into one molded LM component. Such a component would be on the order of 115 kg of composite structure and with the addition of a small number of components would form the entire body structures of a vehicle.

Recently, Beardmore[108] forecast that RTM will assume a much greater role as a composite processing technology for making auto body panels and structural components during the next 10 years. He believes that improved tooling, glass preform techniques,

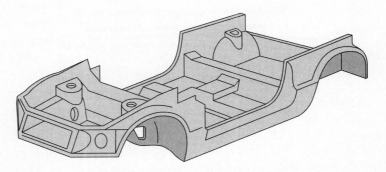

Figure 1.49 Composite body tub structure.[107]

and resins will lead RTM to eventually challenge SMC as the dominant automotive composite process. Beardmore also sees thermoplastic RTM arriving in the near future as a result of research investigating applications and processing techniques for GE's low-viscosity cyclic thermoplastic oligomers. Beardmore also says that automotive original equipment manufacturers (OEMs) will need to decide in the next decade whether to invest in captive RTM capability in order to ensure that their future production and quality needs will be met.

Flexible Resin Transfer Molding (FRTM).   The FRTM process has been under development at Draper Laboratory, Cambridge, Massachusetts.[109] In this process, a two-dimensional flat stack of carbon or glass fibers is placed between two elastomeric diaphragms attached to matching halves of a rigid picture frame assembly (Figure 1.50).

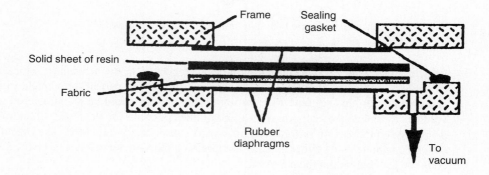

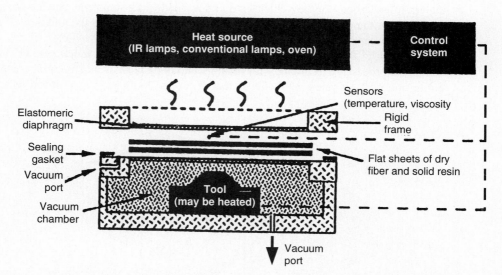

Figure 1.50   (*a*) FRTM process concept—exploded view.[109] (*b*) Schematic of FRTM process.[109]

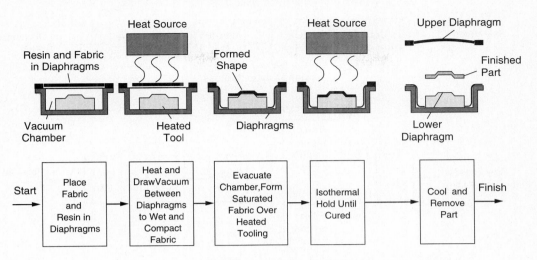

Figure 1.51   FRTM process flow diagram.

The part can then be formed in a number of ways, ranging from application of simple vacuum to the use of segmented matched-metal molds.

FRTM offers an advantage over traditional RTM by reducing the requirements of preform preparation. All that is necessary is to place a flat preform layup and a sheet of resin film—or prepreg if desired—between the diaphragms. Heat is then applied to the diaphragms, and a vacuum is drawn between them to impregnate the preform with resin.

Next, a vacuum in the forming chamber draws the diaphragm-preform sandwich over the molding, shaping the component. The component is then cured and removed (Figure 1.51).

The technology can range from a very simple operation, in which the part is formed with vacuum alone, to the production of exceedingly complex parts, in which fiber volumes and part geometries can be tailored using "pins" or "fingers" placed on one or both sides of the diaphragms. Fiber volumes of 70% with void contents lower than 1% have been demonstrated with the FRTM process. Figure 1.52 compares FRTM to other composite processes.

Continuous Resin Transfer Molding (CRTM).   This process combines the relatively low cost and simplicity of pultrusion with the benefits of RTM, including closed resin injection and enhanced reinforcement.

The technique can be used to make virtually any component with a constant cross section. RTM resins, including vinyl esters, are used in combination with tailored fabrics to produce laminates.

The CRTM method begins with dry reinforcements being continuously pulled from creel systems through alignment guides that form the material into a NNS of the final component. This preform is then drawn into a die where it is first compacted into a final profile dimension; then resin is injected into the die, using either a hydraulic or a pneumatic feed (Figure 1.53).

The injection temperature and pressure control are critical in order to maintain low porosity levels and maximize preform impregnation. The partially impregnated preform

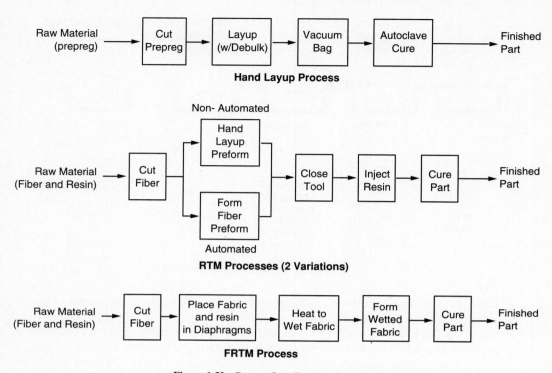

Figure 1.52   Process flow diagrams for several processes.

is drawn through a wet-out, tapered section of the die where full impregnation occurs. The prepreg is then pulled into a multizone cure region of the die.

Key advantages of the CRTM process include elimination of the open resin baths used in traditional pultrusion (and the corresponding elimination of volatile organic compound emissions) and reinforcement volumes higher than those normally associated with pultrusion. The process also enables the relatively rapid production of components normally produced using hand layup and autoclaving, at an estimated 40% cost reduction.

The process has produced laminates for aircraft structural components, including floor beams, and is producing test parts aimed at replacing steel channel sections in the ground transportation sector. Civil engineering applications may represent the most potential.

Vacuum Resin Transfer Molding (VRTM).   Adding vacuum enables RTM to challenge compression molding and autoclaving systems capable of making the best high-performance composites. Texas Instruments (TI)[111] uses vacuum in two ways. First, it mixes resin and hardener under a 91-Pa vacuum just prior to injection. Using an impeller, it agitates the mixture to drive air to the surface where the vacuum removes it. Degassing, which takes about 2 h per tank, also removes volatiles and low-molecular-weight by-products.

TI also places its RTM tools in a vacuum chamber rather than using a vacuum tool.

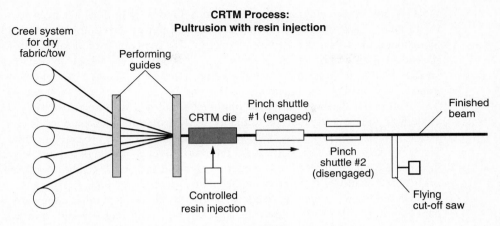

Figure 1.53   CRTM process. Pultrusion with resin injection.[110]

The chamber creates a vacuum that does not vary, even as resin fills the tool. The hard vacuum pulls any air and water vapor off the preform and sucks resin into the mold.

This combination of degassed resin and vacuum keeps voids under 3% and often better. This ensures consistently high structural integrity because voids concentrate stresses that initiate fractures and cause premature failure. It takes only a 2% increase in voids to drain interlaminar shear strength 20% and flexural modulus 10%. TI's VRTM matches equivalent compression molding and autoclaving fiber volumes and voids.

Compared with compression molding and autoclaving, VRTM does not need to apply pressure over the entire skin surface to vanquish voids, simplifying cocuring. TI, for example, fabricates cores and reinforced skins in a single step rather than bonding them after fabrication.

More importantly, VRTM makes composites in fewer, more controllable steps. VRTM users make their core and or preform, add them to the mold and vacuate the air, mix and inject the resin, cure, and, when necessary, machine the final part.

Not only is VRTM simpler, but each process step is also independent and controllable. In VRTM, air evacuation depends on only the vacuum and resin preparation depends on only the resin mixer. Preform production varies with automated fabric weaving and preform placement, whereas core manufacture depends on molding or a machining process. The cure depends on a programmable heat source.

VRTM composites also show excellent resistance to water, solvents, and chemicals. This is largely a function of resin type and surface finish. Rough surfaces pitted with micropores trap water and chemicals and act as tiny reaction chambers that set in motion their own destruction. VRTM yields parts with less than 20-μin. RMS porosity. VRTM can achieve this fine finish repeatedly on all surfaces, depending on the finish of the tool. For high-quality finishes, compression and autoclaving processes depend on uniform resin flow under pressure, which they cannot always maintain.

The main attraction of VRTM, despite its competitive properties, remains cost, where it offers real advantages over compression molding and autoclaving.

VRTM capital equipment costs can run as low as 35% of those of compression molding and autoclaving systems. Because VRTM tooling does not need to withstand high (over 0.7 MPa) pressure, it runs smaller and lighter than compression molding and autoclaving tools. Nor does it take as long to heat and cool VRTM tools. Massive compression molding and autoclave tools require at least 4 h to heat up and cool down; smaller VRTM tools need cycle times of only 0.5–1 hr. Actual cures take as little as 8 min at a sustained cure temperature.

Currently, the process makes primarily small and medium-sized parts. These include high-temperature, high-performance HARM missile antennas, missile fins, for example, that have survived at a 190% rated load at 288°C, and a projectile radome that survived Mach 4 on a test sled track. Prototype optical housings exhibit varying wall thicknesses and net molded lens mounting tabs and flanges. The fabric used was T-300, 3K tow 282 plain-weave graphite molded with cyanate ester resin.

In the future,[112] fuselage sections, wings, stabilizers, and other large parts will perhaps pop out of VRTM molds with numbing regularity. Perhaps our ability to weave complex and cost-effective preforms will grow. The future appears promising.

1.3.4.19 Pultrusion.

Pultrusion is a continuous manufacturing process used to produce high-fiber-content reinforced plastic structural shapes (Figure 1.54). Pultrusion is distinct from filament winding in that filament winding places the primary reinforcement in the circumferential (hoop) direction whereas pultrusion has the primary reinforcement in the longitudinal direction. Accordingly, while good mechanical properties can be achieved in pultrusion in the transverse (crosswise) direction by using special reinforcements, the primary strength occurs in the longitudinal direction. Pultrusion is similar to extrusion except that the raw materials are pulled rather than pushed through the die. The process is ideal for high throughput of constant-cross-section products.[113]

Process Description. The pultrusion process can produce solid, open-sided, and hollow shapes. Almost any length of product is possible. Material utilization is reported to be nearly 95% compared with 75% for manual layup. Accurate resin content can be maintained because of the fixed cross section of the die. As long as the fiber volume passing through the die is constant, excess resin is squeezed out and returned to the resin bath. Inserts made of wire, wood, or foam are readily encapsulated on a continuous basis.

The pultrusion process is continuous, manufacturing 1.5–60 m/h depending on the shape. The machine may be utilized 24 h a day, 7 days a week, with the only scheduled stoppage required to perform routine cleaning, perhaps once every 2 weeks. The operation

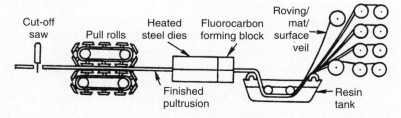

Figure 1.54   Schematic of pultrusion process.

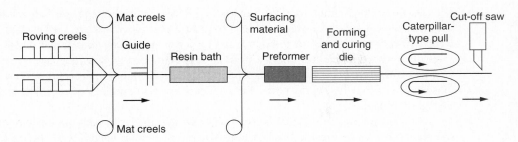

Figure 1.55   Pultrusion machine.[3]

of a pultrusion machine places severe constraints on the raw material and resin mixture combinations that can be used. The continuous nature of the process also creates opportunities and constraints for the quality control system to be utilized in a pultrusion operation.

Equipment.   A pultrusion machine can be conceptually divided into five basic functional areas of operation with each area performing a special task. The machine schematic as shown in Figure 1.55 represents a particular type of machine widely used in the pultrusion industry. The five basic functional areas are listed and described as follows.

Material Feed.   The correct position of the reinforcements within a shape is determined by the design of the composite by production engineers. Pultrusion yields high-strength laminates, but the maximum strength for the shape is attained only when all reinforcements are properly placed in the composite. The process tooling must be engineered to guide the reinforcements into the designed positions within the shape.

1. Longitudinal reinforcement racks (sometimes called roving creels) are the storage areas for the longitudinal reinforcements. These racks have either shelves or horizontal rods (axles).
2. Mat racks (sometimes called mat creels) are the storage areas for reinforcements generally used for transverse properties in pultrusions. These reinforcements in sheet form (mats, stitched fabrics, knitted fabrics) are packaged on a core to be placed on a horizontal rod.

The material feed system for wet resin impregnation consists of a bookshelf-style creel with a capacity for 100 or more packages of roving material. Multiple thread guides or drilled card plates of steel or plastic are used to maintain alignment and minimize fiber breakage. Prepreg tows or tapes can also serve as starting material for pultrusion, but their use is very limited. When prepregs are used, a different creel system is required; the type of system depends on the exact form of the prepreg.

Resin Impregnation and Forming.   Saturation of the reinforcements with the resin mixture is as important as correct fiber placement. Failure to achieve full impregnation of the reinforcements can produce a pultruded shape with lower mechanical properties than intended.

Resin impregnation is achieved by pulling dry fibers through a resin bath equipped with mechanical rollers and wet-out bars to ensure that the fibers are well wetted. The wetted fibers are then pulled through a die or forming bushing to achieve the desired shape. It is also possible to form dry fibers to shape first and then impregnate the shaped fibers by directly injecting resin into the forming die. The latter approach has been used most successfully with hollow parts, although entrapped air and lack of complete wet-out can be a problem.

A good pultrusion practice is to incorporate into the resin bath various structures, sometimes known as breaker bars, to force the reinforcement materials to change direction while passing through the resin bath. These changes in direction serve to slightly spread the longitudinal reinforcements, offering more access to the interior filaments of the bundles for proper impregnation by the resin mix solution. Resin baths are occasionally heated to provide better impregnation of the reinforcements by the resin matrix, but this heating may present problems with certain low-temperature catalysts.

Preforming Area.   Preformers are guides that gently and gradually bend the impregnated reinforcements to form the shape being pultruded. They vary from simple guides at the die entrance to complex guides that are 0.6–1.2 m long. Some reasons for using preformers are (1) the impregnated reinforcements exiting the resin bath are essentially flat, (2) bending them into the proper shape configuration prevents excess stress from occurring during the cure cycle, and (3) by removing excess resin from the reinforcements, one ensures that the maximum fiber volume within the shape is pultruded.

Curing.   One of the most challenging steps in the pultrusion process is the continuous polymerization process that takes place in the die. The shape is cured within the die, which typically can range in length from 30 to 155 cm, although these lengths are not absolute constraints. Dies may be electrically heated or heated with hot oil. Pultruders incorporate one to four heating zones in the die. The choice of the number of heating zones is dictated by several factors, such as the type of resin being utilized, the desired line speed, and the length of the die. Many very successful commercial manufacturers of pultruded products use only one zone, while other manufacturers use multiple zones. There does not appear to be a rule of thumb for choice of the number of heating zones.

Three different methods for curing pultruded components have been developed, and all are in use to varying degrees. Probably the most widely known is the *tunnel-oven process.* The reinforced resin is gelled in the die and fully cured in the free state as it passes through a tunnel oven to produce the desired end product.

The second approach, in which the impregnated reinforcement is drawn through an externally heated, split female die for final shaping and curing, is called *step molding.* When the die is closed, the drawing operation stops until the cure is complete. The process is not really continuous, although the final product appears continuous.

The third pultrusion curing technique is called *die curing* and involves drawing the pultruded material through a die that is subjected to dielectric heating by radio frequency or microwave frequency radiation (Figure 1.56). Curing is rapid with this technique and can be achieved within the confines of the die tube, which is no more than 457 mm long.

The curing process is always the rate-determining step in the pultrusion process. Curing time is controlled by the curing technique, type of resin, and thickness of the

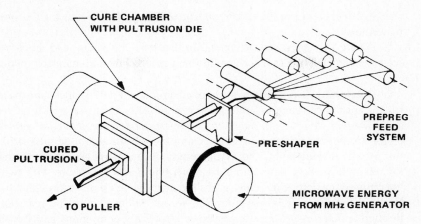

Figure 1.56   Schematic of pultrusion equipment that uses die forming and microwave curing.[113]

part. For polyester resin parts that are 1–76 mm thick, typical pultrusion rates are 0.6–1 m/min.

In some recently completed work Methven[114] found that heat transfer considerations show that microwave heating is more efficient than conduction as the material thickness increases and the thermal conductivity decreases, particularly when the temperature difference between the surface and the core of the material is not great. Microwave heating is therefore ideal for most composite precursors (see Table 1.7) that have low thermal conductivity and where rapid heating without an excessive temperature gradient is required.

One of the most intriguing implications of the data he found was that a pultrusion precursor in a cylindrical single-mode resonant cavity is fully cross-linked in less than a second. On the face of it this is perhaps 10 or 20 times faster than one would calculate from published rate constants at the reaction temperature. In other words, the reaction proceeds faster not because the reactants are at a higher bulk temperature but because of some other effect associated only with microwaves.

He believes, with other authors,[115,116] that this reaction rate enhancement in terms of a nonequilibrated temperature (NET) is obtained in the reaction mixture. According to this scheme, the use of a bulk temperature as a description of a microwave reaction is inadequate (and misleading), and instead each reactant should be assigned a "temperature" that depends on its dielectric loss factor. Moreover, this temperature difference has been

TABLE 1.7.   Properties of Pultrusion Precursors[114]

Property	Aramid	Glass
Permittivity	5.32	5.07
Loss tangent	0.149	0.077
Thermal conductivity (W/m · K)	0.33	0.45

shown to persist during reaction, and it is difficult to account for the observed rate enhancement.

There is a great deal to pursue in this intriguing area, and much of it is likely to have a beneficial effect on both composite manufacture and mainstream kinetics.

Clamping and Pulling.   Many clamping and pulling mechanisms are used in pultrusion machines. They fall into three general categories, referred to as *intermittent-pull reciprocating clamp, continuous-pull reciprocating clamp,* and *continuous belt* or *cleated chain.* In the first category, a single clamp moves through a defined stroke and then returns to the starting point. During the return cycle a pause occurs in the process. The continuous-pull reciprocating clamp approach eliminates this pause with two clamp pullers operating in tandem. One puller grabs the material and pulls it through a complete stroke. While the first puller is recycling, the second puller clamps the material and moves it through another cycle. This approach is effectively continuous and the most popular of the available techniques. The fully continuous approach employs double continuous belts or cleated chains through which the part is passed. This approach is also known as the caterpillar system.[3]

Larsson and Astrom[117] reported on a recently built pultrusion laboratory facility for developing thermoplastic pultrusions. The facility schematic in Figure 1.57 shows a pulling mechanism with a high pulling force capacity and a required continuous pulling motion with closely controlled variable speed. The gripping force of the pulling mechanism should be adjustable. A pulling force capacity of 5000 N and a pulling speed capacity of 10 mm/s are also required.

Cutoff.   Cutting of the pultrusion product is usually done with either a wet or a dry saw. The saw travels with the part to prevent binding of the part in the saw or overheating of the part after stopping in the die during the cut. Saw blades are either continuous-grit carbide or diamond-edged. Either blade works well and provides a reasonable cutting life with all standard reinforcements except aramid. The cutting of aramid-reinforced pultrusions requires frequent blade sharpening or changing (Figure 1.57).

Materials.   As in any technique that uses fiberglass or filament reinforcement, the nature of the process forces constraints on the properties of the raw materials. Conceptually, whether the part is composed of all longitudinal reinforcement or of a combination

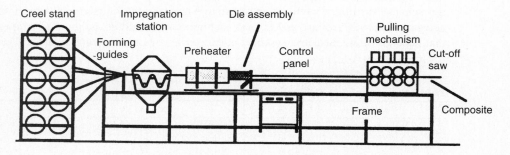

Figure 1.57   Facility schematic.[117]

of longitudinal and continuous-strand mat, different reinforcement schemes mean that different viscosities are required from the resin mixture solution. This in turn requires good consistency in the viscosity of the incoming resin from the resin suppliers. The nature of the pultrusion process is to enter one end of the die as a reinforcement saturated with resin mix and to exit the other end as a solid part in a short period of time. This creates severe requirements for the cure process, which in turn constrains the choice of resin and catalyst raw materials.

Resins.   Polyesters, acrylics, vinyl esters, and silicones have all been used in pultrusion, but 90% of the products use vinyl esters and polyesters because of their favorable costs, desirable properties, and easy handling characteristics. Liquid systems with low viscosity (about 2000 cP) are used to ensure rapid saturation of the fiber bundles. The resin determines the weather resistance, thermal resistance, burning characteristics, moisture sensitivity, chemical resistance, and most of the electrical properties.

The choice of the appropriate viscosity and cure properties for the pultrusion resin is very much a function of the individual pultruder. Some pultruders favor lower-viscosity resins and add more filler to obtain the proper resin mix viscosity, whereas others use a higher neat resin viscosity and less filler. (*Neat* is synonymous with *resin only*—no additives are used.)

While thermoset resins are involved in the vast majority of pultrusion processing, a small number of firms have carved out a niche for thermoplastics in pultruded composites. Thermoplastic Pultrusions, Inc. (TPI), in Bartlesville, Oklahoma, makes custom profiles for medical devices, orthopedic braces, sporting goods, musical instruments, and scientific equipment.

The TPI process utilizes prepreg tapes of glass, carbon, or aramid fibers. Unlike conventional thermoset pultrusion, it cannot use continuous-strand mat. Thermoplastic processing methods include aqueous slurry and powder impregnation. There are three standard polymer matrices: polyethylene terephthalate (PET), nylon 11, and PPS (as well as PEEK).[118]

Reinforcements.   Various grades of reinforcements are available for pultrusion processing, and the reinforcing fibers that are used determine the physical properties of a pultruded product. Because of its cost and physical properties, fiberglass is usually used—particularly E-glass. With fiberglass, the product has highly directional properties. The tensile and flexural strengths in the fiber direction are extremely high, exceeding 78 kg/mm^2, while the transverse properties are much lower. Transverse properties can be enhanced somewhat by incorporating glass mat, glass cloth, or chopped fibers, but with an attendant reduction in the longitudinal property. For this reason, the direction of prime stress must be carefully considered in the design of a pultruded part.

More than 90% of all pultruded products are fiberglass-reinforced polyester. The most common reinforcement is glass roving, which is a glass bundle composed of 1000 or more individual filaments brought together in a single continuous strand and wound on a cylindrical package and is made from E-glass.

The difference in rovings involves the binder, which is a substance attached to the roving to aid internal processing at the roving supplier's plant to permit better adhesion to

Materials		Operating Conditions	Present status of the process
Fibers	Resins		
Glass Rovings	Polyester		Established - - - Already at manufacturing stage.
Glass Rovings	Epoxy		Established - - - Being used for speciality products.
Carbon Rovings	Polyester		Established - - - Limited application
Carbon Rovings	Epoxy		Established - - - Application limited to speciality products.
Glass Preforms	Polyester		Established - - - Limited application
Carbon Preforms	Epoxy		Established - - - Problems still persist but the concern is relative to its intended use.
Thermoset Prepregs			NOT ESTABLISHED - - - Feasible but may require some modification on the resin formulation.

Figure 1.58   Thermoset pultrusions.

the resin matrix. When binders are not compatible with the resin matrix being considered, a frequent result is poor mechanical properties, blisters, and/or generally poor wet-out. Roving that processes well in one resin matrix may not process well in another, and all matrices to be used should be tested independently. Graphite is packaged as an "outside pull" product and is denoted by the number of filaments in the bundle: 12K graphite implies 12,000 filaments.

Another form of reinforcement for pultrusion processing is continuous-strand mat. Continuous-strand mat appears as a sheet of glass in which the primary reinforcement is located in the tranverse direction. The continuous-strand mat must have enough longitudinal strength to be pulled through the resin bath and die. Continuous strand may be purchased either as A-glass or E-glass; only a few companies make any type of continuous-strand mat.

Another form of reinforcement consists of woven or stitched longitudinal reinforcement (both graphite and glass) in which the rovings are placed in specified orientations. Rovings that have been stitched parallel to the line of pull are called 0° rovings; 90° rovings refer to product that has been stitched perpendicular to the line of pull; and +45° rovings refer to product that makes +45° angles with the line of pull.

Cabalquinto[119] of Lockheed examined the automated pultrusion of composites, especially as related to the aerospace industry. His report summarized the general status of the pultrusion technology in Figures 1.58 and 1.59. So far, the only pultruded structures qualified for aircraft use are the fillers and stuffers made from unitapes and tow prepregs.

Other Materials.   These include catalysts, release agents, flame retardants, fillers, UV inhibitors, surface veils, pigments, wetting agents, and other additives.[3]

Materials		Operating Conditions	Present status of the process
Fibers	Resins		
AS4/APC-2 Tow Pregs			Established - - - Already at manufacturing stage.
AS4/APC-2 Unitapes			Established - - - Can be considered as at mfg. stage
IM8/HTA Unitapes			Established - - - Limited application - the matrix is being reevaluated
Radel-X Unitapes			Established - - - Limited application
AS4-PEEK 150 Commingled tows			NO PULTRUSION STUDY
AS4/APC-2 Tow Pregs	Braided or woven		PROPRIETARY PROCESS UNDER DEVELOPMENT.
AS4-PEEK 150 Commingled preforms with off-axial fibers			LIMITED SUCCESS- Process under development.
AS4/APC-2 Unitapes prestacked to form quasi-isotropic laminate			PAPER STUDY - No actual work.

Figure 1.59   Thermoplastic pultrusions.

Mechanical Properties.   Pultruded composites are often preferred to other forms of plastic processing because of the higher mechanical properties available from the pultrusion process. Reviewing individual suppliers' data sheets is often very critical in obtaining an estimate of the mechanical properties available. Although similar reinforcements are used, different pultruders utilize different thermoset or thermoplastic resins and product-forming techniques which may yield considerable variation in the performance of the pultruded part. The designer who works with the pultruded parts must be concerned with the minimum ultimate properties to be expected in the composite, not "typical" values. Typical values do not reveal the ultimate variation within the pultruder's system that may cause extremely low product performance. For example, pultruder A may advertise a typical value of 210 MPa. The use of typical values by pultruder A implies that values less than 210 MPa are possible from pultruder A, while pultruder B states that no values less than 210 MPa are permitted. The designer requires the more reliable presentation of the data.

Applications.   From only 20 pultruders producing 10 million pounds of product in 1970 to over 100–120 pultruders producing well over 200 million pounds of product in 1995 (a figure that could easily grow to 300 million by the year 2000) it is understandable why the pultrusion market and pultrusion applications have grown. This has occurred because of the availability of new composite raw materials, more creative and sophisticated processing techniques, end user and designer awareness of the process and its versatility, and proven success and use of the end product.[120–122]

Because of their precise orientation and construction of the various forms of reinforcements, pultrusions have penetrated a large number of application markets.[123]

Building Construction and Associated Industries.   Because the need exists to
control corrosion or to provide an electrically nonconductive product, pultruded standard
structural building shapes have reached widespread use in the construction industry. One
of the largest volumes of pultruded products today is fiberglass rail for ladders used
mainly by electrical utilities because of their nonconductivity. These ladders are required
to meet stringent structural requirements, and they continue to enjoy excellent growth
each year as fiberglass ladders replace aluminum and wooden products more and more.

Custom shapes in the form of a letter *V* or a *W* are used by many major engineering
firms as mist eliminator blades in the scrubbers of coal-fired power plants to remove corrosive elements from the gas stream. These blades see surge temperatures up to 204°C
and abuse during cleaning operations which demands a high degree of toughness which is
inherent in this pultruded product.

Other high-strength pultruded products are used as support members in a flue gas
air scrubber. Another product that is strong but also requires very consistent electrical
properties is fiberglass pultruded booms used on utility trucks to insulate the worker in
the bucket from high-voltage electric lines.

From a fatigue standpoint, the most dramatic application to date in the nonautomotive area is the fiberglass sucker rod. The fiberglass sucker rod operates continuously in a
tension-tension environment. It can carry an ultimate tensile load in excess of 680 MPa
and in its environment sees an average temperature of 71°C when operating inside the
downhole tubing in an oil well where on each upstroke another column of fluid is brought
to the surface.

The use of fiberglass in pyramid-shaped turrets on the tops of buildings housing
communications equipment, skylights replacing steel in chemical plants with the ability
to last three times longer than steel, covers for sulfur pits containing fumes in oil refineries, stiles in pulp and paper plants with highly corrosive environments, and handrails in a
wastewater treatment plant is possible because pultruded fiberglass shapes are corrosion-
resistant and require minimal maintenance. Various types of pultruded fiberglass grating are used in a wide variety of corrosive environments including wastewater treatment plants, pulp and paper manufacturing, chemical processing facilities, plating plants,
and marine and oil production operations. Finally, pultrsions have been used in an all-
fiberglass impulse generator because of their superior dielectric strength, structural integrity and nonconductivity. Wood was previously employed in the fabrication of these
generators which are used by power companies to simulate lightning when testing switch
gears and other heavy electrical equipment.

Automotive and Trucks and Trailers.   Pultruded roll-up door panels utilize a
custom pultruded fiberglass bunk restraint beam in the sleeper section in an ultraliner series of trucks.

Other applications include battery holddown devices, lightweight bumpers, and alu-
minum-carbon composite driveshafts for which a fiberglass-graphite-proprietary resin
composite is pultruded over a seamless aluminum tube. An isolation barrier between the
aluminum and the graphite eliminates galvanic corrosion, and the composite reinforce-

ment of the tube eliminates the need for a center bearing. By 1994 more than 1 million of these driveshafts for vans and light trucks had been manufactured and sold annually.

Commercial Aircraft and Military.   Eighteen different pultruded fiberglass shapes are used as panel joiners and door framing for modular lavatories, unusually complex and angular, which were among the first fiberglass pultrusions used in commercial aircraft to replace aluminum shapes. They are lightweight and resist the corrosive chemicals used in the lavatory operation and cleaning.

A tow target casing fabricated from a pultruded fiberglass round tube is used in military simulations for fighter pilots to practice firing on enemy aircraft. It is controlled by sending radio commands to a command receiver in the target; therefore low radio interference is a requirement. Because the target is sometimes powered by a jet-fueled combustion chamber located in the aft, the material must also be thermally nonconductive and fire-retardant.

Other Applications.   Additional uses are in fishing rods, I-beams, electrical pole line hardware, tool handles, electric motor wedges, and cables. Finally the lure of pultrusions has even reached out into space. A spiderlike robot has been proposed for space operations.[124] The robot would be operated by a computer, by direct command from an astronaut or from a distant control center. The device has four legs and two arms and a segmented body so that a turret motion is possible. Legs and arms are similar in proportion to those in the remote manipulator system currently used on the space shuttle. The structural elements are made of filament-wound and pultruded composite materials.

The Future.   Pultrusion is already one of the fastest growing sectors of composites processing, and it seems poised for even more rapid acceleration in the next decade as construction and infrastructure applications generate major new opportunities.

Pultrusion appears ready to graduate to a larger scale of part sizes as well. Much of today's research and development (R&D) excitement focuses on thick-walled, structural components for novel marine, power transmission, civil engineering, and high-rise construction projects. These profiles range up to 10.16 cm in thickness and weigh 2.976–81.84 kg/m. They will require new technologies for materials handling, resin delivery, and wet-out of massive stitched fabrics and multilayer preforms.

At the same time, pultrusion process technology is evolving from reliance on "black art" and trial and error to a more scientific understanding of cure dynamics inside the die. Advances in electronic process control, resin injection, and tool design are helping to optimize quality and productivity. And sophisticated new tooling concepts could make possible nonuniform profiles that would never have been considered possible for the pultrusion process.[125]

Better Process Knowledge.   More thorough understanding of the cure process inside the die presents a major opportunity to upgrade pultrusion processing technology, according to J. Sumerak, president of Pultrusion Dynamics, Inc. Using FEA as a predictive tool, Sumerak has created a computer model of die temperature distribution based on part geometry and a die's energy input profile. This computer simulation makes it possible to identify and correct temperature imbalances that impede cure and result in part-tolerance problems.

Design and analysis services use FEA data to position die-heating elements and establish a balanced heat profile.

Most pultruders say one area of process technology that will require improvement is streamlining the collimation of mat, roving, and veil reinforcements upstream of the die. The goal is to reduce scrap, improve line speeds, and shorten changeover times. These challenges resist standardized solutions because each application requires a different package of reinforcements to be gathered and shaped to enter a specific die. So, for now, each processor relies on its own homegrown approaches.

Resin Injection.   Better methods of resin delivery and fiber wet-out are other R&D goals, especially in the pultrusion of larger parts. One evident trend is toward injecting resin into the die rather than pulling reinforcements through a traditional open resin bath. The key advantages of a "closed" resin injection system are reduced emissions and better control of glass placement and integration because the reinforcements enter the die in a dry state.

The disadvantage of injection systems is generally 10–20% lower line speeds. However, line speeds with injection systems have improved significantly during the last 5 years.

Die Technology, Machinery, and Reinforcements.   New innovative technology uses articulated tools and stop-and-go pulling to vary the cross section and contour of pultruded parts at regular or varying intervals (Figure 1.60). The technology utilizes thermal analysis know-how and new process control capability to "manipulate the state of cure" so that the composite remains soft enough to be reshaped in the die. Sumerak[160] claims that the technology lends itself to structural parts that incorporate such features as connection points and interlocking male-female shapes that would otherwise not be possible without machining or secondary bonding.

Machinery is growing larger and achieving greater working envelopes and higher pull strengths. Some machines with comparable pulling force to make 1016-cm-wide panels have been built, and speeds have increased to 72 mm/s.

As pultrusion markets move toward larger and more highly engineered profiles, processors have had to find ways to achieve greater transverse direction strength in their products. Continuous-strand mat has been their main tool so far, and suppliers have worked to improve the mat's wet and dry strength. In addition, some processors have begun to explore the potential of more highly engineered multiaxial fabrics produced by braiding or knitting (also called stitch bonding). Such fabrics can be readily customized to specific applications, allowing higher strength to be provided with less weight of glass than by simply packing in large amounts of mat. For example, composite fabricators can make knitted fabrics in any weight from 43.75 to 420×10^2 g and can bond veils and rovings to one or both sides of the fabric, creating a single reinforcement package that saves creel space. Biaxial fabrics offer new design potential over knitted fabrics of 5 or 6 years ago.

Resins.   Because of its enhanced mechanical, thermal, and corrosion resistance properties, vinyl ester is attractive for many of the new large-part pultruded applications. The tradeoff has been lower line speeds compared with those for polyester.

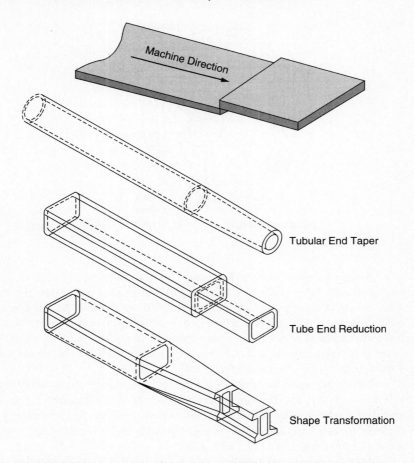

Figure 1.60   New shaping technology using articulated dies and stop-and-go pulling to reform the profile inside the die while it is still soft.[160]

There is heightened interest in pultruding phenolics for low-smoke, flame-retardant applications. The primary drawbacks of phenolics are their water content, which is corrosive to tools, retards line speed, and creates surface porosity. Fabricators have been developing new low-water phenolics as well as phenolics that contain no organic solvents and allow the material to retain mechanical properties after long-term exposure to temperatures up to 260°C.

1.3.4.20 Braiding.   Braiding is a mechanized textile process that dates from the early 1800s. During braiding a mandrel is fed through the center of the braiding machine at a uniform rate and fibers from moving carriers on the machine braid about the mandrel at a controlled angle. The machine operates like a maypole, with carriers working in pairs to accomplish the over-under braiding sequence.

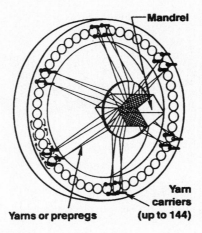

Mandrel

Yarns or prepregs   **Yarn carriers (up to 144)**   Figure 1.61   Braiding.

Since a braiding machine has many carriers, complete coverage of the mandrel with an interwoven layer consisting of rovings lying at plus and minus the braiding can be achieved during one pass. This is compared with multicircuit helical filament winding in which many circuits are required before a single layer is completed. A filament-wound layer also consists of two plies lying at plus or minus the winding angle, but crossovers occur in discrete regions spaced along the length of the filament-wound structure. Figure 1.61 is a schematic of a braiding machine.

Depending upon the thickness of the roving used and the number of carriers, the resulting material is usually less closely interwoven than a standard fabric, comparable to so-called woven roving but more tightly interwoven than filament-wound material. It is difficult to braid an untwisted roving, although the difficulties are more severe with aramid and graphite fibers than with glass.

Since conventional braiding techniques were developed for textiles, there is no standard technique for applying the resin as there is in filament winding. It may be applied after a braided layer is completed on the mandrel, although this step slows down the process to the point where the advantage of more rapid fiber delivery may be lost. An alternative method of resin application involves passing the fibers over a porous ring which delivers a metered amount of resin to each tow. This method provides good fiber wetting and uniform resin content.

Braiding may also be performed on prepregged fibers, and this technique may provide some protection for the fibers, thereby producing braided parts with higher strength.[113]

Because braiding is a rapid reinforcement-forming process, it produces a strong, interwoven tubular or flat structure from glass, carbon, or aramid yarns. The braids can be laid up over a mandrel as described previously, wet or from prepregs and autoclave-cured, or laid up dry and finished by RTM.

Braid Process.   Braids are made on 16-, 32-, 96-, or 144-carrier braiders, most of which are equipped to feed straight axial yarns into the braided structure. Braids are usually produced in tubular form, but a modification of the bobbin track produces instead a flat braid whose width equals the circumference of the normally produced tube. Thus the braider produces triaxial fabric either in tubular or flat form, in which one of the three principal load-bearing directions is parallel to the fabric length and the other two can be varied from nearly parallel to the fabric length to almost perpendicular to it.

Braid Advantages.   Braided textile reinforcement provides several advantages for composite parts. Flexibility of the process permits the braid structure to be designed so that the three yarns lie in the directions that provide the best support to accommodate anticipated stresses. This accomplishes within each layer what, in a conventional fabric layup, requires changing the yarn angles between successive layers of reinforcement. In addition, the interlocked structure of the braid minimizes shear stresses between layers, which in turn reduces the tendency of the composite to delaminate during service.

Braid Applications.   Braiding has become increasingly attractive for composite fabrication because of its high-fiber deposition rates, adaptability to automation, and ability to produce strong, complex shapes. In conjunction with prepreg thermoset or thermoplastic yarns, wet layup, or RTM, braiding has been used to manufacture aerospace (rocket motor exit cones and igniters, ducts, rotor and propeller blades), sports (skis, bicycle frames, tennis racquets), and industrial (pressure vessels, windmill spars) components and assemblies.

Other experimental and developmental applications include composite-intensive vehicles constructed for formula and Baja racing.

A formula SAE racer has thin-walled driveshafts consisting of E-type fiberglass 2D-braided over a collapsible mandrel in an epoxy matrix using shrink tape integrated impregnation. In addition to being a completely original design, these driveshafts are lightweight (eliminating the need for balancing), slightly deformable (assisting with the suspension), and highly resistant to damage. The shrink tape impregnation results in a smooth surface, and the finished shafts are mated to custom-made type 7075-T6 aluminum U-joints.[126]

The body of a mini-Baja buggy is required to be strong, tough, torsionally rigid, and versatile while remaining lightweight. In addition to these performance characteristics, the manufacturing process should remain simple and be able to justify a low cost for mass production. The answer is a monocoque chassis acting as a structural member, designed to exceed the predicted stresses with a safety factor of 3 and made from an E-glass–vinyl ester composite featuring a high strength-to-weight ratio, ease of manufacturing, and low cost. Ho[127] gives a detailed description of the braiding technology used to produce this buggy.

In regard to the tooling used in this program, the finished mandrel was placed on a 144-carrier Wardwell braiding machine. Several layers were braided with E-glass in a $0° \pm 45°$ orientation (to maximize the torsion stability) over the entire body, including the recesses designed for the stiffeners. The stiffeners were then filament-wound until flush with the surface, followed by the remaining layers being braided over the entire body, which allowed an integrated structure linking the skin directly to the stiffeners. The interlacing of the fibers within the braid provides an integrated structure for improved damage tolerance.

This product demonstrates that the construction of a monocoque chassis by judicious selection of preforming techniques can lead to a substantial reduction in part counts.[126–128]

Munjal, Spencer, Rahnenfueher, et al.[129] reported on the design and fabrication of high-quality graphite-epoxy braided composite tubes for space structures. They showed that graphite-epoxy composite tubes (struts), as compared to aluminum tubes, would save over 9200 kg in the weight of the space station truss structure, resulting in significant savings in shuttle launch costs.

Their evaluation of various design, materials, processes, fabrication methods, and tooling concepts indicated that the fabrication of high-quality tubes with Hitco Hitex 46-9A/Shell 9405-9470 graphite-epoxy braided RTM composite material would meet or exceed the critical design requirements, including

- Straightness
- Axial compressive strength and stiffness
- Crush load capability
- Damage tolerance (low-velocity and hypervelocity impact)
- Atomic oxygen protection
- Coating fatigue life
- Thermal cycling—minimal microcracking
- Thermal dimensional stability—low expansion
- Joint bearing strength
- Shear strength
- Vacuum stability—resin outgassing
- Changes in absorptivity and emissivity
- Resistance to plasma, UV, and ionizing radiation environments.

The superiority of braided RTM over filament winding and tape rolling was demonstrated in damage tolerance applications. The superiority of braiding in resistance to microcracking due to thermal cycling was also proven in their work.

Li, Hammad, Reid, et al.[130] studied the bearing behavior of holes formed using different methods in 3D-braided graphite-epoxy composites.

The superior mechanical properties of 3D-braided composites have led to wider applications of these relatively new materials. To meet structural and assembly requirements, holes and slots must often be formed in the composite components, such as in the case of joining parts together using fasteners. These researchers examined three methods of embedding holes in a 3D-braided composite component: cutting or drilling after consolidation, opening by inserting pins during impregnation, and directly braiding into the preform. They drew the following conclusions based on their test results.

1. There are at least two basic failure mechanisms for the holes under bearing load, cracking and yielding. The chance of either one occurring in practice depends on the part geometries and the loading conditions.
2. There is not much difference in performance among the three kinds of holes in the yielding pattern, however, braiding holes demonstrate the best performance in the cracking pattern.

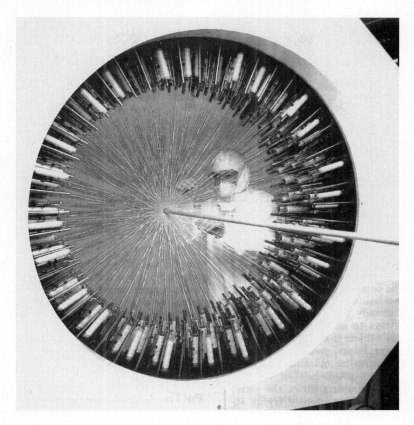

Figure 1.62   3-D braiding equipment for "I," "T," "J," "Z," and cruciforms.[131]

3. In tensile tests, braiding holes demonstrate the highest strength. The average tensile strength of the open hole group is second, and the cutting hole group possesses the lowest strength. These results indicate that cutting of the fibers enhances the chance of crack initiation and propagation, and so the average strength is reduced.

Future of Braiding.   3D textile composites technology was initially started in 1980, however, giant steps were taken in the late 1980s and now into the 1990s (Figure 1.62).[131–137]

Three-dimensional textiles are formed by weaving, knitting, or braiding fibers into thick, skeletal preforms with subsequent infiltration of the binder or matrix materials. Three-dimensional textile composites have greater resistance to delamination and crack propagation than two-dimensional composites, as well as other characteristics that are desirable for certain applications.

Atlantic Research Corporation (ARC) initially introduced braid preforms for CCC rocket nozzles.[132] In the early 1990s it introduced automation into its patented braiding system. With the introduction of innovations in the actuation system, software environ-

Figure 1.63   System for braiding 3-D composite textiles that has been under development for more than 10 years. The use of pneumatic actuation control, an object-oriented programming environment, and a more sophisticated inspection system permit throughput of 12 lb/h of 30.48-cm-wide, 0.635-cm-thick AS-4 carbon fiber-reinforced panel.[132]

ment, and braid inspection system, it was able to cut the cycle time from 6 min per braid to 3 min per braid. This enabled the company to utilize thirty 48-cm-wide, 0.635-cm-thick AS-4 3D carbon fiber-reinforced material for production improvements. The production rate was comparable to that for conventional 2D processes such as weaving and triaxial braiding. The process is illustrated in Figure 1.63.

To move individual fibers during the braiding process, the ARC system uses an array of fiber carriers held in a cylindrical grid arrangement. Fiber carrier travel is accomplished by first shifting columns of carriers in opposite directions and then shifting rings in opposing directions, as shown in Figure 1.64. The fibers are interlocked during the second phase of the shift cycle, during which the fiber carrier direction is reversed.

Peterson[132] offers more details on the programmable logic controller and the automation of this equipment.

1.3.4.21 Weaving.   The goal of preform technology is not only to produce a cost-effective means of making composites but also to find a way of making composites more damage-tolerant. Dexter[138] claims that "the weak link in a laminated composite sys-

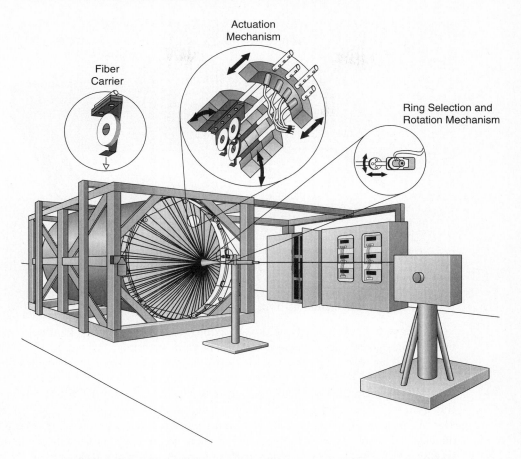

Fiber
Carrier

Actuation
Mechanism

Ring Selection and
Rotation Mechanism

Figure 1.64   The mechanical components of the braiding system are shown in greater detail in the cutaway diagram. Braiding is accomplished by moving the fiber carrier and then rotating the ring and reversing the direction of the fiber carrier. More than 3000 fiber carriers can be operated simultaneously on the largest machines.[132]

tem is that it doesn't have any out-of-plane strength or through-thickness strength." Additionally, he says,

> The only thing that holds it together is the resin. The damage tolerance provided by preforms comes from fiber architecture—from putting fibers through the thickness of the structure. What we're doing with the fiber architecture is either weaving through the thickness, braiding through the thickness, or knitting or stitching through the thickness somehow with a textile process.

The types of textile processes used are braiding (discussed previously), knitting (discussed subsequently), and weaving, each capable of producing a fiber form suitable for reinforcing a composite for structural applications.

Each of the three fabric-forming methods requires a different type of equipment to interlace yarns with each other in forming a fabric structure. Because the fibers are inter-

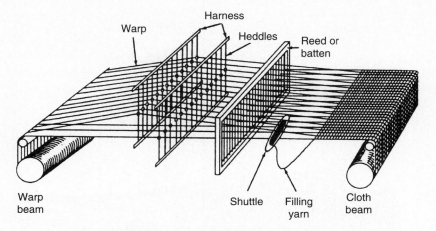

Figure 1.65   Simplified drawing of a two-harness loom.[113]

woven or tied to one another in a textile process—unlike in filament winding and tape laying—the fiber orientation is fixed.

Weaving Basics.   Conventional weaving—in contrast to warp knitting, braiding, or stitching—consists of interlacing two sets of yarns perpendicular to each other. The 0° set is called the *warp,* and the 90° set is called the *filling.* The warp axis is the continuous fabric direction.

Weaving terminology is unique and is best explained with the use of a schematic (Figure 1.65). The warp direction is also called the *machine, long,* or *fiber* direction. The fill or cross-direction is also called the *width direction, weft,* or *woof.* Many two-dimensional weave patterns are possible, but there are five that are used most commonly (Figure 1.66). The stability and open nature of a woven structure make it ideal for layup and matrix penetration.

2-D Versus 3-D.   Although weaving is usually thought of as a two-dimensional process, three-dimensional weaving is often used in composite work. The resulting products are exceptional in performance because the fibers are oriented and arranged for maximum efficiency of reinforcement. Also, a wide range of properties from isotropy to anisotropy are possible with three-dimensional weaving because the types of fibers and number of fiber ends can be varied in each of the three directions. The three-dimensional weaving process is shown schematically in Figure 1.67.

Near-Net Shape Weaving.   NNS weaving[138] is defined by the ability to tailor width and multilayer cross-sectional thicknesses, as well as fiber volumes and orientations. The multilayered turbine blade preform shown in Figure 1.68, for example, features tailored fiber architecture, and its thickness tapers from 2.03 to 0.127 cm.

NNS fiber preforms with conical and curvilinear shapes are woven on looms that produce flat fabrics, and since standard looms can be readily modified to manufacture preforms, weaving provides a cost-effective production method.

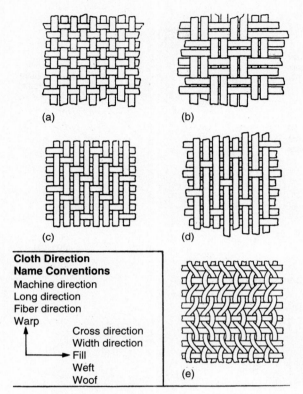

(a)

(b)

(c)

(d)

**Cloth Direction
Name Conventions**
Machine direction
Long direction
Fiber direction
Warp
          Cross direction
          Width direction
          Fill
          Weft
          Woof

(e)

Figure 1.66 Common weave patterns.[113]
(*a*) Box or plain weave. (b) Basket weave.
(*c*) Crowfoot weave. (*d*) Long-shaft weave.
(*e*) Leno weave.

By weaving the reinforcement through the thickness of the plane it all becomes one integrally woven NNS part.

Preforms.   Estimates as to what percentage of fabric preforms are used dry rather than prepregged range upward of 90%. It happens that most people are more comfortable dealing with a NNS part that is dry.

While many fabric preforms are woven, braided, or knitted dry and then resin transfermolded, there are some exceptions. Most notable are the thermoplastic composite preforms made of such materials as PEEK, PEI, and PPS. These are used in commingled yarn form, in which case the resin is already in the preform when it is woven. The preforms can be processed in a press, an autoclave, or conventional equipment and avoid the problems of resin transfer—having to make sure that the resin goes in the right places and that there are no air pockets.

Another exception to dry preforms is the NNS fabric preforms prepregged in the traditional manner. For example, Beech Aircraft[138] uses prepregged carbon fiber-epoxy preforms in Starship wing structures to provide very strong, dependable joints for adhesive bonding.

NNS Preform-Vacuum Resin Impregnation Process.[138,139]   Palmer and Curzio[138] described a manufacturing concept for a dry, semiautomated weaving-stitching operation using multilayers of cloth to produce a NNS preform and with resin vacuum impreg-

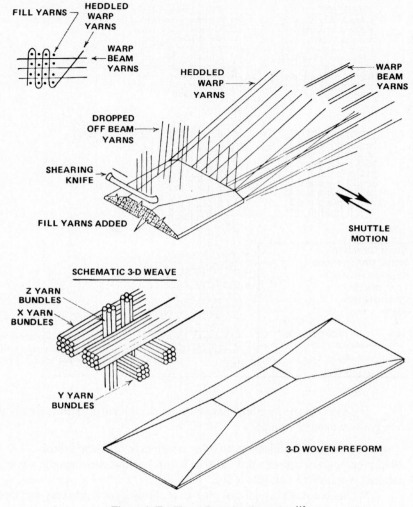

Figure 1.67   Three-dimensional weaving.[113]

nation and a curing process for the near-net preform performed directly in place on the tool.

      They determined that this above approach could

- Reduce layup time
- Reduce detail part count
- Produce a weaving-stitching multilayer preform
- Bypass B-stage resin impregnation costs.

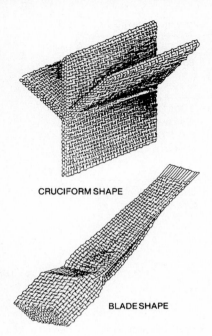

CRUCIFORM SHAPE

BLADE SHAPE

Figure 1.68 Artist's rendering of actual woven shapes.[138]

The described material and processing system is projected to have the capability of both lowering the cost of composite fabrication and improving the structural damage tolerance characteristics.

Dow, Smith, and Lubowinski[140] discussed the NASA Langley Research Center vacuum impregnation molding process that used low-viscosity thermoset resins incorporating various combinations of fibers, and resins, to fabricate composite laminates. The purpose of their work was to improve damage tolerance and economical fabrication because holes, notches, and induced damage severely degrade the structural efficiency of graphite fiber-reinforced epoxy composites. As a result, very conservative performance limits are imposed on the design of composite structures. To obtain full benefit of the weight savings offered by graphite-epoxy materials, improvements are needed in damage tolerance and delamination resistance. Their investigation lists the resins and fiber materials in Figure 1.69.

Their findings concluded that successful fabrication of damage-tolerant structures is the real payoff for the stitching and VIM concepts. Additionally their tests showed

- Outstanding damage tolerance
- Excellent notched compression strength
- Acceptable fatigue behavior
- That strong, closely spaced stitching threads work best
- That Kevlar, glass, and carbon threads provide similar performances.

Finally it should be noted that Zawislak and Maiden[141] and Smith and Dexter[142] have introduced and demonstrated new and advanced weaving and complex woven architectures

Carbon fibers
_____

AS4 (Hercules)
T-800 (Toray)

Dry fabrics
_____

AS4 Uni-weave cloth
fiberglass fill yarn
150-160 gm/sq m
Woven by Textile Technologies, Inc.

Fabric stitching
_____

Lock stitching with
Kevlar, glass and
carbon threads by
Textile Products, Inc.

Molding
_____

Vacuum infusion molding by Douglas Aircraft Co.

Epoxy matrix resins
_____

3501-6 (Hercules)
8551-2 (Hercules)

T-800 Uni-weave cloth
fiberglass fill yarn
150-160 gm/sq m
woven by Toray, Inc.

Chain stitching with
Kevlar thread by
Puritan, Inc.

Figure 1.69 Materials tested and manufacturers.[140]

and concepts indicating that structural preforms can be used to enhance damage tolerance, provide NNS reinforcement, and reduce significant fabrication costs.

Applications.   Swinkels[143] reported on work in which a 3-D woven glass fabric was produced on velvet-weaving machines with glass yarns. It was an integrally woven sandwich fabric with which it was easy to produce a sandwich laminate for all kinds of composite products. The result was a laminate with high strength and stiffness and low weight.[144,145]

He was successful in integrally weaving the vertical glass fibers into the upper and lower plain woven fabric in one process, and as a result the bonding between the core and the reinforcement material was a mechanical one. After impregnation of this 3-D glass fabric a chemical bonding was also achieved. For glass fiber-reinforced products this has all kinds of advantages, such as high impact resistance, high delamination strength, and high stiffness and strength with low weight. For example, in the hollow structure can be used to run hot liquid through in heating up molds for integration of a leak detection system in a graphite-reinforced plastic (GRP) tank.[143]

1.3.4.22 Knitting.   Knitting is a process of interlooping chains of tow or yarn. Knitting does not crimp the tow or yarn as weaving does, and higher mechanical properties are often observed in the reinforced product. Knitted fabrics are easy to handle and can be cut without falling apart.

Types of Knits.   The basic types of knits are 0°, or warp unidirectional, in which fibers run along the length of the roll; 90°, or weft unidirectional, in which fibers run along the width of the roll; bidirectional, with fibers at 0/90°, +45/45°, or other angles; triaxial, with fibers in three directions, +45/−45/0° or +45/−45/90°; and quadraxial, with fibers at +45/−45/0/90°.

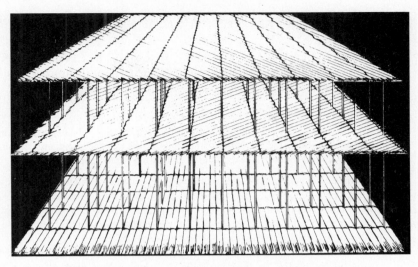

Figure 1.70   Schematic of three-ply 0/+45/–45° stitched fabric.[138]

Knitted fibers are most commonly used to reinforce flat sections or sheets of composites, but complex 3-D preforms have been created by using prepreg yarn.

Broad goods have been processed into ply sets and preforms as well. These materials have axially oriented continuous fibers in the desired direction. Multiple plies of these fibers in the *x-y* direction are stitched in the *z* direction to achieve the balance of drapability and reinforcement properties desired in processing and in performance of the finished parts (Figure 1.70).

Preforms processed from these materials are being used in conventional aerospace production techniques—vacuum bag, autoclave, matched-die molding—and in emerging low-cost production methods such as advanced composite pultrusion and RTM.

A near relative to knitted fabric that has potential applications in custom-engineered structures is a fabric that is a reinforcement consisting of multiple layers of discrete, nonintersecting unidirectional plies interconnected by *z*-direction binder yarns. The patented architecture lays and interlocks the fiber materials in perfectly straight bundles in a no-crimp fashion.

As a result there is a high degree of conformability to complex geometries combined with maximized fiber mechanical property translation.

In developing another new knitting technique that reduces stressing (and breakage) of stiff, high-performance fibers of carbon and ceramic during the stitch buildup phase, engineers at the Swiss Federal Institute of Technology, Zürich, Switzerland, believe that it could lead to the use of knitted fiber-reinforced thermoplastics for low-cost automotive structures. The same workers consider knitting inexpensive because it is fast, easy, and waste-free and because desired shapes are made in a one-step, net shape process. Knits offer freedom of design and have relatively few property limitations, and the machines cost less than weaving machines. The Swiss Federal Institute of Technology has proven the basic relationships between mechanical properties and manufacturing techniques using knitted carbon fibers to reinforce polymethacrylate (PMA) and PEEK matrices.[146]

Future of NNS Fabric Preforms.   In spite of their limitations, NNS fabric preforms have a great future, mainly because of the need for them for structural composites. According to Frank Ko,

> Structural composites are certainly the most exciting area in the decade to come. We have done a lot of work in nonload-bearing and secondary applications for composites, but the growth area is in the structural composites—direct replacement, augmentation of metal. So there's a need for large-scale complex-shape structural composites.
>
> That, of course, makes a large family of continuous-fiber-based composites—"textile-structure composites," as I call them—very attractive candidates. That involves not just braiding but multiaxial knitting and weaving as well. I think there is room for all of these structures in the future, if we know what we want and design accordingly.

New machine concepts are being designed and developed in order to bring down costs and make preforms automatically. Automated braiding machinery has been designed where one can push a button and out comes a preform; there are two types of automated equipment being considered: a Cartesian machine, which is essentially a rectangular braider, and a cylindrical braider that can make surfaces of revolution.

The cylindrical braider, a 304.80-cm cube, has been scaled up from 96 carriers to 6064 carriers. Computer-aided design and manufacturing are also becoming more important in the production of fiber preforms. Although software designed to anticipate stress and strain in a 3-D woven preform are not yet available on the market, such programs are under development.

The key to success will be to use computer programs and in conjunction tap the existing tradition of textile technology and make it a viable manufacturing system for composites.

1.3.4.23 Other newly-developed processes and delivery systems

Automated Tow Placement.   Systems that accurately and efficiently produce advanced materials are not the only fabrication technologies to undergo recent developments in automation, however. Hand layup, which has long been the standard industry approach to positioning preimpregnated tape during part fabrication, also is becoming automated with a fabrication method called tow placement. With the technology under development, engineers believe the process of building up complex-shaped multilayer parts while guaranteeing precise ply orientation can be made less time-consuming, less labor-intensive, and less prone to human error. In its simplest form, automated tow placement uses a multiaxis robot and a tow deposition head to continuously place tow on a tool. The technique has the potential to reduce the production cost of parts by reducing "touch" labor expense and increasing the speed at which material is laid down.

Tow placement for thermoset composites has been under study in the United States for more than 10 years. Recent business alliances have been founded to develop automated thermoplastic fabrication technology for large structures.

Several factors make thermoplastic tow placement desirable for parts manufacturing. When thermoset materials are used, the material and the part are formed simultane-

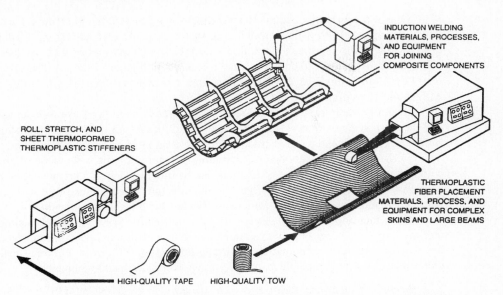

INDUCTION WELDING
MATERIALS, PROCESSES,
AND EQUIPMENT
FOR JOINING
COMPOSITE COMPONENTS

ROLL, STRETCH, AND
SHEET THERMOFORMED
THERMOPLASTIC STIFFENERS

THERMOPLASTIC
FIBER PLACEMENT
MATERIALS, PROCESS, AND
EQUIPMENT FOR COMPLEX
SKINS AND LARGE BEAMS

HIGH-QUALITY TAPE   HIGH-QUALITY TOW

Figure 1.71   The tow placement system processes thermoplastic materials into skins and beams, permitting fabrication of large composite structures.[132]

ously through a cure reaction. Thermoplastics, on the other hand, melt when heated and re-fuse when consolidated to previously placed layers, thereby eliminating the expense and risk of vacuum bag failure during autoclaving or oven curing. In addition, thermoplastics have infinite shelf life, and so users can inventory a supply of material, and then fabricate parts on demand.

Thermoplastic tow placement is not without its challenges, however. With other fabrication methods, it is possible to manufacture parts containing 1% voids from material that has voids of up to 10%. When thermoplastic tow placement is used, the prepreg tow typically must be of a reduced void content to yield parts with the desired quality.

Although the use of thermoplastics eliminates the need for autoclaving, few large parts have been formed using these materials. In several government-sponsored technology programs,[132] the feasibility of forming large structures with a thermoplastic tow placement process by manufacturing aircraft fuselage stiffened panels and skins from AS4 graphite PEKK was demonstrated. The panels measured 152.40 × 213.36 cm and included in situ welded rib stiffeners. The large parts were processed by heating the materials above their 343–371°C melting point and applying pressure to consolidate the layers of tow. A tow placement head is used in in situ consolidation, and so postprocessing is not required. The technique uses an on-line advanced control system and is illustrated in Figure 1.71.

In addition to a low percentage of voids, tow used in such automated placement systems must also have a tight width tolerance so that it can be placed exactly where required.

It has been found that with tow placement, an inspection can be made while manufacturing every layer. This offers the chance to clean up on the thirty-fourth layer before the thirty-fifth layer is placed, for example. In other words, the part can be inspected as it is made. On-line inspection can be fully realized only with in situ processing.

One of the primary concerns surrounding tow placement has been the issue of manufacturing speed. With a filament-winding system capable of handling 32.7 m/min and tow placement moving much more slowly, skeptics have questioned whether the system will ever be fast enough to compete economically with filament winding. The system currently developed moves at 2.4–5.9 m/min, and it is believed that the process eventually will handle 18 m/min.

Recently, thermoset tow placement systems have been used to fabricate sponsons and the aft fuselage structure for the Osprey V-22 helicopter, inlet ducts and stabilators for the F/A-18 E/F, and nacelle structures for the Boeing 767 commercial airliner.

In the future engineers believe that large structures can be manufactured profitably with tow placement. A robotic system that can be programmed to fabricate structures in a wide range of shapes and sizes offers the manufacturer greater flexibility to respond to market demands, and contoured composite structures can be made more simply.[147]

With thermoplastic tow placement, a cost-effective process for fabricating large thermoplastic structures is now available. The first applications are expected to combine the cost advantages of eliminating autoclave processing with the traditional benefits of thermoplastics, for example, high-temperature performance and toughness.

Seeman Composite Resin Infusion Molding Process (SCRIMP)[148].   This technique for comolding composite skins and core in one piece without the need for an oven or autoclave has been used to fabricate glass fiber-vinyl ester arc segments and onionskin that has been glued to concrete columns. This infusion molding of composite exoshells to contain concrete columns offers a low-cost, low-volatile organic compound (VOC) process for large and inherently repeatable infrastructure applications (Figure 1.72).

Engineers have found that parts made by this process can be 30% less expensive to manufacture than those produced with hand layup, and the closed molding system and reduced need for solvents decreases styrene levels and allows no VOC emissions. Worker exposure to wet resins during layup is completely eliminated. The infusion molding proc-

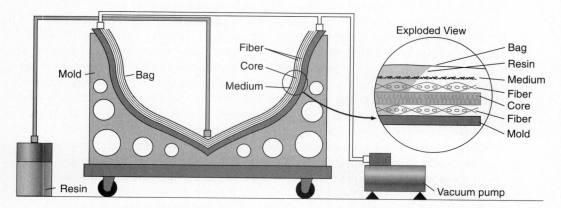

Figure 1.72   In the SCRIMP process, only a one-sided tight vacuum surface is required. In one infusion step, resin eliminates air voids and wets out both skins and core. The use of thick materials can speed up layup; laminate thickness can vary from 0.32 cm to 1.27 cm. (*Source:* Hardcore DuPont.)

ess is cheaper and cleaner than hand layup and offers better structural properties in finished parts. In addition, a work force can be easily retrained to do infusion molding, and no great capital investment in equipment is required.

Thermoformed Thermoplastic Materials[149].   First, a contrast should be made between the thermoforming of unreinforced thermoplastic sheets and continuous fiber-reinforced thermoplastic laminates. When a thermoplastic sheet is thermoformed, the melted sheet thins and stretches to conform to the contours of the mold. The initial sheet thickness and the depth of the draw determine the final part thickness. The surface area of the molded part is usually much greater than the initial surface area of the resin sheet, which results from the material stretching to cover the mold.

The concept of the laminate thinning and stretching is not valid for continuous fiber-reinforced composites. Before the tool closes, the laminate is released from the clamp frame and allowed to lie on the lower tool. Then as the tool closes, the laminate slips into the tool from the edges to cover the contours. The thickness of the laminate does not change during the forming operation. Drapability and conformability are the preferred terms to use when describing the ability of a fabric to form to a contoured surface. As the fabric is forced to conform to the mold contours, the weave pattern distorts slightly and allows the fabric to drape the surface. Movement and slippage of the warp and weft fibers relative to each other account for the ability of a fabric to conform to contours.

The thermoforming process has three key elements:

1. A laminate support frame that carries the laminate into the heat source, supports the laminate during and after the matrix melts, rapidly transfers the melted laminate from the heat source to the forming tool, and then releases the laminate into the lower tool
2. A heat source capable of evenly heating the laminate to its processing temperature in a reasonable period of time
3. A forming tool capable of rapid closing speeds with sufficient clamp pressure to form the laminate.

The schematic in Figure 1.73*a* illustrates the process with a matched mold in the pressure forming stage. First, the laminate is loaded into the clamp frame and placed in the oven for the heating stage of the process. At this point, the laminate is brought up to forming temperature. Once the forming temperature is reached, the laminate is rapidly transferred via the clamp frame to the forming station, at which point the tool is closed and pressure applied. The clamp frame is released just before the upper and lower tools close, allowing the laminate to slip into the mold as required. Mold close times differ for various systems, again varying with laminate thickness.

An alternate technique is shown in Figure 1.73*b*. Vacuum forming, which is conventionally used in thermoforming unreinforced thermoplastics, is also applicable to thermoplastic composites. Although not shown in the figure, additional air pressure can also be supplied to the top of the laminate to assist in forming the part. When vacuum thermoforming, there are limitations on the types of parts that can be formed. Usually, this technique is applicable to gentle curvatures with shallow draws.

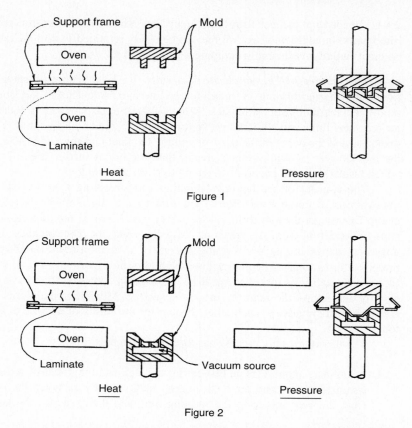

Figure 1.73 (*a*) Schematic illustrating thermoforming with a matched mold in the pressure forming stage.[149] (*b*) Schematic illustrating alternate vacuum forming technique.[149]

With either pressure application technique, plug assist or vacuum, minimizing the time from when the laminate leaves the heat source until the press is fully closed is critical. The laminate begins to cool as soon as it leaves the heat source, and if not formed quickly, it will begin to stiffen and lose its drapability and formability. This transfer time is especially important with thin laminates of only one or two plies (in the range 0.25–0.51 mm). Specific times depend on the laminate thickness; however, most thermoforming equipment can make the transfer and close the tool in 5–10 s.[149]

Electron Beam Curing of Filament-Wound Composites. Beziers[150] reported on the development of a new curing process for composites using an electron beam and x-rays to manufacture filament-wound motor cases. The process, according to Beziers, requires the use of electron accelerators which, at an industrial level, exist in several forms depending on their characteristics. They are principally characterized by two parameters:

- The energy that determines penetration of the radiation (electrons or x-rays), expressed in 10^6 electron volts
- The power, in kilowatts, directly linked to the exposure time.

The electrons and x-rays are obtained from the same accelerator.

He found that radiation curing and use of the combined processes, electrons and x-rays, provides the following advantages.

- Ease of application
- Very short curing time for substantial penetration capacity
- Penetration greater than needed achievable by increasing curing time by using x-rays
- Curing with a small temperature increase, limiting stresses of thermal origin
- Resin pot life much longer than the manufacturing time for filament winding of thermal structures.

However, he also found that there were two disadvantages:

- Cost of equipment
- The special chemistry required.

In his results to date Beziers has been able to cure large structures (maximum length, 10 m; maximum diameter, 4 m).

Figures 1.74 and 1.75 depict production lines that would be largely automated and would take advantage of the fact that matrix resins can be cured by electron beams in addition to heat.

The automation and the continuous nature of the proposed processes would depend on the availability of fabric preforms impregnated with matrix material. The yarns used to make the preforms could be coated with matrix material by any of several commercial processes and then woven into the preforms by use of advanced techniques for producing fabric components with complicated shapes.

Figure 1.74 illustrates a quasi-continuous pressure forming process. Impregnated net shape fabric destined to become layers of composite structural components would be supplied on multiple rolls. The fabric would be unwound from the rolls and drawn through a series of rollers and guides to be assembled into a thicker, multilayer continuous preform. Unit lengths of the multilayer preform would be processed, in succession, through a partial-debulking tool. As its name implies, this tool would compress the unit lengths of preform part way toward their final dimensions by using a combination of mild pressure and mild heating. Because of the mildness of the partial-debulking conditions, the tool and associated equipment should be relatively inexpensive. Typical debulking times would be on the order of 1 min.

Non-heat-curable matrix materials that are electron beam-curable could be solidified without curing to such a degree that the preform could also retain its final shape without support from tooling. The time needed for staging would be on the order of 5 min; be-

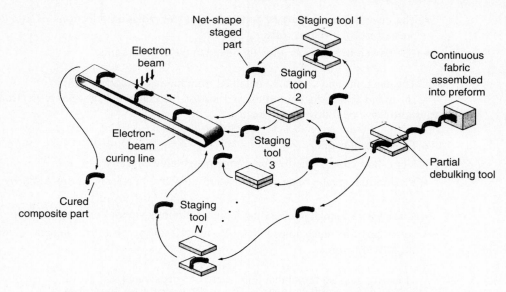

Figure 1.74 This press forming process would produce composite material parts at moderately high rates in production-line fashion instead of by batches as is done now.[151]

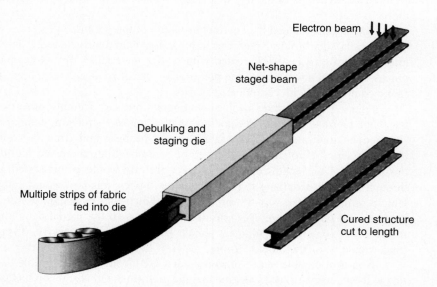

Figure 1.75 This pultrusion process would incorporate features of both conventional pultrusion and of the press forming process shown in Figure 1.74.[151]

cause of the greater duration of the staging cycle in comparison with the debulking cycle, a number of staging tools would be kept running simultaneously to keep up with the output of the partial-debulking tool.

The composite structure would be removed from the staging tool and placed in a chamber where the matrix material would be cured by exposure to an electron beam. Typical electron beam curing times are expected to be on the order of several minutes— significantly shorter than typical autoclave cure cycles. Another advantage is that unlike heat curing, electron beam curing does not cause chemical reactions of the type that forms voids and other anomalies, which can weaken the finished composite parts.

Figure 1.75 illustrates a continuous pultrusion process for making composite I-beams. As in the proposed press forming process, impregnated net shape fabric would be drawn from rolls and assembled into a thicker, multilayer continuous preform. The continuous preform would be fed into a heated die, the cross section of which would taper down to the desired I-beam cross section so that the preform would be gradually debulked to the final cross section as it moved along. The preform would continue along a constant final cross-section length of the die, where it would be partially cured or allowed to solidify. The composite structure would be pulled continuously from the die and cut to length as unit lengths emerged. The staged I-beams would then be electron beam-cured.

Reaction Injection Molding.   RIM may be considered a special case of injection molding of thermosetting material. The basic process utilizes a fast-curing thermoset system in which two reactive components are mixed and then fully cured within a few minutes.

Composite systems known as reinforced reaction injection moldings are made by incorporating a suspension of short fibers in one or both of the prereacted components. Composites may also be formed by preplacing the reinforcement (short- or continuous-fiber preforms) in the mold and then injecting and infiltrating the reacting polymeric components. This is designated *structural RIM*. The concept is now being developed for other types of thermosets, including unsaturated polyester derivatives and polyamide systems.

In the RRIM process, short glass fibers (usually very short *milled* fiber) are mixed into the polyol component. The process is attractive because of the fast cycle times and because tooling costs are significantly lower than for conventional injection molding.

Polyurethane RRIM components have been successfully used for panels and moldings in automotive applications in situations where dimensional stability, rather than load-bearing capability, is the principal requirement. In these circumstances the modest reinforcement confers enhanced stability at moderately elevated temperatures. This enables the moldings to be painted on-line in conventional ovens.

Preform Systems—RTM and SRIM[152].   A new automated net shape fiber preforming system for RTM and SRIM can produce preforms with selectively varied fiber orientations and wall thicknesses in about 1 min according to Monks.[152] This CompForm system uses special binder resins and transcends previous efforts to automate preform fabrication. Here a preform can be engineered with nonuniform wall sections containing various types, thicknesses, and configurations of reinforcements with precise location of specific reinforcement at high-stress points. The system also allows the addition of ribs, closed sections, and cores, and encapsulation of metal, foam, wood, or other materials.

A key element of the system is polyacrylate- and epoxy-based binder resins tailored to be compatible with the composite matrix resin. These binder resins have three times faster cure rates than when CompForm was first introduced. After the fiber broad goods reinforcement (usually mat) is cut to predetermined patterns, it is permeated with the binder by spraying, calendering, or rolling.

Next, single or multiple plies of reinforcement are placed on one half of an epoxy mold mounted in a vertical press and a transparent film is placed over the layup using robots.

The forming press closes, and a $1.02-1.36 \times 10^5$ Pa vacuum is applied to mold the details of the preform shape. When the mold opens again, the half with the preform shuttles into a UV curing tunnel for about 15 s and the short time in the tunnel ensures that there is minimal heating of the reinforcement or mold surface.

After this initial preforming operation, additional sections of reinforcement can be attached to sections of the original preform by robotic application of "energetic stitching" using the same directed UV energy as before.

Long Discontinuous-Fiber Technology (LDF)[153].   Medwin and Coyle[153] were determined to combine roll forming, stretch forming, and press forming into a DuPont LDF thermoplastic material system. The key components of the system are as follows.

1. Highly aligned fibers with random breaks, averaging 55.88 mm in length. Graphite (AS-4 and IM-7), Kevlar, and glass fibers have all been used in LDF systems.
2. A matrix system with a viscosity compatible with thermoforming at the melt point. PEKK, Avimid K3B, and J-2 were all developed for use in thermoforming.

Demonstration aerospace parts that have been made include a V-22 wing rib, a V-22 nacelle door, a F-16 strake door, and an aircraft fuselage frame (shown in Figure 1.76).

Sheet Stamping.   Sheet stamping of glass mat thermoplastics (GMTs) is also called melt-flow stamping because there is considerable flow of material in the dies. The sheet is reinforced with continuous strands of glass fiber randomly deposited in a 2-D pattern. Precut blanks are heated to above the melt temperature of the resin, and the warm sheet is placed in a matched-metal mold in a vertical press where it is compressed in a mold. Ribs and bosses can be molded in, but variable-thickness parts are difficult. Cycle times range from 25 to 50 s for most parts. Resins used are usually polypropylene, but polyester (thermoplastic), PPS, and polycarbonate alloy sheet are also available (Figure 1.77).

Space Composite Manufacturing.   Georgia Institute of Technology researchers have proposed an ingenious system that does not require the use of solid dies in space because composite components for a space station could be manufactured in orbit instead of carrying them aloft after preconfiguration.

How the system would work is best conveyed by describing an example component—a tubular truss member. Tubes of composite preform material such as carbon fiber would be woven on earth around an inner bladder, made of film, which would be slipped over a rigid mandrel. Multiple layers of reinforcing fibers would be braided on top of

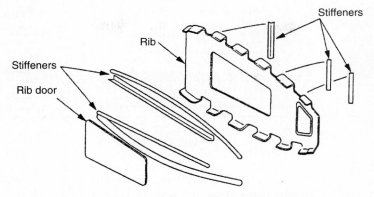

Typical thermoset wing rib and door of V-22

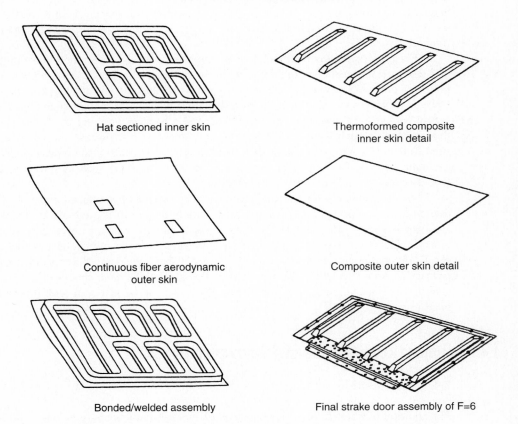

Hat sectioned inner skin

Thermoformed composite
inner skin detail

Continuous fiber aerodynamic
outer skin

Composite outer skin detail

Bonded/welded assembly

Final strake door assembly of F=6

Figure 1.76   V-22 Osprey thermoset wing rib and door, V-22 nacelle door, and F-16
strake door fabricated by LDF technology.[153]

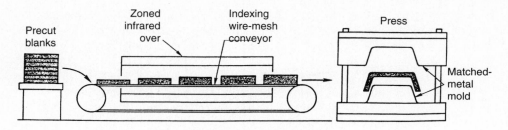

Figure 1.77   Sheet stamping.

each other and stitched together as needed. The preform and the PEEK matrix would be enclosed in an outer containment bag also made of film. Because the completed prepreg would not be consolidated until it reached orbit, it could be slipped off the rigid mandrel and collapsed through a set of rollers for easy transport into space. The flattened prepreg would either be wound onto spools or cut into discrete lengths and placed in cartons for transport.

For prepreg deployment and consolidation in orbit, the researchers have proposed and designed automatic equipment that can be collapsed flat and telescoped toward the prepreg containers (spools or cartons) for ease of transport and deployment in space. Once in orbit, proper lengths of prepreg would be returned to tubular shape by inflating the inner bladder. The bladder would then act like a balloon, pressing the prepreg against the outer bag, which would control the outside diameter and provide a protective outer shell. Inflation would be accomplished by sealing the free ends of the inner bladder with expanding chucks and injecting the bladder with an inert gas such as nitrogen, which would be transported into orbit in liquified form.

The heat required for consolidation would be supplied by concentrating solar radiation on the tube through a set of mirrors that would be part of the automatic deployment and consolidation equipment. While some areas of the tube would receive more heat than others, calculations have shown that the mirrors would provide enough heat to cure all areas. Cooling after consolidation would be accomplished by closing shutters in front of the primary mirrors and letting the heat dissipate. Calculations have shown that the proper cooling rate for obtaining good composite properties can be achieved with this arrangement.

1.4.1 FABRICATION OF GLASS-CERAMIC COMPOSITES

1.4.1 Introduction

The addition of fibers to ceramics has been known for many years to be one approach to developing "tough" ceramics whose performance characteristics retain the best properties of their parent ceramics and have the additional quality of not being susceptible to fracture during impact or under stress in the presence of a notch.

Early experiments performed in England, Germany, and the United States,[154–161] however, have demonstrated that high-performance fibers can be successfully incorporated into glasses to achieve high-strength, tough composite materials. Through the use of carbon fibers to reinforce glasses and glass-ceramics, composites with strengths of over 700 MPa were demonstrated.

The key to the successful development of glass matrix composites lies in the fact that research has been carried out as a direct extension of metal and resin matrix composites efforts. In each case high-elastic-modulus fibers have been incorporated into a lower-elastic-modulus matrix to achieve structural reinforcement. Glass and glass-ceramic matrix elastic moduli are in the range 60–85 GPa, while the reinforcing fibers are generally characterized by an elastic modulus in excess of 210 GPa and in some cases as high as 700 GPa. Another important aspect of the work is that the composites are fabricated in a manner totally analogous to that used for resin matrix composites. This is because the glass matrix can be readily deformed and flowed in its low-viscosity state at elevated temperatures. Not only can glasses be used in this process but also glass-ceramics which have the greatest potential for high-temperature applications. Glass-ceramics provide the unique capability for densifying a composite in the glassy state and then subsequently crystallizing the matrix to achieve high temperature stability. Most, if not all, of the techniques used to fabricate resin matrix composites can be adapted to glass and glass-ceramic matrix composites.

The composites that result from this processing are characterized by high strength, stiffness, toughness, and in general an overall performance similar to that of resin matrix composites except that in this case performance can be maintained up to temperatures as high as 1200°C.

1.4.2 Fiber Reinforcement

The ability to choose from a wide range of both reinforcing fibers and matrices for glass matrix composites plays a predominant role in determining overall composite characteristics. Table 1.8 lists the reinforcements, their properties, and the forms most frequently

TABLE 1.8.   Materials Used to Reinforce Glass and Glass-Ceramic Matrices[162]

Fiber	Form[a]	Diameter (μm)	Density (g/cm³)	E (GPa)	UTS (GPa)	CTE (10^{-6}/°C)
Boron	M	100 to 200	2.5	400	2.75	4.7
Silicon carbide	M	140	3.3	425	3.45	4.4
Carbon	Y	7 to 10	1.7 to 2.0	200 to 700	1.4 to 5.5	− 4 to − 1.8
Nicalon[b]	Y	10 to 15	2.55	190	2.4	3.1
FP alumina[c]	Y	20	3.9	380	1.4	5.7
Nextel 312[d]	Y	10	2.5	150	1.7	—
VLSI-SiC[e]	W	6	3.3	580	8.4	—

[a] M, Monofilament; Y, yarn; W, whisker.
[b] Nippon Carbon Company.
[c] DuPont Company.
[d] 3M Company.
[e] Los Alamos National Laboratory.

used by researchers and engineers to reinforce glasses and glass-ceramic composites.[162,163] When they are divided into three major categories based on fiber form (yarn, monofilament, or whisker) and into more subdivisions based on composition, it is clear that many different levels of tensile strength, elastic modulus, chemical reactivity, electric conductivity, and density are available. Of all the reinforcements listed, the carbon yarns offer the greatest range of mechanical properties and also a potential for lowest cost. This is in large measure due to the already well-established carbon fiber-reinforced polymer composites industry which has created a need for fibers of varying properties in large enough quantities to drive down costs. A wide range of carbon fibers has been found to be compatible with glasses and glass-ceramics for composite fabrication. These fibers, however, lack oxidative stability at elevated temperature, and hence the development of SiC-type fibers in all forms has provided the stimulus to create high-temperature materials that can perform in an oxidizing environment.

1.4.3 Matrices

While reinforcements at first appear to be dominant determiners of composite charateristics, it is the matrices and their wide range of compositions and physical properties that give the materials scientist the control necessary to tailor a successful material. Their compositions not only contribute to such composite properties as elastic modulus, CTE, and high-temperature creep resistance but, more importantly, also provide a means of controlling fiber-matrix reactions during the composite densification step. This reactivity, or lack thereof, in large measure determines the nature of the fiber-matrix interface which subsequently determines composite fracture mode. The matrices in Table 1.9 are a partial list of the general types mostly used to date.

TABLE 1.9. Glass and Glass-Ceramic Matrices of Primary Interest[162]

Matrix Type	Major Constituents	Minor Constituents	Major Crystalline Phase	Maximum-Use Temperature Composite Form (°C)
Glasses				
7740 borosilicate	B_2O_3, SiO_2	Na_2O, Al_2O_3	—	600
1723 aluminosilicate	Al_2O_3, MgO, CaO, SiO_2	B_2O_3, BaO	—	700
7930 high silica	SiO_2	B_2O_3	—	1150
Glass-ceramics				
LAS-I	Li_3O, Al_2O_{3p} MgO, SiO_2	ZnO, ZrO_2 BaO	β-S podumene	1000
LAS-II	Li_2O, Al_2O_{3p} MgO, SiO_2, Nb_2O_3	ZnO, ZrO_2 BaO	β-Spodumene	1100
LAS-III	Li_2O, Al_2O_3 MgO, SiO_2, Nb_2O_3	ZrO_2 —	β-Spodumene	1200
MAS	MgO, Al_2O_3, SiO_2	BaO	Cordierite	1200
BMAS	BaO, MgO, Al_2O_3, SiO_2	—	Barium Osumilite	1250
Ternary mullite	BaO, Al_2O_3, SiO_2	—	Mullite	—1500
Hexacelsian	BaO, Al_2O_3, SiO_2	—	Hexacelsian	—1700

1.4.4 Major Glass-Ceramic Systems

1.4.4.1 Carbon fiber-reinforced glass.

The use of carbon fibers to reinforce glass results in a variety of fiber-glass combinations that can be tailored to achieve a significant range of properties. Because of the availability of high-elastic-modulus fibers, composites with exceptionally high values of axial stiffness have been achieved. Values of 2070–2930 MPa for the ultimate tensile strength are competitive with those of fiber-reinforced resins and can be maintained over a much wider temperature range than in the resin matrix systems.[162] Comparing the retention of flexural strength in argon of both borosilicate glass matrix and two resin matrix composites all reinforced with the same high-modulus (HM) carbon fiber shows that the resin matrix systems are stronger at room temperature. At elevated temperatures the borosilicate system is far superior with eventual strength loss occurring above 600°C as a result of softening of the glass matrix. Similar tests performed in air demonstrate that strength loss is perceptible only above 500°C as a result of oxidation of the carbon fibers.

Carbon fiber-reinforced glass composites also possess several nonstructural performance advantages. Because the carbon fibers have a highly negative axial CTE, when they are combined with a glass matrix, it is possible to achieve a resultant composite CTE of very close to zero. By controlling fiber orientation and fiber content, it has been possible to tailor this behavior to achieve composites whose in-place CTE is nearly zero over a wide temperature range, which makes them candidates to replace monolithic glasses and resin matrix composites in large space-based mirrors and structures.[164,165] Since carbon fibers also possess a certain degree of lubricity when combined with glass, the resultant material exhibits a low coefficient of friction and also has high wear resistance because of the presence of the glass matrix.[166]

High toughness values have been observed for carbon fiber-glass matrix composites. This is due to the high thermal expansion mismatch, high-elastic-modulus ratio, and poor bonding at the interface.[167,168] Thermal expansion mismatches as well as differences in elastic moduli ease load transfer.

The bonding character of the interface plays an important role in generating fiber pullout, large strains to failure, and high work of fracture.[167,168] In carbon fiber-glass matrix composites, poor interfacial bonding allows for a high degree of pullout and large increases in toughness were observed. K_{IC} values for a high modulus strength (HMS) continuous graphite fiber-reinforced 7740 glass were found to be 22 MN/m$^{1/2}$ at 22°C and 16 MN/m$^{1/2}$ at 600°C.[169]

1.4.4.2 Nicalon (SiC)-reinforced glass and glass-ceramics.

When SiC is incorporated into borosilicate or high-silica-glass matrices, these fibers provide high levels of flexural strength[170] accompanied by extremely fibrous fracture morphologies relatable to high crack growth resistance and toughness.

The loss of composite strength at elevated temperatures found for these systems was associated with softening of the glass which in turn resulted in specimens simply deforming under load rather than fracturing. Through the use of glass-ceramic matrices this difficulty was overcome since these compositions could be hot-pressed in a low-viscosity state and then crystallized to achieve a high-temperature composite.

The development of lithium aluminosilicate (LAS-I, II, and III) and (BMAS) matrices (see Table 1.9) in conjunction with SiC_f composites has yielded a material capable of achieving use temperatures from 1000 to over 1250°C. The ternary mullite and hexacelsian glass-ceramic matrices have potential composite use at temperatures of 1500–1700°C.

Prewo and Brennan[171] used a slurry infiltration processing technique to produce a SiC/7740 borosilicate glass matrix composite using monofilaments of SiC. They conducted studies on composites with two levels of fiber: 0.35 and 0.65. A flexure strength of 830 MPa at 22°C was reported which increased to 930 MPa at 350°C and to 1240 MPa at 600°C. This trend was due to softening of the matrix analogous to the behavior of the carbon fiber composites discussed earlier. Weaker strengths were exhibited by the 35% fiber specimens that revealed a value of 650 MPa at room temperature.

Similar strength-versus-temperature behavior was observed in additional experiments[171] involving Nicalon SiC_f in 7740 glass and in 7930 high-silica glass. In all cases, the ultimate strength of the composite at high temperature is limited by the softening point of the glass.

The limited temperature capability of glass as matrix materials led to the use of glass-ceramics. These materials are formed in two steps. First, glass formation processes are conducted followed by a controlled crystallization step to form a fully dense ceramic. This matrix material offers ease of vitreous preparation combined with the high-temperature capability of a crystallized ceramic. Brennan and Prewo[172] reported on composites using a lithium aluminosilicate (LAS) matrix and SiC_f. Flexure strengths were determined by three-point bending, resulting in values of 620 and 370 MPa for unidirectional and 0/90° cross-ply composites, respectively. The unreinforced LAS material showed a strength of 190 MPa. At 1000°C, the strength of the unidirectional composite increased to 900 MPa, again because of the viscoelastic behavior of the glass-ceramics.

Jha and Moore[173] prepared LAS glass-ceramics reinforced with SiC_f to study the effects of matrix attack on uncoated fiber, which reduces fracture toughness. It has been reported[174] that some transitions and refractory metal oxides, when present as one of the ingredients in the matrix glass composition, form a layer of refractory metal carbide. This has been observed particularly with 2–3 wt% Nb_2O_5 in the glass which, during hot pressing and sintering, forms NbC. In this way, nonwettability between the fiber and the matrix has been achieved with some success. Currently, expensive coated SiC and carbon fibers[174] are used to form a barrier between the ceramic matrix and the fibers, and careful selection of materials can therefore reveal the full potential of these new composites. An understanding of the mechanism of the interfacial reaction between the fiber and the matrix is thus important in order to point the way toward cheaper and better materials.

1.4.4.3 Whisker-reinforced glass and glass-ceramics.

In whisker-reinforced glass and glass-ceramic matrix composites (SiC_w and Si_3N_{4w} + 1723 LAS and 7740 borosilicate glass matrices), the whisker composites exhibited room temperature flexure strengths of 140 MPa with linear load deflection curves to the point of fracture and very brittle fracture surfaces with essentially no whisker pullout, in marked contrast to chopped Nicalon–glass-ceramic matrix composites. The strength of a 7740 glass bar tested in flexure under similar conditions was 60 MPa. Thus the whiskers strengthened the glass but did not change its fracture characteristics appreciably.

The reason that the SiC_w do not impart a significant increase in toughness to the composite is that, unlike the case where Nicalon fibers are hot-pressed into these matrices, no carbon-rich, crack-reflecting layer is formed at the whisker-matrix interface. Cracks advancing through the whisker-reinforced composites for the most part encounter whiskers that are strongly bonded to the matrix and so propagate through them with little if any deviation. The consequence is brittle fracture. On the other hand, however, lacking the carbon layer, the whisker composites do not exhibit a loss of properties in oxidative environments.

Lewinsohn[175] reported his work with hybrid whisker-fiber-reinforced glass-matrix composites. When hot pressing methods (slurry winding) were used, less-than-optimum microstructures were obtained because the viscosity of the glass was too high during hot pressing to allow infiltration of the matrix. The fracture toughness of the composites improved with the addition of 20 vol% whiskers. The longitudinal fracture toughness was higher than the transverse, but the transverse fracture toughness of composites containing 20 vol% whiskers was higher than that of those without whiskers. During processing, the loss in toughness in glass-matrix composites without whiskers may have occurred, however, it could be counteracted by increasing the volume percentage of whiskers in the matrix or perhaps by using particulate reinforcements. He showed that the addition of a reinforcement (SiC), via conventional ceramics processing techniques, to the matrix phase of a unidirectionally aligned fiber-reinforced composite improved the transverse properties of the composite.

1.4.4.4 Alumina fiber-glass and glass-ceramics.

Bacon, Prewo, and Veltri[176] conducted a broad study using Al_2O_{3f} in several matrices that focused on the large residual stresses generated by the Al_2O_{3f} in low-expansion glasses because of the large mismatch in coefficients, which may limit the performance of the composites.

Bacon et al. calculated the stresses in various matrices using Lamé equations and found the level of stress and CTE in borosilicate glass 7740 and silica glass to be acceptable when Al_2O_{3f} was used. Strengths observed in the FP alumina fiber-1723 glass composites was 277 MPa, which is much less than that calculated by the rule of mixture (ROM) for a 30% fiber composite (544 MPa). This contradiction was explained by the strong interfacial bonding present in this system and flaws introduced during processing. Composites produced with Al_2O_{3f} still show a considerable increase in strength over that of the unreinforced matrix. Results obtained during three-point bending of a silica glass-Al_2O_{3f} composite resulted in a strength of 187 MPa for a 37% fiber fraction, which is approximately four times the strength of the matrix alone.

Further tests showed the ability of the Al_2O_{3f} to retain strength at high temperatures. Very small losses in strength were recorded over the temperature range 220–1000°C.

Chevron notch tests to determine the toughness of Al_2O_{3f}-glass matrix composites were conducted by Michalske and Hellman.[177] Results showed that although no fiber pullout existed in these systems because of the strong chemical bonding, increases in toughening still occurred because of crack shielding. The mismatch in elastic modulus between the fibers and the matrix shields the fiber from matrix crack extension. This action increases toughness without fiber pullout. Toughness values for the $Al_2O_{3f}/7740$ glass composites were 3.7 MPa m$^{1/2}$.

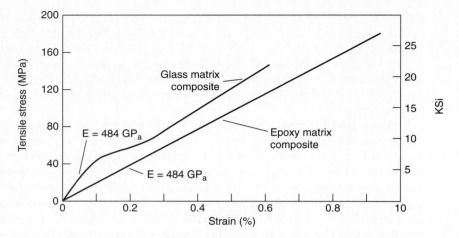

Figure 1.78   Tensile stress-strain comparison for discontinuous carbon fiber-reinforced composites.[162]

The potential for Al_2O_{3f} in reinforcing ceramic matrices is highly dependent upon the incidence of bonding at the interface. Overall, the oxidative stability and corrosion resistance of these fibers is good, however, the formation of strong bonds in matrices such as MgO and SiO_2 greatly reduces the strength and toughness. Furthermore, heating to temperatures greater than 1200°C degrades the fiber by enhancing grain growth and recrystallization. This action limits the high-temperature performance of composites made with these fibers. However, coatings applied on the fibers prior to processing may provide the solution to strong bonding and lead to an increase in strength and toughness.

1.4.4.5 Discontinuous-fiber reinforcement of glass–glass-ceramics.

The use of 2-D and 3-D arrays of discontinuous fibers for reinforcement does not provide the same levels of composite tensile strength and toughness as in the continuous-fiber case. However, significant increases in both properties over those of the matrix alone can be achieved. The tensile stress-strain curves shown in Figure 1.78 compare the performance of epoxy and borosilicate glass matrix composites reinforced with a 2-D array of 1.9-cm-long carbon fibers. As in the previous continuous Nicalon fiber composite comparison, the resin matrix composite behavior is linear, while that of the glass matrix system indicates a matrix microcracking dependence. In this simple tensile case the epoxy matrix composite is stronger. However, in three-point flexure it was found that the glass matrix system could carry a much higher load.[178]

The ability to achieve high toughness for even a 3-D array of discontinuous fibers was also reported when a Nicalon fiber-reinforced LAS composite was fabricated using a chopped-fiber molding compound. As in the case of continuous fiber-reinforced composites, application of load was required to continue crack growth sufficiently to fracture the specimen completely. The appearance of the fracture was quite fibrous and again demonstrated significant crack blunting and diversion.

1.4.5 Fabrication Techniques

One of the most significant attributes of glass matrix composites is the ease with which they can be fabricated. The flow characteristics of the glass at high temperature make it possible to adapt many techniques commonly used to fabricate net shape resin matrix composites to glass matrix composite use. Processes in current use or under study are (1) hot pressing of infiltrated unitape and fabric layups, (2) hot matrix transfer into woven preforms, and (3) hot injection molding of chopped-fiber compounds or preforms.

1.4.5.1 Hot pressing.   The process for making unitape composites is begun by winding slurry-impregnated yarn onto a mandrel to form monolayer tapes as illustrated in Figure 1.79. Usually the slurry consists of water, a water-soluble resin binder, and glass powder. These tapes are cut up to make plies which are then stacked and densified to form the final composite in a hot pressing operation. If the slurry contains a resin binder, it is burned out prior to transferring the ply stacks to the hot pressing molds. For proper densification to occur, pressing must be carried out at a temperature where the glass viscosity is low enough to permit the glass to flow into the interstices between individual fibers within the yarn bundles. Pressure is applied only after the mold temperature reaches the softening point of the matrix glass. The resulting composites are generally greater than 98% of theoretical density at completion of the process.

Highly stressed structural composites generally require unitape ply layup constructions. However, where strength requirements permit, there are practical considerations that favor use of fabric or hybrid fabric-unitape layups, principally because the 2-D structural integrity of fabrics facilitates the molding of complex-shaped parts. The fabrics are prepared by painting on measured quantities of slurry similar in composition to the slurries used in unitape fabrication. The fabric ply layup and hot pressing operations are identical to those used in unitape processing.

Variations of this technique were presented in a paper by Clarke[179] describing *prepregging* and hot pressing for gas turbine engine applications. The essence of the process is to infiltrate a fiber tow with a slurry of powderd matrix, filament-wind the impregnated tow to form prepreg sheets, and then stack and hot press the sheets to form a com-

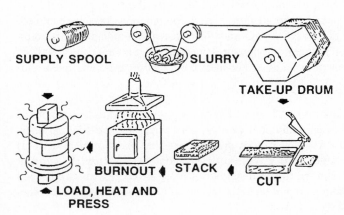

Figure 1.79   Steps in tape layup processing of glass matrix composites.[162]

posite. This ostensibly simple process is in fact a very complex one in which optimization and control of many variables is necessary in order to produce composites with consistently high strength and toughness.

In a joint research project involving Schott Glaswerke (Mainz, Germany), the University of Karlsruhe, and Technical University of Berlin,[180] researchers developed glass and glass-ceramic composites reinforced with continuous fibers of carbon or SiC.

The composites were made using a version of the tape-casting process. The ceramic fiber was exposed to a slurry containing a fine glass powder (which subsequently formed the matrix of the composite), and an organometallic sol-gel (which transformed into an inorganic material when the glass was dried and tempered). The sol-gel addition is used to tailor the composition of the matrix and accurately control fiber-matrix bonding. To produce a glass-ceramic composite, nucleating agents also are added to the batch.

Coated fibers were assembled into wide bands by winding them on reels. To make plates, the resulting tapes were cut, properly oriented, and stacked to form prepregs. Annular shapes were produced by winding prepreg around a cylindrical core. Hot pressing followed at a temperature higher than 1000°C and a pressure greater than 5 MPa.

Schott reportedly has made plates up to 185 mm^2 in area and 40 mm thick or annular shapes having a thickness of 40 mm with an outside diameter of 185 mm and an inside diameter of 10 mm.

Franklin[137] describes another study that examined the interface between uncoated SiC$_f$ and LAS glass powder cold-pressed to produce compacts that were heat-treated in a stream of dry nitrogen gas in the temperature range 1000–1100°C for different lengths of time to form an interface between the fiber and matrix.

1.4.5.2 Matrix transfer molding.   While hot pressing of tape layups and woven structures is the method of preference in many applications, it is not applicable in some cases because of external and/or internal structural geometries. In these more complex situations carefully engineered woven structures can be used. These are first arranged inside a mold cavity, and then fluid matrix is transferred at high temperature into the mold cavity to fill the void spaces around the reinforcement structure. For example, fluid matrix glass is forced into the void spaces in a plain-weave fabric wrapped around a mandrel to produce a thin-walled cylinder.

Although matrix transfer offers the potential for great flexibility in processing net shape composites, it is not applicable to all fiber-matrix combinations that are routinely hot-pressed. This is because fiber matrix glass must be injected at temperatures substantially higher than normal hot pressing temperatures in order to penetrate the interfiber spaces. This in turn can result in excessive fiber-matrix interaction and consequent embrittlement of the composite.

1.4.5.3 Injection molding.   Chopped fiber plus glass frit molding compounds are made by leaving space between individual yarn strands during the tape-winding operation as illustrated in Figure 1.79. The strands are chopped into short (typically 1-cm) segments which are then tumbled into a more-or-less random orientation. In currrent practice, the compound (either in the tumbled state or as a reconsolidated slug) is placed in the reservoir section of a double-chambered die and injected at high temperature into a mold cavity by ram action.

Injection molding of chopped-fiber molding compounds affords versatility for complex shape fabrication equal to that of matrix transfer molding and is suitable for all fiber-matrix combinations that can be hot-pressed. The main disadvantage of this technique is that there is at best very limited control over fiber orientation. Consequently the technique is applicable only to parts that will be subjected to low stress levels. However, in applications for which injection-molded parts are suitable, they offer the toughness characteristics exhibited by the other glass matrix composites.

1.4.5.4 Other processes

Melt Infiltration.   Wolf, Francis, Lin, et al.[181] described how alumina-glass composites were prepared by a melt infiltration process similar to a fabrication method for dental crowns and bridges. Cylindrical alumina samples with green densities ranging from 62 to 72% of theoretical were formed by slip casting followed by sintering at 1100°C for 2 h. A borosilicate glass was infiltrated at 1200°C, resulting in a composite microstructure consisting of fused alumina particles and glass-filled pores. The fracture toughness of the composites, measured by a chevron notch method with a short rod sample, was ~3.8 MPa m$^{1/2}$ and was relatively insensitive to the V_f of alumina in the range 0.62–0.72.

Ultrasonic Sol-gel.   Chiou and Hahn[182] have used an ultrasonic sol-gel technique to process aluminoborosilicate glass and its composite with carbon fiber reinforcement. In this new technique, ultrasonic energy is used in place of alcohol solvent. Gel time is easily controlled by varying the amount of ultrasonic energy, and the resulting gel also exhibits less shrinkage during the densification process. The sonogel-based composite has lower flexural strengths than the alcogel-based composite.

The measured flatwise and edgewise flexural strengths were 241 ± 4.2 and 212 ± 3.1 MPa, respectively. These values are lower than the corresponding average strengths, 320 and 310 MPa, of the alcogel-based composite reported in reference 183. The alcogel-based composite has the same type of glass and the same fiber volume content. The present composite, which is ultrasonically processed, is seen to have only about 80% of the strength of the alcogel-based composite.

The fracture toughnesses K_{IC} were measured to be 9.4 ± 1.1 and 8.3 ± 1.1 MPa m$^{1/2}$ under flatwise and edgewise loading, respectively. These values were almost the same as the corresponding fracture toughnesses, 9.1 and 7.1 MPa m$^{1/2}$, of the alcogel-based composite[183] and were much higher than the typical fracture toughness of 1.5 MPa m$^{1/2}$ for bulk glasses. This increase is the result of effective toughening resulting from fiber debonding, fiber bridging, and fiber pullout.

Once perfected, the new ultrasonic sol-gel technique can offer many advantages over conventional processing. No alcohol is necessary, densification can be done with much less shrinkage, and the process cycle time can be reduced substantially.

1.4.6 Properties

Prewo and Brennan[162] tested three LAS (I, II, and III; see Table 1.9) glass-ceramic matrix–Nicalon fiber composite systems for flexure strength (three-point) as a function of temperature in an inert environment. As in the case of the glass matrix composites, the eventual loss of strength for these composites at elevated temperatures is due to softening of the matrix and thus is highly dependent on the percentage of matrix still left in the glassy state. Testing in air established that composite fracture morphology and strength were strongly related to test environment.

These workers showed that both high composite strength and toughness were achievable only in the presence of a fibrous fracture morphology.

Other composite properties have been determined for the various LAS matrix–Nicalon fiber systems, including fracture toughness, Charpy and ballistic impact, elevated temperature creep, thermal and mechanical fatigue, thermal expansion, and thermal conductivity. Some of these properties have been reported.[184] The results of these measurements have shown that these composite systems exhibit excellent potential for structural use up to temperatures in excess of 1100°C. For example, the flexural strength of a unidirectional LAS-Nicalon fiber composite was measured after thermally shocking the material from an elevated temperature into a water quench and comparing the results to strengths obtained for hot-pressed Si_3N_4 and monolithic LAS quenched under identical conditions. The results of these tests show that the thermal shock properties of this type of composite are excellent.

The effect of fiber orientation on the mechanical properties of LAS matrix–Nicalon fiber composites has also been studied at United Technologies Research Center (UTRC).[185] It is apparent that rather large strains to failure can be obtained in this composite system, particularly for 0°, ±45°, and ±60°-ply orientations. The change from linear to nonlinear behavior in the 0°-oriented composites occurs at the point where the matrix starts to microcrack. Final composite failure does not occur, however, until over 1% strain is reached because of the continuing load-carrying capability of the fibers above the point of matrix failure.

In elevated-temperature creep tests it is apparent that reinforcement of the LAS matrix with Nicalon fibers significantly reduces the creep rate, even for the off-axis fiber orientation.

While additional research is necessary in order to understand the nature of oxidative embrittlement of these composites, the system of Nicalon fiber-reinforced glass-ceramic matrix composites has shown itself to be a viable candidate for a lightweight high-temperature structural material in light of its demonstrated ease of fabrication and attractive mechanical and thermal properties.

Gadkasee and Chyung[186] carried out experiments with a Corning 1723 glass matrix, Nicalon fibers, and Arco SiC_w. They found that the ultimate strength of the composite remained constant up to about 5 wt% whiskers but began to decrease beyond that with increasing whisker percentage. At 24 wt% whisker the ultimate strength decreased substantially. The ultimate strength and the microcrack stress were at the same level, and the composite was essentially a brittle composite. The microcrack stress went through an optimum at approximately 10 wt% whisker loading and had increased by 100% at this level. The drop in ultimate strength at high whisker loading and the optimum in microcrack stress were unexpected phenomena. A detailed study of the composites revealed that they were caused by damage to the fibers induced by whiskers during processing.

Additionally these researchers evaluated the variation in transverse strength and interlaminar shear strength of a 1723 glass matrix nonhybrid composite with Nicalon fibers versus a hybrid plus whiskers. They found there was a several hundred percent increase in both properties with whisker addition. The transverse strength increased from 12 MPa to 51 MPa, and interlaminar shear strength increased from 47 MPa to 134 MPa at 10 wt% whisker addition.

The whisker reinforcement of the matrix thus has been taken advantage of in significantly increasing the fiber-reinforced composite performance. Although principles of hybridization have been proven in glass matrix composites, hybrid glass-ceramic matrix composites are expected to have substantially improved performance over fiber–glass-ceramic composites.

Finally, German researchers[180] found improved ductility and properties with these glass-ceramic materials. They reported that

- Unreinforced glass broke when quenched from 350°C to 20°C. However, SiC_f/Gl can withstand 60 cycles of quenching to room temperature from the tempering temperature of 550°C.

- A low thermal conductivity of 1.7 W/m·K, which is close to that of glass, was found for SiC_f/Gl and C_f/Gl having the C_f running transverse to the diffusion direction. Values up to 25 W/m·K, roughly half that of steel, could be obtained when the C_f were oriented in the diffusion direction.

- The coefficient of sliding friction and the wear factor of C_f/Gl paired with steel were both low–0.1 and ~10^{-6} mm^3/N·m, respectively. They do not vary greatly with fiber direction, and the composite may be self-lubricating.

1.5 METAL MATRIX COMPOSITE PROCESSING

1.5.1 Introduction

The critical need for high-strength, lightweight, high-stiffness materials has in recent years resurrected much interest in continuous and discontinuous reinforced MMCs. These hybrid materials have combined primarily aluminum and titanium, as well as other metal alloys, with a wide variety of reinforcements.

System trade studies have been the primary motivating factor resulting in the renewal of much interest in developing and using MMCs. MMCs, in general, consist of at least two components; one is the metal matrix, and the second is the reinforcement. In all cases the matrix is defined as a metal, but a pure metal is rarely used; the matrix is generally an alloy. The distinction of MMCs from other alloys with two phases or more comes about from the processing of the composite. In the production of the composite, the matrix and the reinforcement are mixed together. This is to distinguish a composite from an alloy with two phases or more where the second phase forms as a particulate and a phase separation such as a eutectic or a eutectoid reaction occurs. MMCs offer a spectrum of advantages over conventional and/or traditional PMCs that are important for their selection and use as structural materials. A few such advantages include the combination of high strength, high elastic modulus, high toughness, and impact resistance; low sensitivity to changes in temperature or thermal shock; high surface durability; low sensitivity to surface flaws; high electric and thermal conductivity; minimum exposure to the potential

TABLE 1.10.   Typical Reinforcements Used in MMCs[195]

Fiber[a]	Diameter (μm)	Tensile Strength (MPa)	Elastic Modulus $(10^6 \text{ psi})^{b}$	Use Limit °F	°C
Boron (C)	100–200	3500	400	1000	538
Carbon graphite pan (C)	7.0	2400–4820	227–390	> 3000	1649
Carbon graphite pitch (C)	5.1–12.7	2067	400–700	> 3000	1649
SiC monofilament	140	4134	400	1700	927
SiC (W)	6.0	3341	500–800	1700	927
FP alumina (C)	20	1378	400	> 3000	1649
Fiberfrax (DC)	2–5	1723	90	2100	1149
3M Nextel 312 (C)	10	1378	150	3000	1649
ICI Saffil (C)	3	2000	300	3000	1649

[a] C, Continuous; W, whisker; DC, discontinuous.
[b] 10^3 psi ≈ 6.89 N/mm^{-2}.

problem of moisture absorption resulting in environmental degradation; and improved fabricability with conventional working equipment.[187–196]

The renewed interest in MMCs has been aided by the development of reinforcement material that provides either improved properties or reduced cost when compared with existing monolithic materials. MMC reinforcements can be generally divided into five major categories: (1) continuous fibers, (2) discontinuous fibers, (3) whiskers, (4) wires, and (5) particulates (including platelets). With the exception of wires, which are metals, reinforcements are generally ceramics. Typically these ceramics are oxides, carbides, and nitrides, which are used because of their excellent combinations of specific strength and stiffness at both ambient temperature and elevated temperature. The typical reinforcements used in MMCs are listed in Table 1.10. SiC, B_4C, Si_3N_4, and Al_2O_3 are the key particulate reinforcements and can be obtained with varying levels of purity and size distribution. SiC_p are also produced as a by-product of the processes used to make whiskers of these materials.

Early studies on MMCs addressed the development and behavior of continuous fiber-reinforced hybrid materials based on aluminum and titanium matrices. Unfortunately, despite encouraging results, extensive industrial application of these composite materials was hindered by (1) exorbitant manufacturing costs associated with the high cost of reinforcement material, and (2) the highly labor-intensive manufacturing process. Consequently, effective utilization of these materials was restricted to military and other highly specialized applications. The family of discontinuous reinforced MMCs includes those with particulates, whiskers, and platelets.

In recent years, particulate-reinforced MMCs have attracted considerable attention on account of[197] (1) availability of a spectrum of reinforcements at competitive costs, (2) successful development of manufacturing processes to produce MMCs with reproducible microstructure and properties, and (3) availability of standard and near-standard metalworking methods that can be utilized to form these materials. Furthermore, use of discontinuous reinforcements minimizes problems associated with the fabrication of continuously reinforced MMCs such as fiber damage, microstructural heterogeneity, fiber mismatch, and interfacial reactions. For applications subjected to severe loads or extreme thermal fluctuations, such as in automotive components, discontinuously reinforced MMCs have been shown to offer near-isotropic properties with substantial improvements in strength and stiffness relative to those available with monolithic materials. However, discontinuously reinforced composite materials are not homogeneous, and material properties are sensitive to the properties of the constituent, interfacial properties, and the geometric shape of the 3-D reinforcement. Overall, the strength of such particle-reinforced metal matrices depends on (1) the diameter of the reinforcing particles, (2) the interparticle spacing, (3) the volume fraction of the reinforcement, and (4) the condition at the matrix-reinforcement interface. Matrix properties, including the work-hardening coefficient, which improves the effectiveness of the reinforcement constraint, are also important.[198]

1.5.2 Materials Selection

In any discussion of materials and their behavior as MMCs, a ROM-type prediction of composite stiffness and strength must be considered and only contributions from the fiber and matrix phases allowed. This rule involves the following assumptions: (1) filaments and matrix are strained equal amounts, (2) the phases act like homogeneous materials and are well bonded, and (3) the constituents have equal Poisson ratios.[199]

1.5.2.1 Reinforcement types.
The following forms of reinforcement types are available: continuous and discontinuous fibers and filaments ranging in diameter from a few micrometers to nearly 300 μm, discontinuous or chopped fibers, whiskers, and particulates and platelets. Among the fibers of greatest interest are B, Al_2O_3, C, SiC, W, and steel wires; among the whiskers and particulates are SiC, Si_3N_4, Al_2O_3, and TiC.

For many years, Al_2O_3 has been recognized as a promising reinforcement in high-temperature applications of MMCs and for many chemically aggressive environmental applications. Below 900°C, it retains most of its elastic stiffness, structural strength, and abrasion resistance. It is also inert to most metals and exhibits excellent oxidation resistance.[200]

All of the technically important metal matrix reinforcing fibers, including Al_2O_3, B, C, and SiC, are intrinsically brittle materials that exhibit significantly large variations in strength. Because of this, their failure strengths can be defined only statistically.[200]

SiC is produced in the form of continuous filaments, whiskers, and particulates. SiC single crystals appear in two useful forms: α-SiC, which is a hexagonal structure, and β-SiC, which has the face-centered cubic form. Whiskers and particulates are produced by various methods.[201–207]

Wire forms of Ni-based amorphous alloys exhibit high strength and good ductility combined with a high corrosion resistance.

B fiber is one of the early fibers produced. It is manufactured by vapor deposition of a W or C core.[208]

1.5.3 Interfaces in MMCs

The interface region in a given composite is extremely important in determining the ultimate properties of the composite. An interface is, by definition, a bidimensional region through which there occurs a discontinuity in one or more material parameters. In practice, there is always some volume associated with the interface region over which a gradual transition in one or more material parameters occurs. The importance of the interface region in composites stems from two main reasons: (1) the interface occupies a very large area in composites, and (2) in general, the reinforcement and the matrix form a system that is not in thermodynamic equilibrium.

An interface is a boundary surface between two phases where a discontinuity in some material parameters occurs. Among the important discontinuities are elastic moduli, thermodynamic parameters such as chemical potential, and CTE. The discontinuity in chemical potential is likely to cause chemical interaction, leading to an interdiffusion zone or chemical compound formation. The discontinuity in the thermal expansion coefficient means that the interface will be in equilibrium only at the temperature where the reinforcement and the matrix were brought into contact. At any other temperature, biaxial or triaxial stress fields will be present because of the thermal mismatch between the components of a composite.

The importance of the interface in composites is recognized on all sides. Therefore, some bonding must exist between the ceramic reinforcement and the metal matrix for load transfer from matrix to fiber to occur. Neglecting any direct loading of the reinforcement, the applied load is transferred from the matrix to the reinforcement via a well-bonded interface. It should be mentioned that in fiber-reinforced composites mechanical bonding is effective mostly in the longitudinal or fiber direction.

Most MMC systems are nonequilibrium systems in the thermodynamic sense; that is, there is a chemical potential gradient across the fiber-matrix interface. This means that given favorable kinetic conditions (which in practice means a high enough temperature or long enough time), diffusion and/or chemical reactions will occur between the components. The interface layer(s) formed because of such a reaction generally have characteristics different from those of either of the components.

Ceramic-metal interfaces are generally formed at high temperatures. Diffusion and chemical reaction kinetics are faster at elevated temperatures.

In general, ceramic reinforcements (fibers, whiskers, or particles) have a CTE smaller than that of most metallic matrices. This means that when the composite is subjected to a temperature change, thermal stresses are generated in both components.

Various processing parameters such as time, temperature, and pressure combined with the thermodynamic, kinetic, and thermal data can be used to obtain an optimum set of interface characteristics in a given MMC.

The extent of the interfacial region in composites can be very great. Mechanical as well as chemical bonding can contribute to the bond strength. A controlled interfacial reaction between a metal and a ceramic reinforcement can provide good adhesion. A diffusion barrier coating between the reinforcement and the matrix is frequently employed to control the extent of interfacial reaction. It is therefore important to understand the thermodynamic, kinetic, mechanical, and microstructural aspects of a metal-ceramic interface in order to use that information to optimize the processing to obtain a desirable set of characteristics in a composite. In particular, such an understanding is important if the composite is to have an ability to sustain high temperatures during processing, as well as long time exposures at operating temperatures.

1.5.4 Processing

Over the years a spectrum of processing techniques have evolved in an attempt to optimize the microstructure and mechanical properties of MMCs.[196,198,200,209–220] The processing methods utilized to manufacture MMCs can be grouped according to the temperature of the metallic matrix during processing. Accordingly, the processes can be classified into three categories: (1) liquid-phase processes, (2) solid-liquid processes, and (3) two-phase (solid-liquid) processes.

1.5.4.1 Liquid-phase processes.   In liquid-phase processes, the ceramic particulates are incorporated into a molten metallic matrix using various proprietary techniques. This is followed by mixing and eventual casting of the resulting composite mixture into shaped components or billets for further fabrication. The process involves a careful selection of the ceramic reinforcement depending on the matrix alloy. In addition to compatability with the matrix, the selection criteria for a ceramic reinforcement include the following factors[221]: (1) elastic modulus, (2) tensile strength, (3) density, (4) melting temperature, (5) thermal stability, (6) size and shape of the reinforcing particle, (7) CTE, and (8) cost. The various possible discontinuous ceramic reinforcements and their properties are summarized in references 198, 200, and 220–222. Since most ceramic materials are not wetted by the molten alloys, introduction and retention of the particulates necessitate either adding wetting agents to the melt or coating the ceramic particulates prior to mixing.

Liquid Metal–Ceramic Particulate Mixing.   Several approaches have been utilized to introduce ceramic particles into an alloy melt.[223] These include (1) injection of powders entrained in an inert carrier gas into the melt using an injection gun, (2) addition of particulates into the molten stream as it fills the mold, (3) addition of particulates to the melt via a vortex introduced by mechanical agitation, (4) addition of small briquettes (made from copressed aggregates of the base alloy powder and the solid particulates) to the melt followed by stirring; (5) dispersion of the particulates in the melt using centrifugal acceleration, (6) pushing of the particulates into the melt using reciprocating rods, (7) injection of the particulates into the melt while the melt is being irradiated with ultrasound, and (8) zero gravity processing. The last approach involves utilizing an ultrahigh vacuum and high temperatures for long periods of time.

In the processes described above, a strong bond between the matrix and the reinforcement is achieved by utilizing high processing temperatures (e.g., $T > 900°C$ for the Al-Al$_2$O$_3$ system) and alloying the matrix with an element that can interact with the reinforcement to produce a new phase and effect "wetting" between the matrix and the ceramic (e.g., Li in an Al-SiC system). Agitation during processing is also essential to disrupt contamination films and adsorbed layers to facilitate interfacial bonding.

To date, liquid-phase processes have reached an advanced stage of development, and SiC or Al$_2$O$_3$ (3–150 μm) particulates are routinely added to a variety of aluminum alloy matrices.[209–212, 223] Among these, the Dural process is perhaps the most advanced in terms of commercial development. This method involves the incorporation of ceramic particulates into a metallic melt through melt agitation. A summary of the mechanical properties of various MMC materials processed by the Dural process is given in reference 211. The results suggest that it is possible to combine up to 20 vol% of either SiC or Al$_2$O$_3$ with various aluminum alloys to obtain MMCs with attractive combinations of properties.

Despite the encouraging results obtained with liquid-phase processes, some difficulties exist. These include agglomeration of the ceramic particulates during agitation, settling of particulates, segregation of secondary phases in the metallic matrix, extensive interfacial reactions, and particulate fracture during mechanical agitation.

Melt Infiltration. In melt infiltration processes a molten alloy is introduced into a porous ceramic preform, utilizing either inert gas or a mechanical device as a pressurizing medium. The pressure required to combine matrix and reinforcement is a function of the friction effects due to the viscosity of the molten matrix as it fills the ceramic preform. Wetting of the ceramic preform by the liquid alloy depends on alloy composition, ceramic preform material, ceramic surface treatments, surface geometry, interfacial reactions, atmosphere, temperature, and time.[218, 219, 224] This approach has been studied extensively and in fact is currently being used commercially to fabricate the Toyota diesel piston, an Al–chopped-Al_2O_{3f} composite material.[225]

Some of the drawbacks of this process include reinforcement damage, preform compression, microstructural nonuniformity, coarse grain size, contact between reinforcement fibers or particulates, and undesirable interfacial reactions.[226]

Melt Oxidation Processes. In melt oxidation processing (i.e., the Lanxide process), a ceramic preform formed into the final product shape by a fabricating technique such as pressing, injection molding, or slip casting, is continuously infiltrated by a molten alloy as it undergoes an oxidation reaction with a gas phase (most commonly air).

Basically a chemical reaction is used to cause Al to infiltrate the ceramic preform. Al_2O_3 and SiC in volumes of 55–60% are used with Al alloy matrices to form NNS parts. An Al alloy ingot containing 3–10 wt% Mg is placed on top of a permeable mass of ceramic material, either Al_2O_3 or SiC. The alloy-ceramic assembly is heated in an atmosphere of nitrogen at temperatures between 1475 and 1835°C. Spontaneous infiltration takes place provided (1) the alloy contains Mg, (2) the temperature is at least 1475°C, and (3) the atmosphere is mostly nitrogen.[227–229]

Because of the nitrogen atmosphere, AlN is formed within the microstructure. The amount depends inversely on the speed of the reaction, with more AlN forming at slower infiltration rates. Infiltration speed is slower at lower temperatures and at a lower nitrogen content of the atmosphere. This is significant because an increase in the amount of AlN increases the stiffness of the composite and reduces the CTE. Therefore, stiffness and CTE can be tailored by controlling the process temperature and the amount of nitrogen in the atmosphere. Also high-temperature oxidation of the molten alloy in the interstices of the ceramic preform produces a matrix material composed of a mixture of oxidation reaction products and unreacted metal alloy.[228] The primary advantage of this process is its ability to form complex, fully dense composite shapes. An attractive combination of mechanical properties has been reported for Al-based MMCs processed using this method.[212]

Squeeze Casting or Pressure Infiltration. Relatively few MMCs have been made by simple liquid-phase infiltration, mainly because of difficulties with fiber wetting by the molten metal noted earlier. However, when the infiltration of a fiber preform occurs readily, reactions between the fiber and the molten metal can significantly degrade fiber properties. Fiber coatings applied prior to infiltration, which improve wetting and control reactions, have been developed and can produce impressive results.[200] In this case, however, the disadvantage is that the fiber coatings must not be exposed to air prior to infiltration because surface oxidation alters the positive effects of coating.

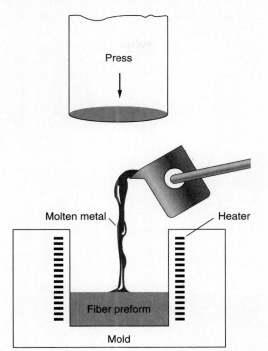

Figure 1.80 Squeeze casting or pressure infiltration process. Liquid metal is forced into a fibrous preform, and pressure is applied until solidification is complete.

Squeeze casting or pressure infiltration involves forcing the liquid metal into a fibrous preform. Pressure is applied until solidification is complete. Figure 1.80 is a schematic of this process. By forcing the molten metal through small pores of a fibrous preform, this method obviated the requirement of good wettability of the reinforcement by the molten metal. Composites fabricated with this method have minimal reaction between the reinforcement and molten metal and are free of common casting defects such as porosity and shrinkage cavities. Infiltration of a fibrous preform by means of a pressurized inert gas is another variant of liquid metal infiltration. Net-shaped components can be produced inexpensively. The process takes place in the controlled environment of a pressure vessel and with a rather high fiber V_f, and complex-shaped structures are obtainable.

1.5.4.2 Solid-phase processes.   The fabrication of particulate-reinforced MMCs from blended elemental powders involves a number of steps prior to final consolidation. Two methods that fall in this category are powder metallurgy (P/M) and high-energy rate processing.

Powder Metallurgy.   Solid-phase processes invariably involve the blending of rapidly solidified powders with particulates, platelets, or whiskers using a number of steps. These include sieving of the rapidly solidified powders, blending with the reinforcement phase(s), pressing to ~75% density, degassing, and final consolidation by extrusion, forging, rolling, or some other hot working method. This technology has been de-

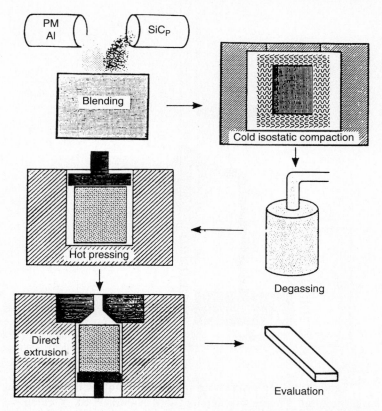

Figure 1.81   Schematic interpretation of the processing route for P/M Al/SiC$_p$ composites.[216]

veloped to various degrees of success by different commercial manufacturers. The Alcoa and Ceracon processes are shown diagrammatically in Figures 1.81 and 1.82, respectively. Consolidation of the MMC preform is achieved by hot extrusion in the Alcoa process, whereas in the Ceracon process final densification is achieved by hot pressing in a pressure-transmitting medium (PTM). The P/M methods have been successfully applied to a large number of metal-ceramic combinations.[232-237] P/M-processed Al-SiC MMCs possess higher overall strength levels relative to those corresponding to the equivalent material processed by a liquid-phase process; the elongation values, however, are lower.

In terms of microstructural requirement, the P/M approach is superior in view of the rapid solidification experienced by the powders. This allows the development of novel matrix materials outside the compositional limits dictated by equilibrium thermodynamics in conventional solidification processes.

Continuous filament-reinforced composites can also be made by the P/M technique. Fiber tows are infiltrated by dry matrix powder, followed by hot isostatic pressing (HIP). Alternatively, matrix alloy powder that has been dry-blended with short whiskers, chopped filament, or particles is hot-pressed to produce a composite material with a ran-

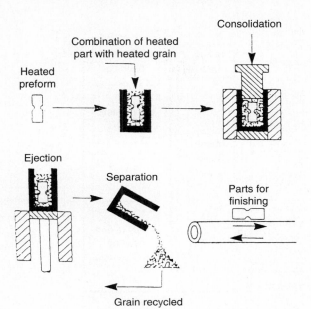

Figure 1.82 Schematic illustration of the Ceracon technique for fabricating P/M MMCs.[230, 231]

dom orientation. This material can become partially oriented by hot mechanical deformation, usually extrusion or hot forming (Figure 1.83).

P/M methods involving cold pressing and sintering, or hot pressing, produce MMCs (Figures 1.81 and 1.82). The matrix and the reinforcement powders are blended to produce a homogeneous distribution. The blending stage is followed by cold pressing to produce what is called a *green body,* which is about 80% dense and can be easily handled. The cold-pressed, green body is canned in a sealed container and degassed to remove any absorbed moisture from the particle surfaces. The final step is hot pressing, uniaxial or isostatic, to produce a fully dense composite. The hot pressing temperature can be either below or above that of the matrix alloy solidus (Figure 1.84).

The P/M hot pressing technique generally produces properties superior to those obtained by casting and by liquid metal infiltration (squeeze casting) techniques. The slightly superior performance of the P/M composite is evident. In studies made by the Brown Boveri Research Centre, Switzerland, the superior mechanical properties obtained by hot pressing an Al/SiC$_w$ composite were attributed to a more homogeneous distribution of whiskers when compared to that obtained with melt infiltration.

Major limitations and constraints on P/M technology are as follows.

- Limited availability of appropriate prealloyed metal powders
- High cost of metal powders
- High cost of hot pressing.

High-energy, High-rate Processes.   An approach that has been successfully utilized to consolidate rapidly quenched powders containing a fine distribution of ceramic particulates is known as high-energy, high-rate processing.[238, 239] In this approach the con-

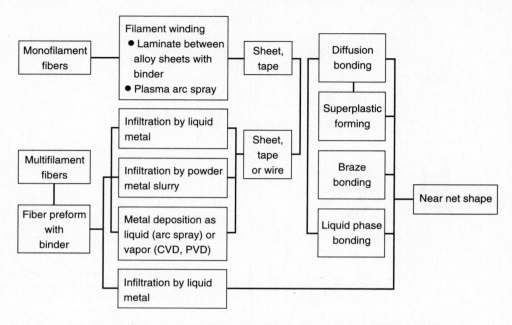

Figure 1.83   Process routes for the production of continuous fiber-reinforced MMCs.

solidation of a metal-ceramic mixture is achieved through the application of high energy over a short period of time. An examination of the literature reveals that both mechanical energy and high electric energy sources can be successfully utilized to consolidate MMCs.[223, 238–240] For example, Marcus et al.[238, 240] were able to consolidate Al/SiC MMCs by heating a customized powder blend through a fast electric discharge obtained from a homopolar generator. The high-energy, high-rate pulse (1 MJ/s) facilitates rapid heating of the conducting powder in a die with cold walls. The short time at temperature approach offers an opportunity to control (1) phase transformations and (2) the degree of microstructural coarsening not readily possible with standard processing techniques. This technique has been successfully used in the manufacture of Al-SiC and (TiAl + Nb)/SiC composites.[223, 240]

Diffusion Bonding.   Diffusion bonding is a common solid-state welding technique for joining similar and dissimilar metals. Interdiffusion of atoms of clean metal surfaces in contact at an elevated temperature leads to welding.[241] The major advantages of this technique are (1) ability to process a wide variety of matrix metals, and (2) control of fiber orientation and V_f. Among the disadvantages are (1) processing times of several hours, (2) cost of high processing temperatures and pressures, (3) ability to produce objects of only limited size. There are many variants of the basic diffusion bonding process, however, all of them involve a simultaneous application of pressure and high temperature. Matrix alloy foil and fiber arrays, or composite wire, or monolayer laminas are stacked in a predetermined order. Figure 1.85 shows the different steps in fabricating MMCs by diffusion bonding. Here primarily the metal or metal alloys in the form of sheets and the reinforcement material in the form of fiber are chemically surface-treated

Process routes for the production of discontinuous fiber, whisker and particulate reinforced composites

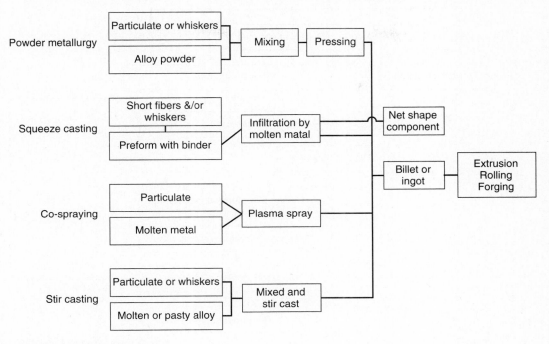

Source: After Harris, University of Nottingham

Figure 1.84 Process routes for the production of discontinuous fiber, whisker- and particulate-reinforced composites.

for effectiveness of interdiffusion. Then fibers are placed on the metal foil in predetermined orientation, and bonding takes place by press forming directly, as shown by the dotted line. However, sometimes the fibers are coated by plasma spraying or ion plating to enhance the bonding strength before diffusion bonding: this is shown by the solid line. After bonding, secondary machining work is carried out. Again the methods of putting together reinforcement fibers and matrix materials depend upon the fiber types. In the case of monofilaments and fiber bundles such as C, SiC/W, SiC, and Al_2O_3, they are wound onto a drum made of a metal having good thermal conduction (Cu, for instance). Then the preform materials of composites are made by applying the matrix material on reinforcements using various coating methods such as plasma spraying, chemical coating, electrochemical plating, CVD, and plasma vapor deposition (PVD). Of these coating methods, plasma spraying[242] is a relatively simple technique of low cost that can produce sheets of large width with good adhesion between the fiber and matrix. Then the preform is press-formed, achieving bonding of fiber and matrix through the application of pressure and temperature either by hot pressing or cold isostatic pressing (CIP), to enhance the density of the composite by removing voids and improve the strength of the composite by introducing some plastic deformation in the metal matrix. Diffusion bonding under vacuum conditions is more effective than under atmospheric conditions.[243–245] Because of

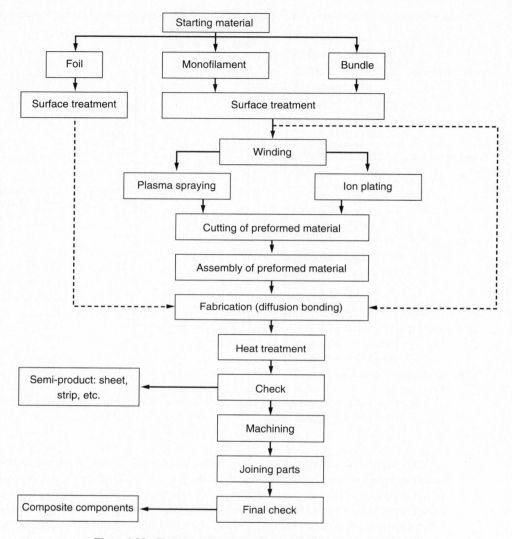

Figure 1.85 Flowchart for composite fabrication by diffusion bonding.[242]

the relatively low temperatures involved, a fiber-coating treatment for solid-phase fabrication is not critical as it is in the case of liquid metal infiltration, but the pressure applied in enhancing diffusion bonding may cause damage.[247–250] The applied pressure and temperature, as well as the time required for diffusion bonding to develop, vary with the composite system. Normally filaments of stainless steel, B, and SiC have been used with matrices such as aluminum and titanium alloys,[251–254] however, this is the most expensive method of fabricating MMC materials.[200, 241, 246, 255, 256]

In this process, the fugitive binder is evaporated and the gas removed. This method enhances composite density by removing voids and improves composite strength by introducing some plastic deformation in the matrix. The basic parameters controlling the forming process are temperature, pressure, pressure holding time, and atmosphere.

Of the several processing methods used for superalloy composites, diffusion bonding of fiber arrays with matrix alloy powder or matrix foil has been the most effective. The ability of the fibers to deform plastically eases the problem of fiber cracking during processing and allows secondary plastic deformation of the composite.

Hot roll diffusion bonding is another variant of the diffusion bonding technique that can be used to produce a composite consisting of different metals in sheet form. Such composites are called *sheet-laminated MMCs*.

1.5.4.3 Two-phase processes.

Two-phase processes involve the mixing of ceramic and matrix in a region of the phase diagram where the matrix contains both solid and liquid phases. Applicable two-phase processes include the Osprey, rheocasting, and variable codeposition of multiphase materials (VCM).

Osprey Deposition.   In the Osprey process, the reinforcement particulates are introduced into a stream of molten alloy which is subsequently atomized by jets of inert gas. The sprayed mixture is collected on a substrate in the form of a reinforced metal matrix billet. This approach was introduced by Alcan as a modification of the Osprey process,[246,257,258] Figure 1.86. The procedure combines the blending and consolidation steps of the P/M process and promises major savings in the production of MMCs.[259,260]

Rheocasting.   In the rheocasting process, fine ceramic particulates are added to a metallic alloy matrix at a temperature within the solid-liquid range of the alloy. This is followed by agitation of the mixture to form a low-viscosity slurry. The approach takes advantage of the fact that many metallic alloys behave like a low-viscosity slurry when subjected to agitation during solidification. This behavior, which has been observed for fractions of solids as high as 0.5, occurs during stirring and results in breaking of the solid dendrites into spheroidal solid particles which are suspended in the liquid as fine-grained particulates.[261,262] This unique characteristic of numerous alloys, known as thixotropy, can be regained even after complete solidification by raising the temperature. This approach has been successfully utilized in die casting of Al-based and Cu-based alloys.[262,263]

The slurry characteristic of the matrix during stirring permits the addition of reinforcements during solidification. The ceramic particulates are mechanically entrapped initially and are prevented from agglomerating by the presence of primary-alloy solid particles. Subsequently, the ceramic particulates interact with the liquid matrix to effect bonding. Furthermore, continuous deformation and breakdown of solid phases during agitation prevent particulate agglomeration and settling. This method has been successfully utilized by Mehrabian and coworkers[263,264] to incorporate up to 30 wt% Al_2O_3 and SiC and up to 21 wt% glass particles (14- to 340-μm diameter) in a partially solidified, 0.40–0.45 V_f solid of Al-5% Si-2% Fe alloy. The majority of the particulates were found to be homogeneously distributed in the matrix.

Compocasting is an application of the rheocasting process in which particulate or fibrous materials are added to the semisolid slurry. The particles or short fibers are mechanically entrapped and prevented from settling or agglomerating because the alloy is

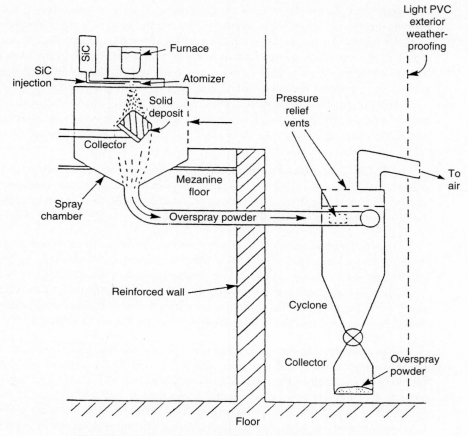

Figure 1.86   Schematic diagram of the modified Osprey technique.[257,258]

already partially solid. With continued mixing, the particles or fibers interact with the liquid matrix to undergo adhesion and bonding. Researchers[255] have combined rheocasting with squeeze casting to produce fiber composite materials (Table 1.11).[265]

Compocasting is one of the most economical methods of fabricating a composite with discontinuous fibers (chopped fiber, whisker, and particulate). This process is the improved version of slush or stir casting.

The technique has several advantages:[255]

- It can be performed at temperatures lower than those conventionally employed in foundry practice during pouring, resulting in reduced thermochemical degradation of the reinforced surface.
- The material exhibits thixotropic behavior typical of stir-cast alloys; it therefore offers NNS forming at low pressure in the semisolid state, resulting in a defect-free product.
- Production can be carried out by conventional foundry methods.

TABLE 1.11.   Comparison of Different Techniques[265]

Route	Cost	Application	Comments
Diffusion bonding	High	Used to make sheets, blades, vane shafts, structural components	Handles foils or sheets of matrix and filaments of reinforcing element
Powder metallurgy	Medium	Mainly used to produce small objects (especially round), bolts, pistons, valves, high-strength and heat-resistant materials	Both matrix and reinforcements used in powder form; best for particulate reinforcement; since no melting is involved, no reaction zone develops, showing high-strength composite
Liquid-metal infiltration	Low medium	Used to produce structural shapes such as rods, tubes, beams with maximum properties in a uniaxial direction	Filaments of reinforcement used
Squeeze casting	Medium	Widely used in automotive industry for producing different components such as pistons, connecting rods, rocker arms, cylinder heads; suitable for making complex objects	Generally applicable to any type of reinforcement and may be used for large-scale manufacturing
Spray casting	Medium	Used to produce friction materials, electrical brushes and contacts, cutting and grinding tools	Particulate reinforcement used; full-density materials can be produced
Compocasting	Low	Widely used in automotive, aerospace, industrial equipment, and sporting goods industries; used to manufacture bearing materials	Suitable for discontinuous fibers, especially particulate reinforcement

However, the process also has some disadvantages:

- Residual pores between fibers cannot be eliminated completely.
- Slurry is apt to fall off on handling, and the method cannot handle fiber for fabricating fiber-reinforced composites.

Variable Codeposition of Multiphase Materials.[196,198]   During VCM processing, the matrix material is disintegrated into a fine dispersion of droplets using high-velocity gas jets. Simultaneously, one or more jets of a strengthening phase are injected into the atomized spray at a prescribed location (Figure 1.87a) where the droplets contain a limited amount of the V_f of liquid. Hence, contact time and thermal exposure of the particulates with the partially solidified matrix are minimized, and interfacial reactions can be controlled. In addition, tight control of the environment during processing minimizes oxidation.[266–268]

In recent studies, Gupta et al.[266] incorporated up to 20 vol% of SiC_p into an Al-Li matrix using VCM. In these studies, injection of the reinforcing phase was accomplished by entraining SiC_p in an inert-gas stream using a suitably designed fluidized bed (Figure 1.87b).

1.5.4.4 Deposition techniques.
Deposition techniques for MMC fabrication involve coating individual fibers in a tow with the matrix material needed to form the composite, followed by diffusion bonding to form a consolidated composite plate or

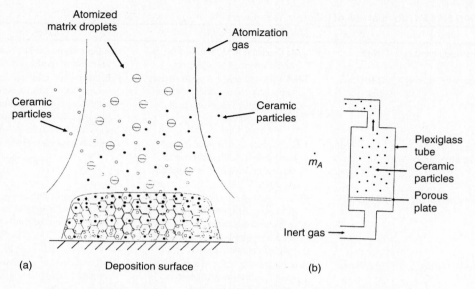

Figure 1.87   Schematic diagram of the variable codeposition process.[266–268] (*a*) Gas flow. (*b*) Fluidized bed.

structural shape. Since the composite is composed of identical units, the microstructure is more homogeneous than that of cast composites. The main disadvantage of using deposition techniques is that they are time-consuming. However, there are several advantages:

- The degree of interfacial bonding is easily controllable; interfacial diffusion barriers and compliant coatings can be formed on the fiber prior to matrix deposition, or graded interfaces can be formed.
- Filament-wound thin, monolayer tapes can be produced which are easier to handle and mold into structural shapes than other precursor forms; unidirectional or angle-plied composites can be easily fabricated in this way.

Several deposition techniques are available: immersion plating, electroplating, spray deposition, CVD, and PVD. Dipping or immersion plating is similar to infiltration casting except that fiber tows are continuously passed through baths of molten metal, slurry, sol, or organometallic precursors. Electrodeposition or electroplating produces a coating from a solution containing the ion of the desired material in the presence of an electric current. Fibers are wound on a mandrel, which serves as the cathode, and placed in the plating bath with an anode of the desired matrix material. The advantage of this method is that the temperatures involved are moderate and no damage is done to the fibers. Problems with electroplating are that voids can form between fibers and between fiber layers, adhesion of the deposit to the fibers may be poor, and only a limited number of alloy matrices are available for this processing. A spray deposition operation typically consists of winding fibers onto a foil-coated drum and spraying molten metal onto them to form a monotape. The source of molten metal may be powder or wire stock which is melted in a flame, arc, or plasma torch. The advantages of spray deposition are easy control of fiber alignment and rapid solidification of the molten matrix. In CVD, a vaporized component decomposes or reacts with another vaporized chemical on the substrate to form a coating on that substrate. The process is generally carried out at elevated temperatures. The product phase can be homogeneously nucleated if it occurs in the vapor phase or heterogeneously nucleated if it occurs on a substrate surface. Metallic, oxide, carbide, or nitride coatings or films can be produced with CVD. These films can be grown in amorphous, single-crystal, or polycrystalline form. When the matrix material is developed on a reinforced preform, such as a woven mat, it is called chemical vapor infiltration. The three basic PVD processes for creating MMCs differ mainly in how the vapor is generated: evaporation, ion plating, and sputtering. Evaporation methods for matrix materials under high vacuum, generally contained in a crucible, include resistive heating, electron beam evaporation, arc evaporation, radiation heating, and laser ablation. Ion plating involves passing the vaporized component through an argon glow gas discharge around the substrate, which ionizes some of the vapor molecules. These ionized atoms subsequently deposit onto the substrate. Sputtering techniques involve vaporization of the coating material (the target or cathode) via a momentum transfer from an ionized argon gas molecule in a glow discharge around the cathode. The primary advantage of PVD is versatility in the composition and microstructure of the coating produced. These techniques can produce coatings with superior bonding to the substrate and excellent surface finishes. In addition, no chemical reaction by-products are created by these processes. The disadvantage is that the processing equipment is complex and expensive.

Spray Forming Technique.   A relatively new and promising development for making particle-reinforced MMCs deserves a separate description. It involves the use of spray techniques that have been employed for some time to produce monolithic alloys. This process uses a spray gun to atomize, for example, a molten aluminum alloy matrix into which heated (to dry them) SiC_p are injected. An optimum particle size is required for transfer. Whiskers, for example, are too fine to be transferred. A preform produced in this way can be up to about 97% dense. The MMC is subjected to scalping, consolidation, and secondary finishing processes, and the product is a wrought material. The procedure is totally computer-controlled and quite fast. It should be noted that it is essentially a liquid metallurgy process. The formation of deleterious reaction products is avoided because the time of flight is extremely short.

A great advantage of the process is the flexibility that it affords in making different types of composites. For example, in situ laminates can be made using two sprayers. One can have selective reinforcement. This process, however, is more costly because of the expensive capital equipment and is akin to the VCM process described earlier.[266–268]

1.5.4.5 In situ processes.

In these techniques, as the name suggests, the reinforcement phase is formed in situ. The composite material is produced in one step from an appropriate starting alloy, for example, Co-NbC or Ni-TiC, thus avoiding the difficulties inherent in combining the separate components as is done in typical composite processing. Controlled unidirectional solidification of a eutectic alloy is a classic example

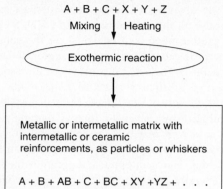

Matrix and reinforcement precursor materials

A + B + C + X + Y + Z

Mixing | Heating

Exothermic reaction

Metallic or intermetallic matrix with
intermetallic or ceramic
reinforcements, as particles or whiskers

A + B + AB + C + BC + XY +YZ + . . .

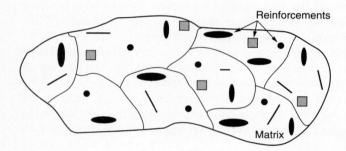

Reinforcements

Matrix

Figure 1.88   Schematic of a SHS process
making a MMC.[241]

of in situ processing. Unidirectional solidification of a eutectic alloy can result in
one phase being distributed in the other in the form of fibers or ribbon. One can control
the fineness of distribution of the reinforcement phase by simply controlling the solidification rate. In practice, however, the solidification rate is limited to a range of 1–5 cm/h
because of the need to maintain a stable growth front which requires a high-temperature
gradient.

The *XD process* is an in situ method that uses an exothermic reaction between two
components to produce a third component. Sometimes such processing techniques are referred to as self-propagating high-temperature synthesis (SHS) processes. Specifically,
the XD process is a proprietary method developed by Martin Marietta Corporation for
producing ceramic particle-reinforced metallic alloy. Generally, a master alloy containing
a rather high V_f of reinforcements is produced by the reaction synthesis. This alloy is
mixed and remelted with the base alloy to produce a desirable amount of particle reinforcement. Typical reinforcements are SiC, TiC, TiB_2, and so on, in an aluminum, nickel,
or intermetallic matrix. Figure 1.88 is a schematic of this process.

Recently there has been considerable interest in the family of processes developed
by the Lanxide Corporation in which metal is directionally oxidized so as to produce a
NNS component containing metal and ceramic.[269,270] Although full details are often unavailable, a typical process involves raising an aluminum melt to a high temperature, with
additions such as Mg, so that the alumina skin becomes unstable and the metal moves by

capillary action into an array of ceramic particles, such as alumina. Many other metal-ceramic systems have been explored.[271] Similar principles are employed in efforts to develop pressureless infiltration processes.[272]

1.5.5 Forms and Reinforcement

A variety of processes have been and are being developed for the manufacture of MMCs. As can be seen in Table 1.12, many fabrication routes are now available by which a reinforcement can be incorporated into a metal matrix. It is important to note from the outset that making the right choice of fabrication procedure is just as important in terms of the microstructure and performance of a component as it is for its commercial viability. However, before looking at the various processing options in detail, it is worthwhile dwelling for a moment on selection of the reinforcement. Clearly, the size, shape, and strength of the reinforcing particles or fibers are of central importance. Often the choice between the continuous and discontinuous options is relatively straightforward in terms of both performance and processing costs. However, within each category there are wide variations in reinforcement size and morphology.

As an example, consider particulate reinforcement. The most convenient form is SiC powder of about 3- to 30-μm diameter which is cheap (largely because of the mature market for its use as an abrasive) and relatively easy to handle. Extensive research has been undertaken on the influence of particle size on properties, especially with respect to failure processes. This work may point toward a preference for a much tighter size range than is currently used. Furthermore, a change from the common angular form to spherical particles might also be expected to offer performance improvements and may lead to superior flow behavior during handling prior to MMC manufacture. There are also grounds for supposing that platelets or ribbons may have certain advantages over equiaxed particles. Alumina has been produced for MMC use in the form of monocrystalline platelets with aspect ratio about 5–25 and diameters on the order of 10 μm. These can be readily assembled into preforms for melt infiltration, often with better control over V_f than is possible with equiaxed reinforcement. However, given the pressures to manufacture MMCs under economically attractive conditions, particularly for Al-based particulate MMCs aimed at bulk markets, the benefits of changing the type or nature of the reinforcement must first be clearly established and understood if such a change is accomplished by increased processing or raw materials costs.

TABLE 1.12.   MMC Fabrication Procedures

	Continuous Reinforcement[a]		Discontinuous Reinforcement		
Processing Route	Monofilament	Multifilament	Staple Fiber	Whisker	Particulate
Squeeze infiltrate preform	(✓)	✓	✓	✓	(✓)
Spray coat or codeposit	✓	✓	x	x	✓
Stir mixing and casting	x	x	(✓)	(✓)	✓
Powder premix and extrude	x	x	✓	✓	✓
Slurry coat and hot pressing	(✓)	✓	x	x	x
Interleave and diffusion bonding	✓	x	x	x	x

[a] x, Not practicable; (✓), not common; ✓, current practice.

1.5.5.1 Preforms and fabrics[241,246,273,274]

Green Tape.   This system consists of a single layer of fibers that are collimated or spaced side by side across a layer, held together by a resin binder, and supported by metal foil. This layer constitutes a prepreg (in organic composite terms) that can be sequentially laid up into the mold or tool in the required orientation to fabricate laminates. The laminate processing cycle is then controlled so as to remove the resin (by vacuum) as volatilization occurs. The method is usually used to wind the fibers onto a foil-covered rotating drum, overspraying the fibers with the resin, followed by cutting the layer from the drum to provide a flat sheet of prepreg.

Prepreg.   This plasma-sprayed aluminum tape is similar to green tape except that the resin binder has been replaced by a plasma-sprayed matrix of aluminum. The advantages of this material are (1) the lack of possible contamination by resin residue and (2) faster material-processing times because the hold time required to ensure volatilization and removal of the resin binder is not required. As in the green tape system, the plasma-sprayed preforms are laid sequentially into the mold as required and pressed to the final shape.[274]

Woven Fabric.   Perhaps the most interesting of the preforms being produced is woven fabric since it involves a universal preform concept that is suitable for a number of fabrication processes. The fabric is a uniweave system in which the relatively large-diameter SiC monofilaments are held straight and parallel, collimated at 100–140 filaments per inch, and held together by a cross-weave of low-density yarn or metallic ribbon.

1.5.6 Application of Processing to MMCs

1.5.6.1 Casting and liquid metal.   In the past few years, a number of potentially low-cost processing methods based on casting technologies have been demonstrated that also have the capability of forming complex-shaped parts. Several composite casting technologies based on liquid metal processing, involving infiltration of continuous or staple fiber preforms, casting of slurries of ceramic particles in molten matrices, or inclusion of ceramic particles in streams of liquid metal by various rapid solidification techniques were investigated and proven economically viable for production. In the future, development of low-cost, stiff inorganic fibers combined with commercially proven metal casting technologies (such as freeze casting, die casting, centrifugal casting, continuous casting, slush casting) for practical composite fabrication of NNS metal composite components will open up new market opportunities for the metals industry (Figure 1.89).

Squeeze Casting.   Fukunaga[275] reported on his fundamental and practical manufacturing process work and how he controlled parameters in fabricating SiC_f (Nicalon)-reinforced aluminum composites and subsequently SiC_w preforms for improved strength and reliability:

1. Molten metal temperature and fiber temperature
2. Infiltration speed
3. Final squeeze pressure
4. Vacuum squeeze casting.

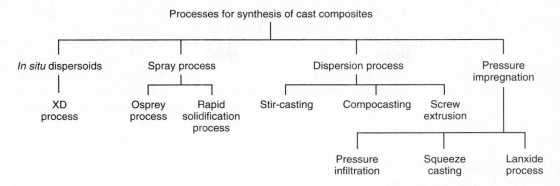

Figure 1.89   Different casting routes for the synthesis of cast MMCs.[323]

Verna and Dorcic[276] indicated in their squeeze casting development of ceramic fibrous preforms that the interface problem involving the fibers and matrix is not as critical as in hot-pressed composites because the porosity present in the preform (used for squeeze casting) is three-dimensional and is filled by liquid melt during casting. Because of the three-dimensional ceramic network in casting, properties of composites are improved considerably compared to those of hot-pressed ceramic-metal composites.

There is growing information indicating that ceramic fiber composites can provide the needed performance and do so at costs that justify their use.

Sample, Bhagat, and Amateau[277] described their development work in the high-pressure squeeze casting fabrication of unidirectional continuous graphite fiber-reinforced aluminum matrix composites with the V_f ranging from 4% to 52%. They demonstrated that during the casting process, the fiber tows were thoroughly infiltrated by the molten metal and rapid solidification was achieved. The fiber-matrix bonding in the squeeze-cast composites was primarily of the adhesion type; and there was very little evidence of an interfacial reaction between the fiber and the matrix. Their studies also included an electrochemical corrosion study of Gr/Al and Gr(Ni)/Al composites in 3.5% NaCl solution and revealed that the presence of nickel coating on the fibers in the fabricated composites significantly reduces damage to the Gr(Ni)/Al composites. This is in contrast to the extensive damage suffered by Gr/Al composites because of the cumulative effects of the dissolution of aluminum, hydration of Al_4C_3 into $Al(OH)_3$, and galvanic coupling between Gr and Al. The latter two mechanisms are almost nonexistent in Gr(Ni)/Al composites.

Waku and Nagasawa[274,278,279] discussed their squeeze casting work with aluminum matrix composites and the hybrid Si-Ti-C-O fiber, which has a fine SiC powder with a mean particle size of about 0.27 μm attached to its surface in order to improve the uniform distribution of fiber in the matrix. The Si-Ti-C-O fiber was synthesized from an organometallic polymer called polytitanocarbosilane (PTC). Their results showed a temperature dependence of the tensile strength of Si-Ti-C-O fiber-reinforced pure Al-based MMCs. Longitudinal tensile strength showed an almost constant value independent of the testing temperature up to 484°C. After 484°C, longitudinal tensile strength decreased with increasing testing temperature. The transverse tensile strength changed in a complicated manner with increased testing temperature, and a rapid drop in transverse tensile strength was observed near 211 and 400°C. The ratios of transverse tensile strength at 211 and 400°C to that at room temperature were about 60 and 20%, respectively.

Chadwick[280] conducted a series of experiments, and his analysis has shown, first, that liquid metal infiltration of preforms in the presence of air can require very high squeeze casting pressures in order to avoid porosity either within the preform itself or in the neighboring matrix phase. Second, he found that that oxide and nitride formation can occur in the matrix phase during infiltration and during subsequent periods of heat treatment.

El Baradie[281] prepared a comprehensive review of a number of manufacturing processes for MMCs. In his review of the squeeze casting MMC process he noted in 1967 the beginnings of B/Al liquid infiltration to SiC_w, in 1974 hot pressing methods, and in 1975 the Japanese patents.

Chadwick[282] also reported on the squeeze casting of magnesium alloys and Mg-based MMCs. He and his associates conducted a series of experiments at Hi-Tec Metals R & D, Ltd. with several different types of fibers that were encapsulated in different metallic matrices. In the results reported, the matrix was AZ91. The fibers used were Saffil, a glass-graphite mixture, and stainless steel wire.

Despite the close packing of the fibers, no problems were encountered in producing the MMCs because the applied pressure was always sufficient to squeeze the liquid magnesium alloy through the fiber bundles and around the individual fibers to preclude any entrapment in the fiber mat and to ensure absolute metal-to-metal interfacial contact. This was true for all different types of fiber arrays. Although the temperature of the magnesium was sufficiently high to melt the fine glass fibers during infiltration, the graphite fibers remained fully dispersed, as well as the high packing density of the unidirectionally aligned stainless steel wire which showed complete penetration of the interfiber spaces by the liquid magnesium alloy.

The simplest mechanical property to be measured for composite materials was their hardness. It is noteworthy that the glass-graphite composite, containing only ~9% graphite, had a hardness very similar to that of the 25 vol% Saffil-reinforced alloy: this is a reflection of the higher modulus of graphite compared with that of alumina. It is also interesting that the temperature dependence of the hardness values is the same for all three composites, implying that the load transfer characteristics are common to all three and that the progressive decrease observed is due entirely to softening of the magnesium matrix alloy.

Creep and fatigue curves were obtained for the 16 vol% Saffil-reinforced alloy and compared with AZ91 tested under the same conditions. The results indicate that the composite material had a creep life an order of magnitude better than that of the matrix phase and a fatigue strength double that of AZ91 alloy alone. This latter result is similar to that already found for aluminum alloy composites containing Saffil pads. Because the fracture toughness values for the high-hardness composites were very low at room temperature (in the region of 10 MPa $m^{1/2}$), these MMCs will find a market only for high-temperature applications where their superior creep and fatigue properties can be fully utilized without fear of an attendant low fracture toughness.

Several investigators, scientists, and metallurgists described their findings on various aspects of squeeze casting and solidification of MMCs, and Rohatgi compiled these reports.[283] Additionally, Rohatgi[284] described the squeeze casting of MMCs—Al-4.5

Cu/SiC$_f$ (0.6 and 10 vol%) and Zn-27Al-3Cu/SiC$_f$ (0.10 and 18 vol%). He pointed out that because the properties of squeeze-cast MMCs depend not only on the processing variables but also on the constituent metal and reinforcement materials, optimization of the composite system based on mechanical property evaluation and microstructural characterization also is important. Tensile and creep testing of cast, hot isostatically pressed, extruded, and heat-treated composite materials (2014 Al/SiC/20$_p$) at and above 150°C showed that the material fails via a mechanism of void nucleation, which initiates within the matrix and at the SiC-matrix interface. The ratio of the tensile properties of squeeze-cast fiber-reinforced composites (Al-3.75% Mg/Al$_2$O$_3$/10$_p$) to the same property for the matrix alloy alone improves even up to 300°C. Fatigue life also increases with increasing fiber content.

Mortensen and Koczak[285] presented an overview of the status of MMC efforts in Japan, especially the squeeze-cast work by UBE Industries (Si-Ti-C-O/Al and Si$_3$N$_4$/Al), Toyota Motor Corporation (Al$_3$Ni/Al pistons in mass production),[286] Toray (C$_f$/Al), Sumitomo Metals Industries and Chemical Company (2014/SiC$_w$), and Mitsubishi Electric Company (Al/SiC$_w$ and C$_f$/Al).

Verna and Dorcic[287] described the squeeze casting of Al$_2$O$_{3f}$ and SiC$_w$ and aluminum alloys 532, 206, 332, and 242.

Brown[288] reported on the squeeze-casting work of J. Cornie of Metal Matrix Cast Composites, which has now been commercialized. Cornie believes he can deliver parts inexpensively enough to compete with conventional ductile cast iron and aluminum on the basis of price as well as performance. The cycle times are currently 10–15 min per mold, and the hybrid-reinforced composite consists of more than 50 vol% ceramic. When the matrix is filled with alumina, the density is reduced 30–40%, a much stiffer part results, and the thickness is reduced 30%. Potential products include rear brake calipers for automobiles, and the assumption is that if you can make a caliper, you can make a lightweight gimbal for an aircraft gyroscope.

Ray and Yun[289] of Alcoa Labs have presented their findings on squeeze-cast Al$_2$O$_3$/Al ceramic-metal composites. In a recent study they formed Al$_2$O$_3$/Al ceramic-metal composites by squeeze casting and compared mechanical properties of the squeeze-cast parts with those of sintered all-ceramic parts.

Ceramic preforms were fabricated with three types of alumina: A-99F 99.5% pure Al$_2$O$_3$ powder, with 0.2 μm average particle size; A-98 99.5% pure Al$_2$O$_3$; and A-96 96% pure Al$_2$O$_3$ powder. Their process consisted of (1) die and preform preheating, (2) molten metal pouring, (3) high-pressure infiltration, and (4) removal of the composites. The alumina preforms were preheated, and squeeze casting was performed in a 1500-ton press. The preheated preform was introduced in a preheated die, and a molten metal (99.99% pure or Al-7Mg-2Si alloy) was poured into the die and high-pressure infiltration carried out. The composite was removed after the infiltration with the help of ejector pins.

The highest hardness, 1566 kg/mm^2, was seen in the squeeze-cast A-99F preform partially sintered at 1300°C for 2 h prior to infiltration with Al-7Mg-2Si. All Al$_2$O$_3$ preforms produced parts with good strength and toughness compared with those of sintered Al$_2$O$_3$ bodies. Flexure strengths of sintered 96% Al$_2$O$_3$ bodies were approximately 320–360 MPa, whereas squeeze-cast bodies showed nearly 33% improved strength. Similarly, the strength and toughness of A-98 and A-99F Al$_2$O$_3$ bodies were significantly higher. The toughness values of squeeze-cast parts were more than twice the values for the sintered bodies. The exceptional strengths and toughnesses of squeeze-cast Al$_2$O$_3$

bodies were believed to be caused by the presence of the metal phase and also because they were fine-grained bodies free of major large defects.

Suganuma, Sasaki, Fujita, et al.[290] fabricated 6061 Al alloy/β-Si$_3$N$_4$/13$_w$ by squeeze casting, and its high-temperature strength and interface microstructure between whiskers and matrix showed that as compared with the 6061 Al alloy, the strength of the composite increased about 20% and elongation reached about 4% at room temperature. The strength beyond 300°C was excellent.

Singh, Goel, Mathur, et al.[291] prepared Al/Al$_2$O$_3$/MgO squeeze-cast composites using a modified "MgO-coating" technique, and their tensile behavior up to 300°C was evaluated.

From their comprehensive tests they concluded the following.

1. Squeezing the stirred slurry in the pressure range 80–140 MPa brings about a distinct improvement in the elevated-temperature tensile properties of the composite. The best properties were displayed by a composite squeezed at 140 MPa and ambient die temperature. Increasing the die temperature systematically causes the properties to deteriorate.

2. The ultimate tensile strength (UTS) of the composite squeezed at 140 MPa and ambient temperature is higher by 52% at a 100°C test temperature compared to that of an ordinary gravity chill-cast composite. This figure improves to 161 and 162% at 200 and 300°C test temperatures, respectively. This implies that the composite retains about 77% of its ambient UTS value at a 300°C test temperature, while the gravity chill-cast composite retains only 44% of its ambient UTS value at the same temperature. The same squeezed composite retains an UTS value of 160.4 MN m^{-2} at 300°C, while the gravity chill-cast composite retains only a 61.1 MN m^{-2} UTS at the same test temperature.

3. Results pertaining to the 0.2% offset yield strength (YS) values of the squeezed composite at different temperatures were more encouraging. The squeezed composite (140 MPa and ambient die temperature) retained 94.4% of its ambient YS value at a 300°C test temperature, while the gravity chill-cast composite retained only 47.6% of its ambient YS value at the same test temperature. The YS of the same squeezed composite was higher by 105% compared to the YS value of a gravity chill-cast composite at a 300°C test temperature. At a 100°C test temperature, the squeezed composite was superior by 53% compared to the gravity chill-cast composite. Therefore, with a progressive increase in test temperature, the performance of the squeezed composite becomes increasingly superior to that of an ordinary gravity chill-cast composite.

4. With a further increase in the squeeze pressure applied, such as 160–240 MPa, the performance of the composite is likely to improve further.

5. Squeezed composites show fully ductile fracture features. No evidence of the presence of porosity and interdendritic unfed regions was observed in the squeeze-cast composites.

Stir-Casting or Compocasting.[246,279]   As mentioned earlier, when a liquid metal is vigorously stirred during solidification by slow cooling, it forms a slurry of fine spheroidal solids floating in the liquid. Stirring at high speeds creates a high shear rate, which tends to reduce the viscosity of the slurry even at solid fractions as high as 50–60% volume. The process of casting such a slurry is called *rheocasting*. The slurry can also be mixed with particulates, whiskers, or short fibers before casting. This modified form of rheocasting to produce NNS MMC parts is called *compocasting*.

The melt reinforcement slurry can be cast by gravity casting, die casting, centrifugal casting, or squeeze casting. The reinforcements have a tendency to either float to the top or segregate near the bottom of the melt because their densities differ from that of the melt. Therefore, a careful choice of casting technique as well as of mold configuration is of great importance in obtaining uniform distribution of reinforcements in a compocast MMC.[292,293]

Compocasting allows a uniform distribution of reinforcement in the matrix as well as a good wet-out between the reinforcement and the matrix. Continuous stirring of the slurry creates intimate contact between them. Good bonding is achieved by reducing the slurry viscosity as well as by increasing the mixing time. The slurry viscosity is reduced by increasing the shear rate as well as by increasing the slurry temperature. Increasing the mixing time provides a longer interaction between the reinforcement and the matrix.

Keshavaram, Rohatgi, Asthana, et al.[294] described the salient features of the solidification microstructures of discontinuously reinforced cast Al/Gl_p composites synthesized using a stir-casting technique. These included formation of an interaction zone at the interface between a cast Al alloy matrix and glass particles. Additionally, the permanent mold castings of Al/Gl_p composites showed a statistically homogeneous particle distribution.

Rohatgi[284] described the stir casting process as an alternative casting method that does not use the application of pressure to produce MMCs. It involves mechanical mixing of the particles in fully molten or partially molten alloys, followed by casting. He claimed he used stir-casting techniques to disperse Gr, B_4C, SiC, Al_2O_3, TiO_2, ZrO_2, and glass particles in cast Al alloys. While these processes lead to some heterogeneity in reinforcement distribution, considerable effort has been made by Alcan International Ltd., Canada, to improve the process sufficiently to make mass manufacturing of SiC/Al alloy shape-cast MMC products feasible.

Other investigators[294] have conducted compocasting experiments to investigate the feasibility of the process as applied to the AZ91 D Mg alloy-SiC_p system. Relationships between processing and macro- and microstructure were examined. Three temperature-time processing sequences were investigated: stirring temperature maintained above liquidus, stirring temperature in the semisolid temperature range, and an imposed temperature rise above the liquidus after stirring in the mushy zone. Stirring temperature and particle size significantly affect spatial particle distribution and porosity level. The easy incorporation and even dispersion of particles in the matrix suggest good wetting of SiC_p by the Mg matrix.

They concluded the following.

1. Mg matrix composites reinforced with SiC_p can be made by compocasting.[295]
2. Particles can be introduced either in the liquid alloy or in the semisolid alloy.
3. The porosity level is much higher with small particles.

4. Stirring the composite above the liquidus subsequent to addition of the reinforcement phase in the mushy alloy produces the soundest and most homogeneously reinforced composite.

5. The composites formed an excellent interface between the SiC_p and the Mg matrix.

6. Tensile tests performed on extruded composites reveal a high modulus level (up to 64.5 GPa compared to 45 GPa for Mg), an improvement of 20% in YS, and a tensile strength equivalent to that of the unreinforced matrix. The ductility level was low in relation to residual agglomerates of particles and associated porosity after extrusion.[296]

Investment Casting.[279,297,298]   Investment casting emerges as one of the most desirable routes to the manufacture of net shape MMC components. As castable composites come of age, superb parts exhibiting superior dimensional stability under thermomechanical loading have been cast on a routine production basis.

The aerospace business has for some time rejected aluminum castings because of the low strengths that are typically achieved; however, with a material that is fiber-dependent and not predominantly matrix-controlled, significant structural improvements now have been made that have revived interest in this low-cost procedure. The investment casting technique, sometimes called the "lost wax" process, utilizes a wax replica of the intended shape to form a porous ceramic shell mold where, after removal of the wax (by steam heat) from the interior, a cavity for the aluminum is provided. The mold includes a funnel for gravity pouring with risers and gates to control the flow of the aluminum into the gage section. A seal is positioned around the neck of the funnel, allowing the body of the mold to be suspended in a vacuum (imposed through the porous walls of the shell mold), and the total cavity is filled with aluminum.

The SiC_f were installed in the mold using fabric, by either first placing the fabric into the wax replica or simply splitting open the mold and inserting the fabric into the cavity after the wax was removed. At present, the latter approach is usually used because of contamination and oxidation of the fibers during wax burnout.

Work is underway to evaluate castable composites targeted for higher-temperature applications. They might displace titanium parts servicing temperatures under 315°C. Development work conducted on composites obtained by preform infiltration adds new capabilities to investment casting. Of particular interest are electronic packages with excellent thermophysical tradeoffs.

Sand Casting.   A process has been developed for making MMCs that blends liquid Mg alloy with SiC_p and Al_2O_{3p}. The method is essentially the same as those that have been developed for aluminum composites in that blending is accomplished via a high-shear process. Major differences in the new process result from increased general reactivity of Mg and the difference in surface chemistry between the Al/SiC and Mg/SiC systems.[299]

The particulate-reinforced Mg MMCs produced by this new technology are lightweight and demonstrate a significant increase in modulus and tensile strength at both ambient and elevated temperatures over those of the unreinforced material, leading to a significant improvement in the operating envelope. In addition, the CTE of these Mg MMCs can be reduced by up to one-third by varying the SiC content, enabling the CTE to be matched to that of other materials.

The process has recently been scaled up to a prototype stage capable of producing 180-kg direct chill billets for subsequent extrusion, foundry ingots, or a bulk supply of liquid composite to be used in the development of sand castings.

To study the effects of alloy type on composite properties, samples of ZM21 (2Zn-1Mn), AZ61 (6Al-1Zn-0.3Mn), AZ80 (8.5Al-0.5Zn-0.3Mn), and ZC71 (6.5Zn-0.7Mn-1.2Cu) wrought Mg alloys reinforced with 12 vol% 9-μm SiC grit were cast in water-cooled steel molds.

Test results indicated that ZC71 had the best tensile properties, and the modulus of the ZC71 composite was found to be 63 GPa. This is a 43% increase over that of the un-reinforced alloy and, if a specific modulus is considered, the value for the composite is 32 MNm/kg, which is higher than that of either unreinforced Mg (24 MNm/kg) or unreinforced Al (26 MNm/kg).

Rotating-bend fatigue testing of the composite showed a fatigue runout at 50 million cycles at 125 MPa, the same value obtained for the unreinforced alloy.

Addition of SiC_p significantly increases the ambient-temperature UTS of ZC71, and the operating-temperature envelope is raised substantially. However, at temperatures approaching 200°C, where the matrix alloy has relatively little strength, the composite is also weakened.

Gravity Casting.   Li, Loh, and Hung[300] described their work using the commercially available A-356 Al alloy, with ~7 wt% Si, 0.4 wt% Mg, and other small amounts of elements since it is the most common type of Al foundry alloy because of its excellent castability. The needlelike Si phase embedded in the Al matrix is the most typical structural feature of this alloy system, and the morphology of the Si phase has been shown to be important in determining its mechanical properties.

Their work describes the microstructural changes caused by the addition of Al_2O_{3p} to molten Al, focusing on the effect of the Al_2O_{3p} on the location, amount, and morphology of the Si or eutectic phase in the matrix. Reasonable Al_2O_{3p} distribution within the matrix Al was achieved by low-temperature mixing and a gravity casting technique under an open atmosphere.

The Al_2O_{3p} was 42 μm on average with irregular morphology and was preheated to 800°C; it was held at this temperature for 30 min and then poured into a semi-solid state molten Al alloy (with the addition of 0.5% Li to improve the wettability of the molten Al and the Al_2O_{3p}) in a clay crucible heated in a conventional conduction furnace, followed immediately by strong agitation of the mixture using a graphite rod. The mixture was put back into the furnace when the temperature dropped too low for agitating. Once it was just melted, the mixture was taken out of the furnace again for further agitation. The procedure was repeated two times, and total agitation time at the lower temperature was about 5 min. When the particles were mixed within the Al, the temperature was increased up to 800°C, and further agitation took place for another 20 min to ensure that they were mixed uniformly within the molten Al alloy. The mixture at 720°C was then poured into a preheated mold (200°C) to form bars of 15-mm diameter and 60-mm length for HIP.

The cast MMC bars were processed by HIP in order to densify the MMCs. The cast and hot isostatically pressed MMCs showed that the 0.2% YS was improved over that of the unreinforced Al alloy.

Centrifugal Casting.[301,302]   Suery and Lajoye[303] reported their concern about the characteristics of the microstructure of Al-Si alloys reinforced by SiC_p produced by centrifugal casting. Attention was particularly focused on distribution of the particles in the casting and on the influence of these particles on the solidification microstructure. By comparing reinforced and nonreinforced zones, it was shown that the presence of the particles changed the dendrite arm spacing, with nucleation of Si on the SiC_p. Segregation of the particles as a result of centrifugal acceleration allowed locally reinforced parts to be produced. It was shown that even better defined zones located anywhere in the part could be obtained using sequential casting and increased the potential of the centrifugal casting process.[293]

Die Casting.[304]   Allison and Cole[286] described the die casting of discontinuously reinforced aluminum (DRA) as an attractive technology that can produce a variety of NNS components at quite high production rates. Because the molten composite is semisolid, it exhibits thixotropic properties, and so die filling is more controlled and uniform compared to the turbulent filling typical of die casting of unreinforced metals. Consequently, in die-cast DRA, the amount of entrapped gas is substantially reduced over that found in typical die-cast materials. DRA die castings are often heat-treatable and may exhibit improved fatigue properties.[305]

The Mg alloy AZ9lHP has been successfully die-cast into NNS with Al_2O_3. By controlling casting conditions such as temperature and shot velocity, consistent distribution of the reinforcement was achieved. In addition to modulus and strength increases, parts cast with this material show 10-fold wear improvement. Corrosion resistance of the composite was excellent, as in the unreinforced hot-pressed alloy. This combination of excellent wear and corrosion resistance allows applications for Mg that are not possible in its monolithic form.

Pressure Casting.[279]   Kobayashi, Yosino, Iwanari, et al.[306] discussed their pressure casting technique for fabricating SiC_w-reinforced Al alloys. SiC_w-reinforced Al alloys fabricated under a pressure of 90 MPa are superior to those fabricated under lower-pressure conditions in mechanical properties. Fracture of SiC_w-reinforced Al alloys is associated with failure of SiC_w and interface decohesion between whiskers and matrix. It was shown that these composites were strengthened by increasing interface bonding between the SiC_w and the matrix; that is, the interface cohesion was strengthened by accelerating the interface reaction adequately. The addition of highly reactive lithium to the Al matrix made the interface cohesion tight and resulted in composites of lower density and greater strength.

Several techniques are now being employed to fabricate intermetallic matrix composite materials. These include P/M, diffusion bonding (DB), plasma spraying, and pressure casting. Because of its advantages over the other techniques, pressure casting is being extensively used to fabricate continuously reinforced MMCs such as Al and Mg alloys reinforced with Gr, Al_2O_3, or SiC_f.

Pressure casting is a relatively simple technique that allows the incorporation of various arrangements of fibers into the matrix. There are no limitations on the diameter of the fibers or the shape and size of the samples. Since pressure casting involves infiltration of preheated fibers by molten metal, it is possible that exposure to high temperatures and to the molten metal may alter the microstructure of the fiber and therefore severely degrade its mechanical properties or result in an extensive chemical interaction between the fibers and the molten metal.

Nourbakhsh, Margolin, and Liang[307] used pressure casting to fabricate composites of Ti-48.4 Al-l Mn intermetallic alloy reinforced with continuous ZrO_2-toughened Al_2O_{3f}, 20-μm-diameter PRD-166. The reactions that took place between the molten alloy and the fibers were dissolution of the ZrO_2 phase from the fiber into the molten alloy and diffusion of Ti from the matrix into the fiber. In addition, the fibers were found to contain entrapped TiAl. Penetration of molten alloy into the fiber was suppressed, and the extent of ZrO_2 dissolution was reduced by lowering the processing temperature.

Vacuum Suction Casting.   Kun, Dollhopf, and Kochendörfer[308] devised a vacuum suction casting process for manufacturing composites of Al reinforced with CVD SiC filaments. Three types of CVD SiC filaments were tested in an Al-10% Si matrix. The wettability and degradation of the filaments during liquid infiltration and the mechanical properties of the composite rods were studied. They found that low infiltration temperatures were achieved as a consequence of good wettability under the specific conditions of the casting process. The uncoated SiC filament showed severe degradation after liquid infiltration, whereas the SiC filaments exhibited an excellent protective response to molten Al. Composite rods consisting of 50 vol% SCS-6 filaments possessed the following characteristics.

Tensile strength	1.6–1.7 GPa
Tensile modulus	208–220 GPa
Compression strength	1.74–1.81 GPa
Compression modulus	212–215 GPa
Shear strength (transverse)	240–270 MPa
CTE	−100 to +20°C, 2.60×10^{-6}/°C
	20–300°C, 3.60×10^{-6}/°C

The tensile strength obtained is higher than that in most data published in the literature for CVD SiC filament-reinforced Al.

They concluded that the vacuum suction casting process was suitable for producing net shape long components with uncomplicated sections. The use of woven fiber preforms was expected to further improve the properties of composites and simplify the process. The simplicity of the equipment, the easy procedures, and the high productivity should make the process very cost-effective in the manufacture of composite parts.

Chill Casting.   Pathak[309] prepared Al matrix high-Al_2O_3 slag (2–16 wt%) particulate composites by impeller mixing and bottom discharge chill casting foundry techniques. In his findings he reported the following.

- Uniformity of slag dispersion in Al matrix depends on slag particle size and weight percent.
- Tensile strength, 0.2% offset YS, hardness, and elongation (percent) increase with the slag content of the composite but only up to a certain weight percent of slag, that is, 4 wt% for 26-μm-particle-size slag and 8 wt% for 105-μm-particle-size slag.

- Al–high Al_2O_3 slag composites wear by adhesion and abrasion, and the wear rate of the composite is less than the wear rate of Al.
- Wear rate increases with an increase in applied load or sliding velocity and has an inverse relation to the hardness of the composite.

Bicasting.   Howmet and General Electric Aircraft Engines engineers[310] developed the bicasting method of selectively reinforcing titanium investment castings with titanium matrix composite (TMC) inserts. The process is based substantially on conventional investment casting and TMC fabrication technology and can be performed at existing facilities. Bicasting was successfully demonstrated on a turbine engine prototype fan frame strut in 1991.

There are four major steps in bicasting: (1) mold manufacture, (2) preform fabrication, (3) casting, and (4) postcasting processing and inspection. Investment casting molds are made conventionally by coating a wax replica of the part (pattern) with multiple layers of a ceramic slurry. The pattern is subsequently melted out of the ceramic mold. TMC preforms are then placed within the mold cavity at the locations that are to be reinforced. Preform geometry, composition, and fabrication method can be tailored to the required mechanical property improvements. Except for modifications that permit preform placement, bicasting molds are identical to conventional titanium investment casting molds.

During casting, the TMC preforms are surrounded by molten metal, which then solidifies, embedding them in the casting. The parts are then processed via conventional titanium practice, including chemical milling, HIP, weld repair, heat treating, and inspection.

In order to take full advantage of the mechanical properties, such as elevated-temperature tensile strength, stiffness, fatigue life, and resistance to impact damage, that TMCs potentially have, future development work will focus on reducing the cost of the TMC by, for example, improving preform design to minimize machining requirements. Other issues to be studied include improving the positional accuracy of the preform in the mold cavity, developing a weld repair process for TMC-containing titanium castings, and improving nondestructive evaluation (NDE) methods and their resolutions.

1.5.6.2 Melt and pressureless infiltration (direct oxidation)

Pressure Infiltration of Preforms.[311,312]   These processes involve unidirectional pressure infiltration of fiber preforms or powder beds to produce void-free NNS castings of composites. Aluminosilicate fiber-reinforced Al alloy pistons made by Toyota have been in use in heavy diesel engines for some years, and a considerable amount of work has been done on pressure infiltration of ceramic particle and fiber reinforced MMCs in industrial applications. The processing variables governing the evolution of microstructures in infiltrated MMCs are fiber preheat temperature, interfiber spacing, infiltration pressure, infiltration speed, and metal superheat temperature.

The process variables in pressure infiltration are the same as those in squeeze casting. If the pressure vessel is not pressurized and the end of the pipe is not vented to a neutral atmosphere but connected to a vacuum line, liquid metal will infiltrate as a result of the vacuum and the process is called *vacuum casting*.

In the United Kingdom, Cray Advanced Materials, Ltd.[313] has developed a patented process for the production of MMCs that is referred to as liquid pressure forming (LPF)

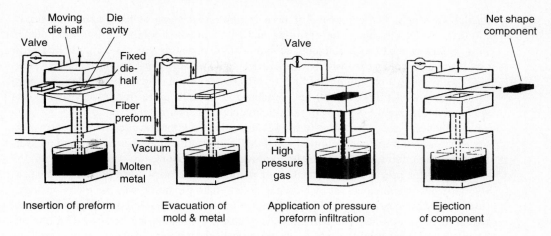

Figure 1.90   Schematic of the liquid pressure forming process.

and has been fully developed for production. It is most suitable for volume-infiltrating fiber preforms with liquid metal under pressure, where it differs from conventional squeeze casting in that the levels of applied pressure are lower and do not lead to preform damage. In essence, the LPF process, illustrated in Figure 1.90, involves a molten metal reservoir located below a heated die in a circuit that can be alternately evacuated or pressurized. A fiber preform is placed in the preformed die, the die is then closed, and a vacuum is applied. A gas is then injected above the reservoir of molten metal, forcing the metal into the die cavity and impregnating the preform. It is claimed that the resultant composites are fully dense, being free of air, dissolved gases, or oxidation products.

A major advantage claimed for this technique is the scope it offers for engineering the final properties of a product. The reinforcing fibers can be preformed either to the full net shape of the component or selectively placed for reinforcement of specific areas. The reinforcement may be selected from steel wire, SiC, Al_2O_3, C, and B_f, and the matrix metal can be an Al, Mg, Pb, Zn, or Cu alloy.

Variations of the above method include the vacuum infiltration process. This technique has been employed successfully as an immersion process for fiber coating in which the fiber is immersed in an organometallic compound solution followed by hydrolysis and pyrolysis. It can be used to deposit a uniform, air-stable SiO_2 coating on the surface of C_f or Gr_f.

There are several advantages in using C_f coated with SiO_2 for manufacturing C/Mg composites. The coating is not only stable in air but is also flexible and convenient to keep and use and improves the wetting between fiber and matrix. As a result, the mechanical properties of C/Mg composite materials are improved.

It was found that because of the chemical reaction that occurs between Mg and the coating, MgO precipitates were produced in a uniform dispersion in the interfacial layer. The interfacial layer is rich in Al, and the reaction between Al and C improves the interfacial bond strength.

Klier, Mortensen, Cornie, et al.[313] have proposed a novel casting process for fabrication of particle-reinforced composites. The two-step procedure consists of (1) the production of a pore-free composite of particle-reinforced metal of high V_f by infiltration, followed by (2) dilution of the resulting melted or remelted composite into more metal. This process is capable of producing economically a pore-free, as-cast composite of particles distributed homogeneously in a metal such as Al or Mg. They demonstrated this process by infiltrating a packed bed of SiC_p (30, 10, and 3 μm in diameter) with Mg, followed by dispersion into more metal, and produced a virtually pore-free, as-cast composite while allowing control of the V_f of reinforcement within the metal.

Blucher[314] described some of his development work that resulted in a simple, versatile technology for the production of cast MMCs. In this method, the transfer and pressurization of the infiltrating metal was by hydrostatic gas pressure.

With this process, fiber, whisker, or particulate preforms were placed in inexpensive throwaway containers into which the molten matrix material was forced by gas pressure. Because the gas pressure on the preform container is quasi-hydrostatic, its strength requirement is minimal, and consequently preform containers, even those with complicated shapes, were easy to prepare. By virtue of the simplicity of the containers and the absence of a requirement for a heavy mechanical system to effect infiltration of the liquid metal into the preforms, the technique is suitable for producing bulk or final-shape composite parts inexpensively. As the processing parameters, such as preform temperature, infiltrating metal temperature, infiltrating pressure, and cooling rate, can be accurately controlled, the process is particularly suitable for limited production.

Cook and Werner[315] have conducted an exhaustive study on the principles, variables, parameters, and so on, of pressure infiltration casting. They claim that this technique is a unique form of liquid infiltration that utilizes pressurized inert gas to force liquid metal into a preform of reinforcement material.

They also claim that there are several methods including

1. *Top fill casting.* Liquid metal is forced downward by a pressurized gas into a preform.

2. *Bottom fill casting.* Liquid metal is forced up a fill tube into a preform by pressurized gas acting on the surface of a melt.

3. *Top pour casting.* A method developed for infiltration of high-temperature alloys where the reinforcement and melt must be prevented from reacting.

The combinations of materials (reinforcement and matrix) include graphite (P100)/Al (6061 alloy), AlN (50 vol%)/Al, graphite (P100)/Cu (OFHC), AlN/Cu, Al (6061 alloy)/TiB_2 (73 vol%), Saffil/Al (6061 alloy), SiC (65 vol%)/Al (6061 alloy), Au/Al_2O_3, and Au-diamond.

Bhagat[316, 317] described a high-pressure infiltration casting (HiPLC) method that he and his colleagues have developed. HiPLC uses rapid application of a relatively high pressure (above 100 MPa) to force-infiltrate molten metal into fiber preforms. Reaction between fiber and matrix metal was negligible, and the cast composites were free of voids, gas porosity, and shrinkage cavities. Experimental results on tensile strength, stiffness, low-cycle fatigue, fracture toughness, damping, and corrosion resistance of the HiPLC composites are summarized as follows:

At a fiber V_f of 27%, planar, random Gr_f-reinforced 6061 Al matrix composites have a Young's modulus of 87 GPa which is about 26% higher than that of the matrix Al alloy.

The low-cycle stress-controlled fatigue of Ni-coated, planar, random Gr_f-reinforced 6061 Al matrix composites resulted in a significant increase in the fatigue life of the HiPIC composites.

The fracture toughnesses of the SiC/6061 Al alloy ($V_f = 0.13$) and Gr/6061 Al alloy ($V_f = 0.12$) were, respectively, 35 and 40% higher than that of the cast Al matrix.

The cast composites have up to an order of magnitude more damping than that of the wrought 6061-T6 Al alloy. The improved damping of the composites is largely attributed to interface-controlled mechanisms.

Finally, coated Gr_f-reinforced Al alloy matrix composites show high resistance to corrosion in 3.5% NaCl solution.

Muscat, Shanker, and Drew[318] described their work on the fabrication of Al/TiC composites in which they infiltrated a porous TiC_p preform with molten Al using the melt infiltration technique in argon at atmospheric pressure and at temperatures ranging from 950°C to 1350°C. The mechanical properties of the composites containing 50–85 vol% TiC exhibited excellent tensile strength values of ~475 MN/m^{-2} and up to 5% elongation.

The elastic moduli of these materials increased with TiC content from 120 GN/m^2 to 290 GN/m^{-2}. Hardness values increased with increasing TiC content and decreased with increasing infiltration temperature. High Vickers hardness values of 800 kg/mm^2 were obtained for composites containing 84% TiC.

Melt Infiltration and Other Infiltration Method Variations. Jiang, Ma, Liu, et al.[319] reported on a new liquid metal infiltration process for fabricating short-fiber-reinforced MMCs via centrifugal force infiltration of fiber preforms with molten Al alloys. They produced composites having fiber V_f values of 4.5, 8.0, 12, and 16% by this method.

They found that the hardness and wear resistance of the MMCs prepared via the centrifugal force infiltration route were almost identical to those of composites obtained via the squeeze infiltration method ($P = 150$ MN/m^{-2}).

The molten infiltration process was markedly affected by the pouring temperature, preheated mold temperature, and time of application of centrifugal force.

Cheng, Lin, Zhou, et al.[320] developed an ultrasonic liquid infiltration technique for the fabrication of carbon fiber-reinforced Al (C_f/Al) precursor wires. They concluded the following.

1. The principal effect of the ultrasound on the infiltration of Al into C_f was caused primarily by the cavitation phenomenon. The acoustic power required for the prediction of cavitation in the practical system used was calculated to be approximately 142 W, which is much greater than the requirement for overcoming the capillary pressure between C_f, that is, about 42 W if a fiber V_f of 70% is assumed.

2. It has been found that C_f can be sufficiently impregnated by molten Al with the appropriate application of ultrasonic waves propagating within a guiding hole in the coupling stub.

3. The C_f/Al precursor wires obtained have an average fiber V_f of 26%. The maximum longitudinal tensile strength of these wires was 605 MN/m^{-2}, which implies a fiber strength transfer efficiency of 0.76.

4. The results of single-fiber tests show that there is no degradation of fiber strength, indicating that no significant chemical reaction occurs between C_f and Al during ultrasonic infiltration. However, it is possible that some fiber breakage caused by strong ultrasound occurs. Microvoids were observed in the C_f/Al precursor wires, which are considered to reduce their strength.

In the United Kingdom, BNF Metals Technology Centre[321] has developed a process for the production of large MMC filament-wound tubes up to 300 mm in diameter. The method was developed to overcome some of the problems associated with the squeeze casting of such large tubes, such as the problem of venting interfilamentary gases from the die cavity and the requirement for a very high degree of control over dimensional tolerance between the die parts in order to avoid leakage of Al or seizing up.

The new process is known as *vacuum infiltration gas pressurization* (VIGP) and uses a closed die containing a filament-wound core which is heated to a temperature above that of the matrix alloy under vacuum. The metal is sucked into the die cavity through a heated tube projecting under the surface of the molten metal held in a crucible. When the die is full, a valve controlling the flow of Al is closed and the die is pressurized by inert-gas pressure to effect infiltration of the preform. Since the die and preform are above the liquidus, premature solidification is avoided. In addition, the pressure is applied isostatically, and so there are no pressure losses due to mechanical friction. When infiltration is complete, rapid controlled cooling can be initiated in order to produce directional solidification toward the source of the feed metal. These features provide a high degree of control over the process.

Gokhale, Lu, and Abbaschian[322] synthesized Nb-based binary IMCs using an electromagnetic levitation processing system. The composites were formed by vacuum-assisted infiltration of the intermetallic melt onto a reinforcement preform. W and Nb filaments of various sizes were used as reinforcements to explore their potential for toughening the intermetallic matrices.

These workers determined that, by analyses, in all cases the interfacial interactions could be explained by phase equilibria. Furthermore, it was found that in these composites, the formation of another intermetallic may not be adequate in preventing long-range intercomponent diffusion. It was observed that both Si and Al embrittled W via liquid metal embrittlement along grain boundaries. As a result of these two observations, it was concluded that in such systems, the use of diffusion barrier coatings may be essential for synthesizing microstructurally stable composites.

Directed Melt Oxidation (DMO) and Pressureless Spontaneous Infiltration. Lanxide has developed a process involving spontaneous infiltration of Al_2O_3 or SiC powder beds by Mg-containing molten Al alloys placed over the beds. The alloy-ceramic assembly is heated in a nitrogen atmosphere at temperatures of 800–1000°C, resulting in spontaneous infiltration of the particles by molten alloys, followed by their solidification to form the composite. The volume percentage of particles that can be incorporated can be generally higher (45–70 vol%) than that possible with stir casting techniques. Some AlN forms during the procedure (depending upon the rate of infiltration) and also influences the properties of the composites.[311] Figure 1.91 is a schematic of the main components involved in the process.

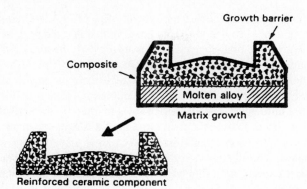

Figure 1.91   Schematic of the Lanxide process for the production of composites.[269,270,323]

The alloy should have a composition such that on melting it wets dispersoid particles and infiltrates into the bed or preform without application of pressure. The resulting composite may be made to net shape or NNS by using a suitable preform of dispersoid. The process variables in this process are (1) infiltration temperature and (2) particle size, apart from the composition of the alloy and the nature of the atmosphere. Here, infiltration takes place almost spontaneously, and thus wettability of the dispersoids by the alloy is extremely important.[269,270,323]

Numerous scientists and researchers have conducted innumerable studies, investigations, and tests to examine and essentially grow various composite compositions by directed melt oxidation (Dimox).

Dhandapani, Jayaram, and Surappa[324] examined the growth and microstructure of $Al_2O_3/SiC/Si(Al)$ composites prepared by reactive infiltration of SiC preforms.

They grew Al_2O_3 matrix composites by directed melt oxidation of an Al-Si-Zn-Mg alloy. Reaction between the melt and the SiO_2 layer on the preoxidized SiC yielded a matrix in which the residual alloy was principally Si. The coarsest particles displayed the least reaction, while porosity increased as particle size decreased. Thus, owing to the enrichment of Si, growth rates were retarded compared to those in free space. The microstructure of the oxide was monocrystalline over distances on the order of interparticle spacing, however, a new type of growth fault that constitutes an inversion boundary in the Al_2O_3, was found. Because of the substantial reaction between the particle and the melt, they concluded that the V_f of the alloy constituent in the final composite depends sensitively on particle size, in contrast to its relative invariance in directed melt oxidation into inert preforms.

Breval, Aghajanian, Biel et al.[325] produced unreinforced AlN/Al ceramic/metal composites by directed oxidation of molten Al alloys in nitrogen. The microstructures of these composites were compared with those of previously studied Al_2O_3/Al composites. In both composite systems the ceramic phase was interconnected and oriented in a columnar structure. The metallic phase showed no significant crystallographic orientation and appeared as both interconnected channels and isolated inclusions. The columns in the nitride system were found to be of micrometer size and to contain subgrains weakly defined by lattice defects, unlike the oxide system where the columnar structure was shown to be of millimeter size and to contain well-defined micrometer-sized subgrains. Finer structures were obtained in both systems via the addition of Ni to the parent alloy.

Antolin, Nagelberg, and Creber[326] evaluated the growth of α-Al_2O_3-metal composites by directed oxidation of molten Al-Mg-Si alloys and determined that the process proceeds through four distinct stages. The first stage encompasses the early heating of the alloy ingot, melting, and continued heating to between 850 and 900°C. In this latter temperature range, the molten alloy surface rapidly oxidizes to form a MgO-covered $MgAl_2O_4$ layer. During further heating and initial soak, the duplex layer slowly thickens (second stage). The start of the third stage, growth initiation, is marked by the spread of a metal-rich zone over the duplex layer. During the final rapid growth stage, the small composite nodules grow and coalesce to form a macroscopically planar growth front which persists until growth is complete. Throughout the growth process, the external surface of the α-Al_2O_3-metal composite is covered by a thin MgO layer. Immediately under this external layer and separating it from the α-Al_2O_3 is a thin layer of molten metal.[327,328,329]

May[329] reported on the Primex pressureless metal infiltration process which utilizes certain process conditions to enhance the wetting of ceramic reinforcing materials by molten Al, thus spontaneously producing a MMC.[330]

The pressureless infiltration of Al into ceramic reinforcing materials is induced by placing Al-Mg alloys (typically 1–10% Mg) in contact with the reinforcing materials at temperatures between 750 and 1050°C in a nitrogeneous atmosphere (10–100% N_2, balance inert) (Figure 1.91). Al_2O_3, SiC, TiB_2, MgO, and AlN particles and Al_2O_3 fibers have been utilized as reinforcing materials, among others. Infiltration rates as high as 25 cm/h have been demonstrated using proprietary modifications to the process. Components up to 7.5 cm in thickness or up to 55 × 90 cm in lateral dimension have been fabricated with no inherent size limits evident.

Johnson and Sonuparlak[331] formed diamond-Al MMCs by the pressureless infiltration process. Diamond particles are unique fillers for MMCs because of their extremely high modulus, high thermal conductivity, and low CTE. The diamond particulates were coated with SiC by chemical vapor infiltration (CVI) prior to infiltration to prevent the formation of Al_4C_3, which is a product of the reaction between Al and diamond. The measured thermal conductivity of these initial diamond-Al MMCs was as high as 259 W/m·K.

The other initial properties were very promising—modulus greater than 400 GPa and a CTE of 4.5–6.8 ppm/K, particularly for electronic applications where stiff, low-CTE, high-thermal-conductivity materials are required.

In Situ Processing

In Situ Production of Dispersoids.   Compound dispersoids such as TiC, TiB_2, TiN and NbB_2 can be produced in an alloy matrix by allowing components to come into contact and react during high-temperature processing in the liquid or solid state. The matrix alloy may include Al- or Cu-based alloys or intermetallic compounds such as aluminides. One of the reacting constituents may remain in solution in the molten matrix alloy, and the other may be added as a fine powder. A typical example is the formation of a TiC-reinforced composite by the addition of Ti or ferrotitanium to molten cast iron. It is also possible to use a gaseous reacting constituent, such as acetylene or methane bubbling, in a molten alloy bath containing a carbide former, and chemical reaction may result in fine dispersion of a carbide. The reinforcements can be of various shapes and sizes depending on the processing conditions.

Systems for in situ production of reinforcements can be designed a priori on the basis of thermodynamic and kinetic studies.

For other types of systems, where there is a reacting gaseous phase or a nondissolving solid constituent, similar thermodynamic considerations may be worked out to determine the feasibility of in situ synthesis of reinforcing compounds. However, the reaction time involved for carrying out such a synthesis is an important variable in determining the cost-effectiveness of a process, and it cannot be predicted from thermodynamic considerations. The detailed kinetic steps involved in each specific system must be examined to determine the overall rate of reaction.

In Situ Formation of Reinforcements.[311]   Directional solidification of eutectic liquids has been used in the past to form fiber-reinforced composites in situ. The XD process is an example wherein Ti and C are added to molten Al alloys to form TiC_p in situ within the matrix. The TiC_p can be varied from 0.1 μm to 10 μm in diameter and from 20 to 75 vol% in a master composite alloy which can be diluted in molten Al to produce 25 vol% TiC. The moduli of these composites can range from 69 to 117 GPa. These alloys have application potentials for brake calipers, connecting rods, and pistons.

In the above-mentioned XD process,[332] the matrix alloy and reacting constituents are mixed in the solid state and ignited to generate a self-propagating reaction throughout the mixture. For the reaction to be self-propagating it has to be exothermic. This process results in a stable submicroscopic dispersion of reinforcing particles in a matrix alloy that generally melts at a temperature as high as that generated by the reaction. Also, the high diffusivity of the reacting constituents in the molten alloy helps to bring them together for further reaction and thereby contributes to an increase in the rate of reaction.

Basics. In situ production of MMCs can create a new class of naturally stable composites for advanced structural and wear applications. Conventional mechanical mixtures of whiskers, fibers, or particles and matrices (i.e., synthetic MMCs) are often not thermodynamically stable. In situ processes for nonferrous and intermetallic systems eliminate interface incompatibility of matrices with reinforcements by creating more thermodynamically stable reinforcements based on their nucleation and growth from the parent matrix phase. By controlling the melt (i.e., matrix alloy design) and reaction gas chemistry, a hierarchical range of carbides, nitrides, oxides, borides, and even silicides have been generated.

The reactions can be categorized generically, in terms of the starting phases, as gas-liquid, liquid-solid, liquid-liquid, and so on. One of the major fundamental scientific challenges lies in controlling the materials synthesis through optimized reaction kinetics and interfacial design.

Various Reactions
Liquid-solid: The liquid-solid reaction method of producing ceramic reinforcing particulates in situ in a matrix is exemplified by the previously discussed XD process.[333–335] In essence, the process is a solvent-assisted reaction wherein the ceramic phase is precipitated in the solvent medium (the matrix) via diffusion of the components. This is best illustrated by the formation of TiB_{2p} in an Al matrix.

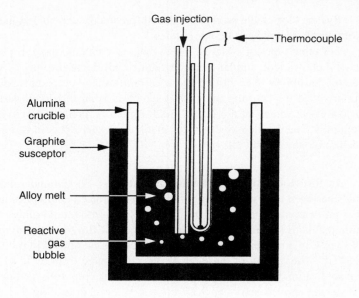

Figure 1.92   Schematic of the liquid-gas in situ reaction process.[335]

Liquid-gas: Liquid-gas reaction processing involves the formation of thermodynamically stable refractory compounds in a nonferrous alloy matrix (e.g., Al, Cu, Mg, Ni, or Ti). The reinforcing phase is the product of a gas-liquid reaction caused by the injection of gas into a reactive liquid metal (Figure 1.92). The rapid reaction between the solute alloying elements and the non-metal-bearing gas introduced into the melt produces a fine refractory dispersion in the matrix alloy.[335]

Mixed salt: In this process, mixed salts of Ti and B are reacted in molten Al to form particles of TiB_2. The by-products of the reaction are decanted, and the Al melt containing borides is cast onto waffles or ingots.

The London Scandinavian Metallurgical Company (LSM)[336] claims that TiB_{2p} of 1–2 μm have been formed by this process; a 2014 alloy with TiB_2 reportedly had a YS of 500 MPa with 5% elongation and an elastic modulus of 90 GPa. The nature of the particle-matrix interface is unclear. Since the process is based on well-established technology, it would be more readily amenable to scaling up to produce commercial quantities of composites, and LSM believes that the products would be technically and economically competitive with existing cast MMCs.

Directed metal oxidation and nitridation: Many researchers consider this innovative technique an in situ process, and it was discussed earlier.[325–328,330,337–339]

Reactive spray forming: Combining spray forming technology for NNS, rapidly solidified products with in situ reaction synthesis to form ceramic particles leads to reactive spray forming. This relatively new processing technique has been demonstrated for intermetallic-based composites such as Ni_3Al reinforced with Y_2O_3.[335]

In principle, by careful selection of alloying additions and the reactive atomizing gas based on thermodynamic principles, it is possible to synthesize a variety of composites containing carbides, nitrides, and oxides.

Self-propagating high temperature synthesis: Few technologies offer cost-effective synthesis of high-performance MMCs, IMCs, and CMCs. A compelling technology developed in Eastern Europe is SHS,[340] which has the advantages of submicrometer reinforcement size, nascent interfaces, economical processing, thermodynamic stability coupled with rapid reaction kinetics, and the ability to produce a high V_f of carbides, borides, and nitrides.

SHS has demonstrated its viability in producing various carbides, borides, silicides, nitrides, and hydride- and oxide-reinforced MMCs, as well as powders. The resultant products have applications such as sheaths for thermocouples, protective coatings, refractory material, nozzles for metal spraying, and line equipment in the chemical industry.

Liquid-liquid: Another technology employing an in situ reaction for the formation of ceramic particulates is the Mixalloy process practiced by the Sutek Corporation. This technique involves the reaction between two metal streams to form the refractory particulates. Two or more high-speed, turbulent molten metal streams are made to impinge upon one another in a mixing chamber, thereby resulting in intimate mixing and a reaction to produce the second phase. The resulting mixture can then be cast in a mold or rapidly solidified via melt spinning or atomization. The technology has been demonstrated[341] for producing Cu alloys reinforced with extremely fine TiB_2 for electrical applications. These Cu composites have good thermal stability with acceptable electric conductivity.

Plasma reactive synthesis: Thermal plasma-enhanced vapor-liquid-solid (VLS) processing is a novel materials synthesis route that utilizes the high thermal energies of a plasma to create reactive chemical species in the presence of heated vapor, liquids, and/or solids. Thermal plasma-enhanced processes either with liquid, gaseous, or particulate precursors have been demonstrated[342–346] to have the potential to synthesize unique composite structures in a wide range of materials and material systems. Examples include Al with AlN, Al_2O_3, or SiC; NiCrTi-based alloys with TiC or TiN; intermetallics such as TiAl, Ti_3Al, $MoSi_2$; and other ceramics with oxides, nitrides, borides, and/or carbides. Figure 1.93 schematically illustrates the principle of the three innovative plasma-processing routes. Particulates, liquids, gases, and solid or liquid surfaces can all be treated with the higher temperatures of plasma jets over 7750°C, allowing melting and reaction between most materials, provided the thermodynamics and the kinetics are favorable. Intermetallic or "ceramic" phases can be formed by injection of the reactive species into the melts at the bulk surface or in the bubbles under the surface, or around melting particles.

In Situ Studies.   Soboyejo, Lederich, and Sastry reported on the mechanical behavior of damage-tolerant TiB_{2w}-reinforced in situ Ti matrix composites.[347]

Khatri and Koczak[348] discussed the formation of TiC in in situ processsed composites via solid-gas, solid-liquid, and liquid-gas reaction in molten Al-Ti.

Valencia, McCullough, Rösler et al.[349] carried out a study demonstrating the feasibility of developing in situ reinforcements for TiAl matrices by solidification processing. Refractory compounds of B, C, and N can be grown as primary phases from the melt in

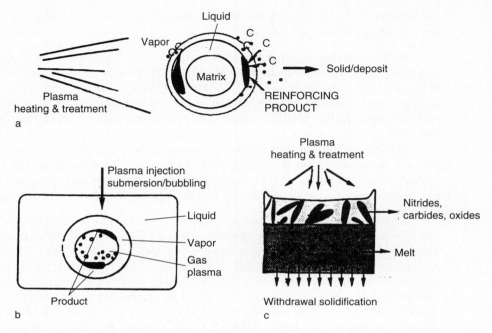

Figure 1.93 Schematic of several reactive plasma processing routes forming composite materials. (*a*) Plasma particulate reactive synthesis. (*b*) Plasma-injected reactive gas synthesis. (*c*) Plasma melt interface-enhanced materials synthesis.[335]

various morphologies and aspect ratios by careful selection of the alloy chemistry and solidification parameters.

Thompson and Nardone[350] reported the development of an approach to fabricating discontinuously reinforced composites with an extremely fine reinforcement size. Specifically, composites were formed in situ by plasma melting an agglomerate consisting of two dissimilar powders that react to form a reinforcement within a metal matrix. TMCs reinforced in situ were successfully fabricated by plasma melting powder agglomerates of Ti alloys and ZrB_2. Mechanical property evaluation of the composites revealed a very high compression strength, increased elevated-temperature strength, and increased elastic modulus relative to the Ti matrix.

Gungor, Roidt, and Burke[351] also reported a plasma melting and deposition approach they used to produce particulate-reinforced MMC systems. Premixed metal powders and ceramic particulates were injected into a plasma jet where they were rapidly heated. These traveling particles were deposited on a metal substrate inserted in the plasma downstream. By melting the metal powders and maintaining the ceramic particulates in a solid form, a composite structure was produced upon impact with the substrate.

Premkumar and Chu[352,353] conducted investigations to show that TiC_p produced in an Al melt by an in situ technique do not show any preferential segregation during solidification.

Using anatose (TiO_2) powders and Al as the reactants, Wang, Liu, Yao, et al.[354] found a new method for fabricating in situ composites, Al_2O_3/Ti_xAl_y. They prepared TiO_2

powder preforms by hydraulic pressing in a well-lubricated die and heated to 738–825°C; then molten pure Al (99.5% purity) was squeezed into the preform using a punch at a certain pressure.

These engineers found that sintering of squeeze-cast bulk is a simple method by which in situ composites consisting of intermetallics Ti_xAl_y and ceramic Al_2O_3 can be made. First, the properties of this kind of composite may be less affected by its porosity because of the high compactness of the squeeze-cast bulks. In addition, the sintering process can be controlled. Different kinds and quantities of intermetallics Ti_xAl_y may be obtained by controlling the sintering temperature and time, and different kinds of material with different properties may be produced. Moreover, the method of controlling the sintering process is much easier than that used to control the reaction squeeze casting process. Hence, in engineering the method appears to be much easier to adopt than other techniques for fabricating in situ composites.

Sahoo and Koozak[355] reported a novel technique for the fabrication of in situ TiC-reinforced Al alloy MMCs. The reacted, cast, extruded, and heat-treated samples exhibited a homogeneous distribution of fine (0.1- to 3-μm) TiC_p in a fine-grained recrystallized Al-4.5 wt% Cu matrix. Elevated-temperature tensile testing indicated that the composite retains its room temperature strengths up to 250°C and compares favorably with composites fabricated by more complex and costly processes. When compared with Al-4.5 wt% Cu alloy processed similarly but without TiC reinforcement, the addition of TiC resulted in YS and tensile strength increases of 130 and 65%, respectively.

1.5.6.3 Solid-phase processing.[2,196,198,200,232–237,265,281,282,285,311,323]

Powder Metallurgy.[356]   The P/M route is the most commonly used method for the preparation of discontinuous reinforced MMCs. Several companies have employed this technique in manufacturing MMCs using either particulates or whiskers as the reinforcement materials. Among them are DWA, Silag, and Novamet. Each of these companies has a unique feature associated with its process or product that is different from the others.

Figure 1.94 is a flowchart of the general P/M route. In this process powders of matrix materials and reinforcement are first blended and fed into a mold of the desired shape. Pressure is then applied to further compact the powder (cold pressing). In order to facilitate bonding between the powder particles, the compact is then heated to a temperature that is below the melting point but sufficiently high to cause significant solid-state diffusion (sintering). Alternatively, after blending the mixture can be pressed directly by hot pressing, however, HIP is helpful in securing high-density material. The consolidated product is then used as a MMC material after a secondary operation. DWA and Silag use the proprietary blending process to combine reinforcement with metal powder, whereas Novamet employs the mechanical alloying (MA) techniques to combine the reinforcement particulate and matrix constituents.

While both DWA and Silag use a proprietary blending process to combine particulate with metal powder, the distinction between the two methods is that Silag uses SiC_w, which are manufactured from rice hulls, as the reinforcement phase rather than particulates.

Since no melting and casting are involved, the powder process for MMCs is more economical than many other fabrication techniques and offers several advantages, some of these being the following.

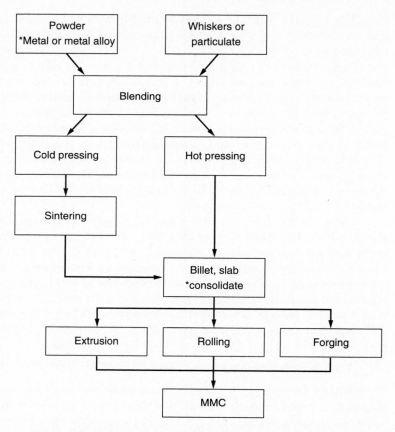

Figure 1.94   Flowchart for composite fabrication by powder metallurgy.[265]

1. A lower temperature can be used during preparation of a P/M-based composite compared to preparation of a fusion metallurgy-based composite. This results in less interaction between the matrix and the reinforcement, consequently minimizing undesirable interfacial reactions, which leads to improved mechanical properties.

2. In some cases P/M techniques permit the preparation of composites that cannot be produced by fusion metallurgy. It has been reported that SiC_w dissolves in a molten Ti alloy matrix, while dissolution can be minimized by using the P/M route. Again it has been shown that SiC_f are highly compatible with solid Al but only fairly compatible with liquid Al.

3. The preparation of particulate- or whisker-reinforced composites is generally speaking easier using the P/M blending technique than it is using the casting technique.

4. Particulate reinforcement is much less expensive than using continuous filaments of similar composition.

This method is popular because it is reliable compared with other alternative methods, but it also has some disadvantages. The blending step is a time-consuming, expensive, potentially dangerous operation. In addition, it is difficult to achieve an even distribution of particulate throughout the product, and the use of powders requires a high level of cleanliness, otherwise inclusions will be incorporated into the product with a deleterious effect on fracture toughness, fatigue life, and so on (Table 1.11).

P/M Applications in Various Types of Reinforcement and Processing (Al/SiC and Al$_2$O$_3$).   Ni, Maclean, and Baker[357] used two P/M routes to manufacture 6061 Al MMCs. Since the acceptance of MMCs for industrial applications depends upon improving properties, they examined two routes that involved (1) blending, vacuum canning, and hot pressing, and (2) blending of elemental powders, liquid-phase sintering, and subsequent hot rolling.

These composites comprised 7.5 or 15 vol% of 7-, 23-, or 45-μm SiC$_p$. From the first method they obtained a near-theoretical density together with a reasonable homogeneous microstructure, however, oxide was observed at the metal powder grain boundaries. The tensile strength was compared to the best data reported in the literature for the same composite made via other P/M routes. The low elongation and prior particle boundary failure were due to limited intermetal particle bonding which could result from the inability to redistribute adequately the surface oxide during the hot pressing operation.

The processing of P/M route 2 was relatively inexpensive and flexible and had the advantage over route 1 of providing composites with superior ductility. Appropriate selection of the initial particle size ratio (Al powder/SiC$_p$) and mixing conditions were very important. Cracked particles were observed after the cold compacting stage but more frequently after rolling. Generally, to achieve optimum mechanical properties, for a MMC with a given particle V_f, the smallest particle size that can develop a uniform distribution of particles is recommended.

Hot rolling of sintered billets minimized their porosity, refined the grain size, broke up clusters, and homogenized the distribution of reinforcement, thereby creating optimum mechanical properties.

In an NDE study Shannon, Liaw, and Harrigan[358] produced SiC$_p$-reinforced MMC billets (335-mm length, 349-mm diameter) by P/M techniques. The billet material consisted of a 6090 Al alloy (1.0 Mg, 0.6 Si, 0.8 Cu, balance Al)/SiC/25$_p$, and during P/M processing the alloy matrix powder and the SiC$_p$ reinforcement powder were blended to form a powder mixture. This mixture was then hot-pressed and consolidated into a composite billet. Following consolidation, the composite billet underwent thermal mechanical treatments to develop final extrusion products. For instance, the composites were extruded into bars or plates.

Geiger and Walker[359] reported on an extensive, intensive study on the strides taken by the manufacturers of DRA composites in moving from the laboratory toward commercialization. Since the reinforcement is discontinuous, DRA composites can be made by P/M with properties that are isotropic in three dimensions or in a plane. Conventional secondary fabrication methods can be used to produce a wide range of product forms, making them relatively inexpensive compared to other advanced composites reinforced with continuous filaments (Figure 1.95).

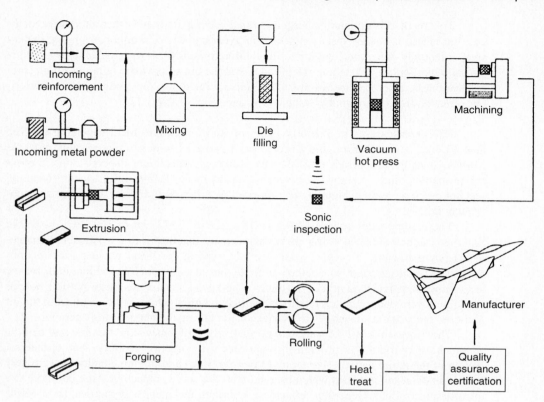

Figure 1.95   Flowchart showing the process for fabricating P/M DRA composites.[359]

Other engineers and developers[360–366] have processed Al matrix composites reinforced with SiC_p, whiskers, or fibers by P/M techniques and found the following.

1. The YS and the UTS increased, while the ductility decreased in the Al/SiC/20_p
   composite with decreasing particle size of the SiC. A faster cooling rate after annealing improved strength. A significant improvement in strength was observed
   with an increase in the V_f of SiC_p.
2. A large increase in the strength of the composite with SiC_w, compared with SiC_p,
   was observed with 10 vol% SiC. This was attributed to the higher aspect ratio and
   the inherent high strength of the whiskers, which resulted in a more efficient load
   transfer.
3. A substantial improvement in strength was possible with Cu modification of the
   SiC_p surface. This treatment probably improved the Al-SiC interface.

Zick[367] reported on the HIP fabrication of Al/SiC/40_p components for insensitive
munitions. These components are one example of how HIP can be employed to fabricate
powdered metals into products with properties tailored for a particular application. Wear-
and corrosion-resistant coatings can be applied in powder form to selective areas of com-

ponents. One application is in hot working tools such as extrusion dies and hot mill rolls. While such coatings are applied by other processes, hot isostatically pressed coatings often show superior bond strength and freedom from porosity. The HIP P/M approach also allows cladding inside long bores, such as in pump bodies and extrusion cylinders.

Premkumar, Hunt, and Sawtell[368] discussed the use of Al composite materials for multichip modules using SiC_p as the reinforcement since the material has emerged as one with an especially attractive combination of physical properties, manufacturing flexibility, and cost. One benefit of these materials is the ability to tailor physical properties through the selection of both reinforcement and alloy variables to match the thermal expansion coefficient of other electronic materials. In addition, the manufacturing flexibility of the various processes, especially P/M using vacuum hot pressing, allows for shape complexity as well as selective reinforcement placement in the component to optimize system producibility. Finally, because raw materials are inherently inexpensive and low-cost production routes have been identified, Al composites may offer a range of cost-effective solutions to emerging problems in electronic packaging and thermal management applications.

Styles, Sinclair, Gregson, et al.[369] discussed the effect of microstructure on mechanical properties of thermomechanical processes for a $2124/SiC_p$ MMC employed to control the grain and subgrain structure of a powder-produced SiC_p-reinforced (3 μm, 17 vol%) Al-Cu-Mg alloy.

Hong and Chung,[370] with support from the Korea Science and Engineering Foundation, conducted an investigation on the effects of processing parameters on mechanical properties of $SiC_w/2124$ Al composites. They found that the effects of P/M processing resulted in the following.

1. The tensile strength of composites increased with increasing vacuum hot pressing temperature because of the local melting and softening of the 2124 Al matrix. The vacuum hot pressing temperature needs to be higher than 560°C, and the vacuum hot pressing pressure needs to be higher than 70 MPa to achieve full densification of the composite.

2. The addition of SiC_w at more than 20% was not helpful in improving the tensile strength because of the inadequate densification from local clustering and the reduced aspect ratio of the whiskers.

3. The increased extrusion temperature and extrusion ratio resulted in improved alignment of SiC with full densification. However, the tensile strength decreased at a higher extrusion ratio of 25:1 because of a reduction in the aspect ratio of SiC_w caused by significant damage to the whiskers during extrusion. The extrusion temperature needs to be slightly above the solidus temperature of 2124 Al matrix in order to avoid fir-tree cracking of the composite.

Tweed[371,372] manufactured 2014 Al alloy reinforced with SiC_p using powder prepared by inert-gas atomization and consolidation by vacuum hot pressing. Billets with 0–40 vol% filler and 0–10% porosity were produced. Tweed found that the thermal conductivity of the poured powder was poor but could be enhanced by cold pressing. During consolidation shear transfer of load to the cylindrical die was reduced by supersolidus pressing in conjunction with low SiC contents. For solid-state pressing, increasing the SiC content from 0% to 20% reduced densification rates by a factor of 2 under identical pressing conditions.

Finally, Ling, Bush, and Perera focused[373–375] their efforts on SiC_p/Al fabricated by four different P/M routes in order to determine the effect of the fabrication method on the mechanical properties of the NNS specimens.

Their study undertook to improve understanding of the effect of fabrication technique on the mechanical properties of Al-based MMCs and compared the mechanical properties of specimens subjected to various postcompaction processing techniques. In particular, these were (1) sintered, (2) cold isostatically pressed and sintered, (3) hot isostatically pressed, and (4) sintered and hot isostatically pressed in the same HIP cycle.

They found that sintered and hot isostatically pressed composites of up to 30 vol% SiC were produced with a significant improvement in ductility and UTS compared to composites obtained with the other fabrication methods. The poor mechanical properties of composites produced by the other methods can be attributed to the weak bonding between adjacent particles and to internal porosity. The measured densities of the fabricated materials indicated that fully dense material was not achieved. For composites with less than a 10% V_f of SiC, the density was relatively independent of the fabrication method adopted.

Microstructural examination of fracture surfaces in representative materials confirmed that sinter-HIP techniques yielded the best composites. For particle loading of less than 10% V_f, ductile failure of the matrix appears to be the limiting factor. At higher V_f values the strength of interfacial bonds, initiation and growth of voids, and particle cracking all play an important role in controlling the mechanical properties.

Other researchers examined the mixing of powder metallurgy Al powder alloys with Saffil (δ-Al_2O_{3f}).

The studies of ter Haar and Duszczyk[376] and Whitehouse and Clyne[377] have been reported as follows.

1. Reference 376 covers a study conducted on dry mixing of air-atomized Al-20 wt% Si-3 wt% Cu-1 wt% Mg alloy and δ-Al_2O_{3f} ("Saffil-milled") in a tumbler mill. The mixing was focused on deagglomeration of clusters of the fiber material and on minimalization of the decrease in fiber length. The general fundamental approach for efficient breakage of solid particles in a ball mill was adopted to disintegrate fiber clusters in the powder metallurgical composites.

   Composites of 5, 10, and 20 vol% δ-Al_2O_{3f} were successfully mixed by ball milling providing geometric mean fiber lengths (aspect ratios) of 38 µm (13), 30 µm (10), and 25 µm (8) at milling times (fractional ball fillings J) of 5 min (30%), 15 min (39%), and 26 min (39%), respectively.

   The mixing process for P/M composites with clustered δ-Al_2O_{3f} can be carried out using the general approach for efficient breakage of solid particles in a ball mill. However, the parameters should preferably be chosen at threshold levels for low-impact operation in order to minimize fiber damage.

2. Reference 377 shows the effect of test temperature and reinforcement shape on cavitation for commercially pure Al and chopped Saffil Al_2O_{3f} composite produced via a powder route followed by extrusion. At room temperature, voids formed at the reinforcement-matrix interface.

Favored void nucleation sites changed from the planar surfaces of reinforcement particles to angularities as the test temperature was increased. Void levels therefore tended to be very low with spherical reinforcement at high temperatures.

The difference in void content with increasing reinforcement content became less pronounced with increasing temperature.

The presence of oxide stringers in the matrix promoted cavitation at all temperatures. At room temperature they also stabilized the voids, preventing rapid coalescence and failure. This stabilizing effect was reduced as the test temperature was increased, leading to reduced ductility for powder route materials.

P/M Applications to Various Types of Reinforcement and Processing (Ti/TiC and Ti/TiB$_2$).   A family of Ti MMCs, CermeTi, with superior properties has been developed by combining cold and hot isostatic pressing (CHIP) manufacturing technology commercialized for monolithic titanium alloys.[378–384]

The manufacture of TMCs utilizing economical particle reinforcement has been demonstrated by the P/M processing route. These composites exhibit homogeneous microstructures with a clearly established particle-matrix interface. TiC-, TiB$_2$-, and TiAl-reinforced Ti/6Al/4V have been produced by P/M with subsequent forging or extrusion at various levels of reinforcement. Properties, particularly room- and elevated-temperature strengths and modulus, offer significant improvement, while useful fracture toughness has been maintained.

These material combinations, with the NNS P/M manufacturing route, offer a viable alternative for economical TMC preparation for both military and commercial applications.

Pauchal and Vela[385] also have reported their P/M efforts to make TiC/Ti composites. Pressing and sintering blends of Ti hydride powder (−325 mesh) and TiC powder ball-milled to 0.8-μin. particle size resulted in nearly fully dense (99% or greater) composites having 35–40 points greater R_c hardness and three to four times greater wear resistance than pure Ti. Just 10 vol% TiC, for example, resulted in a density of 4.53 g/cm^3 and a hardness of R_c 50–52 after sintering at 1200°C for 1.5 h in a vacuum. The TiC (10–30 μin. in diameter) was uniformly dispersed throughout the matrix.

The use of as much as 45 vol% TiC was also studied, but less than 25 vol% was recommended. For one thing, more than 25 vol% formed a network of carbides in a porous matrix. For another, further hardness gains were negligible.

Sliding and rotating mechanical parts—bearing balls and cages, cams, seals, clutches, gears, splines, pump components, and so on—seem to be the most likely potential applications. Others, the researchers noted, include railcar wheels, valves, mixers, and atomizers in the chemical industry, and possibly, orthopedic implants.

Diffusion Bonding.   This method for fabricating MMCs consolidates multifilament yarns preimpregnated with a metal. These preimpregnated ribbons are known as composite wires or impregnated tows, and DB consolidation is achieved by hot pressing, hot rolling, or hot extrusion. This method allows fabrication of composite structures of continuous length.

Diffusion bonding processes are generally considered suitable for producing large products, but this technique has several problems. For example, Al matrix composites produced by hot pressing a prepreg sheet reinforced with B/W and SiC/C_f have excellent mechanical properties. However, it is difficult to produce parts with complex geometries because the large, 100- to 150-µm-diameter fibers lead to low flexibility in the MMC prepregs. On the other hand, Al matrix prepregs reinforced with small carbon fibers (5- to 10-µm diameter) have good flexibility, but unfortunately the composite mechanical properties are poor.[274]

DB/HIP.   Loh, Wu, and Khor[386] used DB by HIP to join Ni powder to Al_2O_3 tubing. Their results showed a maximum shear bond strength of 38.2 MPA which was achieved with a HIP temperature of 1200°C at 200 MPa pressure and 100 min soaking time. This produced a reaction layer of nickel aluminate spinel, and the shear bond strength of the Al_2O_3/Ni interface indicated that it was a function of the HIP processing parameters. In particular, they showed that spinel formation is necessary to *reaction-bond* the ceramic-metal interface. An excessive interphase layer, however, causes a drop in the bond strength.

Shear tests showed the spinel $(NiAl_2O_4)$-nickel junction to be weaker than the spinel-nickel junction.

DB and Filament Winding.   Fabrication of complex-shaped components is commonly achieved by DB of monolayer composite tapes. The tapes may be prepared by a number of methods, but the most commonly used is filament winding where the matrix is incorporated in a sandwich construction by laying up thin metal sheets between filament rows or by the arc spraying technique developed by the NASA Lewis Research Center. In the latter process, molten matrix alloy droplets are sprayed onto a cylindrical drum wrapped with fibers. The operation is carried out in a controlled atmosphere chamber to avoid problems of oxidation and corrosion. The drum is rotated and passed in front of an arc spray head to produce a controlled porosity monotape. When lamination of the fibers between matrix alloy sheets is adopted to produce the monotape, a binder is used to hold the fibers in place prior to consolidation. This technique can also be combined with consolidation by HIP. Filament winding and monotape layup into multiple layers permit close control over the position and orientation of the fibers in the final composite.

A combination of the arc spray technique, DB, and SCS-6 SiC monofilament fibers has produced composites for various prototype applications in the aerospace and defense sectors of the market. In one example, a rocket motor shell has been produced by filament winding of SCS-6 fibers and plasma spraying with an Al alloy. The rocket motor was reported to have been successfully fired and to have withstood the extreme temperatures, pressures, and vibrations associated with missile firing. In addition, the assembled motor, using the MMC shell, weighed about 10% less than a conventional steel motor.

DB by Hot Pressing.[273,387]   The ability to readily produce acceptable SiC_f-reinforced metals is attributed directly to the ability of the SiC_f to (1) readily bond to the respective metals, and (2) resist degradation of strength while being subjected to high-temperature processing.

SCS-grade fibers have surfaces that readily bond to the respective metals without the destructive reactions occurring. For Ti composites, the SCS-6 filament has the ability to withstand long exposure at DB temperatures without fiber degradation. As a result, complex shapes with selective composite reinforcement can be fabricated by the innovative superplastic forming–diffusion bonding (SPF/DB) and HIP process.[388]

DB of SiC/Ti is accomplished by hot pressing (DB) using fiber preforms (fabric) that are stacked together between Ti foils for consolidation. Two methods have been developed by aircraft and engine manufacturers to produce complex shapes. One procedure is based on the HIP technology and uses a steel pressure membrane to consolidate components directly from the fiber-metal preform layer. The other method requires the use of previously hot-pressed SCS/Ti laminates that are diffusion-bonded to superplastic forming operations.

The following is typical of the first fabrication procedure noted above. The fiber preform is placed on a Ti foil, which is then spirally wrapped, inserted, and diffusion-bonded to the inner surface of a steel tube using a steel pressure membrane. The steel is subsequently thinned down and machined to form the "spline attachment" at each end. Shafts have been fabricated for other engine fabricators without the steel sheath. The concept was developed for SPF of hollow engine compressor blades. Here, the SCS/Ti laminates are diffusion-bonded in a press. They are then diffusion-bonded to form monolithic Ti sheets, with "stop-off" compounds selectively positioned to preclude bonding in desired areas. Subsequently, the stack-up is sealed into a female die. By pressurizing the interior of the stack-up, the material is "blown" into the female die to form the desired shape, stretching the monolithic Ti to form the internal corrugations. SCS-6 foil retains its strength with the Ti in the structure at 925°C.

The β-Ti alloy, 15-3-3-3, reinforced with SCS-6 fiber, has achieved superior composite properties, for example, tensile strengths of 1585–1930 MPa[273] by DB and HIP. The HIP technique has been particularly successful in the forming of shaped reinforced parts (e.g., tubes) by the use of woven SiC fabric as a preform. The high-strength, high-modulus properties of SCS-6/Ti represent a major improvement over B_4C-B/Ti composites in which the modulus of the composite is increased relative to the matrix, but the tensile strength is not as high as would be predicted by the ROM.

Foil-Fiber-Foil DB Process.[388,389]   To reach their ambitious goals for civil and military aircraft engines in the next century, researchers agree that TMCs show promise as prime candidates to replace current metallic components. In addition to blades, stators, and shafts, key substitutions include integrally bladed rotors reinforced with TMC cores.

Researchers[388] have focused their attention on the fabrication of MMC rings (Figure 1.96). Sheets of Ti alloy foil are inserted in the spaces between a helix of bundled SiC_f, and the entire assembly is compressed into a slot in a Ti alloy pancake disk. After HIP to consolidate the ring, heat treating relieves thermal stresses and stabilizes the Ti microstructure.

The previously mentioned MMC rings were made from Ti-15V-3Cr-3Al-3Sn and SCS-6 fibers. Chosen for its extensive database, this material represents the MMCs that would be used in an aircraft engine but is not the specific grade. The same material has been successfully used to fabricate integrally bladed rotors as well (Figure 1.97).

1.5.6.4 Spray processes.   A number of processes have evolved in which a stream of metallic droplets impinges on a substrate in such a way as to build up a composite deposit. The reinforcement can be fed into the spray if particulate, or introduced

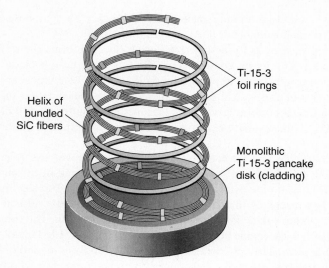

Helix of
bundled
SiC fibers

Ti-15-3
foil rings

Monolithic
Ti-15-3 pancake
disk (cladding)

Figure 1.96   Foil-fiber-foil process for fabrication of MMC rings.[389]

onto the substrate in some other way if fibrous. The techniques employed fall into two distinct classes depending whether the droplet stream is produced from a molten bath or by continuous feeding of cold metal into a zone of rapid heat injection. In general, spray deposition methods are characterized by rapid solidification, low oxide content, significant porosity levels, and difficulty in obtaining homogeneous distribution of reinforcements.

Osprey Process (Melt Atomization).[196,198,200,265]   In the Osprey[323,332] process, a molten metal stream is fragmented by means of a high-speed, cold, inert-gas jet passing through a spray gun, and dispersoid powders are simultaneously injected as shown in Figure 1.98. A stream of molten droplets and dispersoid powders is directed toward a collec-

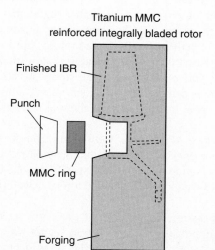

Titanium MMC
reinforced integrally bladed rotor

Finished IBR

Punch

MMC ring

Forging

Figure 1.97   Cross section of an integrally bladed rotor showing the position of Ti MMC core reinforcement.

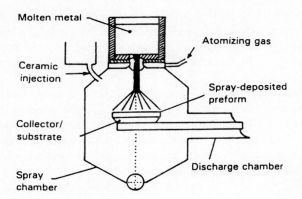

Figure 1.98   Schematic illustration of the Osprey process.[323]

tor substrate where droplets recombine and solidify to form a high-density deposit. Dispersoid particles may combine with droplets during flight, but most particles are generally codeposited. Particle spraying can be independently controlled and may be directed to selected areas. The process depends critically on the ability to control the enthalpy of droplets in the impinging spray, and a droplet should be partially solid when it reaches the substrate. If properly controlled, the process can result in solid deposits in different net shapes of tubes, round billets, strips, or clad products. The grain size of the resulting composite is relatively uniform, and the presence of particles during solidification of droplets refines the matrix microstructure. This process is capable of achieving a high rate of production, and the deposited product can be directly used in hot forming such as forging, rolling, or extrusion. The process variables are (1) temperature of the alloy, (2) speed of the gas jet, and (3) temperature of the substrate.

This process combines the blending and consolidation steps of the P/M process and thus promotes major savings in MMC production.

Spray deposition was developed commercially in the late 1970s and throughout the 1980s by Osprey, Ltd. (Neath, United Kingdom) as a method of building up bulk material by atomizing a molten stream of metal with jets of cold gas, with the effect that most such processes are covered by their patents or licenses and are now generally referred to as Osprey processes.[390–392] The potential for adapting the procedure to particulate MMC production by injection of ceramic powder into the spray was recognized at an early stage and has been developed by a number of primary metal producers.[393,394]

A feature of much MMC material produced by the Osprey route is a tendency toward inhomogeneous distribution of the ceramic particles. It is common to observe ceramic-rich layers approximately normal to the overall growth direction. Among the other notable microstructural features of Osprey MMC material are a strong interfacial bond, little or no interfacial reaction layer, and a very low oxide content. Porosity in the as-sprayed state is typically about 5%, but this is normally eliminated by secondary processing. A number of commercial alloys have been explored for use in Osprey route MMCs.[395–397]

Spray Forming.   Spray forming in the Osprey mode involves the sequential stages of atomization and droplet consolidation to produce NNS preforms in a single processing step. Metallurgically, the as-sprayed material has a density greater than 98% of the theoretical density and exhibits a uniform distribution of fine equiaxed grains and no prior particle boundaries or discernible macroscopic segregation. Mechanical properties are normally isotropic and meet or exceed those of counterpart ingot-processed alloys. A major attraction of the process is its high rate of metal deposition, typically in the range 0.2–2.0 kg/s.

Commercial viability mandates close tolerances in shape and dimensions, as well as consistency in microstructure and product yield. This requires an understanding of and control over the effects of several independent process parameters, namely, (1) melt superheat, (2) metal flow rate, (3) gas pressure, (4) spray motion (spray scanning frequency and angle), (5) spray height (distance between the gas nozzles and the substrate), and (6) substrate motion (substrate rotation speed, withdrawal rate, and tilt angle).

To optimize spray forming in terms of the microstructure and properties of the product, it is necessary to model the process. This has been done by dividing the complete sequence of events into six discrete steps or submodels[398,399]; these submodels, which may be mathematical or conceptual, address atomization, the spray, droplet consolidation, preform shape, solidification in the preform, and development of the microstructure (Figure 1.99). Collectively, the submodels constitute an integral model of the process since the output from one stage is used as the input to another stage downline.[397]

Singer[400] found that spray-formed metals had a finer structure, near-zero segregation, and improved properties resulting from rapid solidification when compared with conventional materials. A further benefit of spray forming is that it gives greater freedom to add any second or third phase, liquid or solid, by entraining a stream of the added phase into the atomized spray before deposition. Singer made MMCs by spray forming and was able to incorporate both of these benefits in one operation.

His work showed that the spray forming route to composites offers an inexpensive product that is usually intermediate in cost between that of products produced by melt stirring and P/M. It also has the advantage of avoiding or reducing solution of the added phase, is applicable to most systems, and has the added advantage of having a fine, non-segregated structure and good mechanical matrix properties. His studies covered the production of Al + 4% Si + 1% Cu alloy/10 μm SiC/15$_p$ into spray-formed billets that were extruded into a wide variety of shapes and forms.

McLelland, Atkinson, Kapranos, et al.[401] reported on their findings in some preliminary studies on the application of thixoforming to MMCs. Thixoforming has potential as a processing route for MMCs that would minimize some of the problems associated with liquid metal routes, for example, segregation of reinforcement particles and detrimental interfacial reactions. In addition, it is potentially cheaper than the powder-mixing routes to particulate MMCs.

The Osprey spray codeposited Al-7% Si-0.5% Fe-0.4% Mg alloy/9 μm SiC/15$_p$, which was thixoformed with a good degree of success. They found that with increasing softness prior to thixoforming there was an improvement in the extent of die filling and in the surface finish, and the solidified front of material filling the die cavity became more continuous. Further, there was improvement in the surface finish as the dies heated up prior to thixoforming.

Finally, the SiC$_p$ flowed with the semisolid alloy so that the particulate volume percentage was maintained throughout the shaped material.

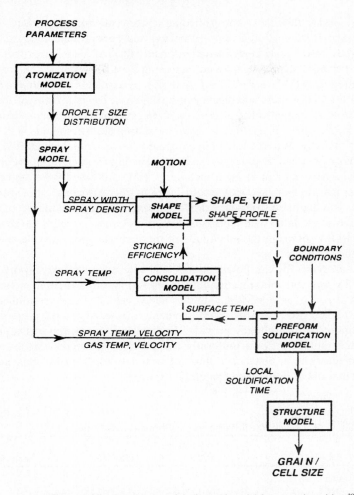

Figure 1.99   Components of the integral model for spray deposition.[397]

Codeposition and Atomization.[265,281,402]   Spray atomization and codeposition processes have received considerable attention in the synthesis of discontinuously reinforced MMCs. This methodology involves the mixing of reinforcements and matrix under thermal conditions such that the matrix contains both solid and liquid phases. In principle, such an approach avoids the extreme thermal excursions, with concomitant degradation in interfacial properties and extensive macrosegregation, that are normally associated with casting processes. Furthermore, this approach also eliminates the need to handle fine reactive particulates normally associated with P/M processes. To investigate the utility of this process in the preparation of MMCs, several $SiC_p/6061$ Al composites were prepared by Wu and Lavernia.[402]

From their tests,[402] the room-temperature mechanical properties may be summarized as follows: First, the YS and the UTS of spray-atomized and deposited materials were higher than those corresponding to conventionally processed 6061 Al. Second, the YS and UTS values of spray-atomized and codeposited 6061 Al/SiC$_p$ materials were slightly lower than those of unreinforced 6061 Al. Third, the ductility and UTS of the 6061 Al/SiC$_p$ decreased with an increasing V_f of SiC$_p$. Finally, the results showed that annealing the MMCs at 560°C for 22 h prior to extrusion resulted in a slight increase in YS and UTS. The most significant improvement in the properties of the MMCs, relative to those of the unreinforced alloys, was in the magnitude of the elastic modulus. In addition, the results show a clear trend toward increasing elastic modulus with increasing SiC$_p$ V_f.

Finally, Wu and Lavernia compared their room-temperature mechanical properties to those obtained by other investigators. The higher strength of the spray-deposited materials, relative to that of ingot metallurgy (IM)-processed 6061, agrees with the results of other investigators and may be attributed to microstructural refinement resulting from the relatively high rates of heat extraction present during solidification. The overall increase in elastic modulus with increasing SiC$_p$ V_f is consistent with the results anticipated from the ROM, although the magnitude of the increase appears to be somewhat smaller than anticipated.

Nayim, Raskin, Polak, et al.[403] showed in their reported work at Ben Gurion University that spray atomization and codeposition was a feasible technique for the fabrication of Al-Li based MMCs, with the "added value" of rapid solidification processing benefits. The Al$_2$O$_3$ and SiC reinforcements in the Al matrix promoted grain refinement additional to that already provided by the spray deposition technique. Solidification occurred rapidly to form uniformly refined microstructures. Mechanical properties were comparable to or better than those of material processed by other methods (Table 1.13), and thermal stability was apparently achieved.

TABLE 1.13.   Mechanical Properties at Room Temperature

Sample ID	Yield Strength (MPa)	UTS (MPa)	Elongation (%)	Elastic Modulus (GPa)
Spray-atomized Al-2024 + Al$_2$O$_3$	214.0 ± 5.0	323.5 ± 2.5	8.5 ± 1.5	78 ± 1.5
Spray-atomized Al-2024 + Li + Al$_2$O$_3$	239.0 ± 21.0	318.0 ± 6.0	4.5 ± 2.0	79.5 ± 2.0
Spray-atomized Al-2024 + SiC	273.3 ± 4.0	342.3 ± 2.0	6.5 ± 2.0	75.0 ± 2.0
As-cast Al-2024	241.5 ± 4.5	339.0 ± 2.0	8.8 ± 1.0	73.0 ± 0.5
As-cast Al-2024 + Li	215.0 ± 5.5	259.0 ± 3.0	4.3 ± 2.0	74.0 ± 1.0
IM Al-2024[b]	400	530	23.0	72.8
IM Al-2.8Li-0.3Mn[c]	261	390	7.0	ng
RS atomized + SHT, 525°C/1 h + 190°C/3 h[d]	269	372	7.0	ng

[a] ng, Not given.
[b] P. Meyer and B. Bubost. In *Al-Li Alloys III,* ed. E. A. Starke and T. H. Sanders, 39, 42. TMS-AIME, Warrendale, PA, 1987.
[c] I. G. Palmers, R. E. Lewis, and D. D. Crooks. In *High Strength Powder Metallurgy Alloys,* ed. M. K. Koczak and G. J. Hilderman, 383. Metallurgical Society of AIME, New York, NY, 1982.
[d] G. Chanani, G. Han Narayan, and I. J. Telesman. In *High Strength Powder Metallurgy Alloys,* ed. M. J. Koczak and G. J. Hilderman, 345. Metallurgical Society of AIME, New York, NY, 1982.

Thermal Spray.[246]   Thermal spraying involves the feeding of powder, or in some cases wire, into the hot zone of a torch where it is heated and accelerated as separate particles by the rapid expansion of a gas. The heat is usually generated by an electric arc or by gas combustion. Particularly good process control is possible with plasma spraying, which is now a mature technology with considerable potential for future development.[404] Another promising technique for spraying MMCs is the use of high-velocity oxy-fuel (HVOF) guns, which are already in established industrial use for the production of surface coatings. There has recently been interest in composite deposits, which are readily produced with twin powder feeding arrangements. In additon to the production of bulk MMC material in this way, there is also considerable interest in graded composition MMC coatings exhibiting both good wear resistance (ceramic-rich region) and strong adhesion to a metallic substrate (metal-rich region).

Thermal spraying differs in several respects from melt atomization processes. Deposition rates (usually less than 1 g/s) are slower, but particle velocities (~200–700 m/s) are higher, particularly for HVOF guns. In general, the substrate and deposit tend to remain relatively cool, and the quenching rates for each individual splat are very high (greater than or equal to 106 K/s). Porosity levels are typically several percent, but can be below 1%. Other types of composite material produced by plasma spraying include laminated metal-ceramic layer structures and fiber-reinforced laminas manufactured by spraying onto an array of monofilaments. There is also interest in producing long-fiber MMCs by simultaneous thermal spraying and filament winding. It has been suggested that a RF plasma torch may be better suited to this type of composite production than a direct current (DC) system, as droplet impact velocities are lower and there is less damage to the fibers.

Steffens and Kaczmarek[405] described their thermal spray method for fabricating both monotapes and multilayer fiber-reinforced MMCs. In this technology different continuous fibers were used, while various matrix materials were applied by thermal spraying. Commercially available stainless steel fibers 0.10 mm in diameter were used to reinforce the thermally sprayed matrices. Sheets of composites with fiber volumes up to 40% and theoretically unlimited dimensions could be fabricated by this process.

Before producing fiber-reinforced composites, a special computer-controlled wrapping unit permitting the processing of different types of continuous fibers was designed. This unit met the requirements of high precision and good reproducibility needed to produce composites with defined fiber volume. The constructed wrapping unit was able to space fibers ranging from 40 μm up to several millimeters with small tolerances. Steffens and Kaczmarek used a computer-controlled plasma system and a two-axis motion unit for the torch, which allowed all significant spray parameters to be controlled and kept within close tolerances. A detailed description of the production technique for the fiber-reinforced composites with sprayed matrices is given in other publications.[406,407]

In their initially reported work Steffens and Kaczmarek[405] claimed that the tensile strength of the constructional matrices exceeded that of the matrix several times. In addition, provided proper selection, the spraying process itself has no influence on the mechanical and chemical properties of the fibers. The theoretical estimation of the tensile strength of composites was compared with the values obtained experimentally. It was found that the ROM seems to be applicable to composites with thermally sprayed matrices, but more work should be done to prove this theory.

In future research, besides further fiber-matrix combinations, Steffens and Kacz-

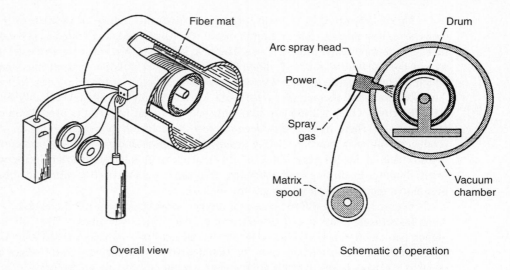

Figure 1.100   Schematized NASA Lewis Research Center patented arc spray monotape technique.[408]

marek plan to investigate the influence of thermal exposures in different atmospheres on the fiber-matrix interface and on composite properties. Moreover, determination of the mechanical properties involving the fatigue strength of composites with thermally sprayed matrices also seems to be very helpful in industrial applications.

Arc Spray.[408,409]   Titan and Grobstein[408] reported the use of a patented arc spray monotape technique for W/Nb composites. This process, schematized in Figure 1.100, consists of wrapping the reinforcing fibers on a drum, placing the drum in an airtight chamber, evacuating the chamber, and backfilling it with argon. Next, wires of the matrix composition are brought together in the arc spray gun, an arc is struck between them, and pressurized argon blows the molten matrix covering the fibers; the resulting structure is a monotape consisting of one layer of fibers in a matrix. Layers of monotapes can then be hot-pressed or hot isostatically pressed, producing a fully densified consolidated structure with negligible fiber-matrix interfacial reaction.

Plasma.   Hall and Ritter[410] studied the structure and mechanical behavior of the fiber-matrix interface in Ti alloy-SCS-6 SiC MMCs. The composites examined in their investigation were fabricated by plasma spraying nuclear metals plasma-rotating electrode powder (PREP) onto a drum wound with SCS-6 SiC_f. Two metal matrix compositions were prepared: Ti-14 wt% Al-21 wt% Nb (Ti-1421) and Ti-6 wt% Al-2 wt% Sn-4 wt% Zr-2 wt% Mo (Ti-6242). A plasma torch hydrogen level of 3% was used for the Ti-1421 fabrication. Four layers of the fiber-containing tapes were then stacked unidirectionally between 0.03-mm Mo foil in a steel can and hot isostatically pressed for 3 h at 1000°C and a pressure of 103 MPa. No pressure was applied until the materials had reached the HIP temperature.

After HIP, the plates were removed from the cans by dissolution in a 50% HNO_3/50% water solution. Samples for heat treating were wrapped in Ta foil and sealed in a quartz capsule which was backfilled with dry argon. The yttrium is used to absorb oxygen first and therefore not allow any to contaminate the base material plates. The yttrium getters or gathers up the oxygen. After heat treatment, the tubes were removed from the furnace and allowed to air-cool.

In their results Hall and Ritter concluded that the effect of composite fabrication and heat treatment on the coating structure was relatively small.

Khor, Yuan, Boly et al.[411,412] prepared an Al-Li/SiC_p composite by plasma spray atomization (PSA) as an alternative to P/M processes because of their reinforcement size limitation (>5 μm).

The technique consists of melting the metal matrix and ceramic reinforcement simultaneously in a high-temperature plasma flame (~2000–10,000°C). The molten droplets are fragmented into very fine composite particles when they impact on a solid rotating disk. The powder collected is characterized using standard metallographic techniques, a scanning electron microscope (SEM), and a laser diffraction particle size analyzer. Results have shown that this process produces fine Al-Li/SiC composite particles in the size range 1–35 μm. SEM observation of polished cross sections showed that the particles consisted of submicron SiC_p dispersed within single Al-Li alloy particles. Quantitative analysis of individual composite particles using wavelength dispersive x-ray analysis (WDX) showed that the concentration of SiC was low in the large particles (>20 μm) but high in particles <10 μm. It was also concluded that the plasma gas composition (Ar/He) has a negligible effect on the production of fine particles.

Pank and Jackson[413] have reported on the development of an induction plasma deposition (IPD) processing method for the fabrication of Ti 6242/SiC material. Large-scale nominally 40 V_f Ti 6242/SCS-6 MMC monotapes were fabricated with precisely controlled fiber spacing using the IPD process. Longitudinal room-temperature tensile results for Ti 6242/SCS-6 MMC fabricated by IPD yielded UTS values between 1540 and 1932 MPa and modulus values in the range of 217 GPa. Fatigue properties for IPD-processed Ti 6242/SCS-6 MMCs with well-controlled fiber spacing exhibited a 2× improvement over material with poor spacing or touching fibers.

Valente and Bartuli[414] described a newly developed plasma spray process for the manufacture of long-fiber-reinforced Ti-6Al-4V composite monotapes.

The plasma spray technique, carried out in an inert atmosphere, was used to deposit the metal matrix onto previously arranged continuous fibers. They found that major benefits were due to a controlled operating environment (the entire process was performed in a neutral gas atmosphere) and to a high solidification rate of the melted material. The deleterious formation of brittle compounds at the interface was limited, or totally avoided, by the precleaning process in operation in the deposition chamber and by the high solidification rate of the sprayed droplets. Preliminary work on diffusion bonding of the monotape was carried out with satisfactory results, even though further improvements in the plasma spray process and the hot pressing procedure for the above-mentioned MMCs are still needed.

Plasma spraying, normally used as a coating technique, was modified to produce a long composite monotape. This required a suitable arrangement of the fiber, placed on a cylindrical substrate, and the identification of suitable operating conditions.

Valente[415] reported some other work on continuous-fiber SiC/Ti and SiC/Al matrix composite monolayers where he used a vacuum plasma spray (VPS) process which needs further optimization work but is a very promising fabrication method for continuous fiber-reinforced MMCs. Sampath and Herman[416] have compiled a review of the work covering plasma spray processing. The technique, which is a droplet deposition method that combines the steps of melting, rapid solidification, and consolidation into a single step allows the processing of free-standing bulk NNSs of a wide range of alloys, intermetallics, ceramics, and composites while still retaining the benefits of rapid solidification processing (RSP).

1.5.6.5 Rapid solidification processing.[323,417,418]

In rapid solidification processing of composites, a jet of liquid alloy-particle slurry impinges under pressure on a water-cooled Cu wheel and the resulting flake powders are collected. The flakes are of thickness 40–60 μm, length 6–8 mm, and width 0.5–0.7 mm. Two critical processing parameters are the speed of the quenching Cu wheel and the amount of material impinging on this wheel. If the wheel turns too quickly, the slurry will not attach to the wheel long enough to produce a powder. On the other hand, if the wheel is too slow, flakes will be too thick and will be unacceptable for subsequent processing. The powder is put into cans and consolidated into billets and/or extruded to form a dense composite with higher YS, UTS, and ductility.

The advantages of this procedure are as follows. RSP has been used in the development of many engineering alloys to produce microstructures that enhance high-temperature properties including strength and corrosion resistance. The microstructures of these alloys are typically characterized by improved compositional uniformity, refined microconstituents, a high level of supersaturation, and retained metastable phases. Extensive refinement of cast dendritic structures is achieved by increasing the cooling rate or dendritic growth velocity during solidification. Increased compositional uniformity is achieved because of the shorter diffusion distances between regions of microsegregation. In addition, precipitates that form in the microsegregated regions tend to be finer and more uniformly distributed.

Application of rapid solidification techniques to ceramic particulate-reinforced MMCs allows the use of high-strength, high-temperature RSP alloys as matrix materials. This permits optimization of matrix-phase properties while using large reinforcement particles to enhance fracture toughness without sacrificing strength.

Processes.   There are three RSP techniques considered to be viable processes: atomization, melt-spun ribbon, and melt-spun flake. Melt spinning is considered the most promising approach because sufficiently high solidification rates are achievable for the flake thickness produced (~40–60 μm). Flakes (~6–8 mm in length and 0.5–0.7 mm in width) having a low surface area-to-volume ratio have been produced by this method. These flakes are suitable for direct consolidation into a composite product without further size reduction.

Cost reduction is also a potential advantage of the melt-spun flake process over conventional P/M methods. For example, the powder blending required with the latter is eliminated in melt spinning, and conventional metalworking operations, such as extrusion, can be used for final consolidation. The time-consuming, equipment-intensive procedures required with P/M processing, such as vacuum hot pressing, also are eliminated. The low-yield atomization process used to produce high-quality metal powders could thus be replaced by a technique employing prealloyed MMC material at a substantial saving in raw material costs.

Two critical parameters of melt spinning are quenching-wheel speed and the amount of molten material impinging on the wheel. Wheel speed, determined by the adherence of molten metal to the Cu wheel, influences the solidification rate.

The amount of material impinging on the wheel is determined by the orifice diameter and the pressure applied to the crucible within the melting chamber. The optimum chamber pressure is that required to supply a uniform stream of metal to the wheel. For example, with a SiC/Mg composite melt, optimum pressure is approximately 27 kPa using an orifice diameter of 1.3 mm. With these parameters, the SiC ceramic particles are uniformly distributed within the quenched flakes.

These flakes have been easily consolidated into billets, and encapsulated in a metal can, and directly extruded to form a fully dense composite with a homogeneous microstructure. Processing temperatures are sufficiently low to prevent deterioration of the matrix alloy properties. Forging, rolling, and other metalworking operations also can be used.

The melt-spun ribbon process results in a product that requires further processing by attrition to reduce its size; this can potentially result in cross-contamination and further oxidation.

The atomization process can achieve high solidification rates if the atomized particles are very small. Unfortunately, the large surface area accompanying small particles results in a greater fraction of surface oxides, which are detrimental to the ductility and fracture toughness of the consolidated material.

Process Comparisons.   A comparison of a conventional Mg alloy with SiC-reinforced (nominally 15 vol%) Mg-6Zn alloy MMCs made by IM and rapid solidification techniques shows that MMC materials have significantly higher YS and UTS values. The RSP material also has a much higher tensile elongation than that of either IM or P/M material, which indicates improved microstructural homogeneity. Elastic moduli are approximately the same for all materials, which is not unexpected because the modulus is controlled by particle V_f as opposed to processing conditions.

The benefits of RS are demonstrated further by the property improvements of SiC-reinforced 5090 Mg alloy. Rapidly solidified 10 vol% SiC-reinforced 5090 has a YS nearly 20% greater than that of the rapidly solidified Mg-6Zn alloy with little loss of ductility. The most dramatic property improvement for an A356 Al alloy/SiC/15$_p$ composite is the increase in tensile elongation. A modest increase in strength is also observed, even though A356 is a casting alloy and is not designed to benefit from RSP.

1.5.6.6 Specialized processes[418,419]

Vapor Deposition.   PVD offers a route for the production of new alloys and microstructures, with advantages such as extremely fine grain sizes, extended solid solubilities, and freedom from segregation. Recent developments in high-rate electron beam evaporation have shown that a wide range of new alloys impossible to produce by conventional IM can be processed by bulk PVD.[418]

Weimer[419] reported on unique facilities for producing continuous, ultrathin Gr/Al composite precursor tapes by the vapor deposition techniques that have recently been developed to fabricate thin-gauge MMC structures. Additional processes were quickly developed for other fiber-reinforced composites, including Gr/Mg, Gr/Cu, and SiC/Ti, as well as for fiber-matrix diffusion barriers.

Vapor deposition methods (VDMs) have the potential to offer a virtually unlimited selection of matrix compositions and fibers, as well as simple incorporation of compliant layers, diffusion barriers, oxidation inhibitors, and bonding agents. In the VDM process for composite materials, the precursor material (prepreg) is a flexible CMC or MMC tape or tow in which each fiber has become a filamentary microcomposite carrying precisely the amount of matrix material required for consolidation to the final V_f desired in the structural component. The extraordinary versatility of this technology comes about because the precursor materials are produced in the form of continuous ultrathin sheets, tapes, or tows that are pliant and lend themselves easily to angle-plying, braiding, weaving, and winding.

The process for manufacturing VDM MMC precursor tapes utilizes fibers drawn from a creel in the form of a tow typically containing several thousand fibers and spread into approximately a monolayer tape. Fiber surfaces are plasma-cleaned by passage through an argon glow discharge. The tape is then coated with a metal matrix by magnetron sputtering, after which it is taken up on a reel and made available for subsequent processing into consolidated composite structures. The system may be configured to deposit as many as six different sequential coatings on the fibers in a single pass.

Weimer claims that by precisely controlling the deposition process, his company has been able to make unidirectional, or quasi-thotropic, and quasi-isotropic laminates with single plies only 0.01–0.0008 mm. Ultrathin Cu matrix composites reinforced with pitch-based fibers have been produced for thermal management applications. By angle-plying the sheets, one can tailor CTEs as required. Recently, Weimer's firm, Cordec, produced a Gr/Cu radiator for attachment to a Ti heat pipe and has also made a Ti/6Al/4V composite reinforced with Nicalon SiC_f. A 14-ply sheet of the composite with 40 vol% fibers is only 0.06 mm thick.

Laser Beam.   Okumura and his associates[420,421] reported that their MMC plates fabricated with a laser beam produced sound thin plates with no damage to the fibers from heating. The mechanical properties of the MMC plates showed degradation according to fiber selection, Figure 1.101.

The materials used were Al-based composites reinforced with Nicalon SiC_f or Torayca C_f, which were preformed into wires by a liquid filtration technique. The wire was 0.5–0.6 mm in diameter and consisted of 500–3000 fibers and commercial pure Al. To make the prefomed sheets, preformed wires were arranged side by side and fixed on a base plate; then Al powder was plasma-sprayed on them in air so that the resulting fiber V_f would be approximately 26–46%.[421]

The conclusions from this work were as follows.

1. The deformability of SiC/Al preformed wire was very large, and strength was maintained even when the ratio of deformation was 0.6. However, when the SiC/Al preformed wires were deformed with high-pressure rolling, strength reduction owing to fiber breakage was observed.

2. The elasticity of SiC/Al composite sheets fabricated with a laser beam achieved a value of 90% or higher than that calculated by a simple ROM. However, the strength of the composite sheet was about 0.6 GPa and was lower than the calculated value.

3. When the C/Al preformed wires were heated with a laser beam at lower than 584°C (600 J in input energy), no thermal damage was observed in the wires.

4. The deformability of C/Al preformed wire was very small, and strength reduction in the wires was observed at small deformations.

5. C/Al composite sheets fabricated by the laser roll bonding process had poor mechanical properties owing to fiber breakage under high-stress rolling.

6. In the case of SiC/Al it could be considered that the laser roll bonding process is a useful technique for the fabrication of MMCs. However, fabrication of MMC using C/Al preformed wires has to be performed under conditions where the temperature is as high as possible and the rolling pressure is as low as possible.

Baker et al.[422] successfully undertook the novel use of laser beams in the design of surface in situ metal-ceramic composite formation. These researchers created wear-resistant surface MMCs in both Al alloys and Ti via the incorporation of preplaced SiC_p by using a 5-kW CO_2 laser.

Thixotropic Metal Molding.[423]   A new process that combines ceramic-reinforced alloys and thixotropic metal molding is being applied to Mg MMCs.

This technology, which resembles thermoplastic injection molding, doesn't allow molten Mg to contact the atmosphere and thus prevents corrosion and is a safe process.

In the new process, Mg pellets are forced through a screw-type extrusion barrel. The screw's shearing action raises the metal temperature to 580°C, about 30–40°C below its melting point. The metal enters the mold as a transitional slurry that fills the cavity with virtually no entrapped gases or inclusions. Temperature, pressure, and screw speed are much higher for Mg than for plastics, but the injection molding principle is the same.

Other advantages of the thixotropic process include low thermal stress, which lengthens tool life, cuts shrinkage, and improves tolerances in workpieces; high production rates; less scrap; and a pleasanter working environment for machine operators. Composites undergoing evaluation include AZ91 Mg with 22 wt% SiC and 27 wt% Al_2O_3. One automotive manufacturer indicates that it has made fiber-reinforced Mg alloy pistons—90% Mg and 10% Al_2O_3. It also claims that the part has five times the abrasion resistance of an Al piston and is about 40% lighter.

Mechanical Alloying.   Asabe, Nakanishi, and Tanoue[424] described the manufacture of a MMC by MA. Their research covered oxide dispersion-strengthened (ODS) alloys which have greatly enhanced high-temperature strength as a result of dispersion of stable thermal ultrafine oxide particles. In the MA process they investigated, ultrafine oxide particles were dispersed forcefully from a powder composed of metal powder and ultrafine oxide powder put into a closed tank with steel balls and stirred forcefully in the solid state using a high-energy ball mill.

They prepared a Y_2O_3 ODS 11 Cr ferritic steel by MA and evaluated the effects of Mo, W, Ti, Nb, and V additions on high-temperature strength. Their results showed the following.

1. The addition of Mo and W greatly enhanced the high-temperature strength as a result of solution strengthening.

2. The addition of Ti greatly enhanced the high-temperature strength owing to the formation of TiO_2-Y_2O_3 complex oxides and refinement of these particles in the MA process.

3. The addition of Nb and V showed little enhancement of the high-temperature strength because of the weak refinement of the oxide particles.

4. The combination of solution strengthening by addition of Mo and W and dispersion strengthening (due to much finer dispersed particles) by addition of Ti enabled them to obtain ferritic stainless steels with a creep rupture strength of about 400 MPa for 1000 h at 750°C, which surpassed that of austenitic stainless steels.

Dynamic Compaction.[425–429]   In explosive compaction, pressures up to 1000 GPa (which last for several microseconds) can be generated by the energy released from the explosive charge. Explosive compaction of powders involves particle deformation at very high speeds. Deformed particles are mechanically locked and welded together, producing compacts of high density. It has been suggested that the interparticle bonding in dynamically compacted materials is a result of localized energy deposition and melting at the particle interfaces which occur as the particles undergo shear and deformation during densification. Intense localized plastic deformation near particle boundaries and interparticle friction convert most of the shock energy into heat. This route has the potential to be an economical method of producing composites from powder mixtures. It also minimizes deleterious interactions between the reinforcement and the matrix alloy.

Using the above-mentioned technique, Sivakumar[425] and his associates examined the compaction of an Al and 20 vol% SiC_p mixture. They found that the compacts produced by explosive compaction were comparable to conventionally produced compacts in terms of properties.

Murakashi and his associates[426,428] used dynamic compaction to fabricate Al-Li powders reinforced by SiC_w. They found the method to be one of the most attractive techniques because it can eliminate exposure to high temperatures for a long time. In this procedure recovery of compacted specimens and control of flyer velocity were important variables in controlling their work which is still in progress.

Tong and his collaborators[427] processed SiC_p-reinforced TMCs by shock wave consolidation. These materials are difficult to produce using traditional methods because of the high reactivity of Ti with most reinforcement materials. However, with shock consolidation, fully dense composite compacts with 100% theoretical density that are free from interfacial reactions and macroscopic cracks have been obtained.

Elimination of prior Ti particle-particle boundary effects in as-consolidated compacts by proper postconsolidation annealing results in a significant increase in the tensile strength and ductility of both Ti and its composite compacts.

Properly annealed 10% V_f SiC_p-reinforced Ti composites show improved tensile strength over that of unreinforced Ti matrix while their tensile ductility remains excellent (>5%).

Choi, Mullins et al.[429] processed fine fibrous TiC by the SHS method and employed it to fabricate Al matrix composites. Two consolidation methods were investigated: (1) combustion synthesis of TiC_f/Al composites directly using Ti powders and C_f ignited simultaneously with varying amounts of matrix metal powder, and (2) combustion synthesis of TiC using Ti powders and C_f followed by consolidation into different amounts of metal matrix powder, Al, via HIP. In the first method, when the amount of Al in the matrix was increased, the maximum temperature obtained by the combustion reaction decreased and propagation of the synthesis reactions became difficult to maintain. In the second method, following separation of the individual fibers in the TiC product, dense composites containing the SHS products were obtained by HIP of a mixture of TiC_f and Al powders. From this one can obtain structural MMCs reinforced with TiC_f.

Specialized Processes

Method of Internal Crystallization (MIC).   Fibrous composites are normally fabricated by inserting preformed fibers into a matrix and trying to tailor mechanical or physical properties of the material by a proper choice of fiber arrangement, fiber V_f, structure and properties of the interface, and so on. As a rule, this method satisfies all the needs fairly well, but in many cases, particularly when heat-resistant composites are involved, it leads to complications that cause composite experts to refrain from being involved in technically very attractive projects. So the need for alternative methods of composite fabrication obviously exists. Mileiko and Kazmin[430] described a process based on the fibers growing from the melt within the volume of the matrix. The matrix should have prefabricated continuous cylindrical channels filled with the melt of the fiber material. The technique was described using as a model a composite with a Mo matrix and single-crystal sapphire fibers. It was shown that the productivity of oxide fiber fabrication based on the process described could be some orders of magnitude higher than that based on the well-known Czochralsky and Stepanov methods. The strength of the single-crystal sapphire fibers obtained was studied, as well as the high-temperature creep strength of composites containing such fibers.

The high-temperature creep strength of composites with continuous single-crystal oxide fibers cannot be very high because single crystals are characterized by the presence of slip systems with low slip resistance.

Liquid Hot Pressing.   Rabinovitch, Daux, and Raviart[431] designed equipment and fabricated Al or Mg MMCs reinforced by long C_f or by liquid or semiliquid hot pressing. Semiliquid-phase infiltration led to a multilayered material having alternate unreinforced matrix-rich zones and composite layers. High tensile strengths (mean value 850 MPa) were obtained for M40/AZ61 Mg-based composites fabricated via the liquid route.

Sol-Gel.   Breval and Pantano[432] prepared, by sol-gel methods, Ni-Al_2O_3 ceramic-metal composites containing up to 50% Ni and consisting of Ni inclusions with a continuous polycrystalline Al_2O_3 microstructure. The Ni inclusions were randomly oriented and found to be both inter- and intragranular. The microstructural evolution depends upon the hot pressing temperature.

Rapid Infrared Forming (RIF).   Warrier and Lin[433] described the use of RIF in controlling interfaces in TMCs. Control of the fiber-matrix reaction during composite fabrication is commonly achieved by shortening the processing time, coating the rein-

TABLE 1.14   Tensile Properties of TMCs[433]

Material	Tensile Strength (MPa)[a]	Modulus (GPa)[a]
P-55 carbon fiber	1730	379
SCS-6 fiber	3950	400
Ti80 (as-cast)	1290	118
Ti85 (as-cast)	1600	120
Ti80/SCS-6 (23 vol%)	1734	195
Ti85/P-55 (40 vol%)	1100	200
Ti-6Al-4V	890	110
Ti-6Al-4V/SiC (35 vol%)	1455	225
Ti-6Al-4V/SiC (35 vol%)	1690	186
Ti-15V-3Cr-3Al-3Sn/SCS-6 (38–41 vol%)	1951	213
Ti-2411[b]/SCS-6 (35 vol%)	1183	184
Beta 21-S[c]/SCS-6 (0–90 vol%)	1680	81

[a] Test results from this work.
[b] Ti-2411 has a nominal composition of Ti-24Al-11Nb.
[c] Beta-21S has a nominal composition of Ti-15Mo-2.7Nb-3Al-0.2Si.

forcement with relatively inert materials, or adding alloying elements to retard the reaction. To minimize the processing time, a RIF technique has shown that it is a quick, simple, low-cost process to fabricate Ti alloy matrix composites reinforced with either SiC_f or C_f. As a result of short processing times, typically on the order of 1–2 min in an inert atmosphere for composites with up to eight-ply reinforcements, the interfacial reaction is limited and well controlled. Composites fabricated by this technique have mechanical properties (tensile strength and modulus) that are either comparable to or, in several cases, superior to those made with conventional diffusion bonding techniques. The Ti80/SCS-6 composites fabricated by RIF were tested, as well as some unreinforced alloys (Table 1.14), by Warrier and Lin. The composites fabricated by RIF exhibited a substantially higher strength and modulus than their constituent alloys, which were comparable to those of some of the reported SCS-6-reinforced Ti-alloy matrix composites.

The flexural strengths and moduli of Ti85/P-55C/35_f and Ti85/P-55C/40_f were also tested by three-point bending. Ti85/P-55C/40_f composites showed strengths and moduli on the order of 1000–1100 MPa and 200–220 GPa, respectively. The Ti85/P-55C/35_f composites had comparatively lower strengths and moduli, typically on the order of 700 MPa and 160 GPa.

These results show that high-strength composites can be fabricated by the RIF technique. The RIF process produces stronger composites because the interfacial reaction, which often leads to fiber degradation, is limited.

Electron Beam Evaporation.   Ward-Close and Partridge[434] have developed a fiber-coating process (electron beam deposition) used to produce continuously reinforced advanced MMCs with up to 80% V_f of SiC_f.

These researchers concluded the following.

1. Electron beam evaporation was used to coat ceramic fibers with matrix alloy. During hot consolidation, the coating deformed and bonded to form the matrix of the composite.

2. No chemical reaction was observed to occur between SiC_f and Ti alloy during coating; alloy coatings up to 65 μm in thickness adhered well to SiC_f, and coated fibers were handled and bent without damage.

3. Fibers coated with Ti-Al-V α /β Ti alloy, intermetallic compound Ti_3Al, or RSP Al alloy Al-4.3Cr-0.3Fe were consolidated into fully dense composites with no evidence of matrix cracking. MMCs produced from fibers coated with dispersion-strengthened alloy Ti-7Al-3V-2Y or TiAl showed some matrix cracking.

4. It was found that after consolidation, compared with other methods, the matrix-coated fiber process produced MMCs with a good fiber distribution and no fiber-fiber contact.

5. The Young's modulus of a Ti alloy MMC with 21% SiC was found to be near the theoretical prediction, but tensile strength was significantly less than predicted. Loss of strength was attributed to the formation of a brittle reaction layer between the fiber and the matrix during hot consolidation.

1.6 CERAMIC MATRIX COMPOSITE PROCESSING[435,436]

1.6.1 Introduction

Ceramic matrix composites, currently 35% of the total market, will grow even faster at an annual rate of 14.5% to $400 million in the year 2000. Quality reinforcements are still not available at competitive prices, and composite components are more expensive to produce in the United States because of high labor costs and environmental regulations. However, reductions in cost are being pursued by developing and improving composite and fiber processes.

Ceramic composites are attracting increasing attention because of the broader diversity of, and especially improvement in, the properties they can frequently provide. Properties, and hence the benefits of composites, typically depend upon achieving a specific range of microstructures. Most modern ceramic composites of course go well beyond traditional ceramics in composition, microstructure, and properties.

The choice of processing of CMCs is based upon the character of the composite, its size and shape, and its cost. Design or tailoring of composite microstructures to achieve improved or new properties presents processing challenges. A combination of the fineness of most dispersed phases and the high V_f often desired commonly make achieving a high density and degree of homogeneity a challenge. These challenges of ceramic composite processing have resulted in a significant shift and development in the emphasis of processing technologies in comparison to the processes used with other ceramic materials.

The ensuing sections will cover the processing of both ceramic particulate and fiber composites, the latter including the use of continuous and short fibers, which in turn includes both chopped fibers and whiskers (Table 1.15).

1.6.2 Composite Processing

Processing of ceramic particulate composites typically has overall similarity, and frequently detailed similarity, to more traditional ceramic processing. The most generic difference and greatest challenge is obtaining homogeneous mixing of the dispersed phase, and obtaining good densification can also be a problem.

More significant differences occur in processing continuous-fiber composites than in normal monolithic ceramic processing; that is, fiber orientation and architecture, and achieving desired size and body shape of the reinforcement in the composite matrix. While the great majority of processing studies on ceramic fiber composites to date have been on unidirectional fiber composites, in practice these have limited uses. Some work on different fiber architectures has been done, but there needs to be a great deal more since this will be required for most applications, as indicated by the use of other composites, for example, PMCs and CCCs. Using tapes of uniaxially aligned fibers or various cloth weaves and laminating them with controlled orientation between the layers are important. The type of weave or the type of fiber within each layer may be varied to balance different processing and property opportunities. Felted mats of chopped or short fibers may also be used. There are also expected to be specialized applications of multidimensional weaves, braids, and so on (Table 1.16).[437]

An important factor in achieving any of the above fiber architectures with fine fibers is the use of fiber tows. Handling of fiber tows thus becomes a major factor in fiber composite processing. The most common approach in introducing the matrix is to put fiber tows (as well as larger individual filaments) through a bath that is the source of the matrix or its precursor and then lay this up directly by filament winding or another related technique. Most commonly, the bath is a slurry or slip of matrix particles, but it may also be, for example, a sol or a preceramic polymer. A challenge is to make the distribution of matrix precursor between tows similar to that within the tows. Cloth may also be infiltrated and then laminated but may often present greater challenges in controlling the quantity and uniformity of matrix (or precursor) infiltration.

1.6.3 Fiber-Matrix Interface[438]

The key to successful processing of CMCs has been shown, experimentally and theoretically, to be control of the fiber-matrix interfacial region.[438] Similar statements can be made about MMCs and CCCs, though some of the particular interface requirements are somewhat different. For high toughness in fiber-reinforced CMCs it is essential to produce and maintain a desirable (typically low) level of interfacial shear strength to permit fiber debonding during the fracture process. Similar requirements are present in brittle matrix fiber-reinforced polymers such as Gr/Ep, while ductile MMCs such as Al with B, Gr, or SiC_f require strong fiber-matrix bonding to utilize fully the high matrix toughness. Composites such as CCCs, in which there is intrinsically minimal fiber-matrix bonding, exhibit very high toughness but also have extremely low interlaminar and transverse tensile and shear strengths. In several high-toughness CMC systems[439] the desirably low

TABLE 1.15. Some Processes for Continuous Fiber-Reinforced CMCs

Processing Method	Advantages	Disadvantages	System		
			Fiber	Matrix	Temperature Range [1]
I. SLURRY INFILTRATION					
a) Glass ceramic matrix	Commercially developed. Good mechanical properties.	Limited max. temperature due to matrix. Needs to be hot pressed, expensive. Formation of complex shapes is difficult.	Graphite Nicalon	Glass-ceramic Glass-ceramic	800–1000°C 800–1000°C
b) Ceramic matrix					
1. Sintered matrix	Potentially inexpensive. Could produce complex shapes.	Shrinkage during sintering cracks matrix. (No promising results to date.) Temperature limit due to glassy phase.		Alumina SiC Si_3N_4	800–1400°C 800–1600°C 800–1500°C
2. "Cement bonded matrix"	Inexpensive. Ability to produce large complex shapes. Low temperature processing	Relatively poor properties to date.	Graphite Nicalon "New" fibers	cements	800–1400°C
3. Reaction bonded matrix	Good mechanical properties. Pressureless densification.	Has required hotpressing of Si powder in silicon nitride system prior to reaction bonding. Simple shapes only (to date)	Nicalon "New" fibers	Si_3N_4 SiC	800–1500°C 800–1600 °C
II. SOL-GEL & POLYMER PROCESSING	Good matrix composition control. Easy to infiltrate fibers. Lower densification temperature.	Low yields. Very large shrinkage. Would require multiple infiltration/densification steps. No promising results reported.	Nicalon	Nonoxides Alumina Silicates	800–1200 °C 800–1400°C
III. MELT INFILTRATION					
a) Ceramic melt	Potentially inexpensive. Should be easy to infiltrate fibers. Lower shrinkage on solidification.	Although attractive in principle, no good ideas demonstrated. High melting temperatures would damage fibers.	Graphite Nicalon "New" fibers	Alumina Oxides	800–1100°C 800–1100°C

TABLE 1.15. Some Processes for Continuous Fiber-Reinforced CMCs *(continued)*

Processing Method	System			Advantages	Disadvantages
	Fiber	Matrix	Temperature Range [1]		
b) Metal melt, followed by Oxidation	Graphite Nicalon "New" fibers	Alumina B_4C SiC	800–1200°C 800–1200 °C 800–1200°C	Potentially inexpensive. Cermet type material.	Difficult to control chemistry and produce all ceramic system. Difficult to envision in use for large, complex parts for aerospace applications.
IV. CHEMICAL VAPOR INFILTRATION a) General approach	Nicalon Nextels	SiC HfC Nitrides Oxides Borides	800–1600°C 800–1800°C	Has been commercially developed. Best mechanical properties. Considerable flexibility in fibers and matrices. High quality matrix, very pure. Little fiber damage. In situ fiber surface treatment. Ability to fill small pores.	Slow and expensive. Requires iterative process. Never achieve fill density. Capital intensive.
V. Lanxide	Graphite Nicalon	Alumina AlN TiN ZrN	800–1200°C 800–1200°C 800–1200°C 800–1200°C	Ability to produce complex shapes. Properties dominated by ceramic. Very pure grain boundaries. Systems include: AlN/Al TiN/Ti ZrN/Zr Flexible Properties. Toughness to 20 MPa/m for fiber reinforced system.	Slow reaction and growth kinetics. Long processing time & high temp. limits chemistry. Wetting and reaction are limitations, e.g. B_4C.

[1] Temperature limit depends on fiber. Currently all systems are limited to ≈ 1200°C with available fibers.

TABLE 1.16.   Continuous-Fiber Ceramic Composite Techniques and Forms[437]

Forming Approach	Uniaxial Shapes with Fiber as Tow or Tape	Planar Shapes with Fiber as Mat or Cloth	Three-dimensional Shapes with Fiber as Three-Dimensional Object
Matrix precursor supported on fiber	Ceramic powder coating, align, consolidate Organic precursor coating, align, consolidate	Ceramic powder coating, layup, consolidate Organic precursor coating, layup, consolidate	Organic precursor coating, pultrude, consolidate
Fiber suspended in matrix or matrix precursor		Hot pressing Isopressing (pressureless sinter) Cold pressing (pressureless sinter) Reaction bonding Polymer pyrolysis	Hot pressing Isopressing (pressureless sinter) Cold pressing (pressureless sinter) Hot isostatic pressing Reaction bonding Polymer pyrolysis
Fiber infiltrated by matrix precursor	Align fiber, wet with sol-gel, dry, fire Direct melt oxidation to form matrix	Layup cloth, wet with sol-gel, dry, fire Layup cloth, gel-cast, dry, fire Layup cloth, chemical vapor infiltrate Direct melt oxidation to form matrix	Preform fiber shape, wet with sol-gel, dry, fire Preform fiber shape, gel-cast, dry, fire Preform fiber shape (thin), chemical vapor infiltrate Preform fiber shape, direct melt oxidation to form matrix

value of interfacial shear strength has been determined to be on the order of 2–3 MPa. Obtaining and maintaining high strength also requires preservation of fiber properties through control of interdiffusion, reactions, recrystallization (of the fiber), and so on. The interface (or interphases) must serve these various functions, that is, control interfacial strength and prevent fiber-matrix reactions not only during processing and fabrication but also during service at high temperatures and in aggressive environments. A number of materials have been used with success for interface control insofar as processing and ambient temperature properties are concerned. These include amorphous C and NbC layers formed in situ on Nicalon$_f$ in Compglas; CVD BN coatings on Nicalon$_f$ in SiO_2, ZrO_2, $ZrSiO_4$, and $ZrTiO_4$ matrices; amorphous C coatings on Si_3N_4 continuous fiber (HPZ) and Nicalon$_f$ in aluminosilicate matrices; and BN/SiC bilayer CVD coatings on Nicalon$_f$ in $ZrSiO_4$ and $ZrTiO_4$ matrices.

1.6.4 Processing Methods[435,439,440]

Processing methods can be broken down into two broad groups: powder consolidation and chemically based methods (Table 1.17). However, within each of these are limitations, as well as several variations, especially in the chemically based methods that further complicate the choices.

1.6.4.1 Powder-based methods.[435,439–441]   Powder consolidation methods, primarily by hot pressing, have been most extensively used to date for both glass-based and crystalline matrices. The most common way of introducing such powders is to draw fibers (or fiber tows) through a slurry (or slip) or a sol (Figure 1.102). Infiltration of a

TABLE 1.17. Main Processing Routes for CMCs

Processing Route	Matrices
Chemical vapor infiltration	Carbides, nitride carbon, oxides, borides
Viscous phase hot pressing (2D preforms)	Glasses, ceramic-glasses
Sol-gel route (2D, 3D preforms)	Oxides
Polymer precursor route (3D preforms)	SiC, Si_xN_y $Si_xC_yN_z$
Liquid metal infiltration	$Si \rightarrow SiC$
Gas-metal reaction	Oxide (Al, nitrides (Al, Zn, Ti)
Solid-state hot pressing	SiC, Si_3N_4
Prepreg curing and pyrolysis	SiC, Si_3N_4
Hot pressing (2D preforms)	Oxides

fiber preform may also be used, but inhomogeneous distribution of the powder can be a problem since the fiber preform may act as a filter, especially in thicker cross sections. Hot pressing is a logical choice for the processing of composites using such a powder-infiltrated fiber system after suitable drying, calcining, and so on. Hot pressing allows achievement of low to zero porosity levels, is applicable in principal to all ceramic materials, and draws heavily on powder technology, which is the predominant basis of conventional ceramic processing.

The limitations of hot pressing ceramic fiber composites are twofold. First, there are the basic limitations of hot pressing itself, namely, that it is mainly applicable to rather simple shapes, for example, plates, blocks, and cylinders, and is not a particularly low-cost process. The other basic limitation stems from the temperatures commonly required to achieve low levels of porosity. Such temperatures, which commonly must be 100–200°C higher than for hot pressing the matrix alone, present limitations with regard to both reaction between the fibers and the matrices and degradation of the fibers, particularly in view of the limited fiber-temperature capabilities. A potentially important way to reduce processing temperatures, as well as to provide much greater versatility in shape, is

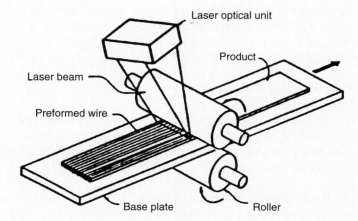

Figure 1.101 Schematic view of an apparatus for the laser process for MMC fabrication.[420]

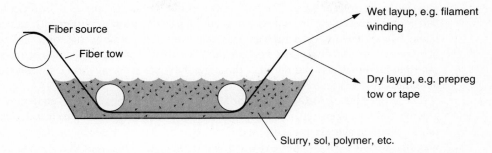

"SLURRY' PROCESSING

'Slurry' -liquid, powder, binder, additives

Figure 1.102   Schematic of the fluid method of coating individual fibers or filaments, or infiltration of tows via a slip, sol, or other fluid source of matrix.[440]

to use HIP with such composites. However, this typically requires a simple, effective canning or sealing method to be practical.

Hot pressing produces the highest densities, that is, the lowest (e.g., near zero) porosity of any of the processing methods. However, substantial densification is obtained at the expense of higher processing temperatures, for example, 1400–1600°C, which seriously limits the type of fiber used as well as the various fiber-matrix combinations. In addition to the temperature issue, other limitations are cost, size, and shape capability. While hot pressing has been applied to date predominantly to unidirectional fiber composites, it should be applicable to 2D composites, for example, cloth laminates, assuming that adequate infiltration of the matrix between the fibers within the cloth can be achieved. Its applicability to three- and higher-dimensional composites will be more limited because of possible fiber damage via buckling and from interference with densification caused by the fibers in the axial direction. Reducing hot pressing temperatures by obtaining higher initial matrix densities and finer matrix particulates and the use of densification aids will all be useful.

Most Al_2O_3/ZrO_2 composites are hot-pressed, however, stresses believed to result from gradients in reduction may have been the cause of delamination and cracking in large zirconia-toughened alumina (ZTA) plates and thus may limit scaling up of hot pressing to large bodies.[435]

While some of the other ZrO_2-toughened composites using oxide matrices show good sinterability, for example, β-Al_2O_3, many of them have been hot-pressed, often in connection with reaction processing. Fabrication with a nonoxide matrix, for example, Si_3N_4 or SiC,[435] again requires hot pressing, for example, at 2000°C for SiC.[435] The high temperatures and reducing conditions of such pressing can cause reactions with the ZrO_2, such as the formation of oxynitride with Si_3N_4, and resultant oxidation problems. However, some systems, such as TiB_2/ZrO_2, show good promise despite the requirements of hot pressing at 2000°C for 3 h.

Hot pressing has been used even more extensively for composites containing nonoxide dispersants. Thus while there has been some reasonably successful sintering of Al_2O_3/TiC composites, most production has been by hot pressing, commonly around 1600°C.[435] Clearly, hot pressing is most practical for simple, flat shapes such as cutting tools and certain wear parts. The practicality of hot pressing is also greater for composites that require processing in nonoxidizing environments. Such environments reduce or remove two of the important advantages that sintering normally has over hot pressing, namely, the use of lower-cost heating facilities for the lower temperatures required for most air-fired material and the high throughput available for air firing, for example, tunnel kilns.

Composites of Al_2O_3 (or BeO or MgO) with a wide range of SiC content have been commercially hot-pressed for use as controlled microwave absorbers in traveling wave tubes. These materials typically show strengthening and toughening due to SiC, with maxima at ~50% SiC. Other oxide-carbide composites have received limited study, for example, hot pressing of Al_2O_3/30 vol% WC to give 97–98% of theoretical density at 1600–1700°C. Some fabrication of Al_2O_3 + diamond composites has also been undertaken by high-pressure (6-GPa) hot pressing of 1-μm powders at 1300°C for 1 h.[435]

Except for some limited sintering, Al_2O_3/BN particulate composites have typically been made by hot pressing. Earlier composites used large (e.g., 100-μm-diameter) BN flakes similar to the large graphite flakes employed in refractory composites. However, more recent developments have concentrated on the use of much finer (2–5-mm-diameter) BN flakes. A SiO_2/BN hot-pressed composite has also been developed and sold commercially. It offers good thermal shock resistance along with a low dielectric constant and reasonable conductivity. Considerable development has been carried out, first on Al_2O_3/BN and then on mullite/BN. Strengths as high as 350 MPa have been achieved with good thermal shock resistance for Al_2O_3/BN. There has also been some promising reaction processing of mullite-BN.[435]

While hot pressing of such oxide matrix composites produces greater densification, it is limited in the densities achievable at higher levels of nonoxide additions despite the general ease with which oxides may be hot-pressed.[435]

A wider variety of all nonoxide composites has been made by hot pressing than by sintering. While some of these composites have increased strength over that of the matrix itself, some have lower strength, which often reflects, at least in part, the nature of the additive. Thus in a few cases strength decreased with larger additive particle size, for example, for 9- and 32-μm SiC_p added to Si_3N_4. Also, the use of weaker second-phase particles, such as Gr or BN (hexagonal), typically decreases strength but provides a large increase in thermal shock resistance. Some composite strengths, whether higher or lower than that of the matrix alone, are also probably limited by nonoptimizing processing. Thus, for example, in a more detailed study of SiC/TiC composite processing, Janney[442] showed that both the type of carbon addition (carbon black or phenolic resin) and the extent of ball milling could significantly affect mechanical properties.

Two other methods of processing ceramic fiber composites using powder or sol-derived matrices are HIP and sintering. HIP clearly offers the potential of achieving high densities at somewhat lower temperatures than are required for hot pressing. It is applicable to a much broader range of shapes than is hot pressing, though potentially more limited in ultimate size than is hot pressing. Until some more practical method of canning is developed, composite HIP applications will probably be quite limited. The method will also experience the same, or possibly more, severe limitations in terms of multidimensional fiber architectures than does hot pressing.

Sintering has a lower cost and, possibly, greater shape and size versatility than HIP or hot pressing. However, several important qualifications and uncertainties must be noted. First, since the cost of a green ceramic fiber composite compact is typically much higher than that of a conventional green ceramic compact because of higher fiber costs and higher composite body formation costs, the cost differential between sintering and hot pressing of ceramic fiber composites may be substantially less than that between sintering and hot pressing of conventional ceramics. Second, sintering faces inherent problems in densifying ceramic fiber composites, as well as intrinsic limitations, for example, as a result of temperature limits based on fiber-matrix interactions and fiber-temperature limitations. Such densification limitations are likely to be least for chopped-fiber and whisker composites and become progressively greater for continuous unidirectional, 2D, and multidimensional fiber architectures. Encouraging results have been achieved in the sintering of some ceramic fiber, as well as ceramic whisker composites. However, it is expected that intrinsic densification problems are likely to present serious challenges in terms of both the variability and reproducibility of sintered composites and that these problems are likely to become more serious with increasing composite size and shape complexity.

Several types of particulate composites have also been hot isostatically pressed, including partially stabilized zirconia (PSZ), ZTA, and Al_2O_3/TiC. While most of these have been hot isostatically pressed after sintering, some have undergone HIP after hot pressing. Typically, HIP results in somewhat higher and more uniform strengths. HIP is generally carried out under nonoxidizing conditions; nevertheless conditions are usually less reducing than for hot pressing and hence there is less concern regarding the use of phases that are subject to reduction, such as ZrO_2. Sintering (or hot pressing) to closed porosity allows HIP without any canning and therefore is much more practical. This allows processing on an industrial scale as shown by its wide use for WC-based cutting tools. Such HIP of bodies sintered to closed porosity greatly reduces isolated pores and pore clusters which are often a major factor in lower-strength failures. Fabrication of large numbers of small components in a single run is the most economical method of using HIP.

Some study of particulate composites having two dispersed phases has been made, again mainly by hot pressing or by sintering and HIP. The two main methods are (1) substitution of a second hard material (e.g., TiN, SiC, or B_4C) for some of the TiC in Al_2O_3/TiC, and (2) substitution of ZrO_2 in some Al_2O_3-carbide or nitride composites or addition of carbide or nitride materials to oxide-ZrO_2 composites.

1.6.4.2 Chemical-based methods.[435,439,440]   A variety of reactions can be used to produce composites under reaction processing. Many reactions are used in preparing ceramics, ranging from powder preparation (e.g., decomposition of salts, gels, polymers, and other organometallics) to direct production of a composite (i.e., CVD or polymer pyrolysis as discussed later). However, what is of interest here are reactions that produce composite powders for consolidation into composites or, more commonly, reactions involving a powder compact that yields a composite product in conjunction with heating, possibly with pressure, such as in hot pressing or HIP. The following are some examples.

- Autocatalytic reaction processes (Al_2O_3/B_4C)
- Displacement reactions (Al_2O_3/TiN)
- Eutectic melt crystallization (Al_2O_3/ZrO_2)
- Reactions of organometallic compounds (SiC/SiO_2)
- Sol-gel and polymer reaction techniques (Si_3N_4/SiC)
- Melting-phase infiltration techniques (Si/SiC)
- Direct melt oxidation (Al_2O_3/Al)
- Gas-phase infiltration/deposition (BN, SiC/SiC).[443]

Clearly, there are various routes to reaction technology, with reaction sintering (or bonding) of Si_3N_4 or SiC from SiC + C compacts, respectively, being a major component. Processing of composites such as Si_3N_4-bonded SiC and SiC-bonded B_4C made by infiltrating compacts of SiC + Si and B_4C + C, respectively, with N_2 or NH_3 gas or molten Si at a high temperature are important examples of the industrial application of this technology for refractory, wear, and armor use. More recently, reaction-formed Si_3N_4 or SiC matrices have been investigated for fiber and whisker composites. While these entail SiC at high temperatures where it is very reactive, and hence may also attack fibers or whiskers, it has the advantage of sintering a matrix without shrinkage and the attendant limitations.

The potential for reaction procesing may go well beyond use of lower-cost raw materials and possible mixing advantages. Sintering (or hot pressing or HIP) reactant compacts and then reacting them may provide opportunities for controlling composite microstructures by controlling nucleation and growth of the new phases formed by the reaction, that is, providing some, possibly all, of the opportunities for microstructural control offered by crystallization of glasses.

Another class of reactions that have attracted considerable recent interest are those often described by terms such as self-propagating high-temperature synthesis, which is referred to by such acronyms as SHS, SPHTS, SHTS, and SPS. The substantial exothermic character of these reactions results in a high-temperature reaction front that actually sweeps through a compact of the reactants once the reaction is ignited at some point. While such reactions can be used to produce single-phase compounds, for example, Ti + C to yield TiC, and Ti + B yield to TiB_2, they can also be used to produce composites directly. The classical thermite reaction $Fe_2O_3 + 2Al \rightarrow 2Fe + Al_2O_3$ is an example of a metal ceramic composite product. However, there are a variety of reactions that can yield all ceramic composite products, for example,

$$10Al + 3TiO_2 + 3B_2O_3 \rightarrow 5Al_2O_3 + 3TiB_2$$

$$4Al + 3TiO_2 + 3C \rightarrow 2Al_2O_3 + 3TiC$$

Another interesting possibility suggested by SHS reactions is that some of them appear to produce whiskers in situ. Besides the possibility of directly producing a product body, such reactions may be used to prepare powders. This in turn introduces another microstructural issue and opportunity in processing, that is, the size of the composite particles and how this affects densification and resultant composite performance.

Another type of reaction-processed composite is that under development by the Lanxide Corporation. This involves the "growth" of Al_2O_3 by the reaction of molten Al at the interface between the Al_2O_3 product (fed by capillary action along microscopic channels in the Al_2O_3 product) and a gaseous source of O_2 (e.g., air). Since this is a growth process, it can grow around particulates or fibers to form composites. The extent to which such composites can be varied in composition and microstructure is not fully known but appears to be quite broad. The process can also apparently be applied to other matrix materials such as AlN, TiN, and ZrN.

Chemical vapor deposition, or more appropriately chemical vapor infiltration, is potentially one of the most important chemical methods of generating ceramic matrices for ceramic fiber composites, directly analogous to its application in the case of some CCCs. It also has potential for substantially wider usage in CMCs because of its potential for producing some of the largest ceramic fiber composites achievable by any processing method. It clearly can provide a wide variety of shapes and can in principle be utilized with any fiber architecture. It has already been used in producing the largest pieces of ceramic fiber composites made to date, for example, a tube consisting of one layer of braided oxide fibers infiltrated with SiC by the CVI process for flame tubes for refining Al metal. Other examples of products made by CVI are CCC aircraft brake disks, SiC/Al_2O_3 heat exchanger tube sections, a SiC/Al_2O_3 thermocouple well, and SiC/C rocket nozzle and exit cones. While CVI has been utilized almost exclusively for SiC-based matrices to date, of the different chemical approaches, it has potential for being applied to the widest range of matrices. The first of two important facts about CVI is that it can frequently be carried out at temperatures in the 1000–1200°C range, and in some cases below 1000°C. Thus it is applicable to a broader range of composites in view of the fiber-temperature limitation noted earlier.

While it is difficult to obtain very low porosities (e.g., <10%) with this method, it can produce reasonably limited porosities in several matrices of interest and has substantial potential for significantly broadening the range of matrices that can be processed in this fashion. The technique is applicable to a wide range of sizes and shapes, although the residual porosity issue becomes more serious as cross-sectional dimensions increase. Also, while different architectures may change the amount of residual porosity, CVI is in principle applicable to essentially any fiber architecture. CVI of composites actually relaxes some of its limitations: for example, it reduces the requirement for homogeneity and dangers of interrupted deposition and can increase deposition rates (because of simultaneous deposition on all fiber surfaces). These factors enhance the already reasonable cost potential of CVD.

The other major chemical method of producing ceramic matrices for ceramic fiber composites is polymer pyrolysis, which is directly analogous to the polymer pyrolysis method used exclusively in the production of CCCs. The two significant advantages of polymer pyrolysis make it a very important competitor of CVI in generating ceramic matrices. The first is that all but the final pyrolysis step is identical to fabrication of a polymeric composite, thus providing essentially the same versatility in terms of handling, fiber architecture, specimen size, and shape. The other major advantage of this method is the low processing temperature for pyrolysis.[435,440]

In principle, polymer pyrolysis consists of using an appropriate polymer containing the atoms of the desired ceramic product in a form that can be pyrolyzed without excessive polymer backbone losses (i.e., ideally the only losses are of H_2) to obtain the desired ceramic product.[439] Several polymers are now known, and additional ones are being developed, that yield basically Si_3N_4 and/or SiC. Polymers are also being developed to yield products that are basically BN or B_4C. (The term *basically* is used because a stoichiometric compound typically cannot be produced by this process.) Such polymers, in addition to or in combination with various carbon-producing polymers, are of interest in producing ceramic composites because of the high temperature and other capabilities of these materials.

Several factors strongly encourage the use of polymer pyrolysis for fiber—as opposed to particulate—composites. These include cost, shrinkage, and low-temperature processing. Thus lower temperatures, as noted above, allow much wider use with fibers, which are commonly much more limited in their temperature capabilities than most particulates of interest. The large density change, for example, typically from about 1 g/cm^3 for the polymer to over 3 g/cm^3 for the theoretical density of resultant products such as SiC and Si_3N_4, means that one has large shrinkages, substantial porosity, and/or microcracking. With a high density of particulates and a small composite body of simple shape, it may be feasible to achieve useful properties. However, in general, it is expected that the limited strengthening and toughening achievable in particulate composites does not produce attractive composites via polymer pyrolysis. On the other hand, the strengthening and especially the toughening available in fiber composites can yield quite useful properties despite the presence of porosity and microcracking from polymer pyrolysis.

Sol-gel processing and resultant gel pyrolysis can be considered similar in some respects to polymer pyrolysis. The sol-gel route is based on the hydrolysis of alkoxides. The gel can be injected into the fibrous preforms and produces the composite after heating and pressing. This method is attractive in that it avoids the problems caused by the dimensions of ceramic particles and product impurities. However, the greater weight losses and associated density changes of gel pyrolysis compared to polymer pyrolysis impose a severe limitation, as may the oxidizing character of some pyrolysis products. Further, pyrolysis of gels frequently results in a more particulate-type matrix as opposed to a coherent matrix. Thus gel-derived matrices fall under powder-based methods because they usually require (and are favorable to) sintering, for example, by hot pressing.

The polymer pyrolysis method is more restricted in terms of the matrices to which it is applicable by the range of polymers that are available. Although it has some limitation in the cross-sectional thickness to which it can be applied, it is in principle applicable to all fiber architectures and is quite versatile in size and shape. An intrisic limitation is the fact that this method cannot yield a totally dense matrix by itself and in fact commonly results in considerable porosity, although this can be reduced by multiple impregnations. A practical limitation is the fact that present polymers are relatively costly and a major need is a wider supply of candidate polymers.

1.6.4.3 Other methods[435]

Melt Processing.   The high melting temperatures of most ceramics pose serious challenges to melt processing of ceramic composites. The frequently very high liquid-to-solid shrinkages of many ceramic materials on solidification (typically 10–25%) can also be an important limitation. Despite this, there are at least three ways in which some selected ceramic composites may be processed using melting techniques, aided by advances in melting technology. Melt processing of composites commonly draws upon the uniform solutions achieved in the melt phase and subsequent solidification of eutectic structures or precipitation in the solid state on or after solidification. One of the most obvious applications of melt processing is in producing powders for the processing of composites such as the most extensively investigated melt-derived powders, Al_2O_3/ZrO_2[436], and PSZ.[435,439,440] The eutectic structure of Al_2O_3/ZrO_2 melts provides a significantly different and much more homogeneous mixing of the two ingredients than simply mixing of Al_2O_3 and ZrO_2 powders via the conventional P/M approach.

The challenge is clearly to obtain melt-derived powder particles sufficiently small for good sinterability but larger than the precipitate or lamellae. Since the latter are typically less than 1 μm for ZrO_2 toughening, this procedure should be feasible, and initial studies support its use. A possible practical aid in obtaining a balance between particle size and phase dimensions is atomization of a molten stream. Splattering molten particles increases cooling rates, giving finer structures than solidification of bulk ingots, and aids in grinding to the desired particle size.

Rice[435] has also proposed that similar processing of PSZ powders, especially those partially stabilized with Y_2O_3, may offer significant advantages in developing precipitation-toughened PSZ for two reasons. First, the achievement of a solid solution by normal processing requires high temperatures, which results in significant grain growth. Thus grain sizes of at least 50–100 μm are typical in precipitation-toughened PSZ. Second, temperatures needed for solid solution are quite high and are difficult to achieve. This is especially true for PSZ systems other than those using CaO or MgO as a stabilizer. Thus the solid solution temperatures for systems utilizing Y_2O_3 that offer greater potential are well beyond the achievement of most processing facilities.

The second method of melt processing composites is to infiltrate compacts consisting of the second phase. This may be the most limited melt approach in that it is most likely to be successful mainly with particular composites. However, infiltration of B_4C + C compacts with molten Si to make B_4C + SiC composites is one practical example.

The third method of melt processing of composites is solidification of the composite from the melt. The extensive studies on fusion-cast refractories, typically using multiphase compositions (with many involving eutectic structures) and some for very refractory nonoxide systems, are a guide to some of the possibilities. The challenge is to keep grain sizes sufficiently fine and control the amount, the size, and especially the location of solidification pores.

A major method of controlling solidification porosity is directional solidification. Such solidification of eutectics results in unidirectional lamella- or rod-"reinforced" composites wherein the lamella or rod diameter is inversely proportional to the solidification rate. Also important are the crystallographic relation of the two phases and the frequency and nature of growth variations. From a practical standpoint, the restriction of such directionally solidified bodies to cylinders has been a basic limitation. However, the use of crystal growth-shaping techniques may offer important opportunities for some practical shaping.

As expected, directional solidification has been explored mostly in oxide systems, with ZrO_2-containing (especially Al_2O_3/ZrO_2) compositions being very common. However, nonoxide systems have also been studied, such as the ZrC/ZrB_2 eutectic.

Another approach to melt processing of composites is preparing a glass and then crystallizing it. Examples of these "glass ceramics" include ZrO_2-containing glass that allows tetragonal ZrO_2 precipitates to be formed, and it is reported[435] that fracture toughness increased by ~100% with up to 12.5% ZrO_2 obtained in a cordierite-based glass body made by melting at 1650°C and then sintering at 860–1100°C. Other reports show room-temperature fracture toughness increases of ~50% in a $Li_2O/SiO_2/ZrO_2$ glass with ~12 vol% fine tetragonal ZrO_2 precipitates. Particularly promising is the directional solidification of a CaO/P_2O_5 glass ($CaO/P_2O_5 = 0.94$) at 20 µm/min at 560°C which resulted in room-temperature strengths of up to 600 MPa and noncatastrophic failure.

Future Cost-Effective Fabrication Methods. Manufacturers are continuing to develop fabrication methods that are economical and can produce NNS parts. One such method is low-pressure (< 2 MPa) injection molding (LPIM). This procedure can achieve a high green density as a result of maximum particle packing, as well as solid loadings ranging from 80 to 87%. It has been successfully used to make composite parts. Preforms of chopped Al_2O_{3f} have been made with a wide range of size and complexity and good aspect ratios. These preforms have been infiltrated with molten metal.

Another NNS process with potential has been developed for $MoSi_2$ and its composites. Conventional processing requires temperatures on the order of 1700°C and therefore restricts the use of a variety of reinforcements. To overcome this limitation, reactive vapor infiltration was used in which a loosely compacted metallic Mo powder was exposed to a gaseous Si precursor. Preliminary results showed that a reaction front moved progressively inward after a surface $MoSi_2$ layer formed. The layer contained less than 0.4 at% and no excess Si or Mo. None of the test samples silicided between 1100 and 1300°C showed any signs of swelling or surface cracks. In the fabrication of composites, the kinetics of silicide layer formation and control of its density will need to be understood.

Another infiltration method is based on vapor-phase sintering of ceramics in a reactive atmosphere. This produces a material with continuous porosity that can be controlled in amount and pore size. Composites of a 3D interconnecting network were made by infiltrating vapor-phase-sintered Al_2O_3 with Cu. Complex shapes were produced by green machining before infiltration, and the Cu acted as a toughening agent. Such composites show promise as intermediate-temperature-range structural composites.

New in situ reactions are also being developed for making composites. In situ chemical reaction sintering is being developed to reinforce Al_2O_3, Si_3N_4, and SiALONs with refractory nitrides, borides, and carbides. This approach can improve the hardness and toughness of the material as well as provide better sintering capabilities. Other advantages include low cost, unique microstructures, and the potential for electron discharge machining (EDM).

1.6.4.4 Hot pressing.

Ruh, Mendirotta, Mazdiyasni, et al.[444] fabricated high-purity mullite/0–40 wt% SiC_w composites and mullite/30 wt% ZrO_2/0–30 SiC_w composites. Different degrees of ZrO_2 stabilization were achieved by varying the amounts of Y_2O_3 or MgO added. The room-temperature flexural strength of mullite was increased 50% by the addition of a 30% ZrO_2 phase. The flexural strength of mullite and mullite + 30% ZrO_2 was increased 25–50%. The addition of SiC_w increased the thermal diffusivity of mullite, and increasing temperature decreased the thermal diffusivity of all composites.

Tiegs[445] reported on current research at Oak Ridge National Laboratory (ORNL) that is centered on alumina and mullite matrix composites in three main areas: (1) further development of a composite property database, (2) NNS fabrication of composites with high whisker volume contents, and (3) understanding of composite behavior and whisker-matrix interactions to develop new approaches to high-temperature ceramics.

The SiC_w-reinforced ceramic composites were fabricated by conventional ceramic-processing techniques. The ceramic powders and SiC_w were mixed in a liquid medium, dried, and then hot-pressed or pressureless-sintered. According to Tiegs, this approach has been found to provide considerable improvement in achieving a satisfactory dispersion of the SiC_w in the alumina or mullite powders compared to hand mixing.

NNS fabrication by hot press or pressureless sintering was limited by the SiC_w content, and the practical limit for achieving high densities was about 20 vol%. Limited bonding at the matrix-whisker interface was necessary for high-toughness composites. The surface chemistry of the whiskers has been found to be an important parameter in determining this bond strength.

Lee, Kim, and Cho[446] reported that the addition of β-SiC_w was effective in increasing the fracture toughness of hot-pressed α-SiC. The mixtures were hot-pressed in a graphite mold at a temperature between 1900 and 2000°C for 1 h with a pressure of 25 MPa under vacuum. The density of the specimens was determined by an immersion technique.

Most of the whiskers added disappeared during the densification process by transformation into the α-phase. The remaining whiskers acted as nuclei for grain growth, resulting in the formation of large tabular grains around the whiskers. These grains were believed to have formed because of the extreme anisotropy of the interfacial energy between α- and β-SiC. The K_{IC} of the material was improved significantly by whisker addition. The increase in this value was attributed to crack bridging followed by grain pullout as a result of the formation of tabular grains in a fine matrix.

CMCs of Si_3N_4/SiC achieve theoretical density, making them viable candidates for structural applications where use temperatures exceed 1371°C, while MMCs are considered limited to below 1093°C.

Foulds, LeCostaouec, DiPietro, et al.[447] described the fabrication of SCS-6 SiC_f-reinforced Si_3N_4 composites by tape layup as well as powder consolidation processes. Theoretical density was achieved by hot pressing or HIP. The densified composites exhibited linear elastic strength properties in tension comparable to those of monolithic material but with continued load support after matrix cracking. No apparent fiber damage resulted from oxidation exposure, although matrix phase changes modified mechanical behavior. High-cycle fatigue tests conducted in tension demonstrated limited residual loss in strength properties over millions of cycles. Resistance to thermal shock was also demonstrated in a simplified water quench test.

Numerous investigators, as in the examples mentioned above, have reported on the hot pressing of a variety of ceramic reinforcements and matrices. They include the following.

- Bhatt,[448] who reported on SiC–reaction-bonded Si_3N_4 (RBSN) where in the bend mode, the unidirectional composites retained their as-fabricated fast-fracture UTS up to 1200°C in O_2 and in a N_2 environment and displayed excellent thermal stability after a 1000-h exposure at temperatures up to 1400°C. In an oxidizing environment, the composites retained nearly 60% of their as-fabricated strength after a 100-h exposure at 1200 and 1400°C. However, those heat-treated between 400 and

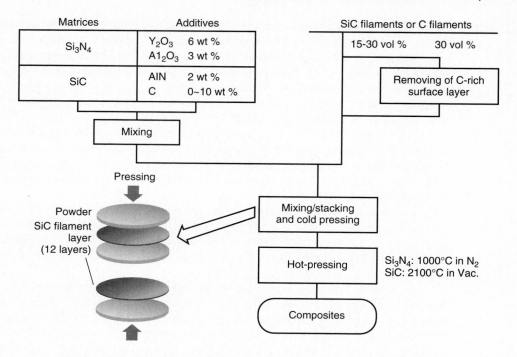

Figure 1.103 Fabrication processes for fiber-reinforced ceramics with a Si_3N_4 or SiC matrix.[449]

1200°C showed strength loss due to oxidation of the interfacial region, the porous Si_3N_4 matrix, and the SiC_f.

- Miyoshi, Kodama, Sakamoto et al.[449] who fabricated CMCs by hot pressing with SiC or Si_3N_4 matrices (Figure 1.103).
- Hirano and Niihara,[450] prepared Si_3N_4 matrix composites containing up to 30 vol% SiC by hot pressing a mixture of submicron-sized α-Si_3N_4 and β-SiC powders. The added SiC powders were located both at grain boundaries and within the matrix grains. The results for mechanical properties and microstructure of the composites are summarized as follows.

  1. The fracture strength reached maximum value at 5 vol% SiC dispersion. Above 10 vol%, the fracture strength decreased slightly with increasing SiC content.
  2. As the SiC content increased, the fracture toughness decreased. This decrease in fracture toughness was attributed to the inhibition of grain growth of rodlike β-Si_3N_4 by SiC dispersions.
  3. The high-temperature mechanical properties were significantly improved by SiC dispersions. The bending strength of the composite containing 30 vol% SiC was almost constant upto 1400°C. The improvement of high-temperature mechanical properties was attributed to the suppression of grain boundary sliding by intergranular SiC_p.

- Kato, Nakamura, Tamari et al.[451] used Al_2O_3-coated SiC_w as the starting material for the fabrication of SiC_w-reinforced Al_2O_3 ceramics by hot pressing. Sintering density and the dispersion state of SiC_w were improved by using Al_2O_3-coated SiC_w. Mechanical strength increased one to five times when Al_2O_3 coated SiC_w were used.

- Goto and Tsuge,[452] who fabricated SiC_w-reinforced Si_3N_4 by extrusion and hot pressing. A unidirectional alignment of the whiskers was achieved through sheet forming by extrusion, and the degree of whisker orientation changed with the thickness of the green sheets. Unidirectionally oriented whiskers increased fracture strength and toughness compared to those of samples with more randomly oriented whiskers. Bridging by whiskers impeded crack propagation when the whisker orientation was perpendicular to the crack plane.

- Tuffe, Dubois, Joraud et al.,[453] who investigated the improvement of mechanical properties of SiC_w-reinforced Al_2O_3 with emphasis on the effects of whisker type and content, hot pressing temperature, and the influence of an interfacial film between whisker and matrix. The introduction of SiC_w significantly improved the fracture toughness, flexural strength, and creep resistance of polycrystalline Al_2O_3. However, these properties were strongly dependent on the size and morphology of the whiskers. Large-diameter whiskers generated extensive microcracking, which lead to a decrease in flexural strength. Also, the presence of C-coated SiC_w substantially increased the high-temperature strain rate by promoting cavitation.

- Duclos, Crampon, and Cales,[454] who also studied microstructure development during hot pressing of SiC_w-reinforced Al_2O_3-based ceramics by transmission electron microscopy (TEM) as a function of composite density. Small, fully densified domains including only Al_2O_3 and ZrO_{2p} were observed from an apparent density of 80% of theoretical density. This corresponded mainly to porosity redistributions in the whisker vicinity. It appeared that grain growth hindrance was most efficient when ZrO_{2p} were gathered in clusters, SiC efficiency being noticeable only at the end of densification.

- Ono, Endo, and Ueki,[455] who reported on their efforts to hot-press monolithic TiC and TiC/Gr composites at temperatures ranging from 1800 to 2100°C and with graphite contents up to 30 wt%. Monolithic TiC was well densified at the hot pressing temperature of 1800°C; significant grain growth and an accompanying decrease in flexural strength occurred with an increase in hot pressing temperature. Conversely, the 10 wt% graphite-containing TiC composite required higher temperatures (1900 to ~2100°C) to reach a degree of densification approaching that of monolithic TiC. Like densification, flexural strength of the composite increased with an increase in hot pressing temperature. Graphite dispersions in the composites inhibited grain growth of TiC. Contrary to the equiaxed nature of the monolithic TiC grains, in the composite system the graphite grains were flattened to thin platelets aligned perpendicular to the hot-pressing axis. At a fixed hot pressing temperature, the relative density of the composites decreased with increasing graphite content.

- Piciacchio, Lee, and Messing,[456] who discussed their work on the fabrication of Al_2O_3/SiC_p composites by hot pressing mixtures of 5–30 vol% SiC with α-Al_2O_3, γ-Al_2O_3, or boehmite (γ-AlOOH) to determine whether grain growth or the α-Al_2O_3

phase transformation could be used to fabricate intragranular particulate composites. Heating at temperatures of 1700–1800°C resulted in only a small degree of grain growth, which indicates that these composites are very stable in a protective environment. The maximum amount of intragranular SiC for an α-Al_2O_3-based composite was 3.6 vol% when a total of 15 vol% SiC was added. It is suggested that there would be more intragranular SiC for conditions resulting in substantially more grain growth such as when liquid grain boundary phases are present or at higher temperatures.

- Brown, White, and Dunlop,[457] who reported that a fine powder consisting of Fe metal and TiC or TiN could be obtained by reduction of the mineral ilmenite ($FeTiO_3$) with carbon under controlled conditions. Hot pressing of Fe/TiC powder caused transformation of ~80% of the α-Fe to Fe_3C, whereas for Fe/TiN powder the extent of transformation to Fe_3C was very minor. The densification process was controlled by the extent to which the ceramic was wetted by the metal matrix and was found to differ considerably between TiN and TiC-based systems. Physical property measurements for hot-pressed specimens showed that density and hardness increased with temperature while porosity was reduced to zero by ~1330°C for Fe/TiN and ~1530°C for Fe/TiC. Vickers hardness values of ~1100 and 1380 were obtained for fully dense specimens of Fe/TiN and Fe/TiC, respectively.

- Cutler, Brinkpeter, Virkar et al.,[458] who fabricated monolithic and three-layered Al_2O_3/15 vol% ZrO_2 composites by dry pressing and casting aqueous slurries. The outer and inner layers of three-layer composites contained unstabilized and partially stabilized ZrO_2, respectively. Transformation of part of the unstabilized ZrO_2 led to surface compressive stresses in the outer layers. These three-layer oxide ceramics with compressive residual stress ranged between 300 and 600 MPa in the outer layers. The outer-layer thickness was controlled in the green state, and this layer protected against damage due to surface flaws or sliding contact. Slip casting was used to improve the uniformity of the interface between layers to allow composites with outer-layer thicknesses of 200–300 µm to be fabricated. The room-temperature strength of Al_2O_3-15 vol% ZrO_2 composites increased from 825 MPa to 1150 MPa, and the strength at 1000°C increased from 320 MPa to 640 MPa. Strength in excess of 1200 MPa at room temperature was achieved by superimposing temperature stress on transformation-induced stress.[458]

- Zhang, Tian, Tong et al.,[459] who reported that a Si_3N_4 ceramic reinforced with 30 wt% 3YZrO$_2$ (ZrO_2 containing 3 mol% Y_2O_3) could be fabricated by using two kinds of processes (hot pressing and pressureless sintering). The ZrO_2/Si_3N_4 composites exhibited high fracture toughness and flexural strength and lower thermal conductivity than Si_3N_4 matrix material.

- Davis, Yang, and Evans,[460] who used hot pressing to fabricate sapphire fibers with thick fugitive coatings of C and Mo into an Al_2O_3 matrix. They demonstrated that the interfacial properties could be controlled with coatings that could be eliminated from the interface subsequent to composite consolidation. However, these fugitive coatings can contribute to the high-temperature strength degradation of sapphire fibers. Such degradation was observed, and in some cases, by selecting appropriate composite processing conditions, such effects could be minimized.

- Shin, Kirchain, and Speyer,[461] who infiltrated and hot-pressed a TaC_p- and SiC_f-reinforced $LiO_2/Al_2O_3/SiO_2$ composite and found that the addition of 0–9 mol% Ta_2O_5 to a $LiO_2/Al_2O_3/SiO_2$ glass-ceramic matrix Nicalon SiC-reinforced composite increased the elastic modulus and UTS of the composite. Reactions at the fiber-matrix carbon-rich interfaces plus the soluble CO gas atmosphere of the hot press converted the TaO to TaC at approximately 1249°C. The improvement in mechanical properties was attributed to TaC_p reinforcement, and suggests a simple glass-ceramic route to the fabrication of particulate-reinforced CMCs.

- Some researchers[435,436,440,444] have obtained dense SiC_w-reinforced mullite composites with up to 50 vol% whisker by tape casting and hot pressing. The tape casting process results in high degrees of SiC_w orientation. The ability to achieve dense composites with as much as 50 vol% whisker is attributed to the higher percolation threshold of aligned whiskers.

1.6.4.5 Slip casting and low-pressure sinter.

Olagnon and Bullock[462] processed high-density sintered SiC_w-reinforced Si_3N_4 composites by slip casting, which allows both complex shape forming and dispersion of fine powders that are readily densified to high density by low-pressure sintering. This consolidation method was used to make sintered SiC_w/Si_3N_4 composites in order to overcome the sinterability decrease normally associated with the introduction of whiskers to monolithic ceramics. The factors influencing green forming and low-pressure (1 MPa N_2) sintering were examined and optimized in order to obtain highly densified composites. In this respect densification to closed porosity for a 20%-SiC composite and to greater than 98% for a 10%-SiC composite was achieved, demonstrating the feasibility of processing SiC_w/Si_3N_4 composites by sintering.

1.6.4.6 Reaction sinter.

Nanocomposite materials in the form of nanometer-sized second-phase particles dispersed in a ceramic matrix have been shown to display enhanced mechanical properties. In spite of this potential, methodologies for producing nanocomposites are not well established. Sakka, Bidinger, and Aksay[463] described a new method for processing SiC-mullite-Al_2O_3 nanocomposites by reaction sintering of green compacts prepared by colloidal consolidation of a mixture of SiC and Al_2O_3 powders. In this method, the surface of the SiC_p was first oxidized to produce SiO and to reduce the core of the SiC_p to nanometer size. Next, the surface SiO was reacted with Al_2O_3 to produce mullite. This process resulted in particles with two kinds of morphologies: nanometer-sized SiC_p distributed in the mullite phase and mullite$_w$ distributed in the SiC phase. Both particle types were immersed in an Al_2O_3 matrix.

1.6.4.7 Pressureless sinter.

Weiser[464] examined the pressureless sintering and densification behavior of model ceramic composites. The effect of the inclusion ratio on densification was studied using SiC_{w+f} in an Al_2O_3 matrix. Composites had maximum sintered density when made from fibers with aspect ratios of about 10. This peak in the density is a result of the density of short-aspect composites being inhibited by inclusion size considerations while long-aspect composites are prevented from densifying by percolation-related effects. Weiser also found that various salts were effective coagulating agents in slip casting Al_2O_3/ZrO_2 composites.

Kim and Lee[465] prepared high-density compacts, up to 88% of theoretical density, of Al_2O_3/SiC_w by a pressure casting and an impregnation technique. Starting with these green bodies, Kim and Lee pressureless-sintered composites of Al_2O_3-20 vol% SiC_w to higher than 95% of theoretical density. They were further densified by HIP up to 99% of theoretical density, resulting in a rupture strength of 680 MPa and a fracture toughness of 4.70 MPa m$^{1/2}$.

Wei and Lee[466] successfully pressureless-sintered SiC/AlN composites by using commercial SiC and AlN powders and the optimum amount of sintering aid. The important parameters during pressureless sintering, including the amount and type of sintering aids, sintering temperature, sintering period, and packing powder were studied. Y_2O_3 was found to be a better sintering aid than Al_2O_3 or CaO. The Y_2O_3 sintering aid reacts with AlN and SiC powders and forms a Y-Al-Si-O-N grain boundary phase to assist densification during pressureless sintering. With 2 wt% Y_2O_3, SiC/AlN composites were pressureless-sintered to high density at 2050–2100°C for 2 h under firing conditions where alpha packing powder was used during firing. At these temperatures, 1–2 h was the optimum sintering period for densification. After sintering, the composites had SiC/AlN solid solution grains and submicrometer SiC-rich grains.

Hoffman, Nagel, and Petzow[467] studied the processing of SiC_w-reinforced Si_3N_4 whereby cold isostatically pressed and slip-cast Si_3N_4 composite materials containing up to 20 vol% SiC_w were densified by pressureless sintering and post-HIP. The final density after pressureless sintering was independent of whisker orientation and green density. Sintering temperatures of 2000°C, higher amounts of sintering additives, and additive compositions with a lower liquid-phase viscosity reduced sintering stresses and increased the final densities. Further densification without any matrix decomposition and whisker degradation was achieved by HIP in a N_2/Ar atmosphere.

1.6.4.8 Slurry.   Freeman, Starr, Harris et al.[468] studied cloth layup fabrication techniques suitable for large, complex components. Cloth fabric layers, in this case made from woven Nicalon SiC_f, were dipped in a slurry composed of submicron (about 0.2 μm average) Si metal powder in methanol. Fabric layers were stacked to a desired thickness and mechanically compacted, allowing excess solvent to pass through a filter while the powder was retained. This compact was then heated in nitrogen gas (N_2/5% H_2) to a temperature of 1200°C, and the Si metal was thereby converted to RBSN. This RBSN was produced in either alpha or beta crystalline form depending on powder size and the kinetics of transport into the powder, as well as impurities present in the material. It was thought[469] that the Si metal vaporized, recondensed as Si_3N_4, and grew via the reaction of Si vapor with adsorbed N_2 on the nitride surface as the nitriding process moved to completion.

In order to solve the major problems of processing whisker-reinforced ceramic composites, such as agglomeration of whiskers and correlation between pH and viscosity, Sato, Ueki, and Shintani[470] carefully investigated a mixed slurry of whiskers and matrix powder. SiC_w and Si_3N_4 powders were dispersed homogeneously by controlling pH in aqueous suspension, and the state was successfully fixed by a sudden change in pH to make the slurry more viscous. The slurry was then filtrated rapidly and dried. In the sintered (hot-pressed) body, fracture toughness was improved more than 75% with whisker addition of 30–40%. Degradation of flexural strength with whisker addition was minimized by the newly developed process.

Iwata, Isoda, and Itoh[471] initiated a study intended to significantly improve the fracture toughness of ceramic materials by incorporating continuous-fiber reinforcement and to demonstrate the fabrication of gas turbine engine components such as shrouds and liners.

Si_3N_4 matrix composites were prepared by stacking precursor tapes consisting of C_f coated with a slurry of Si_3N_4, Al_2O_3, Y_2O_3, perhydropolysilazane, and xylene.

After sintering, the continuous fiber-reinforced Si_3N_4 matrix composites had a 4–5 times higher fracture toughness of 28.1 MPa $m^{1/2}$ and a 250–300 times higher work of fracture of 2.5×104 J/m^2 than typical monolithic Si_3N_4 ceramics.

Roth, Clark, and Field[472] examined the slurry infiltration process for producing SiC/SiC composites for exhaust ducts in aircraft gas turbine engines and where there exists a need for improvement in strength at elevated temperatures. While strength was not generally considered to be a major concern when designing exhaust ducts, the reported strength of a SiC/SiC composite produced by this slurry infiltration was considered low.

The engineers involved in the materials selection process believe that slurry-infiltrated CMCs should exhibit much higher strengths. For strengths of 207 MPa and with all other properties as before, slurry-infiltrated SiC/SiC composite exhaust ducts should have the same utility as superalloy ducts, however, costs must be reduced to be competitive in the future.[473]

Hand layup is probably most applicable for continuous-fiber-reinforced ceramics, with reinforcing fiber cloths placed in the mold and the matrix slurry poured onto the cloth and worked into the cloth by roller, squeegee, or other means. Multiple layers can be built up in this manner, and it is also possible to place compacted whisker mats into the mold. Although hand layup offers great design flexibility and minimal capital investment, it is labor-intensive and would not be an ideal method for mass production.

Vacuum bagging is an extension of hand layup. It involves the placement of a flexible film over the slurry and fibers in the mold. The joints are sealed and a vacuum pulled. The resulting atmospheric pressure on the layup in the mold can help to reduce voids, trapped air, and excess slurry and improve adhesion between layers.

Additional heat and pressure over that used in the vacuum bag method can be applied with autoclave equipment, and pressures on the order of 0.7 MPa can be obtained. Voids can be further reduced below the number observed with vacuum bagging and reinforcement loadings increased by the additional pressure. Undercuts are possible, and cores and inserts can be used. However, autoclave size limits part size, and equipment tends to be expensive.

The fastest growing method of forming PMCs and one that is of considerable interest in forming whisker-reinforced ceramics is injection molding. This procedure can produce complex, greatly detailed parts ranging from small to quite large (such as automobile bumpers). Overall cycle times of less than a minute are common with PMCs. As utilized for whisker-reinforced ceramics, a slurry containing whiskers and matrix particles is injected under pressure into the closed mold. Slurry temperature can be elevated to improve flowability, and the mold cooled to reduce the temperature of the formed part and improve green strength.

1.6.4.9 Chemical vapor infiltration.   There are two CVI processes currently in use: isothermal diffusion-limited infiltration and the forced-flow thermal gradient process (Figure 1.104). The isothermal process was developed in France by the Société Eu-

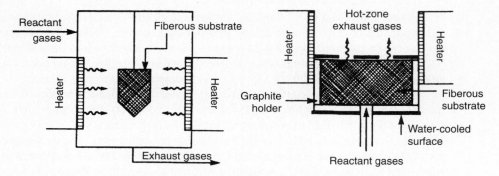

Figure 1.104   Two proven CVI processing techniques for producing CMCs. Isothermal diffusion-limited infiltration (*a*) and thermal gradient forced-flow infiltration (*b*) are relatively low-temperature techniques that also do not stress the ceramic preformed substrate during processing.[474]

ropeenne de Propulsion (SEP). This method produces high-quality composites and places no size or shape limitations on the substrate, which allows for infiltration of NNS parts. However, preferential deposition in the outer regions can lead to premature pore closure, requiring process interruption for diamond machining operations to reopen the pores. Because the SEP process is diffusion-limited, it can require weeks of processing time. The diffusion problem also is aggravated by counterdiffusion of product gases. Japanese investigators have provided an incremental improvement by using pulsed reactant flows. Los Alamos National Laboratory (LANL) also uses an isothermal forced-flow reactor in its work.

The thermal gradient forced-flow process from ORNL[474] intentionally imposes a steep thermal gradient in the direction opposite reactant flow. Cold reactant gases enter under a pressure gradient and flow toward the hot face where deposition reactions occur. As deposition proceeds, the composite thermal conductivity increases in the dense region, allowing the higher-temperature deposition front to approach the cold face. Infiltration times as short as 24 h have been reported. While high-quality, dense composites have been fabricated, regions of high residual porosity are often produced. These may be attributed to the gas flow patterns that develop within the substrate. Furthermore, substrates appear to be limited to relatively simple shapes.

A third method was the development of a CVI technique utilizing a microwave assist (MACVI). A number of processing schemes are possible using combinations of absorbing and transparent material as composite components. This includes the use of an absorbing preform (Nicalon fiber) combined with a transparent matrix (Si_3N_4). Composites 5 cm in diameter and 1 cm thick have been fabricated to densities of 65% theoretical. Processing times for these materials are under 20 h. Higher densities will require additional microwave power now possible with the new reactor. The most effective MACVI scheme will involve the use of a transparent fiber with an absorbing matrix. The hot spot will be initiated by appropriate treatment of the central region of the preform, and to this end Al_2O_{3f} with pretreatments for controlling thermal gradients has been explored. Nextel 610 fibers have been effectively pretreated with a carbon coating, resulting in preferential heating in the interior of the preform. Possible matrix materials include siliconized SiC, doped SiC, Al_2O_3, and ZrO_2.

The CVI process requires the penetration of vapors into a fibrous preform structure and the production of a solid phase from the chemical reaction of the vapor species to form the matrix of the composite.

A model for CVI was used by Tai and Chou[475] in studying the growth of Al_2O_3 from the chemical reaction among $AlCl_3$, H_2, and CO_2 within a SiC_f bundle. Their model considered binary diffusion, chemical reaction of the vapor mixture within the fiber bundle, and deposition film growth. The void space between the fiber bundle and the alumina was assumed as a capillary in this model. Both molecular diffusion and Knudsen diffusion were taken into account sequentially during the infiltration process, and diffusion- and reaction-controlled processes were also considered. Based upon this model, the optimum processing conditions and time required for CVI to form a ceramic-ceramic composite could be predicted.[476–478]

The feasibility of densifying a 3D-braided Nicalon fiber preform with a SiC matrix by the CVI technique was investigated by Burkland and Yang,[479] who produced a toughened structural CMC with a combination of 3D-braided reinforcement and low-temperature CVI. These workers reported that (1) the mechanical properties of strength, fracture toughness, impact resistance, and environmental durability of the CVI-processed ceramic composite demonstrated that a promising structural material could be produced by this process; (2) further information on the characteristics of the fiber-matrix interfaces, elevated-temperature properties, and long-term reliability require development; and (3) further improvements in strength and fracture toughness may be possible by optimizing the processing parameters and tailoring the fiber-matrix interface bonding.[480]

Lasday[481] reported on DuPont's work with continuous-fiber composites whereby ceramic fiber preforms were fabricated in the shape of final parts (Figure 1.105). These

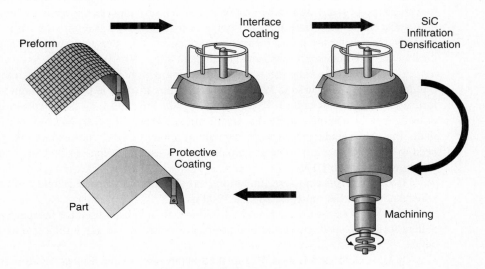

Figure 1.105   CVI process.[481]

preforms were interface-coated and then infiltrated at high temperature with chemical vapors that reacted to form a dense SiC matrix between the fibers. The CVI facilities of the company are capable of producing parts 1.2 m in diameter and 2 m high.

Bashford[482] showed that there are two principal means by which a SiC matrix can be formed with fabrics or preforms of SiC_f: (1) CVI/CVD, and (2) polymer pyrolysis, for example, converting polycarbosilane to SiC.

Both are low-pressure infiltration techniques having advantages and disadvantages, which are summarized in the accompanying table.

Process	Advantages	Disadvantages
CVI/CVD	Even, controlled deposition giving a coherent matrix	Long process times (30–150 h)
	Fiber coatings and matrix built up in single operation	Expensive capital equipment
	Modest reaction temperatures	
	Dopants can be added to matrix	
Polymer pyrolysis	Useful in binding fiber preforms	Repetitive infiltrations and pyrolysis to achieve densification
	Cheaper process than CVI	Long process times with slow heating rates
	Modest process temperatures	Considerable shrinkage on pyrolysis resulting in matrix cracking

Either process gives SiC/SiC composites with theoretical densities of about 85–90% and the remainder porosity. Such composites offer good thermal stability when they're essentially being made of SiC. In terms of engineering uses where the retention of reasonable mechanical properties is sought for prolonged periods in oxidizing environments, the CVI composite is preferable, mainly results of the even, controlled deposition in a single process giving a coherent matrix, which provides the composite with integrity that is less achievable from repeated polymer pyrolysis densification steps. It is felt that polymer pyrolysis may be useful in an initial processing step to provide a bound, handleable preform that can be subsequently densified by CVI.

CVD relies on a chemical reaction occurring within gaseous phases such that material deposits on a hot surface. With CVI the gaseous mixture is passed through fiber preforms where the fibers are 12–15 μm in the case of SiC, and very slow deposition conditions have to be used to control the densification. Because the SiC, Nicalon, and Tyranno fibers that were evaluated begin to degrade at about 1200°C, the reaction temperatures need to be lower than this. The principal reactions used in preparing SiC/SiC composites are shown in Figure 1.106.

CVI composites possess limitations, notably on the thermal stability of the fibers which show an onset of degradation at 1300°C.

SiC/SiC composites prepared by CVI have already been used for prospective bipropellant thrusters, space plane thermal protection systems, and gas turbine engine components.

Kmetz, Suib, and Galasso[483] prepared composites of SiC/Si and SiC/SiC from single yarns of SiC. The use of carbon coatings on SiC yarn prevented the degradation nor-

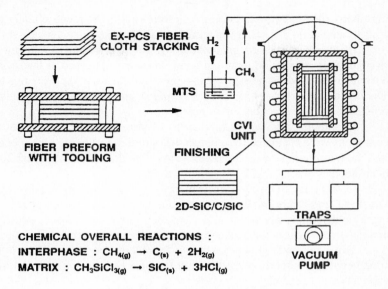

EX-PCS FIBER CLOTH STACKING

FIBER PREFORM WITH TOOLING

MTS

FINISHING

CVI UNIT

2D-SiC/C/SiC

TRAPS

VACUUM PUMP

CHEMICAL OVERALL REACTIONS :
INTERPHASE : $CH_{4(g)} \rightarrow C_{(s)} + 2H_{2(g)}$
MATRIX : $CH_3SiCl_{3(g)} \rightarrow SiC_{(s)} + 3HCl_{(g)}$

Figure 1.106   CVI processing route for CMCs.[482]

mally observed when CVD Si was applied to SiC yarn. However, the strength was not retained when the composite was heated at elevated temperatures in air. In contrast, the strength of a SiC/SiC composite was not reduced after it was heated at elevated temperatures, even when the fiber ends were exposed.[484]

Since most of the work undertaken thus far on CMC materials using CVI processing has been performed on SiC matrix composites, Hoyt and Yang[485] undertook work to demonstrate that fiber-reinforced Si_3N_4 matrix composites can be fabricated using CVI techniques.

The resulting matrix material that they produced using current infiltration technology was amorphous Si_3N_4, however, it crystallized at temperatures above 1250°C. Devitrification occurred with a linear shrinkage of approximately 0.5%, and the crystalline phase that formed was α-Si_3N_4. Since the Vickers hardness of the deposit was similar to that of crystalline Si_3N_4, the amorphous deposit was suitable for use as a matrix material.

The fracture surfaces of composites heat-treated at 1100 and 1300°C revealed virtually no fiber pullout, indicating that the fiber-matrix bonding was strong. Also, shrinkage of the matrix around the fiber during devitrification would induce compressive stress at the surface, creating additional mechanical bonding between the fiber and the matrix. Since a weak bond between fiber and matrix is necessary in a damage-tolerant composite, an appropriate coating layer has to be developed.

The optimization of CVI conditions for infiltrating the composite are needed and the challenge is to balance improved infiltration with the production of high-quality matrix material.

Mitsui Engineering[486,487] reported the development of a fiber-reinforced SiC composite material for large industrial components fabricated using the CVI process.

Initially they fabricated small combustor liners which were produced by first forming fiber-structured porous ceramic (preform) by the filament-winding process, followed by SiC deposition in a small vapor deposition furnace.

In the large preform manufacturing process, a resin displaying greater coupling strength after conversion to an inorganic substance was used as the resin. The fabric was a woven or nonwoven fabric available on the market, and the fiber was SiC, C, or Al_2O_{3f}. The laminate was formed by hand, allowing shaping with flexibility, and the preform was produced in a high-vacuum environment.

In applying CVI to the fabrication of large components, various vapor deposition conditions were evaluated to establish the optimum technique for vapor deposition of SiC into the miniscule pores of the preform. As a result, it is now possible to produce components about 1 m long with a diameter of about 0.7 m.

In preparing the woven fabric made of SiC_f, two sheets of fabric about 1 m long, 50 mm wide, and 8 mm thick were made, and heating tests were repeated several times using a noncooling plate roller in a frequency induction furnace, which showed that the composite can be used in fabricating various components for heating furnaces.

This new material appears highly feasible for commercial use in high-temperature components for aircraft engines and in the area of industrial gas turbines and industrial applications in general.

A new, faster process was developed by Kim, Song, Park, et al.[488] for the fabrication of Nicalon fiber-reinforced SiC composites by combining polymer solution infiltration (PSI) and CVI. The process led to NNS fabrication of fiber-reinforced CMCs and reduced infiltration time. Typical flexural strength and fracture toughness of these composites were 296 MPa and 10.9 MPa $m^{1/2}$ at room temperature and 252 MPa and 9.6 MPa $m^{1/2}$ at 1000°C, respectively. The composites exhibited a load-carrying capability after crack initiation and toughening by a fiber pullout mechanism.[488]

Allaire and Dallaire[489] successfully produced SiC and Si_3N_4 powders and subsequently SiC/Si_3N_4 composites by a one-step DC plasma process in their laboratory from a plasma gas-phase reaction. They followed this initial work with the simultaneous synthesization of two components to form a ceramic-ceramic composite powder in a one-step process, eliminating the powder-mixing step usually required to produce composite materials.

1.6.4.10 Directed melt oxidation.

Manor, Ni, and Levi[490] have shown that Al_2O_3 matrix composites containing SiC_p and an interpenetrating network of metal can be grown by DMO of a slightly modified A380 Al alloy at temperatures ranging from 950 to 1100°C. The relative amount of metal in the oxidation product was found to decrease with increasing processing temperature but also to be relatively independent of the presence of SiC.

The oxidation rate of the A380-M alloy was comparable to those reported for other Zn-bearing alloys under similar processing conditions. It was found that the rate was significantly enhanced by the presence of the SiC preform, with a tendency to increase with decreasing particle size. The underlying mechanism involves an extended oxidation front, arising from a combination of preform wetting by the melt and secondary nucleation of Al_2O_3 on the particle surfaces, perhaps by reaction of the alloy with the SiO_2 layer formed on the SiC.

Finally, the total porosity of the composite was found to increase with increasing Mg content, processing temperature, and/or SiC_p size.

Lasday[481] described the Dimox process for SiC_p/Al_2O_3 which exhibits favorable room- and elevated-temperature properties (Table 1.18). In Table 1.19 are listed the material systems being produced by DMO technology.[481]

The fabrication steps include the preparation of a particulate preform by one of the conventional ceramic-processing routes such as uniaxial or isostatic pressing, injection molding, extrusion, or slip casting. A suspension of ceramic particles is poured into a plaster of paris mold, depositing on its walls. After buildup of the desired wall thickness, the excessive suspension is decanted. When the ceramic layer is dry, the ceramic particulate preform is removed.

The particulate preform is placed in contact with a molten Al alloy in a furnace at a temperature between 900 and 1100°C. Starting at the alloy-preform interface, and with an oxygen (air) atmosphere and a short incubation period, α-Al_2O_3 begins to grow directionally into the preform. The molten alloy, driven by surface energy forces, wicks from the reservoir through an interconnected network of microscopic channels to the Al_2O_3 growth front. Growth is sustained as long as alloy, sufficient temperature, and oxygen are available. The composite formation is complete when the reaction front stops upon contact with a gas-permeable barrier layer. This barrier coating is preapplied to all preform surfaces where matrix growth is to be eventually stopped, thereby enabling component fabrication to the desired shape.

The rate of reaction growth front and microstructure development are influenced by molten alloy composition and element dopants. As mentioned above, filler, filler size, and residual metal in the matrix influence composite properties. The residual Al alloy, for example, significantly contributes to the increasing room-temperature strength and toughness of SiC_p-reinforced Al_2O_3 composites. As the Al alloy increases in temperature to its melting point (~600°C), the strength of the composite with 7 mm filler and larger volume of residual metal decreases because of a reduced load-bearing material cross section. Also, for the same composite, toughness diminishes with temperature increase and softening of the Al alloy because of its decreasing contribution to crack bridging.

The thermal shock resistance of SiC_p/Al_2O_3 is excellent. Creep rates are generally low for these composites, making them attractive for high-temperature industrial heating applications.

1.6.4.11 Sol-gel processing.[484]   Sol-gel processing is an attractive method of producing specialized monolithic ceramic materials. Advantages over conventional ceramic slurry processing include a more homogeneous mixing of components for multiphase materials and reduced sintering temperatures. It can be expected that these attributes will be useful for CMCs as well. In particular, any reduction in the temperature required for matrix densification is more important for ceramic composites, with the possibility of deleterious reactions between the fibers or fiber coatings and the matrix, than for monolithic ceramics.

However, results to date range from poor to only moderately good. The major problem with processing by the sol-gel technique has been matrix cracking and limited densification due to the large shrinkage accompanying the transformation from sol to gel and from gel to a ceramic material. Linear shrinkages of 30% are not uncommon. Although such a large shrinkage can be acceptable in a monolithic ceramic specimen if

TABLE 1.18. Physical Properties of SiC_p/Al_2O_3 Composite[481]

Property	Units	25°C	(73°F)	1000°C	(1832°F)	1550°C	(2822°F)
Density	g/cm³ (lb/in.³)	3.4–3.5	(0.12–0.13)	—	—	—	—
Hardness	Rockwell A	80–90	—	—	—	—	—
Flexural strength[b]	MPa (ksi)	400–500	(65–73)	200–250	(29–36)	175–225	(26–33)
Modulus	GPa (msi)	310–330	(45–48)	—	—	—	—
Poisson ratio	—	0.25–0.29	—	—	—	—	—
Shear modulus[c]	GPa (msi)	120–130	(17–19)	—	—	—	—
Fracture toughness[d]	MPa m$^{1/2}$ (ksi in$^{1/2}$)	7.0–7.5	(6.4–6.8)	3.0–4.0	(2.7–3.6)	2.5–3.5	(2.3–3.2)
Coefficient of thermal expansion[e]	ppm/°C (ppm/°F)	7.8	(3.9–4.5)	—	—	—	—
Thermal conductivity	W/m·K (BTU in h ft² °F)	60–70	(485–555)	15–25	(100–175)	5.5	(38)

[a] Particulate loading of 55 vol%, 5 to 20-μm particle diameter. Loadings can be tailored up to ~75 vol%.
[b] Four-point bend.
[c] Sonic method.
[d] Chevron notch beam.
[e] Average value from 25 to 1400°C.

TABLE 1.19.   Material Systems as Produced by DMO Technology[481]

Silicon carbide particulate/alumina matrix SiC_p/Al_2O_3

Silicon carbide or alumina fiber/alumina matrix SiC_f/Al_2O_3 or Al_2O_3/Al_2O_3

Zirconium diboride platelet/zirconium carbide matrix ZrB_{2p}/ZrC

Aluminum titanate particulate/alumina Al_2TiO_{5p}/Al_2O_3

Silicon carbide or alumina particulate/aluminum metal matrix composite
   SiC_p/Al or Al_2O_{3p}/Al_2O_3

Titanium carbide (TiC)-coated graphite composites

processing conditions are carefully controlled, the nonshrinking fibers in a sol-impregnated composite restrict the possible change in external dimensions and cracking is inevitable.

Steps taken to minimize the shrinkage problem have included hot pressing of the composite specimens. However, this entails the inherent disadvantages of specimen size, shape, and complexity, as well as cost. Another approach by researchers was to incorporate filler particles into the sol material to restrict shrinkage during processing, and this was reasonably successful. Another technique is to perform multiple impregnation-densification cycles. This seems to be a reasonable approach, although costs could increase appreciably.

A number of research groups have investigated the fabrication of CMCs by impregnation of fiber networks with polymeric materials that decompose to nonoxide ceramic materials on pyrolysis. For the most part, the mechanical properties of these materials have been only moderately good. Since in most cases the fibers and the matrix materials have related chemical compositions, suitable fiber coatings will probably be required to prevent undesirably strong bonds from developing between fibers and matrix.

Much of the work has involved the use of either filler particles or multiple impregnation-pyrolysis cycles, both of which address the large mass loss and shrinkage observed in most of the precursors used. It can be anticipated that polymers with greater mass yields and lower shrinkages upon thermal decomposition will be developed, resulting in improved composite properties.

Hot pressing Tyranno fiber networks alone (no matrix material) resulted in fiber deformation into a close-packed hexagonal array and a material with reasonably good room-temperature strength and excellent retention of strength at high temperatures or after heat treatment at high temperatures.

Wu and Messing[491] reported lightweight 20 vol% SiC_w/SiO_2 composites with a reinforced cellular structure that can be fabricated by foaming a sol-gel mixture and sintering to obtain relative densities as low as 10%. Because of the large reduction in drying shrinkage and cracking, the use of reinforcements permits the design and fabrication of significantly larger sol-gel-derived foams. After serving the function of reinforcement during processing, the reinforcement then significantly increases the strength of the sintered SiC_w/SiO_2 composite foams relative to pure silica foams. It is reasonable to propose that whisker or staple fiber reinforcements will result in similar enhancements in the mechanical properties of other lightweight materials. Thus it is anticipated that the structural applications and performance of lightweight cellular materials can be considerably expanded by the use of whisker or fiber reinforcement.

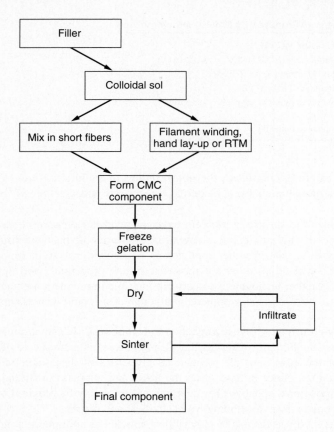

Figure 1.107 Flow diagram of sol-gel processing routes for CMCs.[493]

Feng, Tang, and Zhen[492] have synthesized tetragonal ZrO_2/SiC_w composite powders by the sol-gel method. Analysis revealed that SiC_w were homogeneously dispersed in the composite powder, the matrix being ultrafine metastable tetragonal ZrO_2. The sintering properties of the composite were investigated, and it was found that SiC_w favored an increase in metastable tetragonal ZrO_2. When the SiC_w content was high, the critical stress intensity factor (K_{IC}) decreased.

Floyd, Cooke, and Harris[493,494] have performed a considerable amount of research work using sol-gel processing routes in the production of CMCs as outlined in the flow diagram in Figure 1.107.

Their process involves the use of commercially available, inexpensive aqueous sols combined with the cheaper oxide and C_f. The sol consists of an aqueous dispersion of fine ceramic particles (about 30 nm in diameter) into which somewhat larger filler particles (1–10 µm) are dispersed to increase the solid yield. This liquid sol is then combined with the reinforcing fibers by using the well-known technologies for producing fiber-reinforced plastics, the polymeric resin being replaced by the sol. The sol is converted to a rigid matrix very near to net shape by rapid freezing followed by thawing and then drying off the aqueous solvent. Almost no shrinkage is observed, and certainly none of the cracking inherent in other sol-gel systems during drying and subsequent firing. Because of the high reactivity, much lower sintering temperatures are required, typically less than 850°C for continuous-fiber reinforcement. The inherent porosity within the matrix would normally be reduced to levels below 20% (by volume) via repeated liquid-phase infiltration assisted by a vacuum and small overpressure in order to enhance the mechanical properties.

The main advantages of this process are the ability to form large or small complex-shaped components near to net shape with multidirectional fiber reinforcement by either simple casting for short-fiber reinforcement or filament winding and even hand layup for long-fiber reinforcement. Minimal machining is required, which normally can be completed green (before firing). The V_f of porosity can be varied between 0.2 and 0.8, and the pore size can range from 1 to 20 μm in diameter, giving potential for use in applications requiring a microporous material. Furthermore, the sols and filler powders are relatively inexpensive and widely available. The two main disadvantages are that the sol must contain a minimum quantity of silica and that some residual porosity must remain. The development of other freeze-gelable sols and improvement in the liquid-phase infiltration process will reduce these limitations.

With a fiber V_f of continuous reinforcing fiber above 0.05–0.10, it is the fiber properties that dominate the mechanical performance of the CMC and not the matrix. For applications where the composite must be gastight, glazes have been successfully applied to seal the surface. With short, random fiber reinforcement, however, the mechanical properties of the composite are dominated by the matrix and the residual porosity ensures that maximum strengths of no more than 50 MPa can currently be achieved. A review of the properties of some sol-gel CMCs is presented in Table 1.20.

If production of low-cost CMCs by freeze gelation or other routes is developed, the prospects for their widespread application are likely to be much enhanced. The advantages they offer over conventional engineering materials are those of monolithic ceramics combined with enhanced mechanical performance. It should be noted that in many cases the actual mechanical loading of components is much lower than presently used materials can safely accommodate, the selection of which has been for other design reasons. A typical example of this is in biochemical engineering, where pressures of a few atmospheres and temperatures below 300°C may be encountered. But the prime requirement is for chemical inertness and corrosion resistance. CMCs show potential as substitutes for stainless steels with the added advantage of their lower density, allowing reductions in the size of support structures.

The ability to control the size and V_f of porosity, yet maintain adequate mechanical performance and good corrosion resistance, suggests potential for the application of sol-gel CMCs in a variety of uses requiring a porous, inert substrate. Trial CMC air bearings have been manufactured as replacements for sintered stainless steel porous metallic tubes used in the transmission of delicate film during processing. Porous CMC tubes show significant cost advantages over their metallic counterparts. Other applications include hydraulic oil filtration devices, catalyst supports, precision refractories, including metal casting molds, and low-density, fire-retardant boards.

1.6.4.12 Self-propagated high temperature synthesis or combustion synthesis.

Zeng, Miyamoto, and Yamada[495] successfully synthesized fine Si_3N_4/SiC composite powders in various SiC compositions from 8 to 46 vol% by nitriding combus-

TABLE 1.20. Summary of Mechanical Properties of Typical Freezing in Processing CMCs[493]

Fabrication Route	Fiber	Main Filler Powder	Density (g/cm³)	Dynamic Modulus (GPa)	Flexural Strength (MPa)	ILSS (MPa)	Work of Fracture (kJ/m²)	Maximum Processing Temperature (°C)
1. Filament-wound	FP alumina	Mullite	2.19	—	220	—	2.90	1000
2. Filament-wound	Nextel	Silica	1.91	48.1	202	>25	4.88	700
3. Filament-wound	Nextel	Silica-zirconia	1.95	40.7	104	>20	1.03	1000
4. Filament-wound	Carbon	Low exp glass ceramic	1.70	45.4	400	31.4	10.30	750 (in Ar)
5. Cast	Saffil	Mullite	1.88	30.2	46	—	0.23	1050
6. Cast	Saffil	Silica-zirconia	2.38	29.0	32	—	0.15	1150

Fiber volume fraction: Fiber volume fractions in continuous filament-wound materials are typically 0.15–0.35 (0.55 max), and for short Saffil fiber-reinforced material, typically 0.08–0.15 (0.30 max).

Residual porosity: The residual porosity in continuous filament-wound material is typically 0.2–0.25 and 0.25–0.32 in short Saffil fiber CMCs. This can be increased to about 80%, though with a consequential reduction in mechanical performance.

Thermal shock: Material 3 shows no reduction in mechanical properties after repeated 550°C water quenching. Material 5 exhibits a gradual (near-linear) reduction in performance as the water quench increases to 600°C at which point the flexural strengths have declined by about 30%.

tion of Si and C at room temperature and under 10 MPa N_2 pressure. The powders were composed of α-Si_3N_4, β-Si_3N_4, and β-SiC. The sintered bodies consisted of uniformly dispersed grains of β-Si_3N_4, β-SiC, and a few Si_2N_2O.

The high reaction heat of Si nitridation increases the dissociation rate of Si_3N_4 and assists in the formation of SiC, which is more stable than Si_3N_4 at the reaction temperature. The combustion synthesis is seen as a new route to preparing the unique composites in a simple, rapid, energy-saving way.

Chrysanthou, Saidi, Aylott, et al.[496] examined the production of ceramic composites based on Al_2O_3/TiC by means of SHS. They produced Al_2O_3/TiC composite powders by two reaction routes. The rate of reaction of Ti and carbon black as well as the ignition and maximum temperatures attained were all dependent on the amount of Al_2O_3 present, which moderates the reaction. This reaction was fairly explosive, and fine precipitates of Al_2O_3 in TiC were obtained. The reaction did not follow a self-propagating mode if more than 41% Al_2O_3 was present unless the reactants were inhomogeneously mixed. This led to the conclusion that it is possible to generate enough heat for self-propagation of samples containing 50–70% Al_2O_3 by introducing regions of high Ti and carbon concentration in the reacting powders. The effect of precompaction was to create a continuous grain structure, though at pressures exceeding 2.8 MPa the reaction was not self-propagating but occurred via solid diffusion of carbon through solid Ti.

The aluminothermic reduction of TiO_2 in the presence of carbon black took place through the sample in the form of a wave front. Reaction started at 1310°C by the reaction of TiO_2 and Al, releasing exothermic heat. A second reaction between Ti and carbon black then followed, reaching a maximum temperature of 2100°C. The product microstructure was very different from the one obtained by the previous route. TiC and Al_2O_3 were again intimately mixed, but the Al_2O_{3p} were much larger than in the previous reaction.

Rabin, Korth, and Williamson[497] fabricated dispersed-phase composites of TiC/Al_2O_3 from low-cost reactants by external ignition of powder compacts followed by dynamic consolidation of the hot products using explosives. Near-full densification was achieved using shock pressures as low as 1 GPa. However, care must be taken to allow for the escape of evolved gases during exothermic reaction, and rapid cooling of the consolidated composite must be avoided to prevent cracking.

Subrahmanyam and Rao[498] prepared $MoSi_2$, SiC, and $MoSi_2$/SiC composites from elemental powders Mo, Si, and C by the thermal explosion mode of SHS. The morphology of $MoSi_2$ in the product indicates that it is in the molten state at the combustion temperature. The SiC in the composite shows a very fine particle morphology. They concluded that the thermal explosion mode of SHS can be used very effectively to produce composites of $MoSi_2$ with complete conversion and the desired morphology.

1.6.4.13 Other processing methods.

Aghajanian, Biel, and Smith[499] produced novel AlN-based CMCs using a two-step infiltration and reaction process. Molten Al alloy was infiltrated without pressure into preforms containing Si_3N_{4p} (mixed α and β phases). The resultant composites were heat-treated, causing hard phases (AlN and Si) to form by reaction of the soft Al with Si_3N_4. The ceramic content of the final composites was higher than that of the original preform because of volume changes that occurred during the reaction heat treatment. Initial characterization of mechanical properties demonstrated that the composites had hardnesses, Young's moduli, and compressive strengths that compared favorably with traditionally processed AlN. Flexural strength measurements above the melting point of Si indicated that the composites possessed an interconnected skeleton of ceramic.

Singh[500] has reported some recent development work whereby CMCs with tailored properties have been fabricated from low-cost resin mixtures. The process consisted of the production of a microporous carbon preform and its subsequent infiltration with liquid Si or a refractory metal-Si alloy. Microporous preforms were made by the pyrolysis of a polymerized resin mixture at relatively low temperatures, and low-cost tooling was used to produce complex NNSs. Pore volume and pore size were tailored to control the size and distribution of the final constituents.

The process has been used to fabricate SiC matrix composites. It is suitable for various types of reinforcements such as whiskers, particulates, and continuous fibers and may be applicable to fabrication of composites having 3D architectures. Tailorable properties include strength and toughness, creep, and thermal shock resistance.

These materials are being considered for applications in hot sections of jet engines and on leading edges of reentry vehicles and hypersonic aircraft. In the energy industry, potential applications include radiant heater tubes, heat exchangers, and ceramic burner inserts.

Wu and Claussen[501] successfully fabricated mullite/SiC/Al_2O_3/ZrO_2 composites by the reaction bonding of Al_2O_3 (RBAO) technique using Al/Al_2O_3/SiC powder mixtures milled by tetragonal zirconia polycrystal (TZP) balls.

Reaction-bonded mullite/SiC/Al_2O_3/ZrO_2 composites exhibit superior mechanical properties: a fracture strength of 610 MPa and a toughness of 4.9 MPa were achieved in a sample containing 55 vol% mullite. The small grain sizes (<1 µm) developed during reaction bonding and the t- to m-ZrO_2 phase transformation toughening were responsible for the high mechanical properties.

Finally, HIP significantly improves the mechanical properties of reaction-bonded composites: fracture strength and toughness increased from 490 MPa to 890 MPa and from 4.1 MPa to 5.9 MPa $m^{1/2}$, respectively, for the sample containing 49 vol% mullite.

Pressureless sintering of monolithic ceramics results[502] in substantial shrinkage as the void space in the initially porous starting material is eliminated. Introduction of fibers or platelets inhibits sintering. Thus when CMCs are made, processes other than sintering are used to achieve dense structures or to control the shape and size of the final material. These techniques are either consolidation or infiltration methods (nonreactive and reactive melt infiltration).

Hillig[502] describes his approach to melt infiltration and CMCs. Part of the purpose of melt infiltration is to overcome temperature limitations. The term *melt infiltration* is an abbreviation for pressureless melt infiltration, which is capillary-driven infiltration by a liquid above its melting point into a porous body. When it is cooled, a composite results in which the matrix is fully crystallized.

The advantages of the melt infiltration process are as follows.

- A potential for producing complex shapes with convenience and precision
- High solidus temperatures of the infiltrant
- An expected absence of creep-inducing glassy phases at high temperature.

Because the infiltrants must be heated above their melting temperatures, the processing temperatures tend to be high compared with those of other processes. There are also the usual limitations resulting from considerations of chemical compatibility, matching of thermal expansivities, and possible degradation of reinforcements under the processing conditions.

Melt infiltration can be subdivided into reactive and nonreactive processes. In reactive melt infiltration, the porous preform reacts with the melt, often with considerable heat evolution. In both processes powder pressing, slip casting, or related means are used to form the porous preform to the desired shape.

Particulates, whiskers, and discontinuous and continuous fibers may be incorporated into the preform. The particulate constituents of the preform serve to fix the orientation and spacing of the fibers. Thus the matrix that envelopes the fibers may itself be a particulate composite.

The dry preform is infiltrated with a ceramic melt, and the driving force for the infiltration is surface tension. The melt solidifies when cooled, producing an essentially fully dense composite structure. Volumetric changes during freezing may result in some closed porosity.

The most prominent example of reactive melt infiltration is the preparation of siliconized SiC by infiltrating molten Si or Si alloys into a porous preform that includes C_p and/or C_f. The C reacts with the Si to form SiC.

Composites successfully produced by nonreactive infiltration include those made by infiltrating various silicates, aluminosilicates, or refractory fluorides into porous SiC preforms. Other nonreactive examples are the infiltration of porous AlN by Al and the infiltration of porous SiC by intermetallics in the area of cermets. The examples of nonreactive infiltration are analogous to the well-known powder metallurgical technique of infiltrating porous sintered Fe compacts by Cu.

Krenkel and Schanz[503] developed a liquid impregnation process whereby they fabricated C/SiC_f ceramics by taking advantage of the quasi-ductile damage tolerance behavior of CCC in combination with the temperature stability of SiC.

Because of the high number of processing parameters, a large variation in material behavior can be achieved. First, structural applications showed the feasibility of the process for a fast, low-cost CMC fabrication route. Complex-shaped structures with dimensions limited only by the geometry of the furnaces can be manufactured in NNS technology by liquid impregnation within 2 weeks. Generally, this process offers enormous potential for lightweight structures, not only for applications in the aeronautical industry (Figure 1.108).

According to Stinton, Lowden, Besmann et al.[504] composites reinforced with continuous ceramic fibers can be effectively fabricated by a recently developed forced chemical vapor infiltration (FCVI) process. This thermal and pressure gradient technique can produce thick, relatively simple-shaped composites. Much commercial interest is focused on SiC matrix composites reinforced with Nicalon fibers because high strengths are routinely obtained for flexure specimens that also exhibit the desired composite behavior. Fracture toughnesses of the materials are very high compared to those of monolithic ceramics.

Modeling of the FCVI process has successfully predicted trends in infiltration rates that agree reasonably well with experimental values. The improved understanding of the

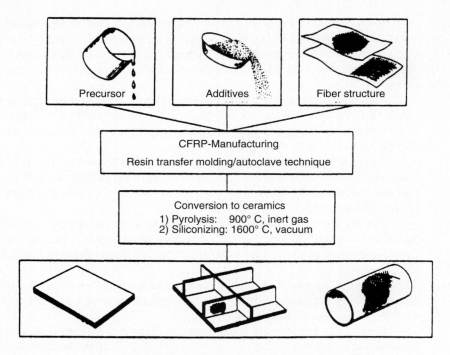

Figure 1.108 Schematic of the liquid impregnation process.[503]

parameters controlling preform permeability and densification resulting from the model allow extension of the process to more complex shapes and alternative reinforcements. The model is used to optimize infiltration conditions to effectively densify voids within fiber bundles as well as between fiber bundles.

Composites reinforced with alternative materials were produced and evaluated to identify materials with promising mechanical properties. Composites produced from SiC_w, SiC_p, or $mullite_w$ exhibited relatively high densities (85% of theoretical density); however, flexure strengths were quite low, and they exhibited brittle fracture. The preforms do provide the anticipated benefit of ease of fabrication compared to hand layup of multiple layers of cloth.

Composites reinforced with continuous fibers of alternate compositions were fabricated during this investigation. Composites produced from Al_2O_3 cloth exhibited relatively low strength, and the fragile and brittle nature of the cloth must be improved before the composites can be fabricated commercially. Composites fabricated from Nextel fibers showed promising results because they exhibited toughening by fiber pullout. The strength of the material was somewhat low but could be improved by optimization of processing conditions and interfacial bonding. The most promising results were obtained from Tyranno fibers. The strength of the material was at least as high as that of Nicalon and exhibited the same strain tolerance.

Within the last decade, a group of technologies collectively called solid free form fabrication (SFF) techniques have emerged for fabricating 3D parts directly from a computer representation of the objects.[505–507] Unlike conventional manufacturing procedures such as casting, injection molding, and forming, this novel approach to manufacturing does not require part-specific tooling. The absence of part-specific tooling in SFF techniques renders them economically attractive in small-scale production runs where tool design and manufacture constitute a significant portion of the total manufacturing cost of the part. Examples of such applications include prototype manufacture for design verification, replacement part production, and tool fabrication.

Selective laser sintering (SLS) is one of the SFF techniques. Lakshminarayan and Marcus[505–507] described a SLS process in which a computer-controlled laser beam was used to create objects directly from the computer-aided design (CAD) data without part-specific tooling. They blended the precursor material alumina and ammonium phosphate powder, and the composite obtained is a potential candidate for fabricating molds for investment casting low-melting-point metals such as Sn and Pb by SLS. In current practice, a wax positive of the part to be cast is fabricated. This wax positive is dipped in a ceramic slurry and then dried. This "dip-and-dry" process is repeated a number of times to build a shell before the wax positive is burnt and the molten metal is cast in the shell. Fabricating the molds directly by SLS eliminates the wax pattern from the investment casting process.[507]

A recently announced new process mixes powders, fibers, or both in a vapor reactant gas stream. The result is less costly fabrication of high-performance ceramic, metal, and carbon composites. The process eliminates the need for a woven preform and yields net shape products with a 0.13-mm tolerance. The high deposition rates permit tailoring the density, composition, and microstructure of the material "as much as 100 times faster than CVI." Once formed, the composite requires only light machining, and an anticipated commercial product would cost about $100/lb.

Reagan and Huffman[508] described a novel, single-step process called chemical vapor composite (CVC) deposition and demonstrated it by fabricating fully dense composite materials without a woven preform. Powders and/or fibers entrained in a CVD reactant stream were codeposited with the matrix material onto a substrate (Figure 1.109). Deposition rates up to 2 mm/h have been obtained. NNS can be CVC-deposited on removable graphite substrates as in conventional CVD. Since a woven preform is not necessary, high-modulus fibers that can withstand high temperatures for extended periods (but cannot be woven or braided) can be used in the CVC process. CVC materials have been deposited with SiC matrices utilizing a variety of ceramic powders (SiC, Al_2O_3, and TiB_2) and whiskers (SiC and Si_3N_4). Tubes 180 cm long and 20 cm in diameter have been made by the CVC process.

Three-point bend tests of SiC powder and SiC matrix have produced the following data.

As deposited	$261 \pm 40 \times 10^6$ Pa
After 300 h at 1400°C in air	$293 \pm 165 \times 10^6$ Pa

CMC parts can be produced via a liquid route instead of conventional powder methods by a new process developed by Sullivan Mining Corporation, Greenwood, Indiana.[509] For example, by reacting a Si-based liquid with NH_3, using supercritical fluid extraction to remove by-products and then heat treating, the Sullivan process can produce a

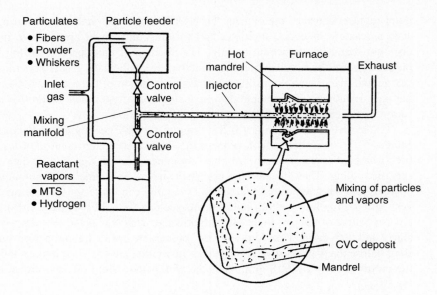

Figure 1.109 SiC CVC reactor.[508]

Si_3N_4 with low density (3.2 g/cm^3), high strength (860 MPa flexural, 73 MPa compressive), and diamondlike hardness. SiC and other ceramics can be manufactured by the Sullivan process if the reactants are changed to include hydrocarbons.

The Sullivan process is capable of producing complex-shaped components for a fraction of the cost of other processes, including pressure sintering, hot pressing, and reaction bonding. When fabricating CMCs, the Sullivan method does less damage to the reinforcing constituents than conventional processes. The liquids used in this process are also able to fill in the spaces between fibers more completely and do not shrink like powders do. Thus more complex shapes are possible. A liquid is also much easier than a powder to handle and pump into forms and makes the process attractive for mass production.

Two basic steps are required for liquid processing of ceramics: (1) reaction of the raw materials, and (2) supercritical fluid extraction of the by-products. This leaves an amorphous ceramic part in a mold and ready for heat treating.

The reaction stage takes place in an autoclave with NH_3 added as a solvent. The NH_3 also supplies N_2 for producing Si_3N_4. Adding H_2 to the autoclave assists in polymerization and helps facilitate the extraction. Other additives, such as metals, help to form second phases with O_2 contaminants.

The extraction step in the Sullivan process is able to separate any undesirable impurities. This makes it possible to use industrial-grade raw materials at a lower cost and still produce high-quality end components. The lower-cost polymeric precursors used as raw materials also instill flexibility and molecular control of chemistry in the final product.

After extraction, the ceramic parts can be heat-treated at temperatures much lower than those of conventional processes, which allows less expensive furnaces to be used.

Overpressure furnaces have been reported to work best, but a variety of furnace types may be used. Further reduction in firing cost by the use of microwave furnaces is being explored.

REFERENCES

1. Blinn, L. 1990. *New Materials Society: Challenges and Opportunities,* New Materials Science and Technology. Bureau of Mines. Washington, DC.

2. Mallick, P. K., and S. Newman. 1990. *Composite Materials Technology: Processes and Properties,* 25–65. Hanser, Munich.

3. Mallick, P. K. 1988. *Fiber-Reinforced Composites: Materials, Manufacturing, and Design,* Mechanical Engineering, vol. 62. Marcel Dekker, New York.

4. Phillips, L. N. 1989. *Design with Advanced Composite Materials,* 38–60. Design Council. Springer-Verlag, London.

5. West, P. 1991. *Adv. Mater.* 13(4):5.

6. Wood, A. S. 1986. *Mod. Plast.* 63:44.

7. Trudeau, E. G., and M. W. Lindsay. 1981. *Proc. 36th Ann. SPI Tech. Conf.* New York, p. 21-B.

8. Schroeter, T. P., and R. K. Leavitt. 1989. *Proc. Adv. Compos. Conf. Exposit.,* September 1989, Detroit, MI.

9. Strong, A. B., and P. Hauwiller. 1989. *Adv. Compos.* 4(5):56–66.

10. New TP composite tapes. *Plast. Technol.* January 1994:13.

11. Pirrung, P. F. 1989. *Adv. Mater. Process.* 136(6).

12. Klein, A. J. 1989. *Adv. Compos.* 4(1).

13. Schwartz, M. M. 1992. *Composite Materials Handbook,* 2nd ed. McGraw-Hill, New York.

14. Jones, S. L. Lockheed unit uses unique tape-laying machine. *Metalworking News* October 23, 1989:10.

15. Leonard, L. 1989. Automating prelayup process steps. *Adv. Compos.* 4(5):49–55.

16. Leonard, L. 1989. *Adv. Compos.* 4(6).

17. Dominy, W. T., Jr. 1993. Cost effective process selection for composite structure. *SME Compos. Manuf. Conf.,* January 1993. SME-EM-93-100. Pasadena, CA.

18. Boey, F., I. Gosling, and S. W. Lye. 1992. High-pressure microwave curing process for an epoxy-matrix/glass-fibre composite. *J. Mater. Process. Technol.* 29(1–3):311–19.

19. Brown, T. J., and C. J. Hooley. 1986. *Int. J. Veh. Des. Technol. Adv. Veh. Des. Ser.* SP6:23.

20. Whelan, A. 1984. *Injection Molding Machines.* Elsevier, London.

21. Johannaber, F. 1984. *Injection Molding Machines: A User's Guide.* Hanser, Munich.

22. Rosato, D. V., and D. V. Rosato, eds. *Injection Molding Handbook.* Van Nostrand Reinhold, New York.

23. Injection molding of high temperature plastics, *NTIS Alert,* 95, (11):10. NERAC, Inc., Tolland, CT, February 1995.

24. Hetzer, R. A. 1983. In *Computer Aided Engineering for Injection Molding,* ed. E. C. Bernhardt. Hanser, Munich.

25. Folkes, M. J. 1986. *Short Fiber Reinforced Thermoplastics.* Research Studies Press, Chichester, U.K.

26. O'Brian, K., H. F. Crincoli, and R. L. Kauffman. 1988. *Proc. 43rd Ann. SPI Tech. Conf.* Cincinnati, OH.

27. Marshall, D. F. 1987. *Mater. Des.* 8(2):77.

28. Cianelli, D. A., J. E. Travis, and R. S. Bailey. 1988. *Proc. 43rd Ann. SPI Tech. Conf.* Cincinnati, OH. Paper also in *Plastic Technol.* April 1988:83–9.

29. Gibson, A. G., S. P. Corscaden, and A. N. McClelland. 1988. *Proc. 43rd Ann. SPI Tech. Conf.* Cincinnati, OH.

30. Gibson, A. G., and A. N. McClelland. 1986. *Proc. 2nd Int. Conf. Fiber Reinf. Compos.* April 1986, Liverpool, U.K. Mechanical Engineering, London.

31. Deans, F. Long-fiber thermoplastics are compounded, molded on line. *Mod. Plast.* November 1993:20–2.

32. Truckenmüller, F., and H.-G. Fritz. 1991. Injection molding of long fiber-reinforced thermoplastics: A comparison of extruded and pultruded materials with direct addition of roving strands. *Polym. Eng. Sci.* 31(18):1316–29.

33. Sasdelli, M. N., V. M. Karbhari, and J. W. Gillespie. 1992. On the use of metal inserts for attachment of composite components to structural assemblies—Review. University of Delaware, Newark. CCM-92-47.

34. Delmonte, J. 1990. *Metal/Polymer Composites.* Van Nostrand Reinhold, New York.

35. Dym, J. B. 1987. *Injection Molds and Molding: A Practical Manual,* 2nd ed. Van Nostrand Reinhold, New York.

36. Carvalho, F. J. 1993. Foreign Industry Analysis: Advanced Composites, Commerce Dept., BXAOFA-93-02, May 1993, 66 p.

37. Fisa, B. 1990. Injection molding of thermoplastic composites. In *Composite Materials Technology: Processes and Properties,* 267–320, ed. P. K. Mallick and S. Newman. Hanser, New York.

38. Kliger, H., and M. Wilson. 1990. *Composite Materials Technology: Processes and Properties,* 179–210, ed. P. K. Mallick and S. Newman. Hanser, Munich.

39. West, P. 1991. *Adv. Mat.* 13(1):2.

40. *NTIS Alert,* Filament winding, NERAC, Inc., Tolland, CT. PB95-852836/WMS: vol 95(3):15.

41. Spencer, B. E. 1988. Tooling considerations for the filament winding process. *SME Prod. Filament Winding Technol. Manuf. Eng. Technol. Conf.,* pp. 151–64, August 1988.

42. Roser, R. 1985. Computer control for filament winding. *Fabricating Composites 1985.* SME, Dearborn, MI.

43. Harper, T. M., J. S. Roberts. 1984. Advanced filament winding machines for large structures. *Proc. 39th Ann. SPI Tech. Conf.* New York.

44. Braun, W. 1986. A control system for filament winding equipment as a prerequisite for automation. *Proc. 1st Int. Conf. Autom. Compos., PRI,* London.

45. Steiner, K. V. 1990. Development of a robotic filament winding workstation for complex geometries. *Proc. 35th Int. SAMPE Symp.,* pp. 757–65, April 1990, Anaheim, CA.

46. Steiner, K. V. 1989. Computer graphics robotic manipulator simulation with collision detection. Master's Thesis, University of Delaware, Center for Composite Materials, Newark.

47. Vaccari, J. A. 1989. More-flexible filament winding. *Am. Mach.,* May:58–9.

48. Franklin, K. M. 1988. Multiaxis filament winding of complex, damage-tolerant structures. *Fiber Tex 1988,* September 1988, pp. 133–7, Greenville, SC. NASA CP 3038.

49. Oldroyd, P. K. 1988. Development and fabrication of V-22 conversion spindle. *Fiber Tex 1988,* September 1988, pp. 331–4, Greenville, SC. NASA CP 3038.

50. Seifert, A., and M. L. Skinner. 1995. An economical solution for automated manufacturing of composite elbows and T-parts. *SME Ann. Conf.,* January 1995, Anaheim, CA. Composites Machine Company, Salt Lake City.

51. Norris, B., and L. Sullivan. 1990. Advanced nacelle structures. *SAE Aerosp. Technol. Conf. Exposit.* October 1990, Long Beach, CA. SAE 901984.

52. Trego, E. 1995. Fiber placement for composite cowls. *Aerosp. Eng.* 15(1):42.

53. McConnell, V. P. 1994. Wish list for repair materials and equipment. *High-Perform. Compos.* July/August:38–43.

54. Munro, M. 1989. Advanced concepts for the filament winding of fibre reinforced polymeric composites. In *Advanced Structural Materials,* ed. D. S. Wilkinson, vol. 9, 295–302. Pergamon Press, Montreal.

55. Rosenow, M. W. K. 1988. Thermoplastic filament wound applications. In *Advanced Structural Materials,* ed. D. S. Wilkinson, vol. 9, 303–10. Pergamon Press, Montreal.

56. Brookstein, D. S. 1986. Braided composites: Attachment considerations. *Proc. CoG/SME Conf. (Compos. Manuf. 5),* January 1986, Los Angeles, CA. EM-111-1 to 9.

57. Campbell, J. H., and J. L. Kittleson. 1991. Winding the tape. *Aerosp. Compos. Mater.* 3(6):21–7.

58. Kittleson, J. L. 1990. Tape winding: A logical progression and alternative to filament winding. *SAMPE J.* January/February:37–41.

59. Stover, D. 1994. Tape-laying precision industrial shafts. *High-Performance Compos.* July/August:29–32.

60. Myers, J. 1994. TP filament winding boosted by technology developments. *Mod. Plast.* August:35.

61. Rudd, C. D., M. J. Owen, V. Middleton, et al. Developments in resin transfer moulding for high volume manufacture. *Proc. 6th Adv. Compos. Conf., ASM/ESD,* pp 301–14, October 1990, Detroit, MI.

62. Rudd, C. D., M. J. Owen, and V. Middleton. 1990. Effects of process variables on cycle time during resin transfer moulding for high volume manufacture. *Mater. Sci. Technol.* 6(7):656–65.

63. Karbhari, V. M., S. G. Slotte, D. A. Steenkamer, et al. 1992. Effect of material, process, and equipment variables on performance of RTM parts. University of Delaware, Newark. CCM-92-17.

64. Spinelli, A. R. 1992. Experimental investigation of flow fronts in resin transfer molding. University of Delaware, Newark. CCM-92-23.

65. Hasko, G., H. B. Dexter, A. Loos, et al. 1994. Application of science-based RTM for fabricating primary aircraft structural elements. *J. Adv. Mater.* 26(1):9–15.

66. Seemann, W. H. 1990. U.S. Patent 4,902,215.

67. Girardy, H. 1992. Resin transfer molding: Answer to problem of industrialization of composites. *SME Compos. Manuf. Conf.,* January 1992, Anaheim, CA. SME-EM-92-114.

68. Pistole, R. D. 1988. Compression molding and stamping. In *Engineered Materials Handbook,* vol. 2, 325–43, ed. J. N. Epel, J. M. Margolis, et al. ASM International, Metals Park, OH.

69. Johnson, C. F. 1988. Resin transfer molding and structural reaction injection molding. In *Engineered Materials Handbook,* vol. 2, 344–57, ed. J. N. Epel, J. M. Margolis, et al. ASM International, Metals, Park, OH.

70. Borstell, H. J. 1988. Hand layup, spray-up, and prepreg molding. In *Engineered Materials Handbook,* vol. 2, 338–43. ed. J. N. Epel, J. M. Margolis, et al. ASM International, Metals Park, OH.

71. Johnson, C. F., N. G. Chavka, R. A. Jeryan, et al. 1987. Design and fabrication of a HSRTM crossmember module. *Proc. 3rd Adv. Compos. Conf., ASM,* pp. 197–218, September 1987, Detroit, MI.

72. Ko, F. K., and C. M. Pastori. 1985. Structure and properties of an integrated 3D fabric for structural composites, pp. 428–39, ASTM, Philadelphia. ST TP 864.

73. *Proc. 2nd Text. Struct. Compos. Symp.,* February 1987, ed. F. K. Ko. Philadelphia.

74. Korngold, J. C., D. E. Larson, A. F., Luscher, et al. 1993. Design for manufacture by resin transfer molding of composite parts for rotorcraft. *SME Compos. Manuf. Conf.,* January 1993, Pasadena, CA. SME-EM-93-103.

75. Horn, S. W. 1990. Complex structural shape preforming process technology. *SME Compos. Manuf. Conf.,* pp. 429–32, January 1990.

76. Hackett, S. C., and P. C. Griebling. 1990. A high performance aerospace resin for resin transfer molding. *Proc. 35th Int. SAMPE Symp.,* pp. 1398–1405, April 1990, Anaheim, CA.

77. Jensen, J. C., D. W. Hearn, and B. R. Colley. 1995. Thermosetting resins for automotive applications. *SME Compos. Manuf.,* February 1995, Anaheim, CA.

78. Bertram, J. L., and P. Mac Puckett. 1994. Getting tough with epoxy resins, *Adv. Mater. Process.* 145(3):29–31.

79. Salem, A. J., K. R. Stewart, S. K., Gifford, et al. 1991. Fabrication of thermoplastic matrix structural composites by resin transfer molding of cyclic bisphenol-A polycarbonate oligomers. *SAMPE J.* 27(1):17–22.

80. Morrison, R. S. 1981. Resin transfer molding of fiber glass preform reinforced polyester resin. *Proc. 36th Ann. Tech. Conf. Plast. Soc.* February 1981, Cincinnati, OH.

81. Harper, J., and H. Sessions. 1992. Fast-curing toughened epoxies for RTM process. *Proc. 8th Adv. Compos. Conf., ASM/ESD,* pp. 95–101, November 1992, Chicago, IL.

82. Chan, A. W., and R. J. Morgan. 1992. Resin impregnation and void formation in enhanced resin transfer molding scheme. *Proc. 8th Adv. Compos. Conf., ASM/ESD,* pp. 69–73, November 1992, Chicago, IL.

83. Sasdelli, M., V. M. Karbhari, and J. W. Gillespie, Jr. 1992. The design and use of molded-in metal inserts and attachments in resin transfer molding. *Proc. 8th Adv. Compos. Conf., ASM/ESD,* pp. 193–202, November 1992, Chicago, IL.

84. Morales, A., and D. Brosius. 1991. Engineered textile preforms for RTM. *SME Compos. Manuf. 10th Conf.,* January 1991, Anaheim, CA. SME-EM-91-109.

85. Fukuta, K., R. Onooka, E. Aoki, et al. 1984. *Proc. 15th Text. Res. Symp.,* ed. S. Kawabata. Textile Machinery Society of Japan, Osaka, Japan.

86. Chou, T.-W., and F. K. Ko. 1989. *Textile Structural Composites.* Elsevier, New York.

87. Brosius, D., and S. Clarke. 1991. Textile preforming techniques for low cost structural composites, *Proc. 7th Adv. Compos. Conf., ASM/ESD,* pp. 1–10, September/October 1991, Detroit, MI.

88. Sharpless, G. C. 1991. Advancement of braiding/resin transfer molding from commercial to aerospace applications, *Proc. 7th Adv. Compos. Conf., ASM/ESD,* pp. 11–21, September/October 1991, Detroit, MI.

89. Jander, M. 1991. Industrial RTM—New developments in molding and preforming technologies. *Proc. 7th Adv. Compos. Conf., ASM/ESD,* pp. 29–34, September/October 1991, Detroit, MI.

90. Horn, S. W. 1991. Advanced in binders for RTM and SCRIM fiber preforming. *Proc. 7th Adv. Compos. Conf., ASM/ESD,* pp. 41–61, September/October, Detroit, MI.

91. Patel, N., M. J. Perry, and L. J. Lee. 1991. Influence of RTM and SRIM processing parameters on molding and mechanical properties. *Proc. 7th Adv. Compos. Conf., ASM/ESD,* pp. 105–11, September/October, Detroit, MI.

92. Owen, M. J., C. D. Rudd, and K. N. Kendall. 1991. Modelling the resin transfer molding (RTM) process. *Proc. 7th Adv. Compos. Conf., ASM/ESD,* pp. 187–202, September/October, Detroit, MI.

93. Peterson, R. C., and R. E. Robertson. 1991. Flow characteristics of polymer resin through glass fiber preforms in resin transfer molding. *Proc. 7th Adv. Compos. ASM/ESD,* pp. 203–8, September/October, Detroit, MI.

94. Loos, A. C., and M. H. Weldemen. 1991. RTM process modeling for advanced fiber architectures. *Proc. 7th Adv. Compos. Conf., ASM/ESD,* pp. 209–16, September/October, Detroit, MI.

95. Wang, H. P., E. W. Liang, and E. M. Perry. 1991. FEMAP-RTM: A resin transfer molding process simulator. *Proc. 7th Adv. Compos. Conf. ASM/ESD,* pp. 217–25, September/October, Detroit, MI.

96. Burgess, K. E. 1993. Integrally woven preforms for RTM. *SME Compos. Manuf. Conf.,* January 1993, Pasadena, CA. EM93-101.

97. *Fiberite's Handbook of Engineered Fabrics,* 1987. Winona, MN.

98. Sharpless, G. 1993. The continuing development of braiding/resin transfer molding for commercial aircraft and aerospace applications. October 1993, *25th Int. SAMPE Tech. Conf.,* Philadelphia, PA.

99. Plastics/composites. *Adv. Mater. Process.* January 1991: 35–40.

100. Beck, W. 1993. Designing the RTM process and product. *SME Compos. Manuf. Conf.* January 1993, Pasadena, CA. EM93-110.

101. Myers, J. 1995. RTM gains speed at lower molding pressure. *Mod Plast.,* February 1995:31.

102. Molina, G., G. Boero, and P. Smeriglio. 1994. Resin transfer molding for car body parts: An integrated approach using CAE methodology. *J. Reinf. Plast. Compos.* 13(8):681–97.

103. Resin transfer molding. *Aerosp. Eng.* June 1994: 36.

104. New composite technique, *Aviat. Week Space Technol.* April 4, 1994:21.

105. Wadsworth, M. 1989. Resin transfer molding of composite structures, *Aerosp. Eng.* December: 23–6.

106. Rock, D. K., and W. S. Mosset. 1990. Composite air intake system for the M1 Abrams tank. *Proc. 35th Int. SAMPE Symp.,* pp. 83–96, April 1990, Anaheim, CA.

107. Chavka, N. G., and C. F. Johnson. 1990. Critical preforming issues for large automotive structures. *Proc. 6th Adv. Compos. Conf., ASM/ESD,* pp. 413–22, October 1990, Detroit, MI.

108. Beardmore, P. 1995. Thermoset and thermoplastic RTM will challenge SMC in automotive. *Plast. Technol.* February: 78.

109. Stover, D. 1992. Near-net preforms and processes take shape. *Adv. Compos.* 7(4):38–42.

110. Innac, J. J. Lower-cost materials, processing reposition composites sector. *Mod. Plast.* June 1994:57–8.

111. Brown, A. S. 1994. Vacuum RTM gains strength. *Aerosp. Am.* September:24–5.

112. Brown, A. S. 1994. Composite makers teach old processes new tricks. *Aerosp. Am.* June:34–37.

113. Mallick, P. K. 1988. *Fiber-Reinforced Composites: Materials, Manufacturing, and Design,* Mechanical Engineering, vol. 62. Marcel Dekker, New York.

114. Methven, J. M. 1994. Microwave heating in the manufacture of composites. *Mater. Technol.* 9(7/8):163–5.

115. Bush, S. F., and J. M. Methven. 1994. A mechanicsm for observed reaction rate enhancement in a microwave applicator based on the concept of a non-equilibrated temperature reaction. *Proc. 39th Int. SAMPE Conf.,* pp. 34–43, April 1994, Anaheim, CA.

116. Lewis, D. A., J. D. Summers, T. C. Ward, et al. 1992. *J. Polym. Sci.* A30:1647–53.

117. Larsson, P. H., and R. T. Astrom. 1991. Design and fabrication of experimental facility for pultrusion of thermoplastic composites. University of Delaware, Newark. CCM 91-32.

118. A thermoplastic pultruder thrives. 1995. *Plast. Technol.* March:84.

119. Cabalquinto, S. C. 1990. Automated pultrusion of composites. SME MM90-144.

120. Johnson, G. S., J. D. Buckley, and M. L. Wilson. 1988. Pultrusion of fiber-reinforced polymethyl methacrylate. *Fiber Tex 1988,* pp. 269–76, September 1988, Greenville, SC. NASA CP 3038.

121. Martin, J. 1988. Pultrusion composites show market growth due to increased technology in three key areas. *Fiber Tex 1988,* pp. 277–84, September 1988, Greenville, SC. NASA CP 3038.

122. Barbero, E. J. 1991. Pultruded structural shapes—From the constituents to the structural behavior. *SAMPE J.* 27(1):25–30.

123. Brown, G., Jr. 1987. Pultrusion—Flexibility for current and future applications. *CoGSME AutoCOM '87,* pp. 220–34, June 1987, Detroit, MI.

124. MacConachie, I. O., and M. L. Wilson. 1988. Application of composites to a space spider robot. *Fiber Tex 1988,* pp. 301–10, September 1988, Greenville, SC. NASA CP 3038.

125. Gabrielle, M. C. 1995. Pultrusion's promise. *Plast. Technol.* March:36–40.

126. Howarth, C. S., R. Balonis, and F. K. Ho. 1990. Design and manufacturing of composite intensive vehicles. *Proc. 35th Int. SAMPE Symp.,* pp. 1544–51, April 1990, Anaheim, CA.

127. Ho, F. K. 1987. Braiding. In *Engineered Materials Handbook,* vol. 1, ed. T. Reinhart, et al. *Composites,* 519–28. ASM International, Metals Park, OH.

128. Hess, J. P. 1988. Braided composite structures. *SME Fabr. Compos. Conf.,* pp. 375–81, September 1988, Dearborn, MI.

129. Munjal, A. K., D. F. Spencer, E. W. Rahnenfueher, et al. 1990. Design and fabrication of high quality graphite/epoxy braided composite tubes for space structures. *Proc. 35th Int. SAMPE Symp.,* pp. 1954–68, April 1990, Anaheim, CA.

130. Li, W., M. Hammad, R. Reid, et al. 1990. Bearing behavior of holes formed using different methods in 3D braided graphite/epoxy composites. *Proc. 35th Int. SAMPE Symp.,* pp. 1638–46, April 1990, Anaheim, CA.

131. Albany braids again. *Aerosp. Compos. Mater.* 4(2):7, 1992.

132. Peterson, C. 1994. Automation developments may offer cost-effective ways to manufacture composite structures. *Adv. Compos.* January/February: 20–26.

133. Florentine, R. A. 1991. The designer of 3D braided preforms and the automotive design engineer—Communications for innovation and profit. *Proc. 7th Compos. Conf., ASM/ESD,* pp. 23–8, September 1991, Detroit, MI.

134. Klein, A. J. 1987. Braids and knits: Reinforcement in multidirections. *Adv. Compos.* September/October:36–48.

135. Brown, A. S. 1993. Weaving new strength into composites. *Aerosp. Am.* 31(9):26–35.

136. Schooneveld, G. V. 1988. Potential of knitting/stitching and resin infusion for cost-effective composites. *Fiber Tex 1988,* pp. 113–12, September 1988, Greenville, SC. NASA CP 3038.

137. Franklin, K. M. 1988. Multiaxis filament winding of complex, damage-tolerant structures. *Fiber Tex 1988,* pp. 133–149, September 1988, Greenville, SC. NASA CP 3038.

138. Leonard, L. 1989. Specialty fabrics: The promise of near-net-shape preforms. *Adv. Compos.* May/June:40–7.

139. Palmer, R., and F. Curzio. 1988. Cost-effective damage-tolerant composites using multi-needle stitching and RTM/VIM processing: An evaluation of stitching concepts for damage-tolerant composites. *Fiber Tex 1988,* pp. 25–51, September 1988, Greenville, SC. NASA CP 3038.

140. Dow, M. B., D. L. Smith, and S. J. Lubowinski. 1988. An evaluation of stitching concepts for damage-tolerant composites. *Fiber Tex 1988,* pp. 53–74, September 1988, Greenville, SC. NASA CP 3038.

141. Zawislak, S. P., and J. R. Maiden. 1988. Advanced weaving concepts for complex structural preforms. *Fiber Tex 1988,* pp. 91–112, September 1988, Greenville, SC. NASA CP 3038.

142. Smith, D. L., and H. B. Dexter. 1988. Woven fabric composites with improved fracture toughness and damage tolerance. *Fiber Tex 1988,* pp. 75–90, September 1988, Greenville, SC. NASA CP 3038.

143. Swinkels, K. 1992. Three-dimension woven glass fabric as reinforcements in composites. *Proc. 8th Adv. Compos. Conf., ASM/ESD,* pp. 125–8, November 1992, Chicago, IL.

144. Mohamed, M. H., and Z. Zhang. 1988. Weaving of 3D preforms. *Fiber Tex 1988,* pp. 193–216, September 1988, Greenville, SC. NASA CP 3038.

145. Hogg, P. J., and D. H. Woolstencroft. 1991. Non-crimp thermoplastic composite fabrics: Aerospace solutions to automotive problems. *Proc. 7th Adv. Compos. Conf., ASM/ESD,* pp. 339–49, September/October 1991, Detroit, MI.

146. Knitted-fiber-reinforced composites. *Automot. Eng.* August 1994:21; *Adv. Mater. Process.,* December 1994:15.

147. Contoured composite structures made more simply. *Production* December 1994:25.

148. McConnell, V. P. 1995. Infrastructure update. *High-Performance Compos.* 3(3):21–5.

149. Krone, J. R., and J. H. Walker. 1986. Thermoforming woven fabric reinforced polyphenylene sulfide composites. *CoGSME Compos. Manuf. 5th Conf.,* pp. 112–24, January 1986, Los Angeles, CA.

150. Beziers, D. 1990. Electron beam curing of composites. *Proc. 35th Int. SAMPE Symp.,* pp. 1220–31, Anaheim, CA.

151. Farley, G. L. 1995. Making composite-material parts at moderate to high rates. NASA Technical Brief, March 1995, pp. 73–4.

152. Monks, R. 1992. Automated system speeds preforming. *Plast. Technol.* August:29.

153. Medwin, S. J., and E. J. Coyle. 1993. Thermoplastic composite parts manufacture at DuPont. SME EM-93-106.

154. Sambell, R. A., D. Bowen, and D. C. Phillips. 1972. Carbon fiber composites with ceramic and glass matrices. Part I. Discontinuous fibers. *J. Mater. Sci.* 7:663–75.

155. Sambell, R. A., et al. 1972. Carbon fiber composites with ceramic and glass matrices. Part 2. Continuous fibers. *J. Mater. Sci.* 7:676–81.

156. Phillips, D. C., R. A. Sambell, and D. Bowen. 1972. The mechanical properties of carbon fiber reinforced Pyrex. *J. Mater. Sci.* 7:1454–64.

157. Phillips, D. C. 1974. Interfacial bonding and toughness of carbon fiber reinforced glass and glass ceramics. *J. Mater. Sci.* 9:1847–54.

158. Levitt, S. R. 1973. High strength graphite fibre—LAS. *J. Mater. Sci.* 8:793–806.

159. Fitzer, E. 1978. Fiber reinforced ceramics. *Proc. Int. Symp. Factors Densification Sintering Oxide Non Oxide Ceram.,* pp. 618–73, Hakone.

160. Sahebkar, M., J. Schlichting, and P. Schubert. 1978. Possibility of reinforcing glass by carbon fibers. *Ber. Deut. Keram. Ges.* 55(5):265–68.

161. Prewo, K. M., and J. F. Bacon. 1978. Glass matrix composites. 1. Graphite fiber reinforced glass. *Proc. 2nd Int. Conf. Compos., AIME,* pp. 64–74, New York.

162. Prewo, K. M., and J. J. Brennan. 1989. Fiber reinforced glasses and glass ceramics for high performance applications. In *Encyclopedia Composites,* ed. S. M. Lee, pp. 97–116. Technomics Pub. Inc., Lancaster, PA.

163. James, P. F. 1995. Glass ceramics: New compositions and uses. *J. Non-Cryst. Solids* 181(1–2):1–15.

164. Prewo, K. M. 1979. Development of a new dimensionally and thermally stable composite. *Proc. Spec. Top. Adv. Compos. Manuf.,* pp. 1–30, El Segundo, CA.

165. Prewo, K. M., and E. J. Minford. 1984. Thermal stable composites—Graphite reinforced glass. *Proc. SPIE—Int. Soc. Opt. Eng.* 505:188–91.

166. Minford, E. J., and K. M. Prewo. 1985. Friction and wear of graphite fiber reinforced glass matrix composite. *Wear* 102:253–64.

167. Zern, C. A. 1989. Strengthening and toughening mechanisms in ceramic fiber-reinforced glass-matrix composites. Ph.D. Thesis, Rutgers University, New Brunswick, NJ. 90-13470.

168. Rice, R. W. 1983. Fundamental needs to improve ceramic fiber composites. *Ceram. Eng. Sci. Proc.* 4(7–8):485–91.

169. Prewo, K. M., and J. F. Bacon. 1979. *Int. Conf. Compos. Mater. Process.,* pp. 64–74.

170. Prewo, K. M., and J. J. Brennan. 1982. Silicon carbide yarn reinforced glass matrix composites. *J. Mater. Sci.* 17:1201.

171. Prewo, K. M., and J. J. Brennan. 1980. High strength silicon carbide fibre-reinforced glass-matrix composites. *J. Mater. Sci.* 15(2):463–8.

172. Prewo, K. M., and J. F. Bacon. 1979. Graphite fiber reinforced glass matrix composites. *SAMPE Q.* 10(4):42.

173. Jha, A., and M. D. Moore. 1992. A study of the interface between silicon carbide fibre and lithium aluminosilicate glass ceramic matrix. *Glass Technol.* 33(1):30–7.

174. Lowden, R. A., and D. P. Stinton. 1987. The influence of fibre-matrix bond on the mechanical behaviour of Nicalon/SiC composites. ORNL TM-10667.

175. Lewinsohn, L. A. 1993. Hybrid whisker-fibre-reinforced glass-matrix composites with improved transverse toughness. *J. Mater. Sci. Lett.* 12(18):1478–80.

176. Bacon, J. F., K. M. Prewo, and R. D. Veltri. 1978. Glass matrix composites. II: Alumina reinforced glass, *Proc. ICCM-2 Conf.,* pp. 753–69.

177. Michalske, T. A., and J. R. Hellmann. 1988. Strength and toughness of continuous-alumina-fiber-reinforced glass matrix composites. *J. Am. Ceram. Soc.* 71(9):725–31.

178. Prewo, K. M. 1982. A compliant, high failure strain fibre reinforced glass matrix composite. *J. Mater. Sci.* 17:3549–63.

179. Clarke, D. A. 1990. Fabrication aspects of glass matrix composites for gas turbine applications. *Int. Conf. New Mater. Their Appl.,* pp. 173–83. IOP Publications, Bristol, U.K.

180. Schott develops glass composites. *Ceram. Ind.* August 1994: 21.

181. Wolf, W. D., L. F. Francis, C.-P. Lin, et al. 1995. Melt-infiltration processing and fracture toughness of alumina-glass dental composites. *J. Am. Ceram. Soc.* 76(10):2691–4.

182. Chiou, S., and H. T. Hahn. 1994. Ultrasonic sol/gel processing of aluminoborosilicate glass and its composite with carbon fiber reinforcement. *J. Am. Ceram. Soc.* 77(1):155–60.

183. Qi, D. Processing and properties of discontinuous carbon fiber reinforced glass matrix composites. Ph.D. Thesis, Penn State University, University Park, PA.

184. Brennan, J. J., and K. M. Prewo. 1982. Silicon carbide fibre reinforced glass-ceramic matrix composites exhibiting high strength and toughness. *J. Mater. Sci.* 17:2371–83.

185. Brennan, J. J., and K. M. Prewo. 1983. Investigation of lithium aluminosilicate (LAS)/SiC fiber composites for naval gas turbine applications. UTRC 83-916232-4, NASC Contract N00019-82-C-0438.

186. Gadkaree, K. P., and K. C. Chyung. 1988. SiC whisker and whisker/fiber reinforced glass and glass ceramic hybrid composites. *Proc. ASM Int. Conf.,* ed. R. A. Bradley, D. E. Clark, and D. C. Larsen, pp. 97–104, Chicago, IL.

187. Fishman, S. G. 1986. *J. Met.* 38(3):26.

188. Flom, Y., and R. J. Arsenault. 1986. *J. Met.* 38(3):31.

189. Flom, Y., and R. J. Arsenault. 1986. *Mater. Sci. Eng.* 77:191.

190. Howes, A. H. M. 1986. *J Met.* 38(3):28.

191. Mortensen, A., M. N. Gungor, J. A. Cornie, et al. 1986. *J. Met.* 38(3):30.

192. Nardone, V. C., and K. W. Prewo. 1986. *Scr. Metall.* 20:43.

193. Mortensen, A., J. A. Cornie, and M. C. Flemings. 1988. *J. Met.* 40(2):12.

194. Manoharan, M., and J. J. Lewandowski. *Acta Metall.* 38:489.

195. Gurganus, T. B., R. G. Gilliland, and W. H. Hunt. 1990. *Ind. Heat.* February:46.

196. Srivatsan, T. S., I. A. Ibrahim, F. A. Mohamed, et al. 1991. Processing techniques for particulate-reinforced metal aluminum matrix composites. *J. Mater. Sci.* 26:5965–78.

197. Divecha, A. P., S. G. Fishman, and S. D. Karmarkar. 1986. *J. Met.* 33:12–17.

198. Ibrahim, I. A., F. A. Mohamed, and E. J. Lavernia. Particulate reinforced metal matrix composites—A review. *J. Mater. Sci.* 26(5):1137–56.

199. Schoutens, J. E. 1982. *Introduction to Metal Matrix Composite Materials, MMCIAC No. 272, MMCIAC.* Kaman Tempo, Santa Barbara, CA.

200. Schoutens, J. E. 1989. Metal matrix composites. In *Encyclopedia of Composite Materials,* ed. S. M. Lee, 175–269. Technomics, Lancaster, PA.

201. Lynch, C. T., and J. P. Kershaw. 1972. *Metal Matrix Composites.* CRC Press, Cleveland, OH.

202. Wagner, R. S., and W. C. Ellis. 1964. *Appl. Phys. Lett.* 4(5):39–90.

203. Wagner, R. S., W. C. Ellis, K. A. Jackson, et al. 1964. *J. Appl. Phys.* 35(10):2993–3000.

204. Wagner, R. S., and W. C. Ellis. 1965. *Trans. Met. Soc. AIME* 233:1053.

205. Milewski, J. V., et al. 1985. *J. Mater. Sci.* 20:1160–6.

206. Galasso, F. S. 1988. *Advanced Fibers and Composites.* Gordon and Breach, New York.

207. Mazdiyasni, K. S. 1990. *Fiber Reinforced Ceramic Composites.* Noyes, Park Ridge, NJ.

208. Kreider, K. G., and K. M. Prewo. 1974. Boron-reinforced aluminum. In *Composite Materials,* vol. 4: *Metal Matrix Composites,* ed. K. G. Kreider. Academic Press, New York.

209. Crowe, C. R., R. A. Gray, and D. F. Hasson. 1985. *Proc. 5th Int. Conf. Compos. Mater.,* ed. W. Harrigan, J. Strife, and A. K. Dringra. TMS, Warrendale, PA.

210. Nair, S. V., J. K. Tien, and R. C. Bates. 1985. *Int. Met. Rev.* 30:275.

211. Hoover, W. R. 1989. Duralcan metal-matrix composites: Data report package. Dural Aluminum Composites Corporation, San Diego, CA.

212. Mehrabian, R. 1988. New pathways to processing composite materials. MRS Symposia Proceedings, vol. 120, 3–20.

213. Steelman, T. E., A. D. Bakalyar, and L. Konopka. 1987. Aluminum metal-matrix composites structural design development. AFWAL-TR-86-3087.

214. Cebulak, W. S. 1986. Advanced metal structures: Review. AFWAL-TR-87-3042.

215. Hunt, W. H., O. Richmond, and R. D. Young. 1987. *Proc. 6th Int. Conf. Compos. Mater.,* ed. F. L. Matthews, N. C. R. Buskell, J. M. Hodgkinson, et al. p. 2209. Applied Science, London.

216. Brown, C. W. 1990. Particulate metal matrix composite properties. *Proc. P/M Aerosp. Def. Technol. Conf. Exhibit.,* ed. F. H. Froes, pp. 203–205.

217. Rack, H. J., T. R. Baruch, and J. L. Cook. 1982. In *Progress in Science and Engineering of Composites,* ed. T. Hayashi, K. Kawata, and S. Umekawa, 1465. Japanese Society of Composite Materials, Tokyo.

218. Clyne, T. W., M. G. Bader, G. R. Cappleman, et al. 1985. *J. Mater. Sci.* 20:85.

219. Cornie, J. A., A. Mortensen, and M. C. Flemings. 1987. *Proc. 6th Int. Conf. Compos. Mater.,* ed. F. L. Matthews, N. C. R. Buskell, J. M. Hodgkinson, et al. pp. 2297–305. Applied Science, London.

220. *Research in Materials Processing at MIT.* Annual Report, 1992, July 1990–December 1991, MIT Press, Cambridge, MA.

221. Rack, H. J. 1993. *Adv. Mater. Manuf. Process.* 3:327.

222. Taha, M. A., and N. A. El-Mahallawy, eds. 1993. *Advances in Metal Matrix Composites,* vols. 79 and 80. Trans Tech, Aedermannsdorf, Switzerland.

223. Rohatgi, P. K., A. Asthana, and S. Das. 1986. *Int. Met. Rev.* 31(3):115.

224. Clyne, T. W., and J. F. Mason. 1987. *Metall. Trans.* 18A:1519.

225. Donomoto, T., N. Miuras, K. Funatami, et al. 1983. SAE Technical Paper 83-052, Detroit, MI.

226. Mortensen, A., J. A. Cornie, and M. C. Flemings. 1988. *Metall. Trans.* 19A:709.

227. Hunt, M. 1989. *Mater. Eval.* October:45.

228. Klimowicz, T. F., and X. Nguyen-Dinh. 1994. *Extrusion of Direct Chill Cast Alumina-Aluminum Metal Matrix Composites,* 127–37. Dural Aluminum Composites Corporation, San Diego, CA.

229. Weinstein, J. 1989. *Proc. Int. Symp. Adv. Process. Charact. Compos. Mater.,* ed. H. Mostaghchi, vol. 17: *CIM/ICM,* pp. 132–42. Pergamon Press, Oxford.

230. Ferguson, B. L., and O. D. Smith. 1984. *Ceracon Process: Metals Handbook,* 9th ed., vol. 7, 537. ASM International, Metals Park, OH.

231. Ferguson, B. L., A. Kuhn, O. D. Smith, et al. 1984. *Int. J. Powder Metall. Powder Technol.* 24:31.

232. Geiger, A. L., and M. Jackson. 1989. *Adv. Mater. Process.* 7:23.

233. Rack, H. J. 1988. In *Processing and Properties of Powder Metallurgy Compos.* ed. P. Kumar, K. Vedula, and A. Ritter, 155. TMS, Warrendale, PA.

234. Rack, H. J. 1987. *Proc. 6th Int. Conf. Compos. Mater.* ed. F. L. Matthews, N. C. R. Buskell, J. M. Hodgkinson, et al., 2382. Applied Science, London.

235. Hunt, W. H., Jr., C. R. Cook, K. P. Amanie, et al. 1987. In *Powder Metallurgy Composites,* ed. P. Kumar, A. Ritter, and K. Vedula. TMS, Warrendale, PA.

236. Rack, H. J. 1988. In *Dispersion Strengthened Aluminum Alloys,* ed. Y. M. Kim, and W. Griffith. TMS, Warrendale, PA.

237. Krisknamurthy, S., Y.-W. Kim, G. Das, et al. 1990. Interfaces in metal-ceramics composites. In *Metal-Ceramic Matrix Composites: Processing, Modelling and Mechanical Behaviour,* ed.

R. B. Bhagat, A. H. Claver, P. Kumar, et al. 145. Minerals, Metals and Materials Society, Warrendale, PA.

238. Marcus, H. L., D. L. Bournell, Z. Eliezer, et al. 1987. *J. Met.,* December.

239. Elkabir, G., L. K. Rabenberg, C. V. Prasad, et al. 1986. *Scr. Metall.* 20:1411.

240. Prasad, C., S. Ranganathan, B. H. Klee, et al. 1988. *Mater. Res. Soc. Symp. Proc.* 120:23.

241. Chawla, K. K. Metal matrix composites. In *Materials Science and Technology,* ed. R. W. Cahn, P. Haasen, E. J. Kramer, vol. 15, 122–179. VCH, Weinheim, Germany.

242. Chou, T. W., A. Kelly, and Okura. 1985. *Composites* 16:187.

243. Moore, J. T., D. V. Wilson, and W. T. Roberts. 1981. *Mater. Sci. Eng.* 48:107.

244. Watanabe, O. 1979. *Proc. 1st Jpn. USSR Symp. Compos. Mater.* p. 226. Moscow University Press, Moscow.

245. Inagaki, J., Y. Terasawa, E. Nakata, et al. 1980. *Proc. 2nd Jpn. USSR Symp. Compos. Mater.,* p. 37; *Tetsu Hagane* 65:1946, 1979.

246. Clyne, T. W., and P. J. Withers. 1984. *An Introduction to Metal Matrix Composites.* Cambridge University Press, Cambridge.

247. Sakai, M., O. Watanabe, and K. Watanabe. *Trans. Jpn. Inst. Met.* 43:181.

248. Moss, M., W. R. Hoover, and D. H. Schuster. 1978. *Proc. Sagamore Army Mater. Res. Conf.,* vol. 21, p. 425.

249. Dolowy, J. F., B. A. Webb, and W. C. Harrigan. 1979. *Proc. 24th SAMPE Symp.,* p. 1443.

250. Debski, R. T., and D. R. Beeler. 1979. *Proc. 24th SAMPE Symp.,* p. 1382.

251. Vinson, J. R., and T. W. Chou. 1975. *Composite Materials and Their Use in Structures.* Applied Sciences, London.

252. Prewo, K. M., and K. G. Kreider. 1972. *Metall. Trans.* 3:2201.

253. Metcalfe, A. G. 1974. In *Composite Materials,* vol. 4, ed. L. J. Broutman, and R. H. Krack. Academic Press, New York.

254. Smith, P. R., F. H. Froes, and J. T. Cammett. 1983. ed. J. E. Hack, and M. F. Amateau, 143. TMS, Warrendale, PA.

255. Abis, S. 1989. *Compos. Sci. Technol.* 35:1.

256. Lock, J. 1990. *Prof. Eng.* April:21.

257. White, J., I. G. Palmer, I. R. Hughes, et al. 1989. In *Aluminum-Lithium Alloys V,* ed. T. H. Sanders, Jr., and E. A. Starke, Jr., 1635–46. Materials and Component Engineering, Birmingham, UK.

258. Willis, T. 1988. *Met. Mater.* 4:485.

259. Singer, A. R. E., and S. Ozbek. 1985. *Powder Metall.* 28:72.

260. Evans, R. W., A. G. Leatham, and R. G. Brooks. *Powder Metall.* 28:13.

261. Spencer, D. B., R. Mehrabian, and M. C. Flemings. 1972. *Metall. Trans.* 3:1925.

262. Mehrabian, R., and M. C. Flemings. 1972. *Trans. Am. Foundrymen Soc.* 80:173.

263. Fascetta, E. F., R. G. Riek, R. Mehrabian, et al. 1973. *Trans. Am. Foundrymen Soc.* 81:81.

264. Mehrabian, R., R. G. Riek, and M. C. Flemings. 1975. *Metall. Trans.* 5:1899.

265. Huda, D., M. A. El Baradie, and M. S. J. Hashmi. 1993. Metal-matrix composites: Manufacturing aspects. Part I. *J. Mater. Process. Technol.* 37(1–4):513–28.

266. Gupta, M., F. A. Mohamed, and E. J. Lavernia. 1990. *Mater. Manuf. Proc.* 5:165.

267. Gupta, M., F. A. Mohamed, and E. J. Lavernia. 1989. *Proc. Int. Symp. Adv. Process. Charact. Ceram. Met. Matrix Compos.,* ed. H. Mostaghaci, vol. 17: *CIM/ICM,* p. 236. Pergamon Press, Oxford.

268. Gupta, M., F. A. Mohamed, and E. J. Lavernia. 1990. In *Heat Transfer Mechanisms and Their Effect on Microstructure During Spray Atomization and Co-Deposition of Metal Matrix Composites,* ed. S. Fishman, TMS Fall Meeting, Detroit, MI, 1990.

269. Nagelberg, A. S. 1986. The effect of processing parameters on the growth rate and microstructure of Al_2O_3/metal matrix composites. In *Processing Science of Advanced Ceramics,* MRS Symposia Proceedings, ed. I. A. Askay, G. L. McVay, and D. R. Ulrich, vol. 155, pp. 275–82. MRS, 1989. DAALO3-89GOO24.

270. Newkirk, M. S., H. D. Lesher, D. R. White, et al. 1987. Preparation of Lanxide ceramic composite materials: Matrix formation by the directed oxidation of molten metals. *Ceram. Eng. Sci. Proc.* 8:879–85.

271. Lewis, D. 1991. In situ reinforcement of metal matrix composites. In *Metal Matrix Composites: Processing and Interfaces,* ed. R. K. Everett and R. J. Arsenault, 121–50, Academic Press, Boston.

272. Shtessel, V. E., and M. J. Koczak. 1994. The production of metal matrix composites by reactive processes. *Mater. Technol.* 9(7–8):154–8.

273. *MMCs: Technology and Industrial Applications.* Techtrends: International Reports on Advanced Technologies. Innovative 128. 1990. Paris, France.

274. Waku, Y., and T. Nagasawa. 1994. Future trends and recent developments of fabrication technology for advanced metal matrix composites. *Mater. Manuf. Proc.* 9(5):937–63.

275. Fukunaga, H. 1988. Squeeze casting processes for fiber reinforced metals and their mechanical properties. *ASM Int. Meet. Cast Reinf. Met. Compos.,* ed. S. G. Fishman and A. K. Dringra, pp. 101–06, September 1988, Chicago, IL.

276. Verma, S. K., and J. L. Dorcic. 1988. Manufacturing of composites by squeeze casting. *Int. Meet. Cast Reinf. Met. Compos.,* ed. S. Fishman and A. K. Dringa, pp. 115–24, September 1988, Chicago, IL.

277. Sample, R. J., R. B. Bhagat, and M. F. Amateau. 1988. High pressure squeeze casting of unidirectional graphite fiber reinforced aluminum matrix composites. *Int. Meet. Cast Reinf. Met. Compos.,* ed. S. G. Fishman and A. K. Dringa, pp. 179–83, September 1988, Chicago, IL.

278. Cook, C. R., D. I. Yun, W. H. Hunt, Jr. 1988. System optimization for squeeze cast composites. *Int. Meet. Cast Reinf. Met. Compos.,* ed. S. G. Fishman and A. K. Dringa, pp. 195–203, September 1988, Chicago, IL.

279. Zhu, Z. 1988. A literature survey of fabrication methods of cast reinforced metal composites. *ASM Int. Meet. Cast Reinf. Met. Compos.,* ed. S. G. Fishman and A. K. Dringra, 93–9. September 1988, Chicago, IL.

280. Chadwick, G. A. 1991. Squeeze casting of metal matrix composites using short fibre preforms. *Mater. Sci. Eng.* A135:23–28.

281. El Baradie, M. A. 1990. Manufacturing aspects of metal matrix composites. *J. Mater. Process. Technol.* 24:261–72.

282. Chadwick, G. A. 1994. *Squeeze Casting of Magnesium Alloys and Magnesium Based Metal Matrix Composites,* 75–8. Hi Tech Metals R&D, Chilworth Research Centre, Southampton, England.

283. Rohatgi, P. 1990. *Solidification of Metal Matrix Composites.* TMS, Warrendale, PA.

284. Rohatgi, P. 1990. Advances in cast MMCs. *Adv. Mater. Process.* 137(2):39–44.

285. Mortensen, A., and M. J. Koczak. 1993. The status of metal-matrix composite research and development in Japan. *J. Met.* 45(3):10–8.

286. Allison, J. E., and G. S. Cole. 1993. Metal-matrix composites in the automotive industry: Opportunities and challenges. *J. Met.* 45(1):19–24.

287. Verma, S. K., and J. L. Dorcic. 1988. *Adv. Met. Proc.* May:48–51.

288. Brusethaug, S., O. Reiso. 1991. Metal matrix composite—Processing, microstructures and properties. *12th Risø Int. Symp.,* ed. N. Hansen et al, pp. 247–55, Roskilde, Denmark. Risø.

289. Ray, S. P., and D. I. Yun. 1991. Squeeze-cast Al₂O₃/Al ceramic-metal composites. *Ceram. Bull.* 70(2):195–7.

290. Suganuma, K., G. Sasaki, T. Fujita, et al. 1992. High temperature strength and whisker/matrix interface of β silicon nitride whisker reinforced 6061 aluminum alloy. *Mater. Trans.* Japan Institute of Metals, 33(7):659–68.

291. Singh, J., S. K. Goel, V. N. S. Mathur, et al. 1991. Elevated temperature tensile properties of squeeze-cast Al-Al₂O₃-MgO particulate MMCs up to 573 K. *J. Mater. Sci.* 26(10):2750–8.

292. Rohatgi, P. 1991. Cast aluminum-matrix composites for automotive applications. *J. Met.* 43:10.

293. Balasubramanian, P. K., P. S. Rao, B. C. Pai, et al. 1990. Synthesis of cast Al-Zn-Mg-TiO₂ particle composites using liquid metallurgy (LM) and rheocasting (RC). In *Solidification of Metal Matrix Composites,* ed. P. Rohatgi, 181–90. Minerals, Metals and Materials Society, Indianapolis, IN.

294. Keshavaram, B. N., P. K. Rohatgi, R. Asthana, et al. 1990. Solidification of Al-glass particulate composites. In *Solidification of Metal Matrix Composites,* ed. P. Rohatgi, 151–70. Minerals, Metals and Materials Society, Indianapolis, IN.

295. Ghosh, P. K., and S. Ray. 1990. Solidification structure in compocast Al(Mg)-Al₂O₃ particulate composite. In *Solidification of Metal Matrix Composites,* ed. P. Rohatgi, 205–12. Minerals, Metals and Materials Society.

296. Laurent, V., P. Jarry, G. Regazzoi, et al. 1992. Processing-microstructure relationships in compocast magnesium/SiC. *J. Mater. Sci.* 27(16):4447–59.

297. Mittnick, M. A. 1990. Continuous SiC fiber reinforced metals. *SAMPE J.* 26(5):49–54.

298. Dumant, X., S. Kennerknecht, and R. Tombari. 1990. Investment cast metal matrix composites. SME EM 90-441.

299. Wilks, T. E. 1992. Cost-effective magnesium MMCs. *Adv. Mater. Process.* 142(2):27–9.

300. Li, Q. F., N. L. Loh, N. P. Hung. 1995. Casting and HIPing of Al-based metal matrix composites (MMCs). *J. Mater. Process. Technol.* 48(1–4):373–8.

301. Divecha, A. P., S. D. Karmarkar, W. A. Ferrando, et al. 1994. Centrifugal casting of reinforced articles, Department of the Navy, Washington, DC. AD-D016 277/6; *NTIS Alert,* 94(18):8.

302. Hunt, M. 1991. Beyond magic: Consistent cast composites, *Mater. Eng.* January:21–4.

303. Suery, M., and L. Lajoye. 1990. Microstructural characterization of Al-Si-SiC composites produced by centrifugal casting. In *Solidification of Metal Matrix Composites,* ed. P. Rohatgi, 171–9. Minerals, Metals and Materials Society.

304. Premkumar, M. K., D. I. Yun, and R. R. Sawtell. 1992. Aluminum composite materials via pressure casting. *Proc. 121st TMS Ann. Meet.,* March 1992, San Diego, CA.

305. Hoover, W. 1991. Metal matrix composite—Processing, microstructures and properties. *Proc. 12th Risø Int. Symp.,* ed. N. Hansen et al., pp. 387–92. Roskilde, Denmark. Risø.

306. Kobayashi, T., M. Yosino, H. Iwanari, et al. 1990. Mechanical properties of SiC whisker reinforced aluminum alloys fabricated by pressure casting method. *ASM Int. Conf.* ed. S. Fishman and A. K. Dringra, pp. 205–12, Chicago, IL.

307. Nourbakhsh, S., H. Margolin, and F. L. Liang. 1990. Fabrication of continuous ZTA fiber reinforced titanium aluminide intermetallic composite material by pressure casting. In *Solidifi-*

cation of Metal Matrix Composites, ed. P. Rohatgi, 103–14. Minerals, Metals and Materials Society.

308. Kun, Y., V. Dollhopf, and R. Kochendörfer. 1993. CVD SiC/Al composites produced by a vacuum suction casting process. *Compos. Sci. Technol.* 46(1):1–6.

309. Pathak, J. P. 1993. Aluminum-matrix high alumina slag particulate composites. *Scand. J. Metall.* 22(5):260–5.

310. Veeck, S. J., G. W. Wolter, C. R. Wojciechowski. 1994. Selective reinforcement of titanium investment castings. *Adv. Mater. Process.* 145(4):37–9.

311. Rohatgi, P. 1991. Cast aluminum-matrix composites for automotive applications. *J. Met.* 43(4):10–5.

312. Production of metal matrix composites by liquid metal infiltration. *NTIS Alert* 95(2):2.

313. Klier, E. M., A. Mortensen, J. A. Cornie, et al. 1991. Fabrication of cast particle-reinforced metals via pressure infiltration. *J. Mater. Sci.* 26(9):2519–26.

314. Blucher, J. T. 1992. Discussion of a liquid metal pressure infiltration process to produce metal matrix composites. *J. Mater. Process. Technol.* 30(3):381–90.

315. Cook, A. J., and P. S. Werner. 1991. Pressure infiltration casting of metal matrix composites. *Mater. Sci. Eng.* A144:189–206.

316. Bhagat, R. B. 1991. High pressure infiltration casting: Manufacturing net shape composites with a unique interface. *Mater. Sci. Eng.* A144:243–51.

317. Öttinger, O., and R. F. Singer. 1993. An advanced melt infiltration process for the net shape production of metal matrix composites. *Zeitschrift fur Metallk.* 84(12):827–31.

318. Muscat, D., K. Shanker, and R. A. L. Drew. Al/TiC composites produced by melt infiltration. *Mater. Sci. Technol.* 8(11):971–6.

319. Jiang, J.-Q., A.-B. Ma, H.-N. Liu, et al. 1991. Fabrication of alumina short fibre reinforced aluminum alloy via centrifugal force infiltration. *Mater. Sci. Technol.* 19(9):783–7.

320. Cheng, H. M., Z. H. Lin, B. L. Zhou, et al. 1993. Preparation of carbon fibre reinforced aluminium via ultrasonic liquid infiltration technique. *Mater. Sci. Technol.* 9(7):609–14.

321. BNF Metals Technology Centre Technical Literature, 1991, Wantage, UK.

322. Rokhale, A. B., L. Lu, and R. Abbaschian. 1990. Interface reactions in melt infiltration processed intermetallic matrix composites. In *Solidification of Metal Matrix Composites,* ed. P. Rohatgi, 115–31. Minerals, Metals and Materials Society, Indianapolis, IN.

323. Ray, S. 1993. Review—Synthesis of cast metal matrix composites. *J. Mater. Sci.* 28(20): 5397–413.

324. Dhandapani, S. P., V. Jayaram, and M. K. Surappa. 1994. Growth and microstructure of Al_2O_3-SiC-Si(Al) composites prepared by reactive infiltration of silicon carbide preforms. *Acta Metall. Mater.* 42(3):649–56.

325. Breval, E., M. K. Aghajanian, J. P. Biel, et al. 1993. Structure of aluminum nitride/aluminum and aluminum oxide/aluminum composites produced by the directed oxidation of aluminum. *J. Am. Ceram. Soc.* 76(7):1865–8.

326. Antolin, S., A. S. Nagelberg, and D. K. Creber. 1992. Formation of Al_2O_3/metal composites by the directed oxidation of molten aluminum-magnesium-silicon alloys. Part I: Microstructural development. *J. Am. Ceram. Soc.* 75(2):447–54.

327. Brown, A. S. 1995. Particulates promise affordable MMCs. *Aerosp. Am.* May:20–1.

328. Ashley, S. 1991. Tailor-made ceramic-matrix composites. *Mech. Eng.* 113(7):44–9.

329. May, C. R. 1990. An economical, net shape process for metal matrix composites. SME EM 90-440.

330. Newkirk, M. S., A. W. Urquhart, H. R. Zwicker, et al. 1986. Formation of Lanxide ceramic composite materials. *J. Mater. Res.* 1:81.

331. Johnson, W. B., and B. Sonuparlak. Diamond/Al metal matrix composites formed by the pressureless metal infiltration process. *J. Mater. Res.* 8(5):1169–73.

332. Christodoulou, L. *In situ* composites by exothermic dispersion, Session II, *ASM International Ann. Mtg.,* Oct 2–6, 1994, Rosemont, IL.

333. Chrisdoulou, L. et al. 1988. U.S. Patent 4,751,048.

334. Brumbacher, J. M., et al. 1988. U.S. Patent 4,836,982.

335. Koczak, M. J., and M. K. Premkumar. 1993. Emerging technologies for the in situ production of MMCs. *J. Met.* 45(1):44–8.

336. Davies, P., et al. *Development of Cast Aluminum MMCs.* London Scandinavian Metallurgical Company, U.K.

337. Newkirk, M. S., et al. 1982. *Ceram. Eng. Sci. Process.* 8(7–8):879–85.

338. Newkirk, M. S., et al. 1986. *J. Mat. Res.* 1(1):81–9.

339. Fareed, A. S., et al. 1990. *Ceram. Eng. Sci. Process.* 11(7–8):782–94.

340. *Int. J. Self-Propagat. High Temp. Synt.* p. 1, vol. 1, 1992.

341. Lee, A. K., et al. 1989. *Multicomponent Ultrafine Microstructures,* ed. L. E. McCandlish, et al. 87–92. MRS, Pittsburgh, PA.

342. Smith, R. W., and Z. Z. Mutasim. 1990. Plasma sprayed refractory metal structures and properties. *ASM Nat. Therm. Spray Conf.,* May 1990, Long Beach, CA.

343. Smith, R. W., and D. Apelian. 1989. Plasma spray consolidation of materials. *9th Int. Union Pure Appl. Chem. Conf. Plasma Chem.,* September 1989, Bari, Italy.

344. Smith, R. W., E. Harzenski, and T. Robisch. 1989. The structure and properties of thermally sprayed TiC particulate reinforced prealloyed powders. *Proc. 12th Int. Therm. Spray Conf.,* June 1989, London.

345. Smith, R. W., D. Apelian, and D. Wei. 1989. Thermal plasma materials processing—Applications and opportunities. *J. Plasma Chem.* 9(1).

346. Smith, R. W., et al. 1989. The structure and properties of plasma sprayed TiC dispersion hardened coatings. *Proc. 2nd Nat. Therm. Spray Conf.,* October 1989, Cincinnati, OH.

347. Soboyejo, W. O., R. J. Lederich, and S. M. L. Sastry. 1994. Mechanical behavior of damage tolerant TiB whisker-reinforced *in situ* titanium matrix composites. *Acta Metall. Mater.* 42(8):2579–91.

348. Khatri, S., and M. Koczak. 1993. Formation of TiC in *in situ* processed composites via solid-gas, solid-liquid and liquid-gas reaction in molten Al-Ti. *Mater. Sci. Eng.* A162(1–2):153–62.

349. Valencia, J. J., C. McCullough, J. Rösler, et al. 1990. Development of TiAl intermetallic matrix composites by solidification processing. In *Solidification of Metal Matrix Composites,* ed. P. Rohatgi, pp. 133–49. Minerals, Metals and Materials Society, Indianapolis, IN.

350. Thompson, M. S., and V. C. Nardone. 1991. *In situ*-reinforced titanium matrix composites. *Mater. Sci. Eng.* A144:121–6.

351. Gungor, M. N., R. M. Roidt, and M. G. Burke. 1991. Plasma deposition of particulate-reinforced metal matrix composites. *Mater. Sci. Eng.* A144:111–9.

352. Premkumar, M. K., and M. G. Chu. 1993. Synthesis of TiC particulates and their segregation during solidification in *in situ* processed Al-TiC composites. *Metall. Trans.* 24A(10):2358– 62.

353. Chu, M. G., and M. K. Premkumar. 1993. Mechanism of TiC formation in Al/TiC in situ metal-matrix composites. *Metall. Trans.* 24A(12):2803–5.

354. Wang, D. Z., Z. R. Liu, C. K. Yao, et al. 1993. A novel technique for fabricating in situ Al_2O_8/Ti_xAl_y composites. *J. Mater. Sci. Lett.* 12(18):1420–1.

355. Sahoo, P., and M. J. Koczak. 1991. Microstructure-property relationships of in situ reacted TiC/Al-Cu metal matrix composites. *Mater. Sci. Eng.* A131(1):69–76.

356. Ghosh, A. 1993. In *Fundamentals of Metal-Matrix Composites,* ed. S. Suresh, A. Mortensen, and A. Needleman, 23–41. Buterworth-Heinemann, Newton, MA.

357. Ni, X., M. S. McLean, and T. N. Baker. 1994. Design aspects of processing of aluminium 6061 based metal matrix composites via powder metallurgy. *Mater. Sci. Technol.* 10(6):452–9.

358. Shannon, R. E., P. K. Liaw, and W. C. Harrigan, Jr. 1992. Nondestructive evaluation for large-scale metal-matrix composite billet processing. *Metall. Trans.* 23A(5):1541–9.

359. Geiger, A. L., and J. A. Walker. 1991. The processing and properties of discontinuously reinforced aluminum composites. *J. Met.* 43(8):8–15.

360. Bhanuprasad, V. V., R. B. V. Bhat, A. K. Kuruvilla, et al. P/M processing of Al-SiC composites. *Int. J. Powder Metall.* 27(3):227–35.

361. Dolowy, J. F. 1986. Increased focus on silicon carbide-reinforced aluminum composites. *Light Met. Age* 44(5–6):7.

362. Hanumanth, G. S., and G. A. Irons. Mixing and wetting in metal matrix composites fabrication. *ASM Int. Conf. Fabr. Reinf. Met. Compos.,* Montreal, Canada, September 19.

363. McDaniels, D. L. 1985. Analysis of stress-strain, fracture, and ductility behavior of aluminum matrix composites containing discontinuous silicon carbide reinforcement. *Metall. Trans.* 16A(5):1105.

364. McKimpson, M. G., and T. E. Scott. 1989. Processing and properties of metal matrix composites containing discontinuous reinforcement. *Mater. Sci. Eng.* A107:93.

365. Rocher, J. P., J. M. Quinisset, and R. Naslain. 1985. A new casting process for carbon (or SiC based) fibre-aluminum matrix low-cost composite materials. *J. Mater. Sci. Lett.* 4(12):1527.

366. Nardone, V. C., and K. M. Prewo. 1986. On the strength of discontinuous silicon carbide reinforced aluminum composites. *Scr. Metall.* 20(1):43.

367. Zick, D. H. 1991. HIP fabrication of Al/SiC MMC components for insensitive munitions. *Ind. Heat.* October:75–7.

368. Premkumar, M. K., W. H. Hunt, Jr., and R. R. Sawtell. 1992. Aluminum composite materials for multichip modules. *J. Met.* 44(7):24–8.

369. Styles, C. M., I. Sinclair, P. J. Gregson, et al. 1994. Effect of microstructure on mechanical properties of thermomechanically processed 2124-SiC_p metal matrix composite. *Mater. Sci. Technol.* 10(6):475–80.

370. Hong, S. H., and K. H. Chung. 1995. The effects of processing parameters on mechanical properties of SiC_w/2124 Al composites. *J. Mater. Proc. Technol.* 48(1–4):349–55.

371. Tweed, J. H. 1991. Manufacture of 2014 aluminum reinforced with SiC particulate by vacuum hot pressing. *Mater. Sci. Eng.* A135:73–6.

372. Parent, J. O. G., J. Iyengar, and H. Henein. 1993. Fundamentals of dry powder blending for metal matrix composites. *Int. J. Powder Metall.* 29(4):353–66.

373. Ling, C. P., M. B. Bush, and D. S. Perera. 1995. The effect of fabrication techniques on the properties of Al-SiC composites. *J. Mater. Process. Technol.* 48(1–4):325–31.

374. Humphreys, F. J. 1991. The thermomechanical processing of Al-SiC particulate composites. *Mater. Sci. Eng.* A135:267–73.

375. Bak, D. J. 1990. Metal matrix woos the military. *Des. News* August 20:72–7.

376. ter Haar, J. H., and J. Duszczyk. 1991. Mixing of powder metallurgical fibre-reinforced aluminium composites. *Mater. Sci. Eng.* A135:65–72.

377. Whitehouse, A. F., and T. W. Clyne. 1994. Effect of test conditions on cavitation and failure during tensile loading of discontinuous metal matrix composites. *Mater. Sci. Technol.* 10(6):468–74.

378. Abkowitz, S., P. F. Weihrauch, and S. M. Abkowitz. 1993. Particulate-reinforced titanium alloy composites economically formed by combined cold and hot isostatic pressing. *Ind. Heat.* September:32–7.

379. Abkowitz, S., and P. F. Weihrauch. 1989. Trimming the cost of MMCs. *Adv. Mater. Process.* 136:31–4.

380. Abkowitz, S., P. F. Weihrauch, and S. Kraus. 1988. Development and evaluation of the metallurgy and shape fabrication of titanium matrix CermeTi composites for high modulus, high temperature applications. Contract N69021-86C-0222.

381. Fujii, H., T. Yamazaki, T. Horiya, et al. 1992. High performance Ti-6Al-4V-TiC alloy by blended elemental powder metallurgy. *Proc. 1st Pacific Rim Int. Conf. Adv. Mater.,* Hangzhou, China.

382. Abkowitz, S., P. F. Weihrauch, and S. M. Abkowitz. 1992. Advanced P/M Ti alloy matrix composites reinforced with ceramic and intermetallic particles. *Proc. 7th Int. TMS/TDA Conf. Titanium,* San Diego, CA.

383. Abkowitz, S., P. F. Weihrauch, S. M. Abkowitz, et al. 1993. Low cost P/M manufacture of titanium alloy components for fatigue critical applications. *Proc. 3rd Int. MPIF Conf. P/M Aerosp.,* San Diego, CA.

384. Abkowitz, S. 1990. Advanced powder metallurgy technology for manufacture of titanium alloy and titanium matrix composites to near net shape. *Proc. Conf. Exhibit. MPIF,* ed. F. H. Froes, pp. 193–201, Princeton, NJ.

385. Pauchal, J. M., and T. Vela. 1990. The promise of TiC/Ti composites. *Am. Mach.* October 1990.

386. Loh, N. L., Y. L. Wu, and K. A. Khor. 1993. Shear bond strength of nickel/alumina surfaces diffusion bonded by HIP. *J. Mater. Process. Technol.* 37(1–4):711–21.

387. Upadhyaya, D., R. Brydson, C. M. Ward-Close, et al. 1994. As received Ti-6Al-4V/σ-SiC fibre composite. *Mater. Sci. Technol.* 10(9):797–806.

388. Mittnick, M. A. 1990. Continuous SiC fiber reinforced metals. EM 90-437.

389. Zaretsky, E. V. 1994. Modeling the future with metal-matrix composites. *Mach. Des.* March 7:124–32.

390. Evans, R. W., A. G. Leatham, and R. G. Brooks. 1985. The Osprey preform process. *Powder Metall.* 28:13–9.

391. Leatham, A. G., A. Ogilvey, P. F. Chesney, et al. 1989. Osprey process-production flexibility in materials manufacture. *Met. Mater.* 5:140:–3.

392. Chesney, P. F., A. G. Leatham, R. Pratt, et al. 1989. The Osprey process—A versatile manufacturing technology for the production of solid and hollow rounds and clad (compound) billets. *Proc. 1st Eur. Conf. Adv. Mater. Process.* ed. H. E. Exner, and V. Schumacher, pp. 247–54, Aachen, Germany. DGM.

393. Willis, T. C. 1988. Spray deposition process for metal matrix composite manufacture. *Met. Mater.* 4:485–8.

394. Kahl, W., and J. Leupp. 1989. Spray deposition of high performance Al alloys via the Osprey process. *Proc. 1st Eur. Conf. Adv. Mater. Process.* ed. H. E. Exner and V. Schumacher, pp. 261–6, Aachen, Germany. DGM.

395. White, J., and T. C. Willis. 1989. The production of metal matrix composites by spray deposition. *Mater. Des.* 10:121–7.

396. White, J., N. A. Darby, I. R. Hughes, et al. 1990. Metal matrix composites produced by spray deposition. *Adv. Mater. Technol. Int.* 58:9–42.

397. Leatham, A. G., and A. Lawley. 1993. The Osprey process: Principles and applications. *Int. J. Powder Metall.* 29(4):321–9.

398. Mathur, P. C., D. Apelian, and A. Lawley. Analysis of the spray deposition process. *Acta Metall.* 37(2):429.

399. Mathur, P., S. Annavarapu, D. Apelian, et al. 1991. Spray casting: An integral model for process understanding and control. *Mater. Sci. Eng.* A142:261.

400. Singer, A. R. E. Metal matrix composites made by spray forming. *Mater. Sci. Eng.* A135:13–7.

401. McLelland, A. R. A., H. V. Atkinson, P. Kapranos, et al. 1991. Thixoforming spray-formed aluminium/silicon carbide metal matrix composites. *Mater. Lett.* 11(1–2):26–30.

402. Wu, Y., and E. J. Lavernia. 1991. Spray-atomized and codeposited 6061 Al/SiC$_p$ composites. *J. Met.* August:16–23.

403. Nayim, S., E. Raskin, M. Polak, et al. 1992. Spray atomization, injection and co-deposition processing of Al-Li metal-matrix composites. *Proc. 6th Int. Alum.-Lithium Conf.,* ed. M. Peters and P. -J. Winkler. vol. 2, pp. 1353–58. Aachen, Germany. DGM.

404. Lugschieder, E. 1990. The family of plasma spray processes—Present status and future prospects. *Proc. 1st Plasma Tech. Symp.,* ed. H. Eschenauer, P. Huber, A. R. Nicoli, and S. Sandmeier, pp. 23–48, Lucerne.

405. Steffens, H. -D., and R. Kaczmarek. 1991. Metal matrix composites made by thermal spraying. *Powder Metall. Int.* 23(2):105–7.

406. Steffens, H. -D., R. Kaczmarek, and U. Fischer. 1988. Fibre-reinforced composite by thermal spraying. *Proc. Nat. Therm. Spray Conf.,* October 1988, Cincinnati, OH.

407. Steffens, H. -D., R. Kaczmarek, and U. Fischer. Production of metal-matrix composites by thermal spraying. *Proc. 12th Int. Conf. Therm. Spraying,* June 1989, London.

408. Titran, R. H., T. L. Grobstein, and D. L. Ellis. 1991. Advanced materials for space nuclear power systems. *AIAA, NASA, OAI Adv. Space Explor. Initiative Technol. Conf.,* September 1991, Cleveland, OH. NASA TM 105171, DOE NASA 16310-16, AIAA 91-3530, E 6464, DE-AI03-86SF-16310.

409. Pickens, J. W. 1994. Fabrication of advanced metal and intermetallic composite materials by the wire arc spray process—An extended abstract. *J. Therm. Spray Technol.* 3(2):204–7.

410. Hall, E. L., and A. M. Ritter. 1993. Structure and behavior of metal/ceramic interfaces in Ti alloy/SiC metal matrix composites. *J. Mater. Res.* 8(5):1158–68.

411. Khor, K. A., Z. H. Yuan, F. Y. C. Boey, et al. 1995. Preparation of Al-Li/SiC$_p$ composite powder by a plasma spray atomization (PSA). *J. Mater. Proc. Technol.* 48(1–4):541–8.

412. Khor, K. A., F. Y. C. Boey, Y. Murakoshi, et al. 1994. Al-Li/SiC$_p$ composites and Ti-Al alloy powders and coatings prepared by a plasma spray atomization (PSA) technique. *J. Therm. Spray Technol.* 3(2):162–8.

413. Pank, D. R., and J. J. Johnson. 1993. Metal-matrix composite processing technologies for aircraft engine applications. *J. Mater. Eng. Performance* 2(3):341–6.

414. Valente, T., and C. Bartuli. 1994. A plasma spray process for the manufacture of long-fiber reinforced Ti-6Al-4V composite monotapes. *J. Therm. Spray Technol.* 3(1):63–8.

415. Valente, T. 1994. Measurement of interfacial properties for aluminum and titanium matrix alloy composites manufactured by vacuum plasma spray. *J. Comp. Technol. Res.* 16(3):256–61.

416. Sampath, S., and H. Herman. 1993. Plasma spray forming metals, intermetallics, and composites. *J. Met.* 45(7):42–9.

417. Rapid-solidification processing improves MMC properties. *Adv. Mater. Process.* November:71–3.

418. Ward-Close, C. M., and F. H. Froes. 1994. Development in the synthesis of lightweight metals. *J. Met.* 46(1):28–31.

419. Weimer, R. J. 1988. Manufacturing MMC precursors by vapor deposition methods. Cordec Corporation and Kaman Tempo, Santa Barbara, CA. MMCIAC 000698, DLA900-85C4100.

420. Okumura, M., S. Murakami, S. Utsunomiya, et al. 1990. Mechanical properties of MMC fabricated with laser beam. *Proc. 35th Int. SAMPE Symp.,* Anaheim, CA, pp. 2153–61, April 1990.

421. Okumura, M., S. Murakami, S. Utsunomiya, et al. 1990. Mechanical properties of MMC fabricated with laser beam. *SAMPE Q.* 21(4):56–63.

422. Baker, T. N., H. Xin, C. Hu, et al. 1994. Design of surface in situ metal-ceramic composite formation via laser treatment. *Mater. Sci. Technol.* 10(6):536–44.

423. English, L. 1990. Magnesium composites molded without melting. *Mod. Met.* August:10–22.

424. Asabe, K., M. Nakanishi, and S. Tanoue. 1991. The development of a metal matrix composite by mechanical alloying. Sumitomo Search N45. Sumitomo Metal Industries, Tokyo, Japan.

425. Sivakumar, K., Y. R. Mahajan, N. Ramakrishnan, et al. 1992. Explosive compaction of Al-SiC composites. *Int. J. Powder Metall.* 28(1):63–8.

426. Murakoshi, Y., T. Sano, F. Y. C. Boey, et al. Dynamic compaction of Al-Li alloy composites using preform. *J. Mater. Proc. Technol.* 48(1–4):407–12.

427. Tong, W., G. Ravichandran, T. Christman, et al. 1995. Processing SiC-particulate reinforced titanium-based metal matrix composites by shock wave consolidation. *Acta Metall. Mater.* 43(1):235–50.

428. Murakoshi, Y., F. Boey, and T. Sano. 1993. Al-Li alloy composites using a dynamic shock compaction technique. *J. Mater. Proc. Technol.* 38B(1–2):351–60.

429. Choi, Y., M. E. Mullins, K. Wijayatilleke, et al. 1992. Fabrication of metal matrix composites of TiC-Al through self-propagating synthesis reaction. *Metall. Trans.* A23(9):2387–92.

430. Mileiko, S. T., and V. I. Kazmin. 1992. Crystallization of fibres inside a matrix: A new way of fabrication of composites. *J. Mater. Sci.* 27(8):2165–72.

431. Rabinovitch, M., J. -C. Daux, and J. -L. Raviart. 1990. Carbon-reinforced magnesium and aluminum composites fabricated by liquid hot pressing. ONERA France. ONERA TP 1990-127.

432. Breval, E., and C. P. Pantano. 1992. Sol-gel prepared Ni-alumina composite materials. *J. Mater. Sci.* 27(20):5463–9.

433. Warrier, S. G., and R. Y. Lin. 1993. Using rapid infrared forming to control interfaces in titanium-matrix composites. *J. Met.* 45(3):24–7.

434. Ward-Close, C. M., and P. G. Partridge. 1990. A fibre coating process for advanced metal-matrix composites. *J. Mater. Sci.* 25(10):4315–23.

435. Rice, R. In *Processing of Ceramic Composites,* ed. J. G. P. Binner, 123–210, Advanced Ceramics Processing and Technology, vol. 1. Noyes, Park Ridge, NJ.

436. Fitzer, E. 1988. Whisker and fiber-toughened ceramics. *Proc. ASM Int. Conf.,* ed. R. A. Bradley, D. E. Clark, and D. C. Larsen, pp. 165–92, Chicago, IL.

437. Karnitz, M. A., D. F. Craig, S. L. Richlen. 1991. Continuous fiber ceramic composite program. *Ceram. Bull.* 70(3):430–5.

438. Lewis, D., III. 1988. Strength and toughness of fiber-reinforced ceramics and related interface behavior. *Proc. ASM Int. Conf.,* ed. R. A. Bradley, D. E. Clark, and D. C. Larsen, pp. 265–73, Chicago, IL.

439. Jamet, J. F., and P. Peres. 1991. *Structures and Fracture Mechanics of Brittle Matrix Composites.* Societe Nationale Industrielle Aerospatiale France, Paris, France.

440. Rice, R. W., and D. Lewis, III. 1990. Ceramic fiber composites based upon refractory polycrystalline ceramic matrices. In *Encyclopedia of Composite Materials,* ed. S. M. Lee, 120–41. Technomic, Lancaster, PA.

441. Anson, D., K. S. Ramesh, and M. DeCorso. 1991. Application of ceramics to industrial gas turbines task 1 technology background. Battelle Institute, Columbus, OH. DOE/CE/40878-1 (DE91013140), Contract AC02-89CE40878.

442. Janney, M. A. 1986. Microstructural development and mechanical properties of SiC and of SiC-TiC composites. *Am. Ceram. Soc. Bull.* 65(2):357–62.

443. Greil, P. 1989. Opportunities and limits in engineering ceramics. *Powder Metall. Int.* 21(2): 40.

444. Ruh, R., M. G. Mendiratta, K. S., Mazdiyasni, et al. 1988. Mechanical and microstructural characterization of mullite-ZrO_2-SiC whisker composites. *Proc. ASM Int. Conf.,* ed. R. A. Bradley, D. E. Clark, and D. C. Larsen, pp. 91–5, Chicago, IL.

445. Tiegs, T. N. 1988. Properties of SiC whisker-reinforced oxide matrix composites. *Proc. ASM Int. Conf.,* ed. R. A. Bradley, D. E. Clark, and D. C. Larsen, pp. 105–07, Chicago, IL.

446. Lee, D. -H., and H. -E. Kim. 1994. Microstructure and fracture toughness of hot-pressed silicon carbide reinforced with silicon carbide whisker. *J. Am. Ceram. Soc.* 77(12):3270–2.

447. Foulds, W., J. -F. LeCostaouec, and S. Dipietro. 1990. Densified SCS-6 SiC fiber reinforced Si_3N_4 composites. *Proc. 35th Int. SAMPE Symp.,* pp. 2163–74, April 1990, Anaheim, CA.

448. Bhatt, R. T. 1988. The properties of silicon carbide fiber reinforced silicon nitride composites. *Proc. ASM Int. Conf.,* ed. R. A. Bradley, D. E. Clark, and D. C. Larsen, pp. 199–207, Chicago, IL.

449. Miyoshi, T., H. Kodama, H. Sakamoto, et al. 1988. Characteristics of hot-pressed fiber-reinforced ceramics with SiC or Si_3N_4 matrix. *Proc. ASM Int. Conf.,* ed. R. A. Bradley, D. E. Clark, and D. C. Larsen, pp. 193–97, Chicago, IL.

450. Hirano, T., and K. Niihara. 1995. Microstructure and mechanical properties of Si_3N_4/SiC composites. *Mater. Lett.* 22:249–54.

451. Kato, A., Nakamura, N. Tamari, et al. 1995. Usefulness of alumina-coated SiC whiskers in the preparation of whisker-reinforced alumina ceramics. *Ceram. Int.* 21(1):1–4.

452. Goto, Y., and A. Tsuge. 1993. Mechanical properties of unidirectionally oriented SiC-whisker-reinforced Si_3N_4 fabricated by extrusion and hot-pressing. *J. Am. Ceram. Soc.* 76(6):1420–4.

453. Tuffe S., J. Dubois, Y. Jorand, et al. 1994. Processing and fracture behavior of hot pressed silicon carbide whisker reinforced alumina. *Ceram. Int.* 20(6):425–32.

454. Duclos, R., J. Crampon, and B. Cales. 1992. Microstructure development during hot-pressing of alumina-based ceramics reinforced with SiC whiskers. *Ceram. Int.* 18(1):57–63.

455. Ono, T., H. Endo, and M. Ueki. 1993. Hot-pressing of TiC-graphite composites materials. *J. Mater. Eng. Performance* 2(5):659–64.

456. Piciacchio, A., S. -H. Lee, and G. L. Messing. 1994. Processing and microstructure development in alumina-silicon carbide intragranular particulate composites. *J. Am. Ceram. Soc.* 77(8):2157–64.

457. Brown, I. W. M., G. V. White, and G. L. Dunlop. 1991. Fe-TiC/N ceramic-metal composites. *Proc. 4th Int. Symp. Ceram. Mater. Components Eng.,* ed. R. Carlsson, T. Johansson, and L. Kahlman, pp. 356–63, June 1991, Goteborg, Sweden.

458. Cutler, R. A., C. B. Brinkpeter, A. V. Virkar, et al. 1991. Fabrication and characterization of slip-cast layered Al_2O_3-ZrO_2 composites. *Proc. 4th Int. Symp. Ceram. Mater. Components Eng.,* ed. R. Carlsson, T. Johansson, and L. Kahlman, pp. 397–408, June 1991, Goteborg, Sweden.

459. Zhang, B. Q., J. M. Tian, X. H. Tong, et al. 1991. Mechanical properties and microstructure of ZrO_2/Si_3N_4 composite. *Proc. 4th Int. Symp. Ceram. Mater. Components Eng.,* ed. R. Carlsson, T. Johansson, and L. Kahlman, pp. 529–35, June 1991, Goteborg, Sweden.

460. Davis, J. B., J. Yang, and A. G. Evans. 1995. Effects of composite processing on the strength of sapphire fiber-reinforced composites. *Acta Metall. Mater.* 43(1):259–68.

461. Shin, H. -H., R. Kirchain, and R. F. Speyer. 1995. Fabrication, microstructure, and mechanical properties of TaC particulate and SiC fiber-reinforced lithia-alumina-silica composites. *J. Mater. Res.* 10(3):602–8.

462. Olagnon, C., E. Bullock, and G. Fantozzi. 1991. Processing of high density sintered SiC whisker reinforced Si_3N_4 composites. *Ceram. Int.* 17(2):53–60.

463. Sakka, Y., D. D. Bidinger, and I. A. Aksay. 1995. Processing of silicon carbide-mullite-alumina nanocomposites. *J. Am. Ceram. Soc.* 78(2):479–86.

464. Weiser, M. W. 1994. Pressureless sintering of ceramic composites. University of New Mexico, Albuquerque. AFOSR-TR-94-0056.

465. Kim, Y. -W, and J. -G. Lee. 1991. Pressureless sintering of Al_2O_3-SiC whisker composites. *J. Mater. Sci.* 26(5):1316–20.

466. Wei, W. -C. J., and R. -R. Lee. 1991. Pressureless sintering of AlN-SiC composites. *J. Mater. Sci.* 26(11):2930–36.

467. Hoffman, M. J., A. Nagel, and G. Petzow. 1989. Processing of SiC-whisker reinforced Si_3N_4. In *Processing Science of Advanced Ceramics,* ed. I. A. Aksay, G. L. McVay, and D. R. Ulrich, MRS Symposia, Proceedings, vol. 155, pp. 369–79. DAAL03-89G0024.

468. Freeman, G. B., R. L. Starr, J. N. Harris, et al. 1991. Characterization of the fiber-matrix interfaces in a SiC-Si_3N_4 ceramic composite system. *Proc. 1st Ann. Conf. Heat-Resistant Mater.,* pp. 299–306, September 1991, Fontana, WI.

469. *Selected Papers from 24th High-Temperature Materials Technology Seminar on Silicon Polymers and Ceramics, August 1993.* JPRS-JST-93-067-L.

470. Sato, Y., M. Ikei, and K. Shintani. 1995. Slurry processing for fabrication of SiC whisker-reinforced Si_3N_4 composites. *J. Mater. Sci.* 30(5):1373–8.

471. Iwata, M., T. Isoda, and T. Itoh. 1994. Processing of continuous fiber reinforced Si_3N_4 matrix composites for gas turbine engine components. *ASME Int. Gas Turbine Aerosp. Cong. Exposit.,* June 1994, New York, NY. ASME-94-GT-396.

472. Roth, R., J. P. Clark, and F. R. Field, III. 1994. The potential for CMCs to replace superalloys in engine exhaust ducts. *J. Met.* 46(1):31–4.

473. Thomson, B. 1994. Preform tape method yields strong ceramic composite components. *Adv. Mater.* January 14, 1994:4.

474. Ceramic-part production made easier by processing advances. *Adv. Mater. Process.* 1991, 139:44–46.

475. Tai, N. -H., and T. -W. Chou. 1988. Modeling of chemical vapor infiltration (CVI) in $Al_2O_3/$SiC composites processing. *Proc. Jt. NASA/DOD Conf.,* ed. J. D. Buckley, pp. 237–46, Cocoa Beach, FL. NASA CP 3018.

476. Chung, G. -Y., and B. J. McCoy. 1991. Modeling of chemical vapor infiltration for ceramic composites reinforced with layered, woven fabrics. *J. Am. Ceram. Soc.* 74(4):746–51.

477. Lu, G. Q. 1993. Modeling the densification of porous structures in CVI ceramic composites processing. *J. Mater. Process. Technol.* 37(1–4):487–98.

478. Tai, N. -H., and T. -W. Chou. 1990. On the deposition mechanism of Al_2O_3 in the CVI process for forming ceramic composites. *J. Mater. Res.* 5(10):2255–6.

479. Burkland, C. V., and J. -M. Yang. 1988. Development of 3-D braided Nicalon/silicon carbide composite by chemical vapor infiltration. *Proc. Jt. NASA/DOD Conf.,* ed. J. D. Buckley, pp. 271–75, Cocoa Beach, FL. NASA CP 3018.

480. Burkland, C. V., and J. -M. Yang. 1989. CVI-processed silicon carbide matrix composites. *Proc. 34th SAMPE Int. Conf.,* pp. 275–83, Anaheim, CA.

481. Lasday, S. B. 1993. Production of ceramic matrix composites by CVI and DMO for industrial heating industry. *Ind. Heat.* April:31–5.

482. Bashford, D. 1990. The preparation of oxidation resistant silicon carbide-silicon carbide composites by chemical vapour infiltration (CVI). *Int. Conf. New Mater. Their Appl.,* IOP Publications, Bristol, UK.

483. Kmetz, M., and S. Suib. 1990. Silicon carbide/silicon and silicon carbide composites produced by chemical vapor infiltration. *J. Am. Ceram. Soc.* 73(10):3091–93.

484. Bilatus, D. 1993. *Fiber and Whisker Reinforced Ceramics for Structural Applications.* Marcel Dekker, New York.

485. Hoyt, J. T., and J. M. Yang. 1991. Chemical vapor infiltration of silicon nitride matrix composites. *SAMPE J.* 27(2):11–7.

486. Mitsui Engr. & Shipbuilding Co. Ltd. 1994. New technology: Japan. *JETRO* 21(11):20.

487. Mitsui Engr. & Shipbuilding Co. Ltd. 1993. Science and technology: Japan. JPRS-JST-93-083-L.

488. Kim, Y. -W., J. -S. Song, S. -W. Park, et al. 1993. Nicalon-fibre-reinforced silicon-carbide composites via polymer solution infiltration and chemical vapour infiltration. *J. Mater. Sci.* 28(14):3866–8.

489. Allaire, F., and S. Dallaire. 1992. Production of $SiC-Si_3N_4$ composites by a one-step d.c. plasma process. *J. Mater. Sci. Lett.* 11(1):48–50.

490. Manor, E., H. Ni, and C. G. Levi. 1993. Microstructure evolution of $SiC/Al_2O_3/Al$-alloy composites produced by melt oxidation. *J. Am. Ceram. Soc.* 76(7):1777–87.

491. Wu, M., and G. L. Messing. 1990. SiC-whisker-reinforced cellular SiO_2 composites. *J. Am. Ceram. Soc.* 73(11):3497–9.

492. GanFeng, T., S. ZhiTong, and W. ChangZhen. 1991. Study of SiC-whisker-reinforced ZrO_2 ceramics using the sol-gel method. *J. Less-Common Met.* 175(2):205–8.

493. Russell-Floyd, R. S., R. G. Cooke, and B. Harris. 1995. Low cost ceramic-matrix composites. *Ceram. Int.* 21:62–5.

494. Russell-Floyd, R. S., B. Harris, R. G. Cooke, et al. 1993. Application of sol-gel processing techniques for the manufacture of fiber-reinforced ceramics. *J. Am. Ceram. Soc.* 76(10): 2635–43.

495. Zeng, J., Y. Miyamoto, and O. Yamada. 1991. Combustion synthesis of SiN-SiC composite powders. *J. Am. Ceram. Soc.* 74(9):2197–200.

496. Chrysanthou, A., A. Saidi, C. E. W. Aylott, et al. 1994. Preparation and microstructure of Al_2O_3-TiC composites by self-propagated high temperature synthesis. *J Alloys Compd.* 203(1–2):127–32.

497. Rabin, B. H., G. E. Korth, and R. L. Williamson. 1990. Fabrication of titanium carbide-alumina composites by combustion synthesis and subsequent dynamic consolidation. *J. Am. Ceram. Soc.* 73(7):2156–7.

498. Subrahmanyam, J., and R. M. Rao. 1995. Combustion synthesis of $MoSi_2$-SiC composites. *J. Am. Ceram. Soc.* 78(2):487–90.

499. Aghajanian, M. K., J. P. Biel, and R. G. Smith. 1994. AlN matrix composites fabricated via an infiltration and reaction approach. *J. Am. Ceram. Soc.* 77(7):1917–20.

500. Singh, M. 1995. Low-cost precursors produce high-tech ceramic composites. *Adv. Mater. Process.* May:11.

501. Wu, S. X., and N. Claussen. 1994. Reaction bonding and mechanical properties of mullite/silicon carbide composites. *J. Am. Ceram. Soc.* 77(11):2898–904.

502. Hillig, W. B. 1994. Making ceramic composites by melt infiltration. *Am. Ceram. Soc. Bull.* 73(4):56–62.

503. Krenkel, W., and P. Schanz. 1992. Fiber ceramic structures based on liquid impregnation technique. *Acta Astronaut.* 28:159–69.

504. Stinton, D. P., R. A. Lowden, T. M. Besmann, et al. 1988. Forced chemical vapor infiltration fabrication of SiC/SiC composites. *Proc. ASM Int. Conf.* ed. R. A. Bradley, D. E. Clark, and D. C. Larsen, pp. 231–41, Chicago, IL.

505. Marcus, H. L., et al. 1990. Ceramic composites fabricated by selective laser sintering. *Proc. 1st Solid Freeform Fabr. Symp.,* August 1990, Austin, TX.

506. Beaman, J. J., et al. 1991. *Proc. 2nd Solid Freeform Fabr. Symp.,* August 1991, Austin, TX.

507. Lakshminarayan, U. 1992. Ph.D. Thesis, University of Texas at Austin.

508. Reagan, P., and F. N. Huffman. Chemical vapor composite deposition. *Fiber Tex 1988,* ed. J. D. Buckley, pp. 247–57. September 1988, Greenville, SC. NASA CP 3038.

509. *CIAC Newsletter,* Center for Information & Numerical Data Analysis & Synthesis—Department of Commerce, Purdue University 2(4):6, 1992.

BIBLIOGRAPHY

ABBOUD, J. H., AND D. R. F. WEST. Ceramic-metal composites produced by laser surface treatment. *Mater. Sci. Technol.* July:725–8, 1989.

ABSON, D. MMC deposits from powder mixtures by friction surfacing. *TWI Bull. 1* 34:12–13, 1993.

Adv. Compos. July/August:38–46, 1989.

AKSAY, I. A., G. L. McVAY, AND D. R. ULRICH. *Processing Science of Advanced Ceramics,* MRS Symposia Proceedings, vol. 155. MRS, Pittsburgh, PA, 1989. DAAL03-89G0024.

ALLEN, L. E., J. R. McCOLLUM, D. D. EDIE, ET AL. Thermoplastic coated carbon fibers for textile preforms. *Fiber Tex 1988,* pp. 367–73, Greenville, SC, September 1988.

ASHLEY, S. Ceramic metal composites: Bulletproof strength. *Mech. Eng.* 112(7):46–51, 1990.

ASM/TMS, *8th Mater. Exposit.* Rosemont, IL, October 1994.

ASSAR, A.-E. M., AND M. A. AL-NIMR. Fabrication of metal matrix composite by infiltration process. Part 1: Modeling of hydrodynamic and thermal behavior. *J. Compos. Mater.* 28(15):1480–90, 1994.

ATWELL, W. H., W. E. HAUTH, AND R. E. JONES. Advanced ceramics based on polymer processing, VI and V2, FR 2/84-2/85; Fiber Technology, AFWAL TR 85-4099 VI and V2, F33615-83C5006, 1987.

Autoclave Curing. Nerac, Tolland, CT, 1994; *NTIS Alert,* 95(7):32.

Automation in Processing Composites: Forming. Nerac, Holland, CT, 1994; *NTIS Alert* 95(2), 1995.

Automation in Processing Composites: Testing and Quality and Control. Nerac, Holland, CT, 1994; *NTIS Alert* 95(2), 1995.

BATES, J. Automated ply cutting cell. *Westec 1990,* Los Angeles, CA, March 1990. SME MS 90-142.

BECKER, W. Developing RTM, aerospace composites and materials. *SME Compos. Conf.,* pp. 12–5, Anaheim, CA, 1992.

BELOY, J., AND S. C. MANTELL. Development of a control system for tape laying of thermoplastic matrix composites. *TMS Mater. Week 1993,* Pittsburgh, PA, October 1993.

BENDER, B. A., D. A. LEWIS, III, W. S. COBLENZ, ET AL. Electron microscopy of ceramic fiber-ceramic matrix composites—Comparison with processing and behavior. *Ceram. Eng. Sci. Proc.* 5:513–29, 1984.

BHAGAT, R. B., A. H. CLAUER, P. KUMAR, AND A. M. RITTER, ED. *Metal and Ceramic Matrix Composites.* TMS, Anaheim, CA, 1990.

BONAZZA, B. R. RYTOX® PPS conductive composites for EMI shielding applications. *Proc. 34th Int. SAMPE Symp. Exhibit.,* pp. 20–34, Anaheim, CA, 1989.

BRAHNEY, J. N. Fiber-lay process. *Aerosp. Eng.* September:33–4, 1993.

BRINDLEY, P. K., P. A. BARTOLOTTA, AND S. J. KLIMA. Investigation of SiC/Ti-24Al-11Nb composite, 1988. NASA TM 100956.

BROOKSTEIN, D. Interlock-braided shapes have high tensile properties. *Mater. Eng.,* June:9, 1992.

BROWN, G. L., JR. Pultrusion—Flexibility for current and future automotive applications. *CoG/SME Autocom '87 Conf.,* pp. 220–34, Detroit, MI, June 1987.

BUCKLEY, J. D., ED. *Proc. Jt. NASA/DOD Conf. Met. Matrix Carbon Ceram. Matrix Compos.,* Cocoa Beach, FL, January 1986. NASA CP 2482.

BUCKLEY, J. D., ED. *Fiber Tex 1988,* Greenville, SC, September 1988. NASA CP 3038.

BURKLAND, C. V., W. E. BUSTAMANTE, R. KLACKA, ET AL. Property-structure relationships for CVI-processed ceramic matrix composites. *Proc. ASM Int. Conf.,* ed. R. A. Bradley, D. E. Clark, and D. C. Larsen, pp. 225–30, Chicago, IL, 1988.

CARLSSON, R., T. JOHANSSON, AND L. KAHLMAN, ED. *Proc. 4th Int. Symp. Ceram. Mater. Components Eng.* Göteborg, Sweden, June 1991. Elsevier, London.

CARNAHAN, R. D., R. F. DECKER, P. FREDERICK, ET AL. New manufacturing process for metal matrix composite synthesis. *ASM Int. Conf. Fabr. Reinf. Met. Compos.,* Montreal, Canada, September 1990.

CARON, S., AND J. MASOUNAVE. Fabrication of MMCs by a bottom mixing foundry process. *ASM Int. Conf. Fabr. Reinf. Met. Compos.,* Montreal, Canada, September 1990.

CARON, S., AND J. MASOUNAVE. A literature review on fabrication techniques of particulate reinforced metal composites. *ASM Int. Conf. Fabr. Reinf. Met. Compos.,* Montreal, Canada, September 1990.

CARVALHO, F. J. Foreign Industry Analysis: Advanced Composites, Commerce Department, BXAOFA-93-02, 1993.

Ceramic reinforces diesel pistons. *Am. Mach.* May:21, 1989.

Characterization and uses of high temperature fibrous ceramics. *Ind. Heat.* April:66, 1993.

CHAN, A. W., AND S. -T. HWANG. Modeling resin transfer molding of polyimide (PMR-15)/fiber composites. *Polym. Compos.* 14(6):524–8, 1993.

CHARTRAND, F., AND G. LESPERANCE. Effects of processing parameters on the microstructure of a 30 W/O SiC$_w$-2080 aluminum composite. Institute for Aerospace Research, Ottawa, Canada, 1992.

CHEN, R., AND X. LI. A study of silica coatings on the surface of carbon or graphite fiber and the interface in a carbon/magnesium composite. *Compos. Sci. Technol.* 49(4):357–62, 1993.

CHOI, R., AND H. -E. E. KIM. Effect of Si$_3$N$_4$-whisker addition on microstructural development and fracture toughness of hot-isostatically pressed Si$_3$N$_4$. *J. Mater. Sci. Lett.* 13(7):1249–51, 1994.

CHRISTODOLOU, L., P. A. PARRISH, AND C. R. CROWE. *Proc. Symp. Mater. Res. Soc.* 120:29, 1988.

CLAAR, T. D., M. K. AGHAJANIAN, J. T. BURKE, ET AL. Particulate-reinforced aluminum matrix composites formed by a pressureless metal infiltration process. *ASM Int. Conf. Fabr. Reinf. Met. Compos.,* Montreal, Canada, September 1990.

CLAUS, S. J., A. C. LOOS, AND W. T. FREEMAN, JR. RTM process modeling for textile composites. *Fiber Tex 1988,* pp. 349–65, Greenville, SC, September 1988.

COMMITTEE ON HIGH-PERFORMANCE SYNTHETIC FIBERS FOR COMPOSITES. *High Performance Synthetic Fibers for Composites.* National Materials Advisory Board, Washington, DC, 1992. NMAB-458.

Composite materials for NASP enter production. *Des. News,* September 7:36–7, 1992.

Composites may cut costs. *Aviat. Week Space Technol.* January 24:53–54, 1994.

COOKE, T. F. Inorganic fibers. *J. Am. Ceram. Soc.* 74(12):2959–78, 1991.

COOPER, T., ET AL. *Aerosp. Mater. Process Technol. Reinv. Workshop,* Wright Air Force Laboratory, Dayton, OH, May 1993. 94-32941.

CORNIE, J. A. Semi-solid slurry processing of Al-matrix composites. *ASM Int. Conf. Fabr. Reinf. Met. Compos.,* Montreal, Canada, September 1990.

CORNIE, J. A., Y. -M. CHIANG, D. R. UHLMANN, ET AL. Processing of metal and ceramic composites. Industrial Liason Progress Report, Massachusetts Institute of Technology, Cambridge, MA, 1986. MIT 4-24-86.

DONALDSON, P. Cooking an aeroplane. *Aerosp. Compos. Mater.* June/July:6–10, 1989.

DREGER, D. R. The challenge of manufacturing composites. *Mach. Des.* October 22:92–8, 1987.

DREVET, B., S. KALOGEROPOULOU, AND N. EUSTATHOPOULOS. Wettability and interfacial bonding in Au-Si/SiC system. *Acta Metall. Mater.* 42(11):3119–26, 1993.

DRORY, M. D., R. J. MCCLELLAND, F. W. ZOK, ET AL. Fiber-reinforced diamond matrix composites. *J. Am. Ceram. Soc.* 76(5):1387–89, 1993.

DU, G. -W., P. POPPER, AND T. -W. CHOU. Analysis and automation of two-step braiding. *Fiber Tex 1988,* pp. 217–33, Greenville, SC, September 1988. NASA CP 3038.

EMBS, F. W., E. L. THOMAS, C. J. WUNG, ET AL. Structure and morphology of sol-gel prepared polymer-ceramic composite thin films. *Polymer* 34(22):4607–12, 1993.

ENGLISH, L. K. SAMPE 1989: The continuing composites chronicle. *Mater. Eng.* July:55–9, 1989.

EVANS, A. G., AND F. A. LECKIE. *Processing and Mechanical Properties of High Temperature/High Performance Composites,* book 5: *Interface Effects,* book 4: *Processing/Property Correlations,* California University, Santa Barbara, CA, 1994.

Fiber-optic sensor improves composite curing. *Mach. Des.* June 22:70, 1989.

Fiber reinforced aluminum composites. *NTIS Alert,* 95(11):10; Nerac, Inc., Tolland, CT, 1995.

FORSBERG, G., AND S. KOCH. Design and manufacture of thermoplastic F-16 main landing gear doors. *Proc. 34th Int. SAMPE Symp. Exhibit.,* pp. 9–19, Anaheim, CA, 1989.

FRIEDMAN, M. The future of composites in advanced applications. *Adv. Compos.* May/June:28–36, 1986.

FROES, F. H., ED. *Proc. P/M Symp. MPIF,* Tampa, FL, March 1991.

FRONING, M. J., G. SCHWIER, J. BECZKOWIAK, ET AL. Carbide/metal matrix coatings by HVOF processes. *TMS Mater. Week 1993,* Pittsburgh, PA, October 1993.

FUNAHASHI, T., K. ISOMURA, A. HARITA, ET AL. Mechanical properties and microstructures of Si_3N_4-BN composite ceramics. *Proc. 3rd Int. Symp. Ceram. Mater. Components Eng.* ed. V. J. Tennery, pp. 968–76, Las Vegas, NV, November 1988.

GABRIELE, M. C. RTM research took center stage at composites conference. *Plast. Technol.* February:24–6, 1995.

GABRISCH, H. -J., AND G. LINDENBERGER. The use of thermoset composites in transportation: Their behavior. *SAMPE J.* 29(6):23–7, 1993.

GALASSO, F. S. *Advanced Fibers and Composites.* Gordon and Breach, New York, 1982.

GHOSH, P. K., AND S. RAY. Influence of process parameters on the reacted layer at particle-matrix interface in Compocast Al(Mg)-Al_2O_3 composite. *ASM Int. Conf. Fabr. Reinf. Met. Compos.,* Montreal, Canada, September 1990.

GILMAN, P. S. The spray deposition of metals and composites. *J. Met.* 45(7):41, 1993.

GOWRI, S., K. NARAYANASAMY, AND R. KRISHNAMURTHY. Recent developments in ceramic-materials processing and applications. *J. Mater. Process. Technol.* 37:571–82, 1993.

GUO, Z. X., AND B. DERBY. Fibre uniformity and cavitation during the consolidation of metal-matrix composite via fibre-mat and matrix-foil diffusion bonding. *Acta Metall. Mater.* 41(11): 3257–66, 1993.

HALL, I. W., J. -L. LIRN, Y. LEPETITCORPS, ET AL. Microstructural analysis of isothermally exposed Ti/SiC metal matrix composites. *J. Mater. Sci. Lett.* 11(13):3835–42, 1992.

HAMILTON, S., AND N. SCHINSKE. Multiaxial stitched preform reinforcements. *SME Compos. Manuf. Conf.,* p. 433A, Pasadena, CA, January 1990.

HAUBER, D. E., AND D. J. HARDTMANN. Recent advances in thermoplastic composite fabrication using rows. *Proc. 35th Int. SAMPE Symp.,* pp. 767–71, April 1990.

HAUGHT, D., I. BALMY, D. DIVECHA, ET AL. Mullite whisker felt and its application in composites. *Mater. Sci. Eng.* A144:207–14, 1991.

HAUPT, N. A real alternative to hand lay-up of composite. *Westec 1990,* Los Angeles, CA, March 1990. SME MS 90-203.

HENRIKSEN, B. R. The microstructure of squeeze-cast SiC_w-reinforced Al_4Cu base alloy with Mg and Ni additions. *Composites,* 21(4):333–38, 1990.

HILINSKI, E. J., J. J. LEWANDOWSKI, T. J. RODJOM, ET AL. New techniques in processing of discontinuously reinforced aluminum (DRA) composites. *TMS Mater. Week 1993,* Pittsburgh, PA, October 1993.

HISCOCK, D. F., AND D. M. BIGG. Long-fiber-reinforced thermoplastic matrix composites by slurry deposition. *Polym. Compos.* 10(3):145–9, 1989.

HOLLAND, D., ED. *Proc. 2nd Int. Symp. New Mater. Their Appl. Inst. Phys. Gr. Brit.,* Bristol, UK, 1990.

HOOVER, W. R. Recent advances in castable aluminum composites. *ASM Int. Conf. Fabr. Reinf. Met. Compos.,* Montreal, Canada, September 1990.

HOUSTON, D. Q., AND C. F. JOHNSON. High-speed resin transfer molding premolds. *Automot. Eng.* 97(5):63–7, 1989.

HU, D. T. P. JOHNSON, AND M. H. LORETTO. Microstructural characteristics of isothermal forging in cast Ti-6Al-4V/TiC composites. *Mater. Sci. Technol.* 10(5):421–24, 1994.

HU, C. -L., AND M. N. RAHAMAN. Factors controlling the sintering of ceramic particulate composites. II: Coated inclusion particles. *Ind. Heat.* November:31–4, 1992.

HU, C. -L., AND M. N. RAHAMAN. SiC-whisker-reinforced Al_2O_3 composites by free sintering of coated powders. *J. Am. Ceram. Soc.* 76(10):2549–54, 1995.

HU, M. -S., J. YANG, H. C. CAO, ET AL. The mechanical properties of Al alloys reinforced with continuous Al_2O_3 fibers. *Acta Metall. Mater.* 40(9):2315–26, 1992.

HUNT, M. Progress in powder metal composites. *Mater. Eng.* January:33–6, 1990.

HUNT, M. Form and function in metal matrix composites. *Mater. Eng.* June:27–32, 1990.

Injection molding of high temperature plastics, *NTIS Alert,* 95(11):10; Nerac, Inc., Tolland, CT, 1995.

IRVING, R. R. Textron unit adding CMCs to its lines. *Metalworking News,* December 11:88–89, 1989.

JANGHORBAN, K. Processing of ceramic matrix SiC-Al composites. *J. Mater. Process. Technol.* 38(1–2):361–68, 1993.

JANGHORBAN, K. Processing of ceramic matrix SiC-Al composites. *Proc. 1st Asia-Pacific Conf. Mater. Process.,* pp. 361–8, Singapore, February 1993.

JANICKI, G., V. BAILEY, AND H. SCHJELDERUP, EDS. *Proc. 35th Int. SAMPE Symp. Exhibit.* pp. 1–1182, 1193–2310, Anaheim, CA, April 1990.

JARFORS, A. E. W., L. SVENDSEN, M. WALLINDER, ET AL. Reactions during infiltration of graphite fibers by molten Al-Ti alloy. *Metall. Trans.* 24(11):2577–83, 1993.

JENQ, S. T., AND J. W. FAN. Transient response of a pultruded fiber reinforced plastic rod subject to impact loading. *Proc. 35th Int. SAMPE Symp.,* pp. 1420–8, Anaheim, CA, April 1990.

JOHNSON, G. S., J. D. BUCKLEY, AND M. L. WILSON. Pultrusion of fiber-reinforced polymethyl methacrylate. *Fiber Tex 1988,* pp. 269–76, Greenville, SC, September 1988.

KANDORI, T., Y. UKYO, AND S. WADA. Directly HIP'd SiC-whisker reinforced silicon nitride, whisker- and fiber-toughened ceramics. *Proc. ASM Int. Conf.* ed. R. A. Bradley, D. E. Clark, and D. C. Larsen, pp. 125–29, Chicago, IL, 1988.

KARBHARI, V. M., AND D. S. KUKICH. Polymer composites technology in Japan. *Adv. Mater. Process.* 144(2):26–9, 1993.

KARBHARI, V. M., AND T. MCDERMOTT. Investigation of felts as low cost preform materials for RTM. *Proc. 25th Int. SAMPE Tech. Conf.,* Philadelphia, PA, October 1993.

KARBHARI, V. M., AND G. POPE. Effects of freezing and moisture on impact properties of glass fiber reinforced RTM parts. *Proc. 8th Adv. Compos. Conf., ASM/ESD,* pp. 135–47, Chicago, IL, November 1992.

KARDOS, J. L. Composite interfaces: Myths, mechanisms, and modifications. *Chem. Tech.* July: 431–4, 1984.

KENNERKNECHT, S. MMC studies via the investment casting process. *ASM Int. Conf. Fabr. Reinf. Met. Compos.,* Montreal, Canada, September 1990.

KHATRI, S., AND M. KOCZAK. Formation of TiC in in-situ processed composites via solid-gas, solid-liquid and liquid-gas reaction in molten Al-Ti. *Mater. Sci. Eng.* 162(1–2):153–62, 1993.

KIM, K. S., AND H. M. JANG. SiC fibre-reinforced lithium aluminosilicate matrix composites fabricated by the sol-gel process. *J. Mater. Sci.* 30(4):1009–17, 1995.

KOLSGAARD, A., L. ARNBERG, AND S. BRUSETHAUG. Solidification microstructures of AlSi7Mg-SiC particulate composite. *Mater. Sci. Eng.* 129(1–2):243–50, 1993.

LACKEY, W. J., J. A. HANIGOFSKY, G. B. FREEMAN, ET AL. Continuous fabrication of SiC tows by chemical vapor deposition. Georgia Institute of Technology, Atlanta, GA, 1993. GIT-A-9056.

LAI, S. -W., AND D. D. L. CHUNG. Aluminum-matrix composites fabricated by liquid metal infiltration of filler preforms prepared using an acid phosphate binder. *TMS Mater. Week 1993,* Pittsburgh, PA, October 1993.

LANGE, F. F., D. C. C. LAM, AND O. SUDRE. Powder processing and densification of ceramic composites. In *Processing Science of Advanced Ceramics,* MRS Symposia Proceedings, ed. I. A. Aksay, G. L. McVay, and D. R. Ulrich, vol. 155, pp. 309–17. MRS, Pittsburgh, PA, 1989. DAAL03-89G0024.

LANGE, F. F., D. C. C. LAM, O. SUDRE, ET AL. Powder processing of ceramic matrix composites. *Mater. Sci. Eng.* A144:143–52, 1991.

LA SALVIA, J. C., D. K. KIM, D. K. HOKE, ET AL. Combustion synthesis/dynamic densification of ceramics, composites, and cermets. *TMS Mater. Week 1993,* Pittsburgh, PA, October 1993.

LASDAY, S. B. Plasma activated sintering combined with resistance heating provides new method for densifying powdered metals and ceramics in minutes. *Ind. Heat.* April:41–3, 1993.

LAUDER, A. J. Manufacture of rocket motor cases using advanced filament winding processes. *Mater. Manuf. Process.* 10(1):75–87, 1995.

LAVERNIA, E. J. Manufacture of particulate reinforced metal matrix composites using spray atomization and deposition. ARO 26439.33 MS, DAAL03-89K0027, 1990.

LEONARD, L. Automated material deposition: A putdown for fabrication costs. *Adv. Compos.* September/October:51–8, 1987.

LEWIS, D., AND M. SINGH. Recent developments and future prospects for XDTM and other in situ composite materials. *TMS Mater. Week, 1993,* Pittsburgh, PA, October 1993.

LI, W., AND A. E. SHIEKH. Structural mechanics of 3-D braided preforms for composites. Part 2: Geometry of fabrics produced by the 2-step process. *Fiber Tex 1988,* pp. 249–58, Greenville, SC, September 1988. NASA CP 3038.

LIU, D. S., AND A. P. MAJIDI. Creep behavior of SiC_w/Al_2O_3 composites. *Proc. 3rd Int. Symp. Ceram. Mater. Components Eng.* ed. V. J. Tennery, pp. 958–67, Las Vegas, NV, November 1988.

LYE, S. W., AND F. Y. C. BOEY. PC-based monitoring and control system for microwave curing of polymer composites. *Mater. Manuf. Process.* 9(5):851–68, 1994.

MABUCHI, M., T. IMAI, K. KUBO, ET AL. New fabrication procedure for superplastic aluminum composites reinforced with Si_3N_4 whiskers or particulates. *Proc. 7th Compos. Conf.,* pp. 275–82, Detroit, MI, September/October 1991.

MACCONOCHIE, I. O., M. L. WILSON, K. H. JONES, ET AL. Application of composites to a space spider robot. *Fiber Tex 1988,* pp. 301–15, Greenville, SC, September 1988.

MAJKOWSKI, T., AND U. KASHALIKAR. Metal-matrix composite parts with metal inserts. NASA Technical Brief, May 1995, pp. 28–29.

MARSH, G. Braving the heat. *Aerosp. Compos. Mater.* February:28–32, 1988.

MARSHALL, I. H. Composite structures. *Proc. 6th Int. Conf. Compos. Struct.* Elsevier, London, 1991.

MARZULLO, A. SiC fibers for cast metal matrix composites. Hartford Graduate Center, Hartford, CT, 1990.

Materials Technologies 1991: The Year in Review. D92-1627, SRI International, Palo Alto, CA, 1992.

MARTIN, J., AND J. E. SUMERAK. Pultrusion composites show market growth due to increased technology in three key areas. *Fiber Tex 1988,* pp. 277–83, Greenville, SC, September 1988.

MEYER, D. W., R. F. COOPER, AND M. E. PLESHA. High-temperature creep and the interfacial mechanical response of a ceramic matrix composite. *Acta Metall. Mater.* 41(11):3157–70, 1993.

Mineral fillers in plastics and elastomers, *NTIS Alert,* 95(11):10; Nerac, Inc., Tolland, CT, 1995.

MITTNICK, M. A. Continuous SiC fiber reinforced metals. *Proc. 35th Int. SAMPE Symp.* pp. 1372–83, Anaheim, CA, April 1990.

MONKS, R., AND M. NAITOVE. Technology update: Advanced composites. *Plast. Technol.* March:48–55, 1991.

MORTENSEN, A. Interfacial phenomena in the solidification processing of metal matrix composites. *Mater. Sci. Eng.* A135:1–11, 1991.

MORTENSEN, A. Metal matrix composites: Composite systems. Industrial Liaison Progress Report, Massachusetts Institute of Technology, Cambridge, MA, 1993. MIT 3-30-93.

MORTENSEN, A., AND I. JIN. Solidification processing of metal matrix composites. *Int. Mater. Rev.* 37(3):101–28, 1992.

MORTENSEN, A., AND V. J. MICHAUD. Infiltration processing of metal matrix composites. Industrial Liason Progress Report, Massachusetts Institute of Technology, Cambridge, MA, 1992. MIT-6-39-92.

MROZ, C. Silicon carbide-whisker preforms for metal-matrix composites by infiltration processing. *Am. Ceram. Soc. Bull.* 72(6):76–8, 1993.

MUSCAT, D., M. D. PUGH, R. A. L. DREW, ET AL. Microstructure of an extruded β-silicon nitride whisker reinforced silicon nitride composite. *J. Am. Ceram. Soc.* 75(10):2713–8, 1992.

NAYIM, S., AND J. C. BARAM. Spray atomization and codeposition of Al-Li based metal-matrix composites. *Proc. 7th Adv. Compos. Conf.,* pp. 267–73, Detroit, MI, September/October 1991.

NELSON, R. W. Fabrication of Polyetheretherketone (PEEK) parts. KCP-613-4473, December 1991. Contract DE-AC04-76-DP00613.

New technology for high-volume molding of TP composites. *Plast. Technol.* December:14, 1993.

NICOLAOU, P. D., AND H. R. PICHLER. Experimental studies of the consolidation of Ti-6Al-4V matrix, SCS-6 fiber composites under different stress states using the foil/fiber/foil technique. *TMS Mater. Week 1993,* Pittsburgh, PA, October 1993.

NIIHARA, K., AND A. NAKAHIRA. Strengthening of oxide ceramics by SiC and Si_3N_4 dispersions. *Proc. 3rd Int. Symp. Ceram. Mater. Components Eng.* ed. V. J. Tennery, pp. 919–26, Las Vegas, NV, November 1988.

NIX, W. D. High temperature deformation processes and strengthening mechanisms in intermetallic particulate composites. AR 11/89-11/90, AFOSR TR 91-2, AFOSR89-0201, 1991.

Oak Ridge research on E-beam composite curing. *SAMPE J.* 30(4):20–1, 1994.

OGUMI, Z., T. IOROI, Y. UCHIMOTO, ET AL. Novel method for preparing nickel/TSZ cermet by a vapor-phase process. *J. Am. Ceram. Soc.* 78(3):593–8, 1995.

OLDROYD, P. K. Development and fabrication of V-22 conversion spindle. *Fiber Tex 1988,* pp. 331–47, Greenville, SC, September 1988.

PAN, N. A detailed examination of the translation efficiency of fiber strength into composite strength. *J. Reinf. Plast. Compos.* 14(1):2–28, 1995.

PENNINGTON, J. N. So you're planning to cast composites? *Mod. Met.* June:26–9, 1994.

PETROVIC, J. J., AND R. E. HONNELL. $MoSi_2$ particle reinforced-SiC and Si_3N_4 matrix composites. *J. Mater. Sci. Lett.* 9(9):1083–4, 1990.

PHILLIPS, D. C. Interfacial bonding and the toughness of carbon fibre reinforced glass and glass ceramics. *J. Mater. Sci.* 9:1847–54, 1974.

PICKLES, C. A., AND J. M. TOGURI. The plasma arc production of Si-based ceramic whiskers. *J. Mater. Res.* 8(8):1996–2003, 1993.

PITCHUMANI, S., P. A. SCHWENK, A. K. KORDON, ET AL. An expert system approach to manufacturing preforms for infiltration-processing of ceramic- and metal-matrix composites. *Proc. Adv. Mater.* 4(3):155–65, 1994.

PLUNNETT, K. P., C. H. CA'CERES, C. HUGHES, ET AL. Processing of tape-cast laminates prepared from fine alumina/zirconia powders. *J. Am. Ceram. Soc.* 77(8):2145–53, 1994.

PLUNNETT, K. P., C. H. CA'CERES, AND D. S. WILKINSON. Tape casting of fine alumina/zirconia powders for composite fabrication. *J. Am. Ceram. Soc.* 77(8):2137–44, 1994.

POWELL, R. E., AND H. -Y. YEH. Mechanical characterization of hibridized braided composite structures. *J. Reinf. Plast. Compos.* 14(2):164–94, 1995.

PRATT AND WHITNEY. Titanium matrix composite turbine engine component consortium (TM-CTEC), West Palm Beach, FL. Contract F33615-94-2-4439.

Proc. Adv. Compos. Conf., ASM/ESD, Detroit, MI, September/October, 1991.

Proc. 8th Adv. Compos. Conf., ASM/ESD Conf., Chicago, IL, November 1992.

Proc. 7th Tech. Conf. Am. Soc. Compos. Technomic, Lancaster, PA, 1992.

PUJARI, V. K., AND D. M. TRACEY. Processing methods for high reliability silicon nitride heat engine components. *J. Eng. Gas Turbines Power,* 117(1):156–60, 1995.

Pultrusion of reinforced plastics. *NTIS Alert* Jun 1, 1995, p 11, Nerac, Inc., Tolland, CT,

RAJ, R. Workshop on high temperature metal-ceramic composites. FR 5/90-4/91, AFOSR TR 91-0025, AFOSR90-0230, 1990.

RAWAL, S. P., L. F. ALLARD, AND M. S. MISRA. Microstructural characterization of interfaces in diffusion-bonded and cast Gr/Mg composites. *AIME Symp. Proc. Interfaces Compos.,* ed. S. Fishman and A. K. Dhingra, pp. 211–26, New Orleans, LA, March 1986.

RAWAL, S. P., AND M. S. MISRA. Development of a ceramic microballoon reinforced metal matrix composite. *Proc. 34th Int. SAMPE Symp. Exhibit.,* pp. 112–24, Anaheim, CA, 1989.

REHMAN, F. U., H. M. FLOWER, AND D. R. F. WEST. Optimisation of alloy-alumina fibre combination for magnesium based metal matrix composite. *Mater Sci. Technol.* 10(6):518–25, 1994.

ROGERS, J. K., AND M. C. GABRIELE. Composite resins, fibers, machinery debut at SAMPE conference. *Plast. Technol.* July:25–31, 1989.

ROHATGI, P., AND R. ASTHANA. The solidification of metal-matrix particulate composites. *J. Met.* 43(5):35–41, 1991.

ROHATGI, P. K., S. RAY, R. ASTHANA, ET AL. Interfaces in cast metal-matrix composites. *Mater. Sci. Eng.* 162(1–2):163–74, 1993.

ROSENOW, M. W. K. New generation of advanced thermoplastic materials for industrial applications. *8th Ann. ASM/ESD Conf.,* pp. 75–80, Chicago, IL, November 1992.

RUSSELL, K. C., J. A. CORNIE, AND S. -Y. OH. Particulate wetting and particle: Solid interface phenomena in casting metal matrix composites. *AIME Symp. Proc. Interfaces Compos.,* ed. S. Fishman and A. K. Dhingra, pp. 61–91, New Orleans, LA, March 1986.

SARHADI, M., T. A. MITCHELL, AND R. F. J. McCARTHY. Robotic lay-up of high performance composite preforms. *Proc. 14th Int. SAMPE Eur. Conf. Exhibit.,* Birmingham, U.K., October 1993.

SCHUSTER, D. M., M. K. SKIBO, R. S. BRUSKI, ET AL. The recycling and reclamation of metal-matrix composites. *J. Met.* 45(5):26–30, 1993.

SESHADRI, S. G., M. SRINIVASAN, J. MACBETH, ET AL. Fabrication and mechanical reliability of SiC/TiB$_2$ composites. *Proc. 3rd Int. Symp. Ceram. Mater. Components Eng.* ed. V. J. Tennery, pp. 1419–28, Las Vegas, NV, November 1988.

SHARMA, S. C., AND S. R. ARUN. Fabrication and evaluation of the mechanical properties of aluminum alloy-glass particulate composites. *Proc. 1st Asia-Pacific Conf. Mater. Process.,* pp. 361–8, Singapore, February 1993.

SHARPLESS, G. The continuing development of braiding/resin transfer molding for commercial aircraft and aerospace applications. *Proc. 25th Int. SAMPE Tech. Conf.,* Philadelphia, PA, October 1993.

SIKORSKI, K., W. CZEREPKO, M. J. EDIRISINGHE, ET AL. Sintering of Al$_2$O$_3$-ZrB$_2$ composites. *Proc. Adv. Mater.* 4(2):95–103, 1994.

SOEBROTO, H. B., C. M. PASTORE, AND F. K. KO. Engineering design of braided structural fiberglass composite. *Fiber Tex 1988,* pp. 435–40, Greenville, SC, September 1988. NASA CP 3038.

Sol gel processes (June 89-present). NTIS, 1993. N-012638001.

SOMIYA, S., AND Y. INOMATA. *Silicon Carbide Ceramics.* 2: *Gas Phase Reactions, Fibers and Whisker, Joining,* Ceramic Research and Development in Japan. Elsevier, New York, 1991.

SONTI, S. S., E. J. BARBERO, AND T. WINEGARDNER. *J. Reinf. Plast. Compos.* 14(4):390–401, 1995.

SOUTHMAYD, A. D. Tooling for composite parts utilizing CAD/CAM. *SME Automat. Compos. Manuf. Conf.,* pp. 235–45, May 1987.

SPRINGER, G. S. Modeling the cure process of composites. *SAMPE J.* September/October:22–7, 1986.

STEENKAMER, D. A., D. J. WILKINS, AND S. G. SLOTTE. Preform joining technology applied to a complex structural part. *Proc. 7th Adv. Compos. Conf., ASM/ESD,* pp. 443–52. Detroit, MI, September/October 1991.

STIEGLER, J. O., AND J. R. WEIR, JR. Metals and ceramics division progress report for period ending September 30, 1991. ORNL-6705, UC-904, 1992.

STONE, K. L. Automation in composites processing. *SAMPE Q.* July:8–11, 1984.

STRIFE, J. R. Study of critical factors controlling synthesis of ceramic matrix composites from pre-ceramic polymers, FR 10/87-11/90, UTRC R 90-917810-5, F49620-87C0093, 1990.

STRONG, A. B., AND P. HAUWILLER. Incremental forming of large thermoplastic composites. *Adv. Compos.* 4(5):56–66.

STUDT, T. Breaking down the barriers for ceramic matrix composites. *Res. Dev.* August:36–42, 1991.

SWAIN, M. V. *Materials Science and Technology—Comprehensive Treatment: Structure and Properties of Ceramics.* VCH, Weinheim, Germany, 1994.

TANIKAWA, E., S. TAKEYAMA, AND T. SAKAKIBARA. Mechanical properties and fabricability of MMC by roll diffusion bonding process. Fuji Heavy Industries, Ltd., Tokyo, Japan, pp. 449–57.

TAYLOR, A. Fabrication and performance of advanced carbon-carbon piston structures. *Fiber Tex 1988,* p. 375, Greenville, SC, September 1988.

TAYLOR, R. S. Braiding and resin transfer molding: A technical approach to fabricating composite drive shafts. *Proc. 25th Int. SAMPE Tech. Conf.,* Philadelphia, PA, October 1993.

Technology base enhancement program. *NTIS Alert* Jun 1, 1995, p 11; BDM Federal, Inc., Toronto, Canada, 1993.

TENNERY, V. J., ED. Ceramic materials and components for engines. *Proc. 3rd Int. Symp.,* Las Vegas, NV, November 1988.

TRESPAILLE'-BARRAU, P., AND M. SUE'RY. Microstructural and mechanical characterisation of aluminum matrix composites reinforced with Ni and NiP coated SiC particles via liquid processing. *Mater. Sci. Technol.* 10(6):497–504, 1994.

TSAO, I., S. C. DANFORTH. Injection moldable ceramic-ceramic composites: Compounding behavior, whisker degradation, and orientation. *Am. Ceram. Soc. Bull.* 72(2):55–64, 1993.

TSENG, W. J., AND P. D. FUNKENBUSCH. Microstructure and densification of pressureless-sintered Al$_2$O$_3$/Si$_3$N$_4$-whisker composites. *J. Am. Ceram. Soc.* 75(5):1171–5, 1992.

USTUNDAG, E., R. SUBRAMANIAN, R. VAIA, ET AL. In situ formation of metal-ceramic microstructures, including metal-ceramic composites, using reduction reactions. *Acta Metall. Mater.* 41(7):2153–61, 1993.

Vacuum hot-press composites. *Mater. Eng.* March 1992:7.

VERMA, S. K., AND J. L. DORCIC. "Squeezing" production costs from metal-ceramic composites. *Adv. Mater. Process.* May:48–51, 1988.

VIVÉS, C., J. BAS, G. BELTRAN, ET AL. Fabrication of metal matrix composites using a helical induction stirrer. *Mater. Sci. Eng.* 129(1–2) A173:239–42, 1993.

WADSWORTH, J., AND T. G. NIEH. Superplasticity in ceramic and metal matrix composites and the role of grain size, segregation, interfaces, and second phase morphology. *Mater. Sci. Eng.* A166: 97–108, 1993.

WAKU, Y., AND T. NAGASAWA. Future trends and recent developments of fabrication technology for advanced metal matrix composites. *TMS Mater. Week 1993,* Pittsburgh, PA, October 1993.

WALKER, J. A. The treatment and preparation of silicon carbide whiskers for the manufacture of powder metallurgical discontinuously reinforced aluminum composites. *Adv. Compos. Mater. Conf.: P/M in Aerospace and Defense Technologies,* vol. 1. Seattle, WA, November 2–3, 1989, pp. 207–211.

WANG, N., Z. WANG, AND G. C. WEATHERLY. Formation of magnesium aluminate (spinel) in cast SiC particulate-reinforced Al (A356) metal matrix composites. *Metall. Trans.* 23A(5):1423–30, 1992.

WARRIER, S. G., AND R. Y. LIN. Silver coating on carbon and SiC fibres. *J. Mater. Sci.* 28(18): 4868–77, 1993.

WEEKS, G. P., AND A. H. WHITE. Advances in thermoplastic compression molding material and technology. *8th Ann. ASM/ESD Conf.,* pp. 81–4, Chicago, IL, November 1992.

WEI, J., M. C. HAWLEY, J. JOW, ET AL. Microwave processing of cross-ply continuous graphite fiber/epoxy composites. *SAMPE J.* 27(1):33–9, 1991.

WEST, P. *Adv. Mater.* 13(3):1, 1991.

WILKINSON, D. S., ED. *Proc. 27th Int. Symp.,* Montreal, vol 9. Pergamon Press, New York, 1989.

WOOD, J. Metal matrix composites: Design and innovation. *Powder Metall.* 37(1):19–21, 1994.

Workshop AGARD Struct. Mater. Panel, Antalya, Turkey, April 1993. AGARD-R-794.

WU, M., AND G. L. MESSING. Fabrication of oriented SiC-whisker-reinforced mullite matrix composites by tape casting. *J. Am. Ceram. Soc.* 77(10):2586–92, 1994.

WU, Y., AND E. J. LAVERNIA. Mechanical behavior of Al-Si-X (X = SiC, TiB_2) MMCs processed by spray atomization and deposition. *TMS Mater. Week 1993,* Pittsburgh, PA, October 1993.

YANG, H., AND J. S. COTTON. Microstructure-based processing parameters of thermoplastic composite materials. Part I: Theoretical Models. *Polym. Compos.* 15(1):34–41, 1994.

YANG, H., AND J. S. COTTON. Microstructure-based processing parameters of thermoplastic composite materials. Part II: Experimental results. *Polym. Compos.* 15(1):42–5, 1994.

YE, L., AND K. FRIEDRICH. Processing of thermoplastic composites from powder/sheath-fibre bundles. *J. Mater. Process. Technol.* 48(1–4):317–24, 1995.

YOUNG, W. -B. Thermal behaviors of the resin and mold in the process of resin transfer molding. *J. Reinf. Plast. Compos.* 14(4):310–32, 1995.

YUN, H. M., AND R. H. TITRAN. Tensile behavior of tungsten/niobium composites at 1300 to 1600 K, 1989. NASA TM-103727, DOE/NASA/16310-15, DE-AI03-86SF16310.

ZAKRZEWSKI, G. A., D. MAZENKO, AND S. T. PETERS. *Proc. 34th Int. SAMPE Symp. Exhibit.* Reno, NV, 1989.

2

POSTTREATMENT PROCESS

2.0 INTRODUCTION

Design is a partnership between engineering and manufacturing. The designer is the focal point and leader who defines the requirements for a part and puts them down in graphical form. Manufacturing knows best how to make the part. The designer must have a good understanding of manufacturing methods, and the manufacturing engineer must understand design requirements.

The area that leads to the most problems is selection of the material. Today there are and in the future there will be many different types of advanced composite materials. Each type has its advantages, and each component must be designed using the most advantageous material and posttreatment process (machine, join, form, finish, etc.).

There are of course limitations on what materials can be used in combination with each other. The problems of galvanic corrosion and thermal expansion are well known. Thermal conductivity and electric conductivity are also considerations. Producibility, cost, and material density are other issues.

As the performance requirements for various aerospace, transportation, and civil structures become more varied and demanding, it is often impossible for a single material to be entirely suitable for the structure in question. It is becoming more commonplace for two or more materials to be used in tandem in an application. This combination of materials allows the structures to be capable of meeting multiple performance targets.

When different materials are used within a given structure, there is most definitely a need to somehow drill and/or form and/or machine and/or cut and/or grind and/or join and/or finish. The methods of the past used to accomplish any of these procedures are still available, however, the trend is toward finding less labor-intensive operations and methods geared more toward automation.

2.1 POSTTREATMENT

As more and varied advanced composite materials appear with new matrices, new reinforcement materials, and complex interfaces between reinforcement and matrix, new machining, cutting, drilling, joining, and so on, techniques and methods are invented, developed, evaluated, tested, and applied. With the establishment of these new methods engineers must now be aware of the numerous choices and that they may be the solution to a particular problem.

For example, there were two fundamental ways of joining structures—fastening and bonding. Fasteners include rivets, bolts and their advanced derivatives, and bonding agents (usually adhesives), but in the case of thermoplastics, for example, fusion welding might be used. For most solutions there are usually advantages and disadvantages which have to be carefully weighed before a final choice is made. Drilling holes in monolithic materials is in many cases a well-understood process, but drilling holes in composite materials often has to be considered much more carefully. Particulate composites may behave like monolithic materials, but with fiber-reinforced composites the drilling of holes can be highly detrimental because fibers and laminates are being cut. Making holes may not be easy. In the case of carbon- and aramid-reinforced plastics, the material is frequently very hard and special drills, ultrasonics, and so on, have been used.

Computer information software and database planning networks have been developed, such as American Welding Institute Composite Material Joining Database (AWI-CMJDB),[1] to extend expert information systems to the designer, engineer, and technologist. Consider a joint in a Gr_f/PEEK composite where mechanical fastening (MF), bonding, or fusion welding could be used. In the case of bonding, should an adhesive be added or should a solvent be used? Which adhesives will adhere to the surface and not cause degradation of the matrix or the matrix-fiber bond? In the case of fusion welding, what will be the degree of degradation (strength loss, change in glass-transition temperature, etc.) due to heating? What temperature and heating rate will the graphite fibers be able to stand?

The areas of concern for the designer in machining composites are (1) quality of the machined region or machined component using either conventional or nonconventional cutting methods, and (2) rapid tool wear, especially when using conventional methods.[2,3]

Because of these concerns and the results of conventional machining methods (excessive tool wear and poor quality of machined components), the pursuit of other cutting methods with little or no tool contact with the workpiece has occurred. The advantage of using a noncontact tool cutting method is twofold: on the one hand, the rate of tool wear is low and easy to monitor, hence lending itself well to automation; on the other hand, surface damage due to mechanical action of the tool is negligible because of the noncontact characteristic of the cutting tool. It should be pointed out that these methods may not be capable of inducing shape changes in a workpiece as is possible by traditional methods (milling, turning, etc.); however, some nonconventional approaches have proved to be most effective for secondary processes (drilling, slitting, etc.). The various types of non-conventional methods available for machining materials include the following.

Laser beam machining (LBM)

Water jet machining (WJM) and abrasive water jet machining (AWJM)

Electric discharge machining (EDM)

Ultrasonic machining (USM)

Electron beam machining (EBM)

Electrochemical machining (ECM).

A wide variety of both traditional and nontraditional machining methods are currently available for machining various composites. In selecting a traditional method of machining, it is important to note that because of the excessively rapid tool wear associated with these methods, diamond cutting tools are considered to be the most effective. This can limit the application of these methods in generating complicated shapes and features. The quality of the machined component is also an important factor in the selection of a machining method, and it has been shown that conventional machining can cause severe damage to a component.

Nonconventional methods have a promising future in the machining of composites. Because of advances being made in the production of versatile robots with improved accuracy, some of these methods can be fully automated, which will make them cost-effective.

2.2 PROCESSING OF COMPOSITE MATERIALS

The processing of composite materials can be grouped into two categories: primary and secondary. Even if the primary processing of a composite with unique properties is established, it is the secondary processing that makes an otherwise unique composite unattractive to end users. This is partly because secondary processing of composites and composite structures has been less studied in comparison to primary processing. Two of the key secondary processes are machining and joining.

The economic manufacture of high-precision engineering components with functional surfaces presupposes that the problems of machining can be solved. The machining cost often amounts to 10–100 times the raw material cost. Nontraditional machining processes, mentioned above, are being used to produce complex parts of fiber- and particulate-reinforced composites in a variety of applications throughout the aerospace, automobile, and electronic industries. Nontraditional manufacturing operations could be competitive because of the reduction in the number of rejects and the ease of implementing computer control. Because of steadily improving cutting capabilities, computer control, and adaptive control, these nontraditional methods are ensured an increasingly important role in industries.

The laser is fast replacing other conventional tools in the secondary processing of composites. The ability of a laser to cut intricate shapes and complex patterns and join either similar or dissimilar materials with minimum distortion will make it a more competitive manufacturing tool. Other advantages of lasers in secondary processing are as follows.

- Laser machining and joining processes are capable of a high level of automation and can achieve very high production rates. Automation improves precision, repeatability, and flexibility.
- The noncontact nature of the process makes it an ideal choice for components that are easily distorted and for abrasive materials such as composites that are difficult to join and machine.
- Large mechanical and thermal forces are not exerted on the workpiece.
- High processing speeds can be obtained, and minimum retooling time is required.
- A narrow to negligible heat-affected zone (HAZ) is produced when compared to other thermal processes.
- Fewer moving parts are involved, and thus maintenance costs are reduced.
- A wide range of materials can be cut and joined without requiring a substantial change in the optical components of the system.

In spite of these numerous advantages, laser beam technology has a few disadvantages because of its inherent thermal nature. The main drawbacks in using the laser as a secondary processing tool are the following.

- Production of toxic fumes in the cutting of PMCs
- Inability to cut or join thick sections (i.e., sections thicker than 2.54 cm) because of the unavailability of high-power lasers and large focal length optics (with a large depth of focus).

2.3 MACHINING

Machining is generally considered a finishing process, with specified dimensions, tolerances, and surface finish. The quality of a surface generated by a machining process and its characteristics are important factors when evaluating the service life of a component under dynamic loads.

The methods and techniques for machining composites are related to the type of composite, for example, a thermoplastic or a thermoset matrix, and the type of reinforcement such as continuous or discontinuous and inorganic versus organic.

The machining techniques for thermosets utilize standard metal- and woodworking machining equipment and can be used with modification to increase spindle speeds along with reduced feeds. Standard cutting tools are suitable only for short production runs, and therefore presently used tools are WC- or diamond-tipped. Tools used on such composites are required to be sharp not only for clean cuts but to minimize the possibility of delamination.

Although the machining of thermoplastics is a well-known art, it is complicated by the wide property difference among the large number of available materials. The generalized machining hints given below are for the largest majority of thermoplastics, although it must be recognized that anomalies do exist.

Coolants should be used to avoid overheating and subsequent melting of the workpiece.

Machinery should be operated at high speeds.

Liberal clearances should be provided on cutting wheels.

Light cuts and a slow feed of the workpiece should be used.

Turning tools should be ground to provide rake angles that minimize tool cutting and thrust forces.

Slow spiral drills designed for thermoplastics are recommended for use.

Tools should be carbide-tipped or utilize special high-speed steel (HSS) tools designed for composites.

The workpiece must be properly supported to avoid distortion under cutting pressures.

Allowance for plastic memory and shop room temperature should be made for accurate machining.

Tool bits and cutters should be sharp since dull tools increase forces on the workpiece.

For the various traditional machining operations performed on thermosets and thermoplastics some general recommendations are listed in the accompanying table and more specifics follow later in this chapter.

Operation	Thermosets	Thermoplastics
Sawing	Hand saws and circular saws.	Band saws and circular saws—carbide-tipped.
Drilling and routing	Dry with finer finish results with free-flowing lubricants or air blasting. Use a high helix angle and wide, polished flutes of WC with diamond tips.	Specific helix angles, wide, polished flutes, and point angle for drilling. HSS- and carbide-tipped drills. For B/Ep use diamond-coated core drills and either free-flowing or submerged coolants. EDM can successfully drill Al/B.
Tapping and threading	Standard metal taps—WC. High speed for lathe threading with an oil coolant.	Threads can be made using lathes or taps and dies. HSS straight, fluted taps. Air blasting or coolants minimize tap clogging and permit high tapping rates.
Milling	Standard machines with HSS, WC, or diamond. Air or vapor mist coolants.	Standard metalworking lathes and milling machines with HSS tools, although carbide- or diamond-tipped tools are preferred.
Turning	Conventional engine or bench lathes.	For light and heavy cuts slight nose radii on cutting edge.
Boring, facing, and cutoff	HSS-, WC-, or diamond-tipped. Coolants can be used.	—

Operation	Thermosets	Thermoplastics
Sanding and grinding	Belt- and drum-type machines. SiC abrasives.	Wet belt sanders and dry and abrasive disks. Grinding use—SiC or Al_2O_3 grit wheels and free-flowing coolants. Boron composites can be slit with a table saw or a cutter-grinder machine. Diamond cutting wheels are used as well as coolants. Diamond wheels appear to be best, although SiC and Al_2O_3 wheels can be used and coolants are required to prevent matrix thermal degradation.
Shearing and punching	Conventional sheet metal shears-Punch and die block -WC or carbon steels, chrome steels, and tool steel.	For blanking use steel rule dies and a "clicker press." Pierce, punch, and shear on standard metalworking equipment with the workpiece either hot or cold.
Finishing and polishing	Sand blasting, honing, lapping, buffing, and polishing with sand and Al_2O_3 abrasives. Honing—dry or wet with an Al_2O_3 abrasive.	Same techniques and equipment as used on thermosets for polishing.

2.3.1 Sawing PMCs

Cured laminates may be cut by band, circular, or saber saws, or water jet–abrasive water jets (WJ/AWJ), which are discussed further later in this chapter. Clamping of the part is required to eliminate vibration which may cause delamination, and the cutting edges require frequent checking in order to maintain their sharpness. The following recommendations apply to each sawing technique noted above.

1. *Band sawing.* Use a fine, offset stagger-tooth blade (14–20 teeth/25.4 mm). Use a high surface speed of 23.0–32.5 m/s (preferrably 30.6 m/s). The preferred cutting method is to utilize the heel of the cutting tooth blade rather than the hook of the cutting tooth blade for cleaner cuts. Band saw cutting blades should be sharpened first before usage.

   Additionally, most operators prefer to use water as a coolant. Some edge fuzzing may remain but can be removed by light, wet sanding. The band should be run in reverse, such that the heel of the tooth enters the composite first (Figure 2.1). The band saw teeth should also be honed. Honing can be done while the blade is turning.

2. *Circular sawing.* Use a fine, offset tooth blade (60–80 teeth/25.4 mm). Use a high surface speed of 25.5–30.6 m/s, although speed and feed are governed by the thickness of the material.

   Straight cuts on laminates up to 0.32 cm thick can be made with only slight fuzzing on the exit side. The blade should be run in reverse, such that the heel of the tooth enters the composite first.

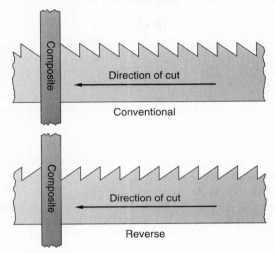

Figure 2.1   Reverse band sawing Kevlar aramid composites.[2]

3. *Saber sawing.* A saber saw blade has been developed that cuts the outermost fibers on both sides of a laminate toward the interior, leaving an excellent cut edge.

   A saber saw blade with five alternating teeth in opposing directions is recommended for cutting at blade speeds of 2500 strokes/min, however, blade speed and feed rates may vary according to material thickness.

2.3.2 Turning PMCs

PCD tooling has been used for diamond turning, boring, and milling tools. Some of the tools have brazed tips, and others accommodate replaceable inserts. For feeds and speeds, it has been recommended to begin with carbide guidelines and then, with experience, increase cutting rates to more optimum levels. A general range of cutting parameters for Gl/Ep and C/Ep materials can be found in Table 2.1.

Kim and his associates[4,5] recently reported a series of quantitative results for the cutting of C/Ep materials and the turning of C/Ep materials in order to obtain the Taylor tool wear constants and surface roughnesses as well as chip formation mechanisms and the optimum cutting speed and feed in the turning of C/Ep materials.

Conclusions from the study were the following.

- The flank wear of the tool increased with an increase in the fiber-winding angle. Also, the dependency of tool wear on the cutting speed was greater when the fiber-winding angle was greater.
- The Taylor tool wear constants n and C for the WC tool (K10) were in the ranges 0.45–1.12 and 55.4–241.5, respectively. The WC tools used in the machining of C/Ep materials were uncoated. Multicoated inserts and polycrystalline (PCD) tools might be more efficient in the machining of such materials.[5]

TABLE 2.1   Machining Parameters for PCD Tooling—Turning, Boring, Facing, and Cutoff Operations[3]

Workpiece Material	Nose Radius (in.)[a]	Side Relief Angle (deg)	Positive Back Rake Angle (deg)	Speed (ft²/min)	Depth of Cut (in.)[a]	Feed Rate (in. rad)
Glass-fiber-plastic composites	0.030–0.090	5–20	0–5	400–3600	0.001 and up	0.001–0.010
Carbon-plastic composites	0.020–0.040	5–20	0–5	500–2000	0.010 and up	0.005–0.015

[a] 1 in. = 25.4 mm.

- The surface roughness was more dependent on the feed and the fiber-winding angle than on the cutting speed.
- The cutting speed and the feed recommended are 20–40 m/min and 0.2 mm/r, respectively, from the point of view of the flank wear of the WC tool (K10) and the surface roughness.

Santhanakrishnan and his associates[6] conducted in-face turning trials on carbon, glass, and Kevlar fiber-reinforced epoxy cylindrical tubes and used HSS and sintered carbide tools.

In addition to recording the surface roughness profiles of the machined surfaces, the morphology of the machined surfaces and the worn-out tool portions was examined and their results and conclusions included the following.

1. The C/Ep surface had the best finish value (R_a) in comparison with Gl/Ep and Kv/Ep surfaces.
2. Because of the crushing and sharp fracturing of C_f, the C/Ep surface exhibited a better surface texture than Gl/Ep. The Kv/Ep surface was poor owing to the fuzziness caused by tough Kv_f.
3. HSS tools with sharp cutting edges appeared to be well suited for machining Kv/Ep, while tool wear was extremely high for Gl/Ep and C/Ep composites.

Ramulu and his colleagues[7] investigated, by turning (cutoff) tests, the use of different grades of PCD inserts in the machining of a Gr/Ep composite material. Their initial tests showed the following.

1. The sharpness of the tool and its microstructure had a great influence on the cutting efficiency of the Gr/Ep material. The coarse-grade polycrystalline diamond (PCD) was more wear-resistant than a finer grade.
2. The flank wear was fairly uniform, and the wear mechanism was found to be due to mechanical processes. Rounding of the edges and cracking and chipping were observed in all grades of PCD. The carbide tool exhibited nonuniform flank wear and became dull within the first 10 s of machining time.
3. The surface roughness studies showed that the surface quality of machined Gr/Ep remains the same or gets even better as the cutting time increases, regardless of the PCD grade.

4. Machined Gr/Ep surface texture was dependent on fiber orientation. The surface exhibited many small holes due to fiber pullout, the presence of cracks, and smearing of the matrix material.

2.3.3 Drilling PMCs

A question normally asked is why it takes so long to drill holes in composite materials. The answer is that the available tooling was not designed for cutting composites. Gl/Ep and Gr/Ep are so abrasive that initially only WC was used; however, PCD has been very successful (Table 2.2). Drill tips were designed for metalworking, the tip heating the metal to provide the plastic flow needed for efficient cutting. Because composites cannot tolerate this heat, production must be slowed down to reduce the heat. Drill designers had to abandon cutting tips with neutral and negative rakes and wide chisel points because a drill with a neutral rake scrapes the material and causes it to resist penetration by the drill tip. The operator must exert pressure to drill the hole, and pressure causes heat buildup.[8,9]

Fully fluted drills with PCD tips on solid carbide provide the strength of carbide and hardness of diamond while allowing innovative geometries. Users must be careful, however, to adjust feeds and speeds for each layer of composite, and employing a torque sensor to find the layers is very helpful.[10–12]

For drilling a graphite composite with an aluminum matrix backing, a PCD double-angle drill is recommended; with ordinary carbide drill technology, the hole finish quality of the graphite layer would be damaged, with chips of aluminum being pulled into contact with the graphite. The double-angle drill allows the shallow center section of the tool to center the tool and prevent wandering, while the larger angle eases the tool into the hole up to its diameter, creating less heat and eliminating interface bell mouthing.

The life expectancy of PCD drills is now well documented. Even without resharpening, which can be accomplished at least two to three times per tool, diamond drills have increased holes in composites per tool by a factor of 100 or more compared to ordinary carbide drills. Even though diamond drills are more expensive than carbide drills, they offer the same flexibility of design. A diamond drill often provides manufacturers the most cost-effective method of optimizing production rates without the need for a large investment in new equipment.[13]

The best way to analyze a drilling operation is to examine the chips. The ideal chip form for composites is a dry, easily moved chip that looks like confectioner's sugar. If

TABLE 2.2   Schematic of EDM Process Setup[7,8]

Tooling Material	Material Thickness (mm)	Hole Diameter (mm)	Speed (m/s)		Feed Rate (in./rad)
			UDFs[a]	MDFs[b]	
Carbide	Up to 12.7	4.8–8.1	0.7	1	0.001–0.002
	12.7–19	4.8–8.1	0.55	0.7	0.001
PCD	Up to 12.7	4.8–8.1	1.625	1.625	0.002–0.0835
	12.7–19	4.8–8.1	1.625	1.625	0.002–0.0035

[a] Unidirectional fibers.
[b] Multidirectional fibers.

the speed of the cutting tool is too high, heat will make the resin sticky and produce a lumpy chip; if the cutting edge is scraping but not cutting the plastic, the chips will be large and flaky. Either type will eventually clog any removal system.

In the past few years, tooling has been developed that has greatly improved drilling operations on PMCs as well as MMCs. Some tool bits are made of particles of WC smaller than 1 mm. Another technique developed incorporates two different disciplines. Ultrasonic-assisted drilling involves the use of a rotary tool on which is superimposed an axial vibratory motion at high frequency. A special adaptor is required to transmit the vibration from a piezoelectric transducer to the tool. Ultrasonic vibration can reduce friction, break chips, and reduce tool wear. It is a particularly useful technique when the matrix or reinforcing fibers are hard, brittle materials. The use of a core drill permits cutting fluids to pass through its center. Ultrasonic machining, though slow, can result in a high finish and accuracy of intricate parts. Hence it is recommended for applications in which intricate shapes of high accuracy and finish are required. This technique has been used to drill and countersink Gr/Ep and B/Gr/Ep holes in prototype aircraft structures such as the B-1 horizontal stabilizer. The drills are water-cooled, and the tools used are all-diamond types—sintered and coated. The machine is versatile in that it can drill, countersink, ream, and counterbore. The ultrasonic drill-countersink is fitted with a sintered-diamond core drill plus an electroplated Ni or diamond sizing band behind the tip to maintain size and concentricity. The countersink surface is also plated so that it can be stripped chemically and replated at low cost to extend tool life. Depending on the application, specially designed, core drill-countersink combination tools ranging in diameter from 4.8 to 12.7 mm have been used.[14] The application of ultrasonic energy to diamond core drills when drilling either B/Gl/Ep or B/Ep hybrids increases drill life. For example, 50 holes were drilled by a portable drill and by a diamond core drill without ultrasonics in a 10.2-mm-thick Gr/B/Ep hybrid material. When ultrasonic energy was applied, a 100% increase occurred (200 holes in a 5.6-mm-thick hybrid).[14–16]

2.3.3.1 Kv/Ep.

Spade drills are best suited for drilling Kv/Ep, leaving very little fuzz and fraying on hole edges. Like carbide drills, these drills have a tendency to burn if they are used too long. Conventional drills can also be used, but firm sacrificial backing must be provided at the exit surface. Twist and flat-ended HSS drills perform quite well on Kv/Ep, especially with a firm backup of the composite to eliminate fuzzing and delamination at the hole exit. A 0.08-mm layer of fiberglass on the top and bottom surfaces of the Kv/Ep composite produces the best holes, leaving clean entrance and exit surfaces. Drill speeds range from 25,000 to 35,000 r/min. A HSS drill is shaped like a twist drill fluted on the end. In a production environment, 45–50 holes have been made before resharpening was required although signs of wear have been found after drilling 5 holes. The best lubricant is water.

2.3.3.2 B/Ep, Gl/Ep, and hybrid drilling.

A titanium diboride coating improves the life of a drill by 877% when used on Gl/Ep.[17] The best way to drill B/Ep and B/Gl/Ep hybrids is with a diamond-impregnated core drill and an ultrasonic machine acting as a drilling-assist tool; however, the cores have a tendency to get stuck in the drills. The best combination is a diamond-impregnated core, a reamer drill, and a countersink tool. Although the cores get stuck, as in the case of ultrasonically powered drills, they can be removed by stopping the drill spindle and leaving the ultrasonic unit on. However, this is a slow process in production applications. The best speed for drilling B/Gl/Ep laminates is 2000–3000 r/min. The best machine spindle feed rates for drilling B/Gl/Ep laminates are 0.14–0.08 mm/s. A 5% commercial surfactant in water is the best coolant when drilling B/Ep and B/Gl/Ep hybrids.[14]

2.3.3.3 Gr/Ep drilling.

WC drills have been successfully employed in drilling Gr/Ep using a backing plate. Various coolants have been tried to prevent drills from breaking, dulling, or burning up. A lubricant called Boelube has worked well, but water is the best. To reduce surface delamination during drilling, a layer of woven glass (0.08 mm) on both sides of the laminate is recommended.

Solid carbide Daggar drills have also proved quite successful in drilling Gr/Ep, producing a clean hole with little or no breakout.[18] Drilling has been accomplished without using backup material. Since the Daggar drill maintains hole tolerance, no reaming is necessary. The overheating problem can still occur in drilling other composite materials such as carbon chopped fibers and glass-filled plastics; using a double-angled drill has worked best.

An approach[19] to tooling for drilling and countersinking Gr/Ep composites has been introduced in the aircraft industry. This advanced tooling system (ATS) has replaced the manual approach using WC Daggar drills where each drill lasted for 40 holes. The ATS consists of the following components.

- Specially shaped PCD drills
- A PCD countersink insert
- A drill countersink tool holder
- A hydraulically operated depth sensor.

PCD was chosen for the cutting tool material in this system because of its outstanding wear resistance. It consists of fine grains of diamond directly bonded to each other and to a WC substrate, and PCD is 100 times more wear-resistant than WC.

A new approach to producing holes when drilling Gr/Ep composites is to add an abrasive slurry. Conventional drill bits alone are blunted very quickly and produce holes with cracked edges when drilling these composites, but when an abrasive slurry is added, clean holes are produced inexpensively and efficiently.[14] The technique was developed for preparing specimens of Gr/Ep for shear testing. A slurry of SiC powder, containing 60% SiC by weight, in water is fed onto the drill. The powder particles become trapped between the cutting edges and the composite and apparently remove the composite material by abrasion. The drill is also dulled, but the researchers who developed the process report that cutting is unaffected.[20] With the slurry, the dull drill cuts as fast as or faster than a sharp one. With the slurry, the holes can be drilled efficiently regardless of the ply orientation.

In the search for an optimal cutting tool material for machining Gr/Ep, PCD inserts have been shown to be more resistant than carbide (C6-grade) inserts.[21,22] In a recent study, Wern, Ramulu, and Colligan[23] drilled Gr/Ep test panels using two different PCD-tipped drills at a constant speed of 4550 r/min, for a feed range of 0.0254–0.254 mm/r in increments of 0.0254 mm/r. Drill A was 12.7 mm in diameter with an axial rake angle of 27°, while drill B had a diameter of 13.9 mm and an axial rake angle of 7°. Apart from the difference in axial rake angle and drill size, the two drills had essentially the same geometry. The surface produced by drill geometry A was superior to that produced by drill geometry B for three reasons. First, when the feed rate is low, the surface produced by drill B was almost four times rougher than that produced by drill A. Drill B also produced a surface with a wider surface damage zone. Last, the probability of the occurrence of deep valleys in surfaces produced by drill B was greater than that in surfaces produced by drill A. Feed rates had a definite effect on the machined surfaces. Between 0.0254 and 0.1778 mm/r, the surface roughness decreased with an increase in feed rate. At feed rates above 0.1778 mm/r, the surface roughness increased with an increased feed.[23]

A study by Malhotra[24] on tool wear thrust and torque using HSS- and WC-coated drills evaluated C/Ep and Gl/Ep materials. He found, in general, that tool wear increased with the number of holes and that flank wear was higher than chisel edge wear. Carbide drills performed much better than HSS drills with both materials. Tool wear, thrust, and torque are much higher in drilling C/Ep as compared to drilling Gl/Ep, and this is due mainly to the higher abrasiveness of C_f compared to Gl_f.

2.3.3.4 Other composite materials.

Hocheng and Puw[25] found that chips from drilling carbon fiber-reinforced ABS indicated that considerable plastic deformation had occurred. Compared to chip formation in thermoset plastics, carbon fiber-reinforced ABS contributes to the improved edge quality in drilling. The edge quality is generally fine except in the case of concentrated heat accumulation at tool lips, which is generated by the high cutting speed and low feed rate. Plastics tend to be extruded out of the edge rather than neatly cut. The average surface roughness along hole walls is commonly below 1 μm for all sets of cutting conditions, and values between 0.3 and 0.6 μm are typical. The HSS drill undergoes only minor tool wear during the tests. Based on these results, they concluded that the carbon fiber-reinforced ABS demonstrated good machinability in drilling.[25–33]

The rules have changed since graphite composites were required to be joined with aluminum, titanium, or both. Taken separately, technologies were available to successfully produce holes in each material, but when these methods were applied to drilling the combinations, the results were generally unsatisfactory.

Air feed peck drilling was considered to overcome the excessive heat generation, degradation of the graphite composite, short cutter life, and frequent cutter breakage encountered with other drilling options. Air feed peck drilling can best be described as a reciprocating action of the drill spindle where the cutter generates small chips, retracts to allow these chips to be removed from the hole, and repeats this action for the time required to complete the hole. Consequently there is no flute clogging, no scrubbing action and heat generation is well within acceptable limits. The benefits of air feed peck drilling (Figure 2.2a) over conventional power feed equipment (Figure 2.2b) in drilling dissimilar combinations are as follows.

- Elimination of flute clogging
- Reduced heat generation

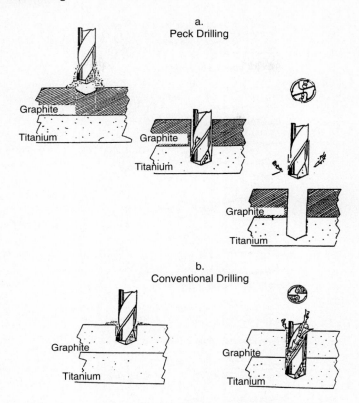

Figure 2.2   (a) Air feed peck drilling.[34] (b) Conventional drilling.[34]

- Increased cutter life
- One-shot capability
- Consistent hole diameter and finish
- Utilization of inexpensive cutters
- Reduced capital expenditure.

The implementation of air feed peck drilling has made the drilling of dissimilar material combinations trouble-free.[34]

Miller[35] reported on a Gr/Ep drilling program that Lockheed Missiles and Space Company initiated to study drill bit geometries and materials, speeds and feeds, and equipment, especially portable air motors, that was anticipated to be used in the *Trident II* (D5) missile body assembly.

Figure 2.3 illustrates 9 drill bit types that were evaluated out of the 17 configurations tested. Results showed that based on the production of more than 5000 holes in Gr/Ep and aluminum with NC drilling and more than 6000 holes in Gr/Ep with a self-feed air motor, the following solid carbide drill geometries produced the best overall results in drilling Gr/Ep alone or in combination with aluminum.

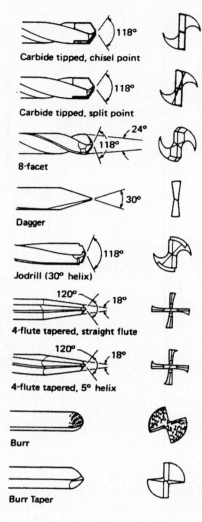

Carbide tipped, chisel point

Carbide tipped, split point

8-facet

Dagger

Jodrill (30° helix)

4-flute tapered, straight flute

4-flute tapered, 5° helix

Burr

Burr Taper

Figure 2.3   Drill bit types used in air motor wear tests.[35]

Drill 1   Eight-facet split point
Drill 2   Jodrill with 30° helix

With these two drill geometries, quality holes were produced in the range 4000– 20,000 r/min.

In summary, a very safe projected drill life of 200–300 no. 11 holes in Gr/Ep up to 6.35 mm thick can be expected when drilling at cutting speeds from 200 to 1000 ft²/min with a solid carbide drill of the recommended geometry.

Development work on the B-2 bomber was reported[36] to show that a computer-controlled drill would reduce hole-cutting time by about two-thirds and avoid the cost of reworking or scrapping damaged composite and metal panels.

An adaptive control drilling system was designed to sense the hardness of different materials and automatically adjust the cutting speed accordingly. By altering drill advancement (or penetration) rates, the time required to cut a single hole in a three-material sandwich or stack-up of graphite composite, titanium, and aluminum on the B-2 was reduced from about 5 min to 1 min 20 s.

Estimates showed that decreasing drilling time by an average of 3 min for a million holes would save 50,000 h, providing a significant cost reduction over current methods.

The system utilized an industry-standard pneumatically driven positive feed drill with new sensors and a microprocessor. An IBM PC/AT personal computer and software tailored to exchange data with the portable adaptive drill completed the system.

The desktop computer was used to load control parameters such as torque limits, feed rates, longitudinal pressure (known as *thrust*), material characteristics, and cutter configurations into the drill's microprocessor or on-board process controller.

Throughout the drilling operation, torque, thrust, spindle speed, and position data are stored in on-board memory and compared to the original programmed parameters for that particular task. This prevents chip overload or a dull cutter from causing hole distortion or other damage to the part being drilled.

Although the adaptive drill costs about twice as much as a conventional power feed or positive feed drill, its speed and consistent high-quality hole production has demonstrated that the tool quickly pays for itself.

Colligan[37] described an advanced wing development that required fastener holes drilled simultaneously through both carbon and titanium. With conventional manufacturing technology, these two materials require different tool geometries, different tool materials, and different drilling speeds for optimum results.

The improved wing called for Gr/Ep skins to be bolted directly to a substructure formed from a combination of titanium and Gr/Ep "I-shaped" ribs and spars. To ensure that the bolt holes in the skin aligned properly with the corresponding holes in the substructure, it was necessary to assemble the wing with temporary fasteners, drill through both the Gr/Ep and the titanium in a single setup, and install the bolts through the holes.

However, the optimum drilling process for Gr/Ep conflicts with that for titanium. Gr/Ep requires a PCD bit drilling at high speed. Titanium, on the other hand, requires a carbide-tipped bit drilling at low speed, and the method originally devised to drill through the Gr/Ep skins and into the titanium spars and ribs was based on a three-step machining procedure.

Because this method is time-consuming and expensive, engineers came up with a new drill bit and a procedure consisting of a single-pass method for drilling all 8000 holes in the wing.

Mehta and Soni[38] investigated the optimum drilling conditions and hole quality in PMR-15 PI/Gr composite laminates as applied to fasteners. Their study covered tool geometry in relation to hole quality attributes. Several solid carbide drill geometries (commonly used in aerospace manufacturing) were evaluated for acceptable hole quality, including the Daggar drill, an eight-facet split point drill, and a four-facet standard master drill. Their performances were compared with that of the NAS 907 HSS twist drill currently used by the U.S. government in composite repair operations.

Researchers, machinists, and tool and manufacturing engineers have studied and attempted to understand the properties of various materials in order to select the best one for a tool for drilling composites. The majority have found that PCD, coated diamond, and solid WC are good materials for round tools used in machining composites. The higher the resin content, the better diamond tooling works. The higher the fiber content, the more success has been found with WC tooling.

The price of WC and coated diamond tooling is about the same. PCD tools cost more, but they last longer and can be resharpened many times with diamond wheels, and so the cost per cut is comparable to that of other tooling materials. WC can be resharpened, and diamond-coated tools can be recoated.

Coated diamond tooling and WC tooling can be run on most commercially available equipment, and users should not assume that they need special equipment to machine composites with these materials. PCD requires a rigid spindle, making it unsuitable for use in hand-held tools, but it can also be run on commonly available equipment, although it performs best at higher revolution rates.

Fabricators can avoid much damage to composites by keeping tools sharp, which minimizes heat damage and delamination. Delamination during drilling can be avoided by using fluted drills rather than abrasive drills. The lip of the tool shears the fiber, producing a clean cut. Experimentation can yield the best drill design for a particular material. Brad point or shear bore WC drills have worked well, and PCD-tipped drills have also been useful. In some cases, a diamond core drill works well on thinner composites, and on heavier materials that have a lot of fiber, WC drills work very well. For larger holes, diamond core drills are recommended for thinner materials, and diamond hole saws for thicker materials.

The most preferred speed for drilling is 5000 r/min according to K. Smith of Greenleaf Corporation.[39] Smith also believes that drill geometry is very important. He claims that an eight-facet point is usually best if entry-side burrs are the major consideration or if the material being drilled has a backing. When exit-side burrs are the major problem, a four-facet split point is best for overall performance. Regardless of the choice of a four- or an eight-facet point, a split point—or web—is usually recommended to reduce the pressure at the drill point.[39,40]

In drilling, tap water or other coolants can be used, but oils or synthetic oils should not be employed when machining C/Ep material because the dust produces a heavy sludge that can damage the coolant system and must be cleaned off the machined parts. Coolants should be applied in a low-pressure stream instead of a spray mist, which creates a sludge that adheres to the part.

It is recommended that brazed diamond or PCD tools be used for C/Ep. After each use, they should be cleaned with a soft wire wheel on a bench grinder. Carbide tools can be used if the turnaround time for a custom diamond tool is too long, but it is recommended that carbide cutters be replaced after each use, and carbide drills after 30 holes. Worn carbide tools cause fraying and delamination. HSS drills should be replaced after each hole.

The surface of the part should be electronically probed before machining to ascertain that sufficient stock exists for cleanup and to ensure that the cutter will not encounter unexpected material thickness that exceeds the axial cutting edge length of the cutter.

During drilling, delamination can result if the drill surges through the workpiece after the point exits the material, and devices that control surging are available. It is always advisable to back up the part with sacrificial material to avoid splintering.

Drilling composites with an aluminum or titanium substructure presents a problem because metal chips trapped in a hole counterbore the back side of the hole. A drill with a four-facet point makes smaller chips that can travel freely up the flutes and out of the hole.

Diamond-impregnated hole saws can be used to make larger holes that are accurate to within 0.76 mm of the saw diameter. The machinist should drill a pilot hole and saw halfway through the material. Then, using the pilot hole, the machinist should complete the hole from the opposite surface. If both sides of the part cannot be accessed, the exit surface should be backed.[39]

In the drilling of reinforced plastics the quality of the surfaces is strongly dependent on the appropriate choice of drilling parameters. Caprino and Tagliaferri[41] carried out drilling tests on Gl-polyester composites using standard HSS tools; drilling was interrupted at preset depths to study damage development during drilling. Their main conclusions were as follows.

The type of damage induced in a composite material during drilling is strongly dependent on the feed rate f. When f is high, the failure modes show features typical of impact damage, with steplike delaminations, intralaminar cracks, and high-density microfailure zones. If sufficiently low f values are adopted, the failures consist essentially of delaminations originating mainly near the intersection between the conical surface generated by the main cutting edges and the cylindrical surface of the hole.

At low feed rates most of the delaminations are induced near the tool exit edge and are generated when the chisel edge and the inner portion of the lips have already left the work material. The delamination located near the back face of the sample are characterized by a large average length, showing a higher tendency to grow.

Finally, Hocheng, and Hsu[42] reported on the use of ultrasonic drilling (USD) of C/Ep and C/PEEK composite materials. In their study they found that the major cutting mechanism of USD was abrasive particles hammering or impacting on the workpiece to remove material in microcraters. The obtained surface roughness increased with grain size, energy of ultrasonic oscillations, and concentration of abrasives and was independent of feed rate of the tool, fiber direction, or matrix plasticity. The experimental results showed that no delamination occurred at the hole edge. The C/PEEK composite showed minor fuzz at the exit edge because of its thermoviscosity. With the aid of liquid nitrogen, the exit edge produced was neat. Grain size and energy of ultrasonic oscillations significantly affected the hole clearance.

They compared three conventional machining processes, namely, abrasive water jet, laser, and ultrasonics. The main cost factors were the investment, economic life of the equipment, salvage value, annual revenue, and expenditures. For the test sample that was selected, AWJ or a laser cut on an average of 15 times faster than ultrasonic machining and for single-hole production was much faster than USM. However, the USM process can readily perform multihole production more economically than the other two processes.

2.3.3.5 Laser beam machining.
LBM is based on interaction of the work material with an intense, highly directional, coherent monochromatic beam of light. Material is removed predominantly by melting and/or vaporization. In the case of resin matrix material, it is also removed by chemical degradation.

The physical processes involved in LBM are basically thermal in origin. When a laser beam impinges on a work material, several effects occur including reflection, absorption, and conduction of the laser beam.

Most metals absorb more readily at shorter wavelengths, hence less power is required to machine these materials at these wavelengths. Therefore, Nd:YAG, with a wavelength of 1.06 μm, is more suitable for machining MMCs than CO_2. In contrast, some organic resins and other compounds have a higher percentage of absorption at higher wavelengths close to that of a CO_2 laser (10.6 μm), and so CO_2 is more appropriate for machining such materials (e.g., aramid-resin composites). As melting begins or the material begins to interact with its atmosphere, the percentage of absorption may change as the process continues. For example, in drilling, the percentage of absorption in part of the drilled hole may be different from its initial value at the surface.[43,44]

2.3.4 Countersinking PMCs

Conventional countersinking tools, from 1750 to 6000 r/min have been used successfully on Gr/Ep, Gl/Ep, Kv/Ep, and Gr/Gl/Ep hybrids, and butterfly countersinking tools modified with serrations have also been used.[45] Carbide countersinks have shown good results with Gr/Ep and Gl/Ep with optimum combinations of speed and relief angles.

In countersinking B/Ep and Gr/Ep hybrids, the feed rate has a pronounced effect on diamond tool life. Low feed rates are recommended. Although the amount of wear on the countersink itself is slight, the angular change approaches the maximum allowable countersink angle tolerance. The pilot on both types of countersinks wears rapidly when B_f are cut. The pilot in the plated tool is replaceable, and the pilot in the sintered tool can be refurbished by Ni plating. Although plated tools wear more rapidly than sintered tools, their lower cost makes plated countersinks more cost-effective.

2.3.5 Grinding PMCs.
Kv/Ep composites can be ground or chamfered by strict adherence to specific conditions. A glazed work surface with a minimum of uncut fibers is obtained. A coolant is necessary to prevent wheel loading, and Al_2O_3 or SiC grinding wheels are satisfactory.

Hand-held grinding tools can be used with diamond-coated wheels in finishing cut edges. The use of diamond-coated wheels in finishing allows the epoxy to dissipate so that it doesn't load the tool. Bonded tools have a tendency to load, but diamond-coated wheels have a lot of space between the particles, which helps prevent loading.[46–49]

2.3.6 Routing, Trimming, and Beveling PMCs

These operations are essentially equivalent, involving the use of hand routers, mechanical Marwin machine routers, and Roto-Recipro machines. Diamond-cut carbide and four-fluted milling cutters have been used to machine Kv/Ep and Kv/Gr/Ep hybrids. Diamond-coated router bits have been used with Roto-Recipro machines to rout and trim B/Ep and B/Gr/Ep hybrids. Speeds for these operations range from 3600 to 45,000 r/mm.[50]

2.3.6.1 Routing.
In manual routing, a higher-torque but lower-speed router (Buckeye) works best. Diamond-cut carbide bits attain higher feed rates with less operator effort than a six-flute configuration. Using a coolant tends to extend tool life and in-

crease cutting force (probably due to sludge formation) with no effect on cut edge quality. Based on operator effort, maximum diametral tool wear would be about 0.04 mm before a tool change would be required. An opposed-helix carbide cutter has successfully routed Kv/Ep cured laminates. The opposed-helix cutter is especially useful in severing laminates into sections, cutting slots and notches, and trimming honeycomb-sandwich panels. The portable version of the cutter has performed better than diamond-cut carbide cutters with Kv/Ep and Kv/Gr/Ep hybrid laminates. Although cut quality is equivalent for both types of cutters, only the opposed-helix carbide cutter can trim 6.4-mm material. For routing and trimming B/Ep and B/Ep and Gr/Ep hybrids, diamond-plated router bits are used. Spray mist or flood cooling is required for all power feed routing of B/Ep; spray mist or air-blast cooling is recommended for hand routing. Power feed is recommended for routing B/Ep thicker than 1.3 mm.

2.3.6.2 Trimming and beveling.   Manual trimming and beveling with a Buckeye router against a guide have been successfully used on Gr/Ep, Gl/Ep, Kv/Ep, and Gr/Gl/Ep hybrids. The router bits were diamond-cut carbide types, and depths of cuts were 1.5 or 3.3 mm and 45° by 6.4 mm. Good cuts have been made on thicknesses up to 6.9 mm.

2.3.7 Milling

To produce a good surface finish and good, sharp corners without breaking the corners off, the edges of the composite block should be machined, leaving a step on the edges to reduce pressure on the laminate during milling. In milling from one side of the composite block to the other, it is always recommended that the block be precut. Doing this prevents potential breakage problems in the lamination.

In milling composite materials such as unidirectional and cross-directional carbon fibers, it is always a good practice to make sure of the direction of the layers. According to Murray,[51] in milling composite materials, good results can be achieved by either plunging into or machining across the direction of the fibers, as well as by machining parallel to the fibers.

End mills of WC with at least four flutes should have either a 45° angle or a generous radius. A four-fluted end mill reduces cutting pressure on the laminate and keeps it cooler. A sharp-tipped end mill dulls very quickly while machining composite materials. Tools should always be in good condition. A dull tool causes overheating of the epoxy-resin, resulting in delamination not only on the surface but also within the layers, which cannot be seen. A dull tool requires more pressure against the composite material while machining, thus causing overheating. In milling operations, using a climbing cut works exceptionally well because it keeps the fibers from separating. This method creates a chopping cut, whereas machining in the opposite direction creates a pulling cut.

The use of a fluorocarbon coolant is recommended because of its cooling efficiency. The coolant is applied as a spray mist during machining; the distance between the spray applicator and the cutter is adjusted so that frost forms on the cutter.

In milling Kv/Ep, conventional fluted cutters are satisfactorily operated at speeds of 0.4 m/s and feed rates of 5 mm/s. In milling Gr/Ep, HSS end mills or carbide cutters can be used provided they are multifluted. Four-fluted end mills are recommended for efficiency and to reduce cutting forces to a point where there is less chance of delamination. All cutters should be sharp; dull ones cause delamination. HSS cutters should always be of the four-fluted positive-rake type; carbide milling cutters should also be of a positive-rake type with chip loads of 0.10–0.15 mm per tooth. Radiused end mills last longer than square-cornered ones. Plunge-cut milling is not recommended unless there is sufficient backup support to prevent delamination.

Milling of a complex shaped hole in graphite composites, AS4/3501-6 tape 0.152 mm thick, laid into $456 \times 610 \times 6.4$-mm thick sheets with 12.7-mm, two-flute, PCD end mills using straight and 30° helical end mills yielded some guidelines on machining parameters for this type of work.[52] For all tools tested, cutting parallel to the fibers produced greater wear than cutting perpendicular. Lower feed rates meant more cutter revolutions to remove the same amount of material, and so the amount of wear increased, indeed doubled in some cases. But lower feed rates produced a better surface finish. The effect of spindle speed on wear was much less significant than that of feed rate, but in the same direction. There was some decrease in wear at high r/min; 11,460 r/min was the maximum speed tested. Cutting speed also showed little effect on the surface finish, which researchers Kohkonen, Anderson, and Strong[52] found surprising. The tools with the least wear were the 30° helical PCD end mills because both types of end mills became dull much more slowly than the carbide end mills. The PCD end mills produced a superior surface finish over longer lengths of cut. Shallower depths of cut (1.27 mm), as expected, produced better surface finishes than greater depths of cut (2.54 mm). Interestingly, the same amount of flank wear (0.127 mm) produced more edge delamination with the carbide end mills than with the PCD end mills. New ceramic matrices whisker-reinforced with SiC single crystals are being applied to materials as cutters in end milling, routing, and drilling tools.

Reporting their findings on PCD tools in machining Gr/Ep composites was the team of Ramulu, Faridnia, Garbini, et al.[53] They found that flank wear was the dominant wear mode and that the coarse grade of a DeBeers Syndite PCD tool was more resistant to flank wear than the finer grades. While the hardness of the three grades of PCD (2, 10, and 25 μm) was the same, the modulus of elasticity of the PCD increased as the diameter of the grain increased. The coarse grade had a grain size averaging 25 μm, the medium was 10 μm, and the fine grade was 2 μm. All the tools showed abrasive tracks or grooves after a short machining time, and all grades also experienced a small amount of edge cracking (cracks from 5 to 40 μm). However, for a given amount of cutting time the abrasion tracks or grooves formed on the flank face of the tools were shortest in the coarse grade. The differentiation in flank wear increased with greater machining times. At 23 min of cutting time the coarse grade of PCD showed 55 μm flank wear, whereas the fine grade reached up to 75 μm flank wear, Figure 2.4.

The texture of the machined Gr/Ep surfaces exhibited many small holes due to fiber pullout, the presence of cracks, and smearing of the matrix material. The surface quality tended to remain the same or even improve as cutting times were increased and was independent of the PCD grade. The R_y value (maximum peak-to-valley roughness height) varied a minimum for the coarse grade of PCD when compared to that of the other grades.

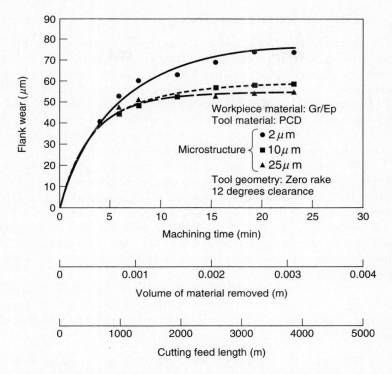

Figure 2.4   Growth of flank wear with increased machining time.

The rough grade, with its lower flank wear, also produced better R_a (surface roughness) values than the other grades.

2.3.8 Cutting PMCs

Cutting is necessary at one or more points in the manufacture of most structural fiber composite parts. Cutting operations can be divided into those that occur before and those that occur after cure.

Materials cut before cure include various mats, fabrics, and other dry reinforcements, as well as prepreg layers. Cutting tools range from manual scissors to automated laser or water jet devices. Between these extremes are power shears, rotary power cutters, reciprocating knives, and steel rule dies for blanking. Which tool is used depends on the material to be cut and the speed and accuracy desired.

As for postcure cutting, traditional tools include various band saws, hand routers, shears, and sanders. These tools are highly labor-intensive and noisy and create serious dust problems. While they are still very widely used, they are steadily being replaced by more advanced cutters. The most important of these are the high-pressure water jet, the abrasive water jet, and the laser.

2.3.8.1 Uncured composites

Reciprocating Knife.   There are two reciprocating-knife cutting machines, both of which incorporate high-speed reciprocating knives that are driven through the material to be cut by a minicomputer- controlled *xyzc* positioning system. In one system, the cutting knife penetrates through the material into closely packed plastic bristles that constitute the surface of the cutting table. The surface is nondegradable and does not require periodic removal.[54] The cutting knife ranges in width from 1.3 mm for the diamond cutter up to 4.4 mm for the carbide cutter. The system cuts in a chopping mode; that is, the knife rises above and plunges through the material onto the table.

The second system[45] can cut desired patterns in a continuous line at high speed. Curves, sharp corners, and notches can also be cut without lifting the knife from the material. The knife can be lifted, as required, to start new cutting lines, to pass over sections without cutting, or to cut holes of any diameter. The system uses a blade 6.4 mm wide and cuts by either chopping or slicing. In the slicing mode, it remains buried in the material after the first stroke (each stroke is 19 mm) and is always at least 3.2 mm below the material being cut. Computer-controlled rotation of the knife about the *c* axis keeps the blade properly positioned at all times.

Cutting test results for both systems indicate that the slicing-and-chopping system can cut a greater number of Gl/Ep and Gr/Ep plies at twice the feed rate of the chopping system. Visually the quality of the edges of laminates cut by both systems is about equivalent.[45,54]

Steel Rule Die Blanking.   Steel rule blanking dies are an economical method for cutting relatively small quantities of composite prepreg materials. If a die requires reconditioning, only a minor expenditure is involved. Normally, steel rule dies are positioned above a flat mild-steel plate to permit blanking on the downstroke. Each die consists of a 3-mm, one-side-beveled, hardened steel strap embedded in a wooden base with a cork stripper plate. This cutting edge configuration gives a higher quality than other standard configurations. Single-ply laminates of all composite materials except Kevlar cut cleanly and easily, requiring only minor die position and pressure changes. Kv/Ep requires more buildup with paper and metal for additional pressure and a better impression than the other materials to achieve a clean separation of the blanked configuration. Multiple cutting is clean and easily performed for all materials except Kevlar prepregs. The criterion used in selecting the maximum number of plies is squareness of the cut. As the number of plies increases, edge squareness decreases, Kv/Ep requires significantly higher pressures since the fibers are difficult to sever, limiting the maximum number of plies cut.

Boszor and Ecklund[55] evaluated the automation of the cutting and stacking of 29 pieces of Gr/PI composite prepreg used for the outer shell of an engine inlet guide vane. The program examined various methods of cutting prepregs that need an automated system. A steel rule die with elastic foam ejector pads that cuts 20 plies of prepreg simultaneously was selected as the most appropriate, flexible, automated system for cutting these unique composite airfoil components.

Laser Beam Cutting.   Continuous-wave 250-W CO_2 lasers have produced cutting speeds up to 169 mm/s in single-ply Kevlar prepregs and up to 127 mm/s in single-ply B/Ep. Although B/Ep laminates, up to four plies thick were also laser-cut at a feed rate of

25.4 mm/s, the slow feed rate resulted in a marginal cut with a significant amount of resin retreat. The best cuts have been made in two-ply B/Ep laminates cut at a feed rate of 51 mm/s. A feed rate of 127 mm/s on single-ply Gr/Ep tape is the most effective cutting rate; for two-ply laminates, 63.5 mm/s is recommended. Although Gr/Ep tape laminates up to three plies thick have been successfully laser-cut, an excessive edge bead of resin develops at 25.4 mm/s. In uncured Gl/Ep laminates, laser cutting has been accomplished at 127 mm/s for single-ply thicknesses and at 38 mm/s for thicknesses up to three plies. Higher feed rates reduced cut quality, apparently because the resin could not vaporize quickly enough.

Woven graphite broad goods were readily cut at 127 mm/s with nitrogen assist gas pressure, and no evidence of heat damage to the fibers was visible. A Gr/polysulfone with its thermoplastic matrix exhibited the greatest matrix damage. In tests, as much as 0.6–0.9 mm of the matrix material has been removed. Cutting rates up to 114.3 mm/s are easily obtained with the laser.[50]

Ultrasonic Cutting.   Ultrasonic cutters are being used on most natural and synthetic fiber materials. With nonthermoplastic materials such as Gl, aramid, and C fibers, ultrasonic machines make clean cuts without displacing fabric or warp ends. On thermoplastic materials, both separation and selvedge sealing are done to prevent the edges from unraveling. And unlike the situation in hot cutting, the selvedged edges are free of welding beads. The ultrasonic process also prevents yellowing because heat, generated by mechanical vibrations between a sonotrode and countertool, occurs on the inside of the material rather than the outside.

Ultrasonic cutters, vibrating at 20,000 strokes/s, use TiC blades at the end of a "horn" to reduce frayed edges in composite materials. These machines can cut PI-based resins, epoxies, and adhesive films.

Dry fabrics, sheet prepregs, and molding compounds continue to offer a cost effective solution for many medium-batch production requirements. It is therefore important that cost-effective methods like the ultrasonic cutting process, which has the unique benefit of cutting through the prepreg and leaving the protective backing layer intact, be adopted. Additionally, a family of needle-based grippers have been developed that are able to pick up cut prepregs off the uncut backing layer and automatically transfer them to a tool at the side of the machine.[56,57]

The above developments have been extended to provide prepreg cutting systems for use in industry, which requires relatively long machines with good access from either side to facilitate the removal of components automatically. The gantry of the machine, which has to carry the ultrasonic head, has been designed such that its bearings are below the surface of the table and yet not too far below the center of mass of the moving gantry to produce bending during the acceleration and deceleration of the machine. One machine is 1280 cm long and 152 cm wide and includes on the gantry a gripper device for reeling out the preprogrammed length of material.

One of the advantages of the ultrasonic process is automatic handling of prepreg parts, where the prepreg material can be cut without significantly cutting into the backing layer. It is this vital principle that enables the gripper devices to peel components off the backing layer while the vacuum system prevents the backing layer from being lifted up from the table.

Kim and Lee[58] conducted a series of experiments using C/Ep composites and ultrasonic vibration cutting equipment and found the following:

1. Below the critical cutting speed (19–165 m/min), ultrasonic vibration cutting results in a better surface than conventional cutting does in the machining of C/Ep.
2. The diamond tool is effective in the machining of C/Ep, both in conventional cutting as well as in ultrasonic vibration cutting.
3. In ultrasonic vibration cutting, the surface quality of the C/Ep is not related to the depth of the cut but is related strongly to the cutting speed and the feed rate.

2.3.8.2 Cured composites.

Sawing.   *Flash,* excess material that must be removed in order to bring the composite part to its finished size, must be removed. Where the flash is more than 1 mm thick or contains a high proportion of reinforcement, a robust cutting tool is required. The most suitable hand tool is a standard hacksaw with a metal cutting blade, and this saw can also be used for general cutting operations on most composites.

To speed up the rate of cutting, some form of power tool is necessary. The jigsaw is a versatile and effective power tool but relatively slow. To achieve a high rate of cut on thermoset composites, a high-speed abrasive disk cutter should be used. A diamond-impregnated cutting wheel is essential to achieve reasonable wheel life. Circular saws can be used to cut composites, but both saw and workpiece must be firmly held to avoid binding. This means that the saw needs to be bench-mounted and the workpiece held in a traversing table, or vice versa, and a diamond-impregnated saw blade is necessary.

Water Jet (WJ) Cutting.   In the past, cured composite parts were trimmed by band sawing, hand routing, shearing, nibbling, and sanding. These processes were labor-intensive and noisy and produced critical dust problems. WJ cutting has reduced the dust and noise problem and produced clean, trimmed edges, eliminating hand sanding. Cured materials follow a rule of thumb: the cut quality improves with increasing nozzle pressure, increasing nozzle orifice diameter, decreasing traverse speed, and decreasing material thickness and hardness. Softer materials are cut with a better edge quality than harder ones. The order of decreasing hardness is B/Ep, Gr/Ep, Gl/Ep, and Kv/Ep, with hybrid materials occupying positions midway between those of the parent materials.[14–16,39,59–66]

Most WJs are based on the concept of a router walking around a part, pressing a template against it and trimming. Some systems feature WJ cutting heads suspended in the air by an articulated counterweighted arm. The user pushes against the template and walks around the part, trimming it as he goes. These systems cut quickly and leave an excellent edge. They do not produce the tearing, fraying, or delamination of some router applications, and they are cleaner. Unfortunately, they are effective only on material thicknesses of up to 2.54 mm.

A basic WJ cutting system consists of a water pressure booster and filtration system, hydraulic pump, nozzle, and catcher. The booster increases incoming water pressure to the level required by the intensifier; filters remove particles that would damage pump seals, check valves, and orifices. Usually, the unit runs at 345 MPa in conventional operation; the water is 12% compressible, and for the first 12% of the stroke of the plunger, nothing comes out. To keep the jet running, compressed water is stored in the accumulator while the pump is making its initial compression (like a storage tank on an air compressor). Traveling at speeds up to 900 m/s, the water moves to the nozzle where it passes through an orifice and emerges as a coherent cutting stream. At this point, the jet is a tight core surrounded by a shroud of light mist.

The WJ cuts when its force exceeds the compressive strength of the material. Depending on the properties of the target material, the actual cutting is a result of erosion, shearing, or failure under rapidly changing localized stress fields. The process produces no thermal or mechanical distortion, but there is a negligible to slight work hardening of metals at the cut surface. After the material is cut, the water or water-abrasive stream is collected in a tank or compact traveling catcher unit.[60]

There are many advantages to cutting with WJs. Listed here are some of the more significant ones.

- Little or no dust or fumes generated in cutting
- No heat-affected zone or thermal deformation
- Little or no exit edge burr or dross to be removed
- A small kerf, meaning better material utilization
- No predrilling for cut starts
- Omnidirectional planar cutting without articulation
- High cutting speeds in many materials
- Minimum delamination of most composites
- High throughput with less material handling
- Elimination of most secondary finishing processes
- Easily adapted to automated contouring systems
- Part programming can be downloaded from Computer-Aided Design/Computer Aided-Manufacturing (CAD/CAM)
- Part cut path easily reprogrammed
- Six-axis cutting capability
- Simple fixturing because of minimum surface pressure
- No tool setup or changeout required
- Ease of integration into FMS
- Just-in-time (JIT) manufacturing ability
- Elimination of costly dies
- Elimination of pattern storage
- Versatility in cutting of different materials.

Studies by Bartlett[64] compared conventional machining equipment (a three-axis programmed mill, a wire electric discharge machine, and two current technology WJ machines. He found that the WJ machines were not as accurate as the conventional equipment. The resolution of the WJ equipment was only +0.13 mm, as compared to +0.005

mm for the conventional equipment, however, the WJ equipment was cost-effective and useful for a variety of composite material applications.

Reference 66 describes a five-axis WJ cutter-router combination that permits programming of all five axes of movement, as well as water pressure and feed rate, with a full-function hand-held programmer.

Finally, Parsons Precision Products Company discussed its use of WJ equipment operating 24 h/day. The CNC WJ unit makes all functions repeatable and removes the human error factor. The computer provides contouring accuracy of + 0.33 mm in each axis over the total travel of the machine. Full control of the WJ function is maintained through the standard RS-274D programming language.

Abrasive Water Jet Cutting or Hydroabrasive (HAJM) Cutting. The process, also known as *hydroabrasive machining,* is identical to WJ cutting except for the injection of an abrasive into the cutting stream.[67,68] To the basic ultrahigh-pressure water-generating system is added recent developments that revolve around entraining abrasives (typically 80- to 100-mesh garnet) into the water stream. Part of the water's momentum is transferred to the abrasive particles which do most of the cutting. This cutting system adds an abrasive hopper, an abrasive metering valve, a specially designed mixing chamber, and a focusing nozzle.[15,16,69–72] Several attributes make AWJs superior to more conventional cutting methods in a variety of applications:

- A low operating temperature avoids the heat-affected zones and distortion created by laser and plasma arc systems. Thus there is no need for expensive grinding, annealing, or secondary machining of heat-affected surfaces to obtain final dimensions.

- The jet cuts with little lateral or vertical force and is virtually vibration-free. This minimizes mechanical stress at the material edge and eliminates the need for costly part support fixtures.

- Abrasive water jets cut in any direction, and the small, lightweight cutting head allows them to take full advantage of robotic manipulation which is often limited by the weight and geometry of traditional tools. Integrating AWJs with computer numerical control (CNC) manipulators combines omnidirectional cutting capability with the flexibility and repeatability of computerized cutting programs.

- Because the cutting head does not contact the workpiece, AWJs are highly maneuverable and excellent for cutting precise right angles and special shapes. In more conventional methods, the cutting path is restricted by blade or tool geometry.

- The 0.76- 2.5-mm-thick water stream minimizes width of cut, or kerf, which maximizes material use. AWJs can cut small holes, narrow slots, and closely spaced patterns, and small parts can be nested within larger ones. Unlike a saw, which produces a rough cut and leaves material that must subsequently be machined away, an AWJ often requires only a single pass to produce the final dimensions and finish.

Hand-held AWJ cutting is too dangerous, however. Entraining the garnet abrasive into the stream allows the cutting of much thicker materials than lasers can handle. AWJ cutting employs erosive action rather than friction and shearing to produce a finished edge. Low operating temperatures do not affect material being cut at speeds of 381–762 mm/min and thicknesses of 0.5–25 mm, in most cases, and some difficult-to-cut materials up to 100 mm in Gr/Ep or MgB_4C. Moreover, there is less dust, especially in cutting Gr/Ep. While AWJ is not completely clean, it leaves no more than a small puddle on top of the part, whereas routers spray particles out at high velocity and even the most elaborate vacuum systems have trouble keeping the dust down. With the abrasive cutter, a small point catcher follows the head, collecting residue and dissipating kinetic energy. In addition, abrasive jets operate with low cutting forces. Typically, a workpiece has about 4.4 N of force on it. This can be increased by cutting faster, but not above about 35.6 N on composites. At that level, the tooling cannot handle any real weight. For example, on its 777 airplane, Boeing uses universal tooling for the composite tail sections it cuts with abrasive jets; with a router, it needs different tooling for every part (Tables 2.3 and 2.4).

Another advantage of water jets is that they can produce inside corners with 0.76-mm diameters. These tight corners still have some rounding to eliminate stress risers. Routers usually need tool bit diameters of 6.3 mm, and this sacrifices speed. It is suggested that the largest cutting head possible be used. The bigger the cutting head, the larger the stream, but one can cut up to 50% faster with a large head.[73,74] To reduce tooling and labor costs associated with net trimming of composite components, systems have been designed and implemented that combine the flexibility and accuracy of robotics with the productivity of AWJ cutting. The system is comprised of a large, six-axis gantry robot that uses specially developed AWJ end effectors to trim the edge-of-panel (EOP) and integral stiffener blades. These end effectors employ compact catchers to contain the spent system and thereby eliminate the need for large catcher tanks commonly used in cutting. The robot is programmed off-line to perform trimming on large, complex contoured panels while a vision system monitors kerf width.[75]

AWJ technology has become a much more reliable and proven technology over the past 10 years. On the cutting equipment side, the work life of AWJ cutters has been extended and the wear-out rate been made more predictable.[76] Sapphire orifices now last 80–100 h, and their wearout rate can now be predicted, making it possible to anticipate and compensate for changes in the size of the water-abrasive stream through appropriate computer programming. Nozzle life, too, has been increased to a maximum of 100 h or more. On the robotics (or motion equipment) side, improvements in programming have made the equipment far more accessible and user-friendly. The technology has gone from a very hands-on, operator-dependent technology to a hands-off, accepted manufacturing process. This has been demonstrated on the B-1B production line with a sytem that can locate and cut precisely any of 100 aerospace components regardless of a given part's orientation or placement within the work envelope, thus eliminating the need for special material-loading and -handling fixtures. Unlike traditional blades or saws, the PACER abrasive jet[67] can cut omnidirectionally and, in addition, the system's light weight makes it easily adaptable to robotic manipulation.

Groppetti and Cattaneo[77] demonstrated the high effectiveness and potential for multilayer C/Ep composites in their modeling of the mechanism of delamination occurring during multilayer C/Ep composite machining by means of HJM and HAJM. They showed that a predictive model of edge damage caused by WJs and AWJs can be useful in evaluating machining conditions to predict the machining results in terms of kerfing conditions without time-consuming and expensive experiments.

TABLE 2.3   Cutting Data[a]—AWJ[60]

Material	Thickness (in.)[b]	Nozzle Speed (in./min)[c]	Edge Quality
Metals			
Aluminum	0.130	20–40	Good
Aluminum tube	0.220	50	Burred
Aluminum casting	0.400	15	—
Aluminum	0.500	6–10	—
Aluminum	3.0	0.5–5	—
Aluminum	4.0	0.2–2	—
Brass	0.125	18–20	Good or small burr
Brass	0.500	4–5	—
Brass	0.75	0.75–3	Striations at 1 +
Bronze	1.100	1.0	Good
Copper	0.125	22	Good
Copper-nickel	0.125	12–14	Fair edge
Copper-nickel	2.0	1.5–4.0	Fair edge
Lead	0.25	10–50	Good to striated
Lead	2.0	3–8	Slower = better
Magnesium	0.375	5–15	Good
Armor plate	0.200	1.5–15	Good
Carbon steel	0.250	10–12	Good
Carbon steel	0.750	4–8	Good to bad edge
Carbon steel	3.0	0.4	Good with small nozzle
4130 carbon steel	0.5	3.0	—
Mild steel	7.5	0.017–0.05	—
High-strength steel	3.0	0.38	—
Cast iron	1.5	1.0	Good edge
Stainless steel	0.1	10–15	Good to striated
Stainless steel	0.25	4–12	Good to striated
Stainless steel	1.0	1.0	65–150 rms
15-5 PH stainless	4.0	0.3	Striated
Inconel 718	1.25	0.5–1.0	Good
Inconel	0.250	8–12	Good to striated
Inconel	2–2.5	0.2	Good to fair
Titanium	0.025–0.050	5–50	Good
Titanium	0.500	1–6	65–150 rms
Titanium	2.0	0.5–1.0	125 rms
Tool steel	0.250	3–15	125 rms
Tool steel	1.0	2–5	—
Nonmetals			
Acrylic	0.375	15–50	Good to fair
C-glass	0.125	100–200	Shape dependent
Carbon-carbon composite	0.125	50–75	Good
Carbon-carbon composite	0.500	10–20	Good
Epoxy-glass composite	0.125	100–250	Good
Fiberglass	0.100	150–300	Good
Fiberglass	0.250	100–150	Good
Glass (plate)	0.063	40–150	Good

TABLE 2.3   Cutting Data[a]—AWJ[60] (continued)

Material	Thickness (in.)[b]	Nozzle Speed (in./min)[c]	Edge Quality
Glass (plate)	0.75	10–20	125 rms
Graphite-epoxy	0.250	15–70	Good to practical
Graphite-epoxy	1.0	3–5	Good
Kevlar (steel-reinforced)	0.125	30–50	Good
Kevlar	0.375/0.580	10–25	Good
Kevlar	1.0	3–5	Good
Lexan	0.5	10	Good
Phenolic	0.25–0.50	10–15	Good
Plexiglas	0.175/0.50	25	
Rubber belting	0.300	200	Good
Ceramic matrix composites			
Toughened zirconia	0.250	1.5	—
SiC fiber in SiC	0.125	1.5	—
Al_2O_3/CoCrAly (60%/40%)	0.125	2	—
SiC/TiB$_2$ (15%)	0.250	0.35	—
Metal matrix composites			
Mg/B$_4$C (15%)	0.125	35	Fair
Al/SiC (15%)	0.500	8–12	Good to fair
Al/Al$_2$O$_3$ (15%)	0.250	15–20	Good to fair

[a] In trying to provide data on waterjet and abrasive waterjet cutting we have collected material from diverse sources. But we must note that most of the data presented is not from uniform tests. Also, note that in many cases data was largely absent on such parameters as pump horsepower, water jet pressure, abrasive particle rate of flow or type or size, and stand-off distance. So these cutting rates vary widely in value—from laboratory control to shop floor ballpark estimates. Many of the top speeds cited either represent cuts made to illustrate speed alone, without regard to surface quality, or may reflect data from machines with very high power output.
[b] 1 in. = 25.4 mm.
[c] 1 in./min = 0.4233 mm/s.

McMullin[39] discusses another problem—embedding of garnet from the abrasive jet in the cut edge of the material. Although this is not a problem in most applications, it was unacceptable to a medical device manufacturer trying to produce a part to be implanted in the body. The manufacturer ended up cutting out a NNS with the WJ cutter and finishing the edges by hand to remove the garnet.

As mentioned previously, the Boeing 777 tail fin production line[50,78] boasts three AWJ cutting system installations. They are used to machine composite production skin panels for the 777 empennage, horizontal stabilizer, and vertical fin. Six panels in all are used to form the horizontal stabilizer and vertical fin. Also trimmed are the 777 spars, ribs, and T-beams.

The above components are equivalent to a wing on a Boeing 737 aircraft, and the total system is controlled by a computerized gantry robot. A program instructs the robot to trim the periphery of parts using water mixed with abrasive garnet. One tool is needed to cut all the parts, one pass is required to complete the cuts, and the machines perform their own inspection.

These machines have two independent z-axis masts on the gantry. One is equipped with the cutting system which performs six-axis machining. The second is fitted with a

TABLE 2.4   Cutting Data[a]—WJ[60]

Material	Thickness (in.)[b]	Nozzle Speed (in./min)[c]	Edge Quality
ABS plastic	0.087	20–50	100% separation
Aluminum	0.050	2–5	Burr
Cardboard	0.055	240–600	Slits very well
Delrin	0.500	2–5	Good to stringers
Fiberglass	0.100	40–150	Good to raggy
Formica	0.040	1450	—
Graphite composite	0.060	25	—
Kevlar	0.040–0.250	50–3	Fair, some furring
Lead	0.125	10	Good, slight burr
Plexiglas	0.118	30–35	Fair
Printed circuit board	0.050–0.125	50–5	Good
PVC	0.250	10–20	Good to fair
Rubber	0.050	2400–3600	Good
Vinyl	0.040	2000–2400	Good
Wood	0.125	40	Fair

[a] In trying to provide data on water jet and abrasive water jet cutting we have collected material from diverse sources. But we must note that most of the data presented is not from uniform tests. Also, note that in many cases data was largely absent on such parameters as pump horsepower, water jet pressure, abrasive particle rate of flow or type or size, and stand-off distance. So these cutting rates vary widely in value—from laboratory control to shop floor ballpark estimates. Many of the top speeds cited either represent cuts made to illustrate speed alone, without regard to surface quality, or may reflect data from machines with very high power output.
[b] 1 in. = 25.4 mm.
[c] 1 in./min = 0.4233 mm/s.

probe which performs coordinate measuring with the aid of a customized software package running on the machine's computer.

Two operators run the machine, one for the cutting system and one for the computer software program. Three high-resolution closed-circuit TV cameras, one at each end of the machine and one mounted on the gantry, help the operators monitor the action.

Recent studies[79] on AWJ machining have concluded the following.

1. Material removal when AWJ machining Gr/Ep occurs by brittle shearing mechanisms which fracture and micromachine the constituents of the composite material, and this mechanism cuts wear.

2. Mechanisms of material removal below the initial damage zone in AWJ machining of both ductile and brittle materials do not change with cutting depth or cutting parameters despite the distinct macro features of the machined surface only because of their effect on jet energy.

Now that WJ cutting systems can operate with as many as six axes of motion, they are capable of cutting three-dimensional complex contoured parts, catching the kerf, water, and abrasive residue, and in some cases providing inspection via coordinate-measuring machine (CMM) technology. For flat parts or parts with a slight contour, two- or three-axis machines can be used. A manufacturer reports that it can invert and mount the articulated arm that comprises the sixth axis of motion onto a vertical servo-driven track positioned above the part to constitute a seventh axis.[76,80,81]

The above AWJ technology and CAD work together in an advanced aerospace cutting tool that allows designers to download cutting instructions directly from design computers.

With the new robot system, technicians at Lockheed Aero Systems Company (LASC) in Georgia position the material to be cut on the cutting platform. Targets placed on the piece automatically guide the water nozzle into position to begin the cutting operation. The operator chooses the precision cutting instructions for any particular job from a computer menu and then enters the command to begin the process.

The six-axis robot-controlled WJ cutter, which uses garnet powder as the abrasive, delivers a stream of water at supersonic speed through a hole 0.25 mm in diameter. The only thing that limits what can be cut with this new system is the part configuration.

AWJ versus WJ.[71,82,83]   In several investigations and studies, Ramulu and Arola[82] found that the surface characteristics of WJ- and AWJ-machined unidirectional Gr/Ep composite showed the following.

- The principal material removal mechanism present in WJ machining of unidirectional Gr/Ep composite is material failure associated with microbending-induced fracture and out-of-plane shear. WJ machining exploits the weaknesses of the composite's mechanical properties.
- AWJ machining of unidirectional Gr/Ep composites consists of a combination of material removal mechanisms including shearing, micromachining, and erosion. AWJ machining is found to be a more feasible machining process for unidirectional Gr/Ep because of its material removal mechanisms and superior quality surface generation (Table 2.5).

Geskin, Tisminetski, et al.[84] demonstrated that AWJ cutting of composites with continuous matter distribution is similar to the cutting of conventional materials. The specific feature of composite machining is a narrow range of process variables that ensure optimal process results. AWJ machining can be used routinely for cutting structural-type composites, which is possible only with reduced requirements for surface waviness. However, the results of their study showed the feasibility of developing a reliable technology for the shaping of these materials. WJ was best suited for rough cutting of materials, not for generating finished surfaces. Jet-based machining allows the generation of a wide variety of flawless surfaces with acceptable topography. Improvements flow and robot control will result in the development of AWJ-based technologies for manufacturing complex parts into the next decade.[84]

Abrasive Liquid Jet Machining (Cutting).   Machining with abrasive liquid jets entails the removal of material by the erosive action of impacting particles. Relatively high levels of particle kinetic power density are delivered to the workpiece. The key to obtaining these high levels of kinetic power is the use of liquid (as opposed to gas) jets to accelerate the particles.

There are two main methods of forming abrasive liquid jets:

TABLE 2.5   AWJ Process and Machining Parameters[83]

AWJ Parameters	Machining Parameters			
	Linear Cutting	Drilling	Milling	Turning
Pressure				Rotational speed
Water jet diameter	Angle	Angle	Traverse rate	Direction of rotation
Mixing tube length	Stand-off distance	Stand-off distance	Lateral increment	angle
Mixing tube diameter	Traverse rate	Dwell time	Number of passes	Traverse rate
Abrasive material	Number of passes	Pressure profile	Number of sweeps	Initial diameter
Abrasive size	Material thickness	Material thickness		Final diameter
Abrasive flow rate				Depth of cut
Abrasive condition				
	Machining Requirements (Dependent Variables)			
	Traverse rate	Diameter	Volume removal rate	Turned diameter
	Surface finish	Drilling time	Depth control	Surface finish
	Width of cut	Hole shape		Machining time

1. *Entrainment.* Abrasives are entrained and accelerated using a high-pressure liquid jet. Entrainment systems typically employ a high-velocity water jet to entrain abrasives and form an AWJ to perform the cutting.

2. *Direct pumping.* Abrasives are premixed with a suspending liquid to form a slurry. The slurry is pumped and expelled through a nozzle to form an abrasive slurry jet (ASJ).[85]

Abrasive liquid jets can be used in many types of machining applications. Because of the large number of parameters involved in abrasive liquid jet machining, optimization is an important task. To help provide a basis for such optimization, a comparative study of entrainment AWJ systems and directly pumped ASJ systems was undertaken by Hashish[85] and the following conclusions were made.

- High-pressure entrainment AWJ systems (up to 345 MPa) are more effective in material removal than low-pressure directly pumped ASJ systems (up to 70 MPa) in the use of hydraulic power and energy consumption.

- Directly pumped ASJ systems are generally more efficient than entrainment AWJ systems and produce greater power density for impacting particles. At $r = 1$ and $k = 2$, directly pumped ASJs are at least eight times as dense in power as entrainment AWJs. For $k = 10$, directly pumped ASJs are approximately 24 times as dense in power as entrainment AWJs.

Laser Beam Cutting (LBC).   Laser beam cutting has become an industrially accepted production process mainly because of the flexibility of the technique and the high-quality cuts that can be made. With the development of composite materials and the introduction of a reinforcement phase, engineers found that the fiber type had a significant influence on the cut quality, carbon fiber-reinforced materials proving the most difficult to cut. The main problem with carbon fiber-reinforced composites in laser cutting was fiber-resin separation.[62,86,87]

Caprino, Taliaferri, and Covelli[88,89] have cut Gl/Ep composites using a CO_2 laser and have proposed that an analytical model can be usefully employed to predict the depth of the kerf and the maximum cutting speed as a function of the travel direction of the beam.

Mukherjee and his associates[52] at Michigan State University studied the three major PMCs using various combinations of laser input power (800–1500 W) and cover gas pressure (0.07–0.55 MPa). For cutting, the maximum speed was 0.17 m/s. They also used several slower speeds for Gr/Ep composites. A cover gas pressure of 0.55 MPa produced the least charring and the most efficient cutting.

> *Gl/Ep.* The cutting speed was much higher than with conventional machining, and more complex cutting geometries could be programmed with the laser.
>
> *Kv/Ep.* With this material, the extent of fiber damage and edge fraying was much lower than with conventional machining. There was no visible delamination with either straight- or cross-cutting.
>
> *Gr/Ep.* Unlike the other materials, Gr/Ep did not laser-machine well. Because the ablation temperature of graphite fibers is high, extensive ablation, burning, and thermal damage occur long before the graphite is cut. The graphite-reinforced composite also absorbs a lot of incident laser energy. Finally, it has high thermal anisotropy, which produces a preferential temperature gradient along the direction of easy conduction regardless of cutting geometry or speed.

Another series of tests performed by Viegelahn, Kawall, Scheuerman, et al.[90] were conducted to determine the basic cutting perimeters for laser cutting of fiberglass composites of different thicknesses with basic tensile strength determinations made from the different thicknesses.[90]

2.4 JOINING PMCs

Joints in components or structures incur a weight penalty, are a source of failure, and cause manufacturing problems; whenever possible, therefore, a designer avoids using them. Unfortunately it is rarely possible to produce a structure, component, assembly, or product without joints because of limitations on material size, convenience in manufacture or transportation, and the need for access. Such considerations apply equally to joints between metallic components and between composite components.

Fortunately the main methods used for joining metallic parts, mechanical fastening and adhesive bonding, are also applicable to composites provided care is taken to allow for the charateristics of composites. Welding is also a possibility for thermoplastic composites, and this technique is being developed for load-carrying joints.

The "Achilles heel" of many structural designs is the joint configuration and method of joining. Thermoplastic composites offer design advantages because they can be joined in more ways than thermosetting materials, which are limited to fasteners and adhesive bonding.

The meltability of thermoplastics and their rapid forming consideration cycles compare very favorably with the lengthy cure and postcure cycles typically needed for thermosets. Previous work has shown that thermoplastic composites can be joined by local consideration of an interface, using a range of techniques. These include ultrasonic welding, induction bonding, dual resin bonding (amorphous thermoplastic film), resistance heating, and focused infrared energy. These and other processes are discussed further later in this chapter. However, the joining technology developed for thermoplastic composites is still very much in its infancy. Rapid and reliable methods must be developed if thermoplastic composites are to become cost-effective. Other requirements include adaptability to various joint configurations, on-line inspection capability, and adaptability to automation within a production environment.

The joining technique used on a particular composite depends on the application and the material's composition. For instance, composites used in aircraft are usually joined by a combination of mechanical fasteners and adhesives, whereas those used in automobiles are often joined only with adhesives.

2.4.1 Mechanical Fasteners

By combining adhesive bonding with mechanical fastening, a partnership was formed that has resulted in the development of surface-mounted fasteners. For composite materials the benefits are twofold: the structural integrity of the material is preserved by the elimination of holes, and structural strength is enhanced by the reinforcing effect of the base plate, which acts as a "doubler" around the point of attachment. In the many fastening applications found in modern composite structures, the adhesive-bonded, nonpenetrating fastener is an attractive alternative method for attaching secondary systems and components.

Fastener development is evolutionary. New fastener designs are usually adaptations of existing devices modified in response to a specific user's need. Thin sheet metal and composite materials pose special problems in fastener use, and this has given rise to entire families of inserts and blind fasteners for the aerospace industry and other industries using these materials.[91–94]

2.4.1.1 Fastener selection.
Rivets, pins, two-piece bolts, and blind fasteners made of titanium, stainless steel, and aluminum are all used for composites. Several factors should be considered when specifying fasteners for composite materials:

Differential expansion of the fastener relative to the composite

The effect of drilling on the structural integrity of the material, as well as delamination caused by fasteners under load

Water intrusion between the fastener and the composite

Electrical continuity of the composite and arcing between fasteners

Possible galvanic corrosion at the composite joint

Weight of the fastening system

Fuel tightness of the fastening system, where applicable.

Aluminum and stainless-steel fasteners expand and contract when exposed to temperature extremes, as in aircraft applications. In carbon fiber composites, the contraction and expansion of such fasteners can cause changes in clamping load. Potential clamping changes should be determined before the fastening system is chosen so joint design can be modified accordingly.

Fasteners for composites should have large heads to distribute loads over a large surface area. In this way, crushing of the composite is reduced. Fasteners should also fit closely to reduce the chances of fretting in the clearance hole. Interference fits (where the fastener is slightly larger than the hole) may cause delamination of the composite. Sleeved fasteners can limit the chances of damage in the clearance hole and still provide an interference fit. Fasteners can also be bonded in place with adhesives to reduce fretting.[95–98]

Because composites have lower thickness compression strengths than metals, blind fastener heads should expand more for proper fastening. The pull-up length of the fastener should precisely match the thickness of the panels being joined to avoid local crushing. The fastener should not expand within the hole, which also causes delamination. A larger clamp area may prevent pull-through. In addition, holes must be drilled cleanly, minimizing splintering and fraying and consequent fiber separation.[91]

When carbon fiber composites are cut, fibers are exposed. These fibers can absorb water, which weakens the material and adds weight to the structure. Sealants can prevent moisture absorption in the clearance hole, but this complicates the process and adds to the cost. It also defeats any effort made to maintain electrical continuity between the composite fibers and the fasteners. Sleeved fasteners can produce fits that reduce water absorption and provide fuel tightness.

2.4.1.2 Bolts and other mechanical fasteners.
Some of the considerations in mechanical fastening are material selection, brittleness of composites, and ensuring the strength of the laminate against shear and bending stresses in the fasteners.

Materials Selection.   Metallic fasteners have succeeded very well with fiberglass, boron, and Kevlar fiber composites. However, graphite reinforcements present problems. Some metallic fasteners in contact with graphite composites are prone to corrosion. Aluminum is the worst in this respect; stainless steel is somewhat better. Nickel and titanium alloys show excellent compatibility with graphite but are expensive. This kind of corrosion especially concerns aircraft designers because of their growing use of graphite. Other than relying on the more expensive metal fasteners, two main solutions are possible:

- Insulate the graphite from the metal by surrounding an aluminum or steel fastener with fiberglass or an adhesive with *scrim*. (Scrim is a low-cost reinforcing fabric made from continuous-filament yarn in an open-mesh construction.) Although this approach prevents corrosion, it complicates fastener installation.

Use composite fasteners rather than metallic ones. Gl/Ep and Gr/Ep do not corrode galvanically, and they save weight. A Gr/Ep fastener is five times lighter than a stainless steel fastener of similar size, and Gl/Ep is four times lighter. Tests made by one helicopter manufacturer indicate that fiberglass composite fasteners are sound for light to medium loads, although they are somewhat degraded by humidity.

In the end, however, design loads and stresses dictate the choice of fasteners. A high-performance aircraft structure may require titanium fasteners despite the cost.

Brittleness.   Most composites are quite brittle, which means that their strains to failure are low. Thus stress concentrations (and strains) caused by bolts themselves can lead relatively easily to failure of the surrounding composite.

One way to understand this is to compare composites with metals. Suppose that a bolt hole in aluminum or steel is slightly too small. When the bolt is inserted, stresses arise in the surrounding metal. If these stresses are great enough, the steel or aluminum will yield, shifting some of the load to adjacent bolts. At the same time, stress is reduced around the original hole. Brittle composites, on the other hand, do not have yield points. If one bolt is tighter in its hole than the others, the bearing stress at that hole will simply remain higher. When loading increases this stress beyond the ultimate stress, the composite does not yield but rather fails. All of the load then suddenly falls on adjacent bolts, perhaps causing failure there as well.

One way to alleviate this problem is to "soften" the composite in the bolt area by using staggered fabric plies to reduce brittleness. Another approach is to coat the bolt with uncured resin before inserting it. The resin flows into any open areas between the bolt and the composite. When cured, it helps to distribute loads more evenly. The cured resin is also less brittle than the surrounding fibers, such as graphite. Finally, bonded metal inserts can be used so that the bolt is actually in the metal, not the composite. However, this approach is costly and adds weight.

One way not to attack the problem is to reinforce the hole with more high-modulus fibers such as graphite. This only increases brittleness around the hole, making the joint more likely to fail under bolt-induced stresses.

Laminate Strength versus Stresses in Fasteners.   As in metallic joints, prime considerations in composite joint design are bearing, tensile, and compressive strengths in the laminate, and shear and bending stresses in the fasteners.

Advanced composites can be delicate materials to fasten. Ordinary fasteners are not designed to protect the fibrous panels and work within the strength limitations. These materials are sensitive to radial expansion of the holes, which creates undue stresses, and to pressure on the back or blind side, which can result in fastener pull-through. The pressure required during installation can cause buckling or delamination, and the problems that can occur during use include delamination, crushing, and pull-through.

2.4.1.3 Lightning strike resistance.   The use of advanced composites for airframe manufacture, particularly the joining of components, has not been without its difficulties. One of the more interesting uncertainties is the behavior of composite structures when struck by lightning.

An inherent obstacle that engineers must overcome in using mechanically fastened composites is the dissipation of electric energy caused by lightning strikes. It is important to dissipate such energy to avoid damage to the composite structure. Moreover, controlled energy dissipation becomes extremely critical in and around fuel tanks, where electric arcing can have catastrophic consequences. In composite material structures, when a lightning strike attaches to a metallic fastener, the current must be dissipated through the carbon fibers that run perpendicular to the fastener and parallel to the skin surface. For this to occur, the fastener must be in intimate contact with the carbon fibers through an interference fit. In the past, fasteners installed in interference in a Gr/Ep advanced composite structure tended to cause damage to the structure by crushing the laminates, bending down the reinforcing fibers, and delaminating the laminate buildup.[94]

A fastener has been developed, Hi-Lex,[99] that is manufactured from Ti-6Al-4V for light weight and excellent corrosion compatibility with carbon fiber-reinforced composite structures. It is generally installed in the "bare" condition, that is, without subsequent surface treatments or coatings to inhibit the conduction of electric current. When the fastener head is fully seated in the joint, the driving hex shears off smoothly flush with the surface of the fastener head. The collar is then attached with a clockwise rotation on right-hand threads. After the collar is seated, the torque is continued until the wrenching flats on the collar separate at a predetermined torque level. This results in a complete installation with the proper preload or clamp-up for beneficial joint strength and a long fatigue life.

In addition to lightning strike conductivity, there are other structural benefits of this unique method of obtaining an interference fit. To obtain high structural efficiency in bolted composite joints, each fastener must accept its correct share of the bond. Because composites fail at a very low level, proper load sharing between fasteners is important and is not obtainable with loose fits.[93]

2.4.1.4 Special considerations for fasteners.   Several factors deserve special consideration when mechanical fasteners are used with composites.

Installation of fasteners requires special care to ensure joint quality and long fatigue life. High clamp-up pressure is beneficial, especially when using clearance fit holes. Clamping forces must be distributed over a sufficient area to avoid exceeding the composite's compressive strength. Where assembly gaps are initially present in a structure, there must be enough of a clamp load to close them without damaging the structure. Assembly gaps cannot exceed the fastener's grip range, however. As a general rule they are dealt with by using temporary fasteners, some of which offer increased bearing areas for use with composites. Generally, to increase joint performance (particularly fatigue resistance), the fastener should be installed in an interference fit hole (i.e., one that is slightly smaller than the fastener diameter) or a hole-filling fastener should be used that expands radially during installation.[100,101]

Drilling and machining can damage composites. The number of defects, such as delamination, resin erosion, and fiber breakout, that are allowed in any structure depends on the application. For instance, because joint failure in carbon fiber composites is caused primarily by localized bearing stress rather than overall stress, delamination is a much more serious defect than fiber breakout in a carbon fiber composite application. Drilling techniques and the tools selected are determined by the resin, the fiber or fiber combinations in the resin, the way the fibers are configured, and the composite metal composition of the structure.[95–98]

When a composite is joined to a metal and the metal is on the blind side, the principal considerations are the desired mechanical properties of the joint, fastener material and strength, galvanic compatibility, fastener design, hole preparation, composite material thickness variations, lightning strike resistance, and fastener installation.

Critical joint parameters in composite materials are tensile pull-through, static lap shear, and lap shear fatigue.[101]

2.4.1.5 Materials.

Titanium alloy Ti-6Al-4V is the favorite fastener material for joining carbon-reinforced materials. It has ultimate tensile and shear strengths of 1104 and 650 MPa, respectively. For enhanced mechanical properties, Inconel 718 or A-286 corrosion-resistant steel materials can be cold-worked to develop a tensile strength up to 1517 MPa and a shear strength up to 758 MPa.

Multiphase alloys MP35N and MP159 can be used to produce fasteners that have minimum tensile strengths of 1800 MPa and are also resistant to stress corrosion cracking and hydrogen embrittlement. Fasteners made of these materials are available in diameters up to 38 mm. Among the applications for MP35N are as wheel bolts for use on the space shuttle orbiter developed by the National Aeronautics and Space Administration (NASA). The alloy has also been used to produce fasteners for critical joints on the space shuttle.

A carbon-reinforced composite requires that fastener materials be near carbon on the electromotive scale. An aluminum alloy or steel alloy fastener would quickly corrode in the presence of carbon and an electrolyte. Salt spray tests show that titanium, austenitic stainless steel, and certain multiphase and Inconel alloys are the best fastener materials.

A fastener must be compatible with all the materials in the joint. For example, when an uncoated titanium fastener is used to join a carbon-reinforced material with aluminum, the titanium fastener should be coated with aluminum for galvanic protection.

Programs such as the NASP also present new and unique challenges to fastener engineers. While the severe environmental conditions that will be encountered by the NASP are not completely understood, engineers agree that atmospheric reentry profiles will require fasteners to withstand very high temperatures (possibly in excess of 1370°C). These fasteners must also be able to operate at extremely low temperatures for extended periods of time.

Researchers are currently testing fasteners made from materials such as Haynes alloy 230 (good to approximately 1000°C), and Ti1100 high-temperature titanium (good to about 600°C) for certain NASP applications. Other possibilities include using coated refractory alloy fasteners, however, compatibility of the coating with the CCC structures of the NASP is a concern. Engineers are also examining experimental materials such as titanium aluminides and SiC-reinforced titanium and aluminum for use in fasteners. Fasteners made of CCC materials may resolve material compatibility problems, however, much research and development still remain to be done.

In answer to the exacting weight and strength requirements for composite fastening on the European fighter aircraft (EFA), a new Ti-Matic blind bolt, an all-titanium blind fastener, has been developed. In this bolt are combined the desirable strength-to-weight properties of titanium and other composite-friendly features to produce a light, versatile product capable of replacing many existing fastener types and offering considerable weight savings.

By their nature, blind fasteners offer significant productivity gains over two-piece threaded fasteners and also involve a lower skill factor in installation, thus resulting in a lower installed cost. The grip overlap capability of the Ti-Matic allows the operator to use one fastener throughout, where with other types different grip sizes may be needed. The Ti-Matic's minimum blind side clearance also enables the designer to reduce the thickness or depth of a structure and thereby reduce weight and airframe drag if the situation permits.

A few companies are making fasteners out of advanced composites for use with advanced composites, and some are making a variety of fasteners for use with composite materials:

- The Huck Comp GP lock bolt titanium fastener[92] is reported to save about 1.75 kg of weight for every 1000 fasteners.
- The Composi-Lok blind bolt[91,92] is designed for composite-to-composite and composite-to-metal applications where the blind bearing side is against the composite sheet. The fastener ensures structural integrity under severe vibration.

A Live Lock fastener[92] is recommended for access panels, engine cowlings, and any removable composite sections, whereas Comp-Tite[92] fasteners are primarily intended for composite-to-composite fastening applications, such as in empennage and wing sections.

Fastener firms have now developed a group of fasteners, Comp-Fast,[92] that can be customized by design and manufactured to handle various applications. They can be used to fasten composites to metals in skin fastening, structural panel fastening, and in applications along with rivetless plate nuts.

Preliminary efforts have been made toward automatic installation in composites, and automation is expected to become even more prevalent. An automated system for drilling and installing Huck Comp fasteners was reported[92] in which 70,000 lock bolts were used in the V-22 Osprey Vertical Take-off and Lift (VTOL) aircraft. The system is expected to have a machine cycle time of only about 9 s.

Another firm, Monogram,[92] has developed end effectors for robotic installation of its blind bolts. The fairly simple pneumatic tools for the rotary-actuated blind bolts weigh about 15.5 kg, making it easier for the robotic system to maintain placement accuracy. According to Monogram, the installation of blind bolts is easier to automate because they require access to only one side.

Cherry-Textron and several other firms[92,102] have introduced an all-composite fastener. The ACP pin-and-nut fastening system of Cherry-Textron is produced in eight composite materials: (1) PEEK long carbon fiber, (2) PEEK short carbon fiber, (3) VECTRA-glass fiber, (4) PEI-glass fiber, (5) PI- glass fiber, (6) epoxy-carbon fiber, (7) PI-carbon fiber, and (8) epoxy-glass.

The performance of specific composite lightweight fasteners depends on the resin and fiber combination. A fastener consisting of AS-4 fiber and epoxy resin has a 275-MPa shear strength. Manufacturing fasteners out of composite materials is still a relatively recent venture that presents an entirely new set of challenges for fastener engineers.

Thermoplastic composite fasteners are being developed for special applications such as space vehicles. These fasteners are formed by compression molding continuous fiber-reinforced bar stock.[103] The manufacturing procedure forces the fibers into the threads, producing a completely reinforced fastener. The fasteners exhibit excellent strength to weight ratios, weighing one-third that of a titanium fastener. The shear strength is higher for a single lap than a double lap because of the different combinations of shear and moment. In the single-lap shear test, the section of maximum shear contains no moment, or represents a pure shear condition. In the double-lap test a moment occurs at the maximum shear, adding stress and thus reducing the strength.

2.4.2 Adhesive Bonding

Theoretically, all composites could be adhesively bonded. However, many manufacturers avoid adhesive bonds where joints undergo large amounts of stress, and so fasteners are still specified for many joints. In addition, the large size of some structures and components precludes use of the special layup tooling and curing equipment needed for most adhesive applications.

The use of adhesives technology to replace or support more conventional joining methods is currently at the forefront of the industry. This is especially true for composite bonding, with particular reference to large structures and unprepared surfaces.[103–106] One of the major considerations is the surface of the composite, which may have to be treated to obtain adequate initial wetting and adhesion and/or long-term durability of the adhesive-composite interface. Both thermosetting and thermoplastic matrices are affected. A second major consideration is the design of the adhesive joint, particularly when bonding anisotropic materials such as multidirectional fiber composites. For example, when lap joints are stressed, they give rise to high tensile or peeling stresses that act normal to the main fiber direction. This is parallel to the direction in which the composite has relatively low strength. By paying careful attention to the design of the ends of typical lap joints, these peeling stresses could be minimized and the strength of the joint greatly increased.

There are overwhelming reasons to adhesive-bond composites to themselves and to metals, and, conversely, there are substantial reasons for using mechanical methods such as bolts or rivets (Table 2.6).

Whenever bonding is planned, adhesive selection is of primary concern. Two-part, medium-viscosity, nonslumping, room-temperature-curing adhesives are preferred because of their user-friendly properties. The most desirable properties are (1) a 1:1 ratio of two different color components that combine to give a third distinct color, thus signifying a complete mix, and (2) a quick cure for rapid handling strength. However, the adhesive must have a pot life long enough to allow time to complete the application. There is typically a tradeoff between a rapid cure and a long pot life.[107]

Epoxies, urethanes, polysulfides, and acrylics are adhesives that have been used successfully in structural applications. Ultimately, the choice of adhesive depends on the performance characteristics required to make an effective attachment. Characteristics such as installation environment, end use environment, cure time, pot life, application equipment, desired strength level, optimal failure mode, and cost are all factors to consider when identifying an appropriate adhesive (Table 2.7).

When adhesive bonding, surface preparation is the major concern in producing a satisfactory joint. The extent of surface preparation required for a reliable adhesive bond depends on the material being bonded and the adhesive used. From a manufacturing

TABLE 2.6   Reasons for and against Adhesive Bonding

For	Against
Higher strength-to-weight ratio	Sometimes difficult surface preparation techniques cannot be verified 100% effective
Lower manufacturing cost	Changes in formulation
Better distribution of stresses	May require heat and pressure
Electrically isolate components	Must track shelf life and out time
Minimize strength reduction of composite	Adhesives change values with temperature
Reduce maintenance costs	May be attacked by solvents or cleaners
Reduced corrosion of metal adherend (no drilled holes)	Common statement, "I won't ride in a glued-together airplane."
Better sonic fatigue resistance	

standpoint, it is desirable to perform the least amount of surface preparation possible that will result in a reliable bond. There are standard procedures that always enhance bonding regardless of adhesive selection. Wiping the substrate and the fastener base with a solvent is considered minimum surface preparation. Abrasion of the surfaces with a scuff pad, sandpaper, or grit blasting prior to wiping with a solvent provides surface conditions that are excellent for successful adhesive bonding.

The studies on adhesives, surface preparation, test specimen preparation, and design of bonded joints reported for the PABST program[108] give much more credibility to the concept of a bonded aircraft and provide reliable methods of transferring loads between composites and metals or other composites.

2.4.2.1 Designs for joining composites.
"Successful bonding" implies consistent strength and long-term reliability. In the bonding of surface-mounted fasteners, strength is derived from the bond line. To achieve consistent strength, bond line thickness must be controlled so that repeatable joints with similar performance characteristics can be achieved.[109]

Although high stiffnesses and strengths can be attained for composite laminates, these characteristics are quite different from those of ordinary materials to which we generally want to fasten composite laminates. The full strength and stiffness characteristics of the laminate often cannot be transferred through the joint without a significant weight

TABLE 2.7   Maximum Service Temperature of Adhesives

Adhesive Type	Typical Maximum Service Temperature (°C)[a]
Epoxy	93
Epoxy-polyamide	149
Epoxy-nitrile	121
Epoxy-novolac	176
Epoxy-phenolic	204
Polyimide	300
Silicone	300

[a] Short-term, dry.

penalty. Thus joints and other fastening devices are critical to the successful use of composite materials. The specific design of a laminate joint is much too complex to cover in detail in this book; the published state of the art is summarized in references 110, 111, and 112.

Bonded versus Bolted Joints. Advanced composites introduce design problems different from those common to metal and molded-plastic construction, particularly where components must be fastened together. Because most types of composite joints, whether they are adhesively bonded or mechanically fastened, involve some cutting or machining of strength-providing fibers, joint configurations require careful planning to minimize the possibility of failure. The planning must consider weaknesses in in-plane shear, transverse tension, interlaminar shear, and bearing strength relative to the primary asset of a lamina, strength and stiffness in the fiber direction.

Bolted Joints. Because most engineered structural composites are made with thermosetting resins, they cannot be joined by welding methods as thermoplastics and metals can. The choice for composites lies between mechanical methods and adhesive bonding. Each technique produces a joint with significant differences in production and function, and each has its advantages and limitations. Mechanical joints are generally not as adversely affected by thermal cycling or humidity as bonded joints are. They can be disassembled without destruction of the substrate, and they are readily inspected for joint quality. Mechanical joints also require little or no surface preparation of the substrate and do not require white glove, clean room production conditions.

On the less favorable side, mechanical methods add weight and bulk to a joint, and because holes must be machined in the substrate members to accommodate bolts or screws, the members are weakened. In addition, the stress concentrations produced by mechanical fasteners can cause joint failure.

Bonded Joints. The performance and safety requirements of the various applications for advanced composites differ widely, and design approaches differ accordingly. For example, the importance of a failure in a carbon fiber composite tennis racket frame in no way approaches that of a failure in a composite helicopter rotor blade. Thus the consequences of component failure determine the type of design data needed and the nature of the testing required. The special difference that sets apart orthotropic materials, such as advanced composites, from isotropic materials is that damage from drilling and machining makes them susceptible to interlaminar shear, delamination, and peeling.

For this reason, bonded joints are usually preferred over mechanically fastened joints in attaching advanced composite components to each other or to metal joints. More specifically, lap joints and strap joints are usually best because they require no machining of the adherends. The advantages of adhesive-bonded joints include light weight, distribution of the load over a larger area than in mechanical joints, and elimination of the need for drilled holes that weaken the structural members. The drawbacks are that they are difficult to inspect for bond integrity, can degrade in service from temperature and humidity cycling, and cannot be disassembled without destruction of the joined members. In addition, bonded joints require rigorous cleaning and preparation of the adhering surfaces.

Bonded Bolted Joints.   These joints perform better than either bonded or bolted joints alone. Bonding reduces the normal tendency of a bolted joint to shear out, and bolting decreases the likelihood of a bonded joint's debonding in an interfacial shear mode.

A number of decisions must be made before bonding to a composite is attempted:

1. Surface preparation technique
   Composite: Peel ply, use manual abrasion, or, if possible, cocure.
   Metal: For aluminum, phosphoric anodizing is the accepted, preferred airframe method. For titanium, several types of etches are available. For steel, the time between surface preparation and bonding is of major concern.

2. Type of adhesive
   Film adhesive: Reproducible chemistry and frequent quality assurance testing (because resin is premixed by the manufacturer) require special storage and generally cannot accommodate varying bond line thicknesses.
   Paste adhesive: Longer shelf life and fewer or no changes in storage accommodate varying bond line thicknesses, but quality assurance is performed after the structure is bonded.

3. Cure temperature of adhesive. Unless they are cocured, the maximum cure temperature of the adhesive should be below the composite cure temperature. The cure temperature or the upper use temperature of the adhesive may dictate the maximum environmental exposure temperature of the component. There are several cure temperature ranges:
   Room temperature to 107°C: Generally for paste adhesives for noncritical structures.
   107 to 140°C. For nonaircraft critical structures. Cannot be used for aircraft in general because moisture absorption may lower the heat deflection temperature (HDT) below the environmental operating temperature.
   177°C and above. For aircraft structural bonding. A higher cure temperature may mean more strain discontinuity between adhesive and both mating surfaces (residual stresses in the bond).

4. Joint design. The primary desired method of load transfer through an adhesive-bonded joint is by shear. This means that the design must avoid peel, cleavage, and normal tensile stresses. One practical application of a method that avoids all but shear stresses is the use of a rivet-bonded construction. The direction of the fibers in the outer ply of the composite (against the adhesive) should not be 90° to the expected load path.

2.4.2.2 Joint configuration.

There are three basic types of bonds—primary, secondary, and tertiary. A primary bond is a bond between an uncured laminate and a fresh laminate, a secondary bond is a bond between an uncured laminate and a cured laminate, and a tertiary bond is a bond between two cured laminates.

Whatever the bond type, the designer can choose from a great variety of joint configurations. Among the most widely used are the lap and the strap joints shown in Figure 2.5. Joints featuring straps and doublers impose a slight weight penalty (although not as

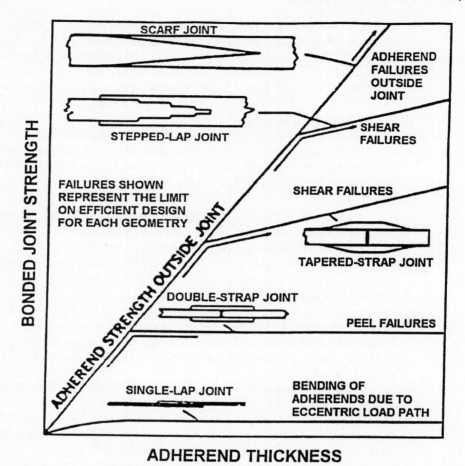

Figure 2.5   Relative uses of different bonded joint types.[112]

much as a bolt). They are unacceptable only when a smooth, uninterrupted surface is necessary.

Also widely used are the scarf joints and other configurations shown in Figure 2.5. Scarf joints, by definition, involve tapered joining surfaces. In this way, they represent a design middle ground between lap joints and the end-to-end attachment of butt joints.

There are also two kinds of joints—mechanical and bonded. *Mechanical joints* are created by fastening the substrates with bolts or rivets. *Bonded joints* use an adhesive interlayer between the adherends. Adhesively bonded joints can distribute the load over a larger area than mechanical joints can, require no holes, add very little weight to the structure, and have superior fatigue resistance.

Kim and associates,[113] using the above evaluation, selected an adhesive-bonded lap joint for work on a torque transmission joint. Included in their investigation were the effects of adhesive thickness and adherend surface roughness on the fatigue strength of a

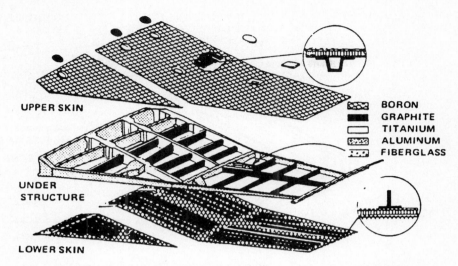

UPPER SKIN

BORON
GRAPHITE
TITANIUM
ALUMINUM
FIBERGLASS

UNDER
STRUCTURE

LOWER SKIN

Figure 2.6   Breakout of a fiberglass composite main rotor blade for a commercial helicopter.[114]

tubular adhesive-bonded single-lap joint. From fatigue experiments, it was found that the optimal arithmetic surface roughness of the adherends was about 2 μm and that the optimal adhesive thickness was about 0.15 mm. Based on these two values, prototype torsional adhesive joints for the power transmission shafts (0.66 mm) of an automobile or a small helicopter were manufactured and statistically tested under torque. The tests were performed on the single-lap joint, the single-lap joint with scarf, the double-lap joint, and the double-lap joint with scarf. One adherend was a high-strength C/Ep composite, and the other was S45C carbon steel. From the experiments, it was found that the double-lap joint was the best among the joints in terms of torque capacity as well as cost of manufacture.

Joint Configuration Applications.[114]   The following examples show how the criteria and theoretical and analytical data developed for adhesive bonding by design engineers are applied with the assistance of specialists in stress, weight, and materials to actual parts.

Figure 2.6 shows a bonded fiberglass composite main rotor blade for a commercial helicopter.

Figure 2.7 illustrates an adhesive-bonded wing torque box in which B/Ep and Gr/Ep composites were used.

Figure 2.8 shows composite rudders of Gr/PI, B/PI, and B/Ep for the F-4 aircraft.

2.4.2.3 Surface preparation.[115]   In recent years, researchers have found that plasma treatment,[116–122] corona discharge pretreatment,[123] chromic-sulfuric acid etch and gas plasma,[124] laser radiation,[124] fluorine pretreatment,[124] and other methods prepare composites and other substrates for adhesive bonding.

Researchers at Lockheed Aeronautical Systems Company (Burbank, CA) com-

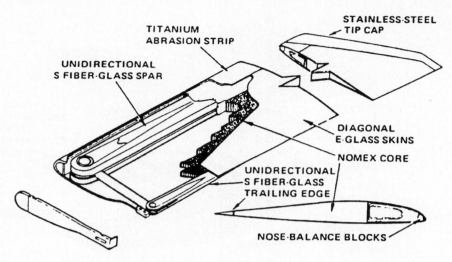

Figure 2.7   Composite flight test wing for the F-15 (McDonnell-Douglas). B/Ep and Gr/Ep composites were used in the adhesive-bonded wing torque box.[114]

pleted studies indicating that treatment of PEEK/Gr composites with chromic-sulfuric acid etch, gas plasma, or a Kevlar peel ply provided good surface preparation for epoxy adhesive bonding.[124]

Scientists at the Ministry of Defense in Israel used argon fluoride (ArF) excimer lasers and 10.6-μm CO_2 lasers to improve the bonding of engineering plastics such as polycarbonate and PEI. The radiation dissociates some surface molecules into reactive fragments. Researchers believe the ArF laser massively alters surface chemistry, whereas the CO_2 laser mainly softens the thermoplastic surface.[124]

Researchers at Air Products in the United Kingdom have been investigating the pretreatment of polyethylene, polypropylene, and acetal with fluorine to improve adhesive joint strengths. Using fluorine-nitrogen and fluorine-oxygen-nitrogen mixtures for pretreating low- and high-density polyethylene films, they found that large increases in adhesion were possible with all mixtures containing fluorine.[124]

Plasma Treatment.   In recent years, researchers[116,117] have found that surface activation by plasma treatment prepares plastic and other substrates for adhesive bonding. The article to be treated is placed in a chamber that is evacuated by a vacuum pump. A specific gas, such as oxygen, argon, or air, depending on the treatment desired, is flowed through the chamber and drawn out the exhaust. The gas is then excited with radiofrequency (RF) energy produced by applying power to the chamber electrodes. This strips electrons from the gas molecules, forming free radicals which react with the surface and create sites for bonding. Water, oils, and other organic materials on the surface are also attacked and broken down by free radicals. Their volatile remains are swept away by the vacuum system, leaving the surface ultraclean. The glow associated with plasma treatment is a result of stripped electrons falling back to a lower energy state, emitting light as in neon, fluorescent bulbs, and lightning.

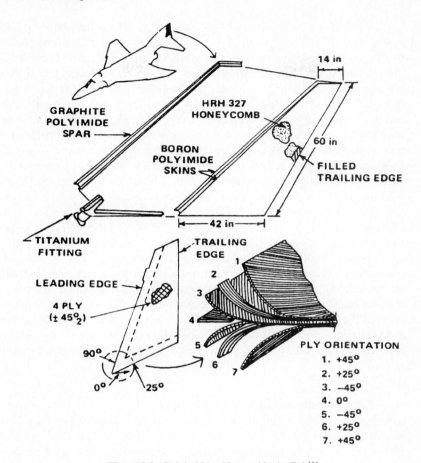

Figure 2.8   Polyimide rudder used in the F-4.[114]

The effect of plasma treatment is to reduce the contact angle (a function of surface energy) of the substrate being treated. This reduction in contact angle results in improved wetting of adhesives on the treated surface and higher bond strengths. When the treated substrate is exposed to air, the contact angles begin to increase again, and so adhesive should be applied to treated surfaces as soon as possible after plasma treatment.

Gas plasma treatment is a good surface preparation for the adhesive bonding of plastics because it involves no solvents, chemical fumes, disposal problems, or heat, which could harm the plastic surface. The plasma reacts to a depth of 0.01–0.1 μm (100–1000 Å).[118] Joining C/PEEK composites is an important technological issue,[125] and surface preparation is particularly critical.[122] A comparison of different surface preparation methods showed that plasma treatment, albeit with particularly "hard" parameters (a long treatment time and the use of a CF_4/O_2 mixture), scored remarkably well.[125] Plasma treatment is particularly interesting because it is reasonably uniform over the whole part, compared with flame or corona treatments, and can be made environmentally-friendly by an appropriate choice of gases.[126–130]

Occhiello et al.[122] assessed the efficiency of plasma treatments in improving the adhesion of C/PEEK composites with epoxy adhesives. C/PEEK composites were treated by plasma to improve adhesion with conventional epoxy adhesives. Oxidizing (oxygen, air) and inert (nitrogen, argon) mixtures were used. An interesting finding was that for plasma treatments lasting more than 5–10 s (much less than the 5 min quoted by Silverman and Griese[125] or the 2 min reported by Yoon and McGrath[127]) there is no real increase in pull strength with treatment time. Also, RF power seems to be a rather mild requirement since 20-W treatments provide excellent adhesion. Oxygen produces consistent improvements in adhesion even at very low treatment times. Argon is most efficient for treatments lasting at least 30 s, and nitrogen is somewhat in between. Moreover, the ultimate pull strength is close for all gases, a behavior reminiscent of that observed with other oxidation-resistant polymers, such as polyarylsulfones, which was attributed to surface cross-linking induced by plasma treatment.[129] Another important factor could be mechanical interlocking, favored by easier penetration of the adhesives into surface imperfections made hydrophilic by the treatment.

Plasma treatments introduced oxygen-containing, and in some cases nitrogen-containing, functionalities at the surface of the composite and increased wettability. While untreated samples showed no pull strength and adhesive failure, plasma-treated samples reached high pull strengths and cohesive failure even after very short treatment times (< 30 s). Pull strength values were relatively insensitive to the nature of the treatment gas and plasma parameters; furthermore, aging did not affect adhesion but in some cases improved it. These effects suggest the occurrence of plasma-induced cross-linking of the surface layer.[122]

2.4.2.4 Adhesives and processing.

Research on high-temperature resin systems has intensified during the past few years. Structural applications include (1) engine nacelles involving long-time exposure (thousands of hours) to temperatures in the 150–300°C range, (2) supersonic military aircraft involving moderately long exposure (hundreds of hours) to temperatures of 150–200°C, and (3) missile applications involving only brief exposure (seconds or minutes) to temperatures up to 500°C and above.[16,112,131]

Adhesive bonding can be carried out at the composite layup stage with the adhesive curing during the laminate process schedule. This is described as *cocuring*. Alternatively, adhesives may be used solely at the assembly stage, where finished components are permanently joined together or to other parts of the overall structure. This is known as *secondary bonding*. The choice between cocuring and secondary bonding as a method requires consideration of a number of parameters at the design stage as well as during manufacture. These are shown in Table 2.8.

One significant advantage of using advanced composites is that single components can be produced with complex geometries in one operation. This reduces the number of parts in comparison with a metal assembly of equivalent performance.

With careful design, composite layups are created so that the part can be manufactured as a single entity. This can be achieved by

- Combining the various layups without an adhesive
- Combining the various layups with an adhesive layer.

TABLE 2.8   Comparison of Cocuring and Secondary Bonding as an Assembly Technique[112]

Cocure	Secondary Bonding
Parts to be joined are both composites.	Parts to be joined are composite, metal, or a combination of both.
Required mechanical performance of joint region can be met by either composite resin matrix or adhesive.	Adhesive is compatible with all adherends.
Adhesive is compatible (chemically plus cure) with composite.	Required mechanical properties of joint region can be met by an adhesive.
Composite layup has been designed for cocuring.	Composite layup has been designed for adhesive bonding.
Adjacent plies of preferred orientation without compromising strength or stiffness of composite requirements.	Surface preparation methods and equipment are available.
Tooling is available for complex geometry required and is cost effective.	Jigs and tools are available.
Process machinery (layup and consolidation) is available for total component size.	Process machinery is available.
Dimensional tolerances on component can be met.	Adhesive cure schedule is known not to degrade adhesives.
The thickness of the "adherends" in the joint overlap is acceptable.	Inspection and text requirements can be met.
Wide variations in thicknesses within a single component may result in a change in the composite cure schedule.	—
The total component size can be handled without subsequent damage.	—
Inspection and test requirements can be performed on complex tool component size.	—

Both these techniques are known as cocuring. It should be remembered that differences exist between these methods and are discussed in detail in reference 112.

A general relationship exists between the viscosity of an adhesive and the optimum bond line thickness range for that adhesive. A low-viscosity adhesive has a thin optimum bond line thickness, and a higher-viscosity adhesive requires a thicker bond line to achieve optimum performance characteristics such as bond strength and long-term bond reliability.

Because of fatigue considerations, whenever possible it is preferable to bond rather than mechanically fasten composite structures. For this reason, the increased use of high-temperature PMCs has necessitated the development of comparable, equally heat-stable adhesive systems.[131,132]

Epoxy adhesives based on multifunctional resins are available that exhibit excellent strength retention at temperatures up to about 225°C. Where long-term aging is required, epoxies are generally limited to applications requiring continuous service at temperatures no higher than 175°C. The adherends involved are most commonly epoxy matrix composite structures. Newly developed epoxy adhesives for use on supersonic military aircraft have excellent strength retention at temperatures up to 215°C, but strength then drops rather sharply to 6.9 MPa at 260°C. After 3000 h aging at 215°C, the adhesive retains approximately 80% of its original lap shear strength, and this is more than sufficient for most military aircraft. Other multifunctional epoxy adhesives are available for use in applications such as engine nacelles, where resistance to heat aging for very long periods of time is required.[133–135]

Epoxy-phenolic adhesives rival PI adhesives in their ability to withstand short-term exposure to extremely high temperatures. For this reason, they are admirably suited for use on missiles, and they are generally preferred to PI adhesives because of their lower cost and ease of processing. Bond pressures of 0.28 MPa and cure temperatures of 150°C are usually adequate for most applications.

Condensation reaction PI adhesives are high-temperature adhesives based on condensation reaction PI precursors that have been marketed for over 15 years. Adhesives of this type are supplied in both liquid form and as solid films. In addition, they may be filled or unfilled. Condensation reaction PI adhesives require high cure temperatures, 260°C or even higher, depending on the expected service temperature. Processing is further complicated by the highly volatile content of these adhesives. One processing advantage that they have over addition reaction PI adhesives is that they can be cured under pressure at 175°C and then postcured without pressure at higher temperatures. This permits the use of low-temperature autoclaves and conventional bagging materials. Inexpensive ovens can be used for postcuring the bonded panels.[136]

Addition reaction PI adhesives make it possible to obtain bond lines with extremely low void contents. Adhesives of this type are generally supplied as supported film containing sufficient alcohol to impart tack and drape to the adhesive. Unlike the condensation reaction type, addition reaction PI adhesives remain thermoplastic after imidization and solvent removal occur. Another difference is that addition reaction PI adhesives must be held under pressure throughout the cure cycle, and so relatively costly high-temperature presses or autoclaves are required for fabricating panels. Composites made from addition reaction PI adhesives are generally fabricated using 1.4 MPa to ensure low void content in the laminate.

BMI adhesives fill a niche between high-temperature epoxy and PI adhesives. Unmodified BMI resins are hard, brittle solids, and they must be plasticized in order to impart drape to the adhesive films based on them. However, unlike PI resins, they can be plasticized without resorting to the use of solvents.[110,111] Cure temperatures vary, depending on the BMI resin employed and the specific monomeric plasticizers selected. In general, good results can be obtained using a cure of 2 h at 175°C under a pressure of 0.28 MPa, followed by a 2- to 4-h postcure at 200–225°C. The strength retention of BMI adhesives is surprisingly good up to about 300°C, but in applications requiring long-term exposure to high temperatures, BMI adhesives are not expected to be durable beyond 200–225°C. Almost limitless variations in formulation are possible with BMI adhesives.

Structural Adhesives. Advances in structural adhesives technology are revolutionizing methods of manufacture and assembly in virtually every modern industry for many reasons. Structural adhesives offer significant advantages over traditional fastening methods: increased basic strength, greater durability, and remarkable resistance to water, outdoor environments, solvents, impact, corrosion, temperature extremes, and thermal cycling. Structural adhesives also impart greater "give" than mechanical fastening methods and provide a better appearance, one that is smooth and uncluttered by the heads of rivets and bolts. They offer improved stress distribution and even insulate and protect against galvanic corrosion.

Structural adhesives are applied in almost any way imaginable: spraying, brushing, rolling, silk screening, or as a bead, depending on the adhesive's physical properties and the design of the bond line. They offer the user the ability to increase the performance and quality of new and existing products by using advanced materials (composites), new processes, and new quality control techniques.[137,138] Two of the new processes are rapid adhesive bonding (RAB)[139,140] and dual-resin bonding (D-RB).[137,141]

Rapid Adhesive Bonding. Most adhesive bonding takes place in heated platen presses or autoclaves that have substantial thermal mass, limiting the heating and cooling rates of the work considerably. A few minutes to achieve a viscosity low enough to obtain flow and wetting of the adherend surfaces is sufficient for thermoplastic adhesives. Because of the cost savings possible, adhesive bonding processes could be accomplished in minutes, using advanced induction heating technology at kilohertz frequencies. These concepts are designated as RAB for components and field repair.[139] The technology utilizes a combination of aerospace adhesives and induction heating to produce lightweight, compact, energy-efficient prototype equipment.

RAB equipment is based on a self-tuning solid-state power oscillator, which may be powered from a variety of sources, feeding kilohertz power to a ferrite toroid induction heater. The toroid geometry produces a uniform, concentrated magnetic flux into the specimen or component to be bonded, causing eddy currents to flow in a ferromagnetic susceptor and/or paramagnetic adherends. These currents heat only the bond line or its vicinity, and the equipment operates at 30–80 kHz. The power required to heat a 6.5×10^2-mm bond area to temperatures above 427°C within ~1 min is 300 W. Maintaining the bond line temperature at 177–427°C typically requires less than 200 W. Since no large fixtures are being heated, the cooling rate of the work is rapid, typically less than 2 min from the bonding temperature to below the glass-transition temperature of the adhesive, at which time the bonded component may be removed from the RAB equipment.

Researchers[139,140] have also designed a machine for seam welding composites by the toroid induction method. Bonding tests were conducted with the toroid induction welder, and the results of these tests were compared to the strengths of bonds made using a conventional bonding technique.[16]

Dual-Resin Bonding. A process that uses a low-melting-point analog of the high-temperature thermoplastic used in the composite matrix, D-RB involves melting and bonding the analog without melting the composite.[116] The Thermabond process utilizes PEEK composites which are coated with PEI film during part consolidation prior to joining. Because the amorphous PEI has a T_g of 210°C, parts can then be fused together without sophisticated tooling at temperatures well below the PEEK melt temperature of 340°C. For this process to be successful, adhesion of the amorphous layer to the composite is essential. This is accomplished by selecting an amorphous polymer that is compatible with the composite matrix resin and/or by migration of some of the fibers from the surface of the laminate into the amorphous layer.

Several aircraft companies have evaluated this system for several thermoplastic materials. Lockheed reports[116] lap shear strengths of 28–35 MPa for this dual-resin system with an additional layer of neat PEEK film at the interface. Joining was accomplished at 260–282°C at pressures of 0.33–0.60 MPa. After wet conditioning, lap shear strengths, when tested at 149°C, exceeded 21 MPa. These joints survived three lifetimes of fatigue cycling before the fixtures broke.

One weakness of D-RB is reliance on an amorphous interlayer that can be attacked by hostile solvents. Lap shear strengths with Gr/PEEK laminates, for instance, decrease by about one-half after a 24-h immersion in dichloromethane. However, development work is currently evaluating a thermoplastic interlayer resin that can be crystallized in an annealing process after D-RB has taken place.

In another program[142] the performance of C/PEEK subassemblies joined using a D-RB approach was observed. The program focused on the properties of joints between square tubes, formed from AS4 carbon fiber-reinforced polyether ether ketone (APC-2/PEEK) prepreg tape, and half-cruciform joints woven from commingled AS4 carbon fiber-PEEK yarn. The joints were assembled using a bonding polymer, PEI, which is miscible in the PEEK matrix. The conclusion from the tensile lap tests was that the bond strength was greater by ~50% than the failure stress recorded. The average lap shear strength value was 16.29 MPa, a conservative estimate of the actual bond strength. This work indicates that the use of PEI/PEEK as a miscible combination in the bonding of structural PEEK matrix composites results in excellent joint performance with minimal susceptibility to surface contamination during bonding. D-RB offers an efficient method for rapidly assembling space structures from standardized components. Further, the characteristics of thermoplastic matrices permit joining techniques with a potential for reconfiguring the structure to accommodate design changes without scrapping existing components. This would also allow reconfiguration of structures in space without the need to place additional materials into orbit. Finally, the results indicate that the joints showed good reproducibility in bond strength over the entire test temperature range. However, use of the lower-melting-point PEI bonding material decreases the upper service temperature of the resulting structure.

2.4.3 Fusion Bonding

The chemical adhesion that holds a material together, cohesion, results from the strong valence attraction between substrates that are caused to flow together, such as when metal surfaces melt and flow together during welding. In cohesive bonding, there is an intermingling of molecules between substrates. In thermoplastics and thermoplastic composites (TPCs), the fusion process involves heating the polymer to a viscous state and physically causing polymer chains to interdiffuse, forming a weld. Welding techniques can provide bond strengths equal to those of the parent polymer.

There are three groups of fusion welding techniques that deal with the basic problem of localizing heating to the bond line so that part configuration and structural integrity are maintained in the process (Table 2.9).

The first technique has heat directly applied to the individual surfaces to be joined. The surfaces are quickly brought into contact, held under pressure, and allowed to cool, forming a bond. Methods used to heat the surfaces include heated tools and platens, hot gas, and focused infrared and laser beams.

In the second method, parts are first put in contact with each other, and then heat is applied at the bond line. In the case of spin welding, for instance, two concentric parts are assembled and one is held stationary while the other is rotated. Heat is generated by friction welding the parts. Other examples include vibration, ultrasonic, induction, and resistance welding, all of which have been evaluated with advanced TPCs.[116]

A third series of welding processes that are both quick and efficient includes induc-

TABLE 2.9   Thermal, Friction, and Electromagnetic Methods for Welding Thermoplastics

Thermal	Friction (Mechanical)	Electromagnetic
Hot gas welding	Spin welding	Resistance (implant) welding
Extrusion welding	Vibration welding (100–250 Hz)	Induction welding (5–25 MHz)
Hot tool welding	Ultrasonic welding (20–40 Hz)	Dielectric heating (1–100 MHz)
Infrared welding		Microwave heating (1–100 GHz)

tion, microwave, dielectric, and implant resistance welding. The technique uses a sacrificial heating element placed between two APC-2 laminates. By applying pressure and passing an electric current through the element, heat is generated and a bond produced.

Induction welding is based on the principle that a magnetic field passing through a conductor generates eddy currents which dissipate in the form of heat. If the conductor is also magnetic (graphite fibers, for instance, are conductive but not magnetic), hysteresis losses also dissipate as heat. In practice, a thin layer of metal-filled adhesive or a thermoplastic film-laminated metal screen is placed at the joint line. This metal susceptor localizes heat in the joint area once the magnetic field is applied. Lap shear strengths of 42–49 MPa have been reported for Gr/PEEK laminates welded in this fashion.

Some PEEK, PPS, and PEI composites of graphite without metal susceptors have been successfully induction-welded[116] with lap shear strengths of 38.5–48.3 MPa by using APC-2 laminates and PEEK film at the interface. Some Gr/PPS composites have been welded with lap shear strengths of 23.8–30.8 MPa, while a Gr/PEI composite was joined with lap shear strengths of 31.5–41.3 MPa.

Resistance-welded Gr/PEEK composites have generated lap shear strengths up to 52.5 MPa.

2.4.3.1 Thermal welding

Focused Infrared Energy (FIRE).   FIRE, a novel noncontact fusion bonding technique, has been introduced to industry in a semicommercial phase. It is said to be particularly suited for parts with problematic geometries and for high-performance TPCs, such as carbon fiber-reinforced PEEK and polyphenylene sulfide composites. It is targeted for applications that require joints as strong as the composite itself.[143–145]

The FIRE technique overcomes many problems by using reciprocating focused infrared heaters that progressively apply a precise amount of intense heat to bond lines on two separated bond surfaces which are pressed together after heating to produce a strong welded joint (Figure 2.9). During reciprocation, the surface temperatures of both bond faces increase with each stroke. However, because the infrared beam is moving, it is not in any one position long enough to burn the resin. This also allows the surface temperature to increase to the melt point without increasing the internal temperature of the composite to the point that the matrix would be damaged by slippage or fiber separation.

Temperature sensors can be positioned to directly monitor temperature at the bond lines and to regulate the power applied to the infrared heaters. There is no distortion of the bond lines or part surfaces and no adhesion of melted material to the heaters. The temperatures of both surfaces can be controlled individually during welding, which permits assembly of thick and thin parts as well as thin-walled parts.

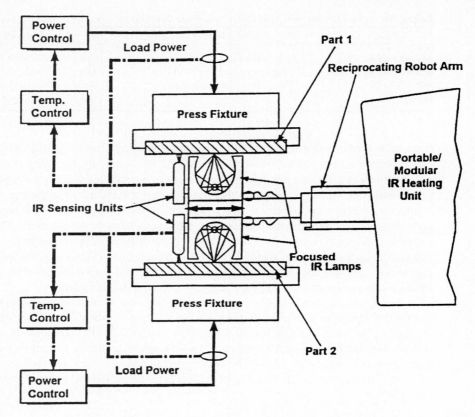

Figure 2.9 The FIRE process consists of moving a pair of linear, focused infrared lamps back and forth along the bond lines of the material surfaces that are to be joined. The reciprocation rate of the lamps is selected by the melt-flow temperature of the material. The temperatures of the surfaces are monitored to control the power applied to the lamps. When the required bonding temperature is reached, the lamps are retracted and the two surfaces are pressed together to make the bond.[143]

FIRE can join any combination of straight or circular flat parts. The process can bond curved lines for applications such as fuselages or three-dimensional parts, and it also accommodates continuous welds. A wide range of materials can be welded, from very rigid, high-temperature advanced composites to low-modulus elastomers. Strong welds have been produced in glass, carbon, and aramid fiber-reinforced PEEK, PEI, thermoplastic PI, PPS, polyamide-imide, polysulfone, and liquid crystal polymers (LCPs).

The reciprocation zone controls bond line width, and wide bond lines are made possible by using multiple lamp fixtures placed side by side. For example, it is possible to heat a bond line area 51 cm wide and 3.75 m long by using four sets of focused infrared lamps, 1.2 m long and spaced approximately 10 cm apart, back to back.

Good wetting of a clean, release agent-free bond line is essential for successful use of FIRE with advanced composites. Treatment with cleaning solutions of plasma is preferred to sanding or abrading the surface, which may remove resin in the two facing adherend plies and loosen the fibers. A fiber-free film of resin tape should be applied to one part surface, but resin-rich surfaces on the composites to be joined may eliminate the need for this step. A benefit of FIRE is that the high temperature of the reciprocating beam (up to 650°C) burns off surface contaminants that may be trapped in asperities in the interface layers.

Heated Tool and Platen Welding.   Heated tool welding, more commonly called *hot plate welding,* is the simplest welding technique used with plastics. It is popular both in mass production and for large structures.

A heated plate is clamped between the surfaces to be joined until they soften. The plate is then withdrawn, and the surfaces are brought together under a controlled pressure for a specific period. The fused surfaces are allowed to cool, forming a joint that normally has at least 90% of the strength of the parent material.[146] Work conducted on hot plate welding of APC-2 indicated that welds could be achieved with a lap shear strength of up to 50 N/mm^2.

Hot plate welding large components with wall thicknesses above 25 mm currently presents difficulties.[146] The Welding Institute (TWI) has designed and built a test bed machine that has an electrically heated hot plate and hydraulically applied axial force in common with commercial equipment, but these parameters can be varied over wide ranges. The maximum hot plate temperature is 650°C, and force is adjustable from 1kN to 150 kN so that all current thermoplastic high-temperature composites can be welded in sizes up to the dimensions of the hot plate (650 × 650 mm). Typical areas of current research include joining large-diameter (up to 630 mm) advanced thermoplastic composite parts for aerospace use.

Hot Gas Welding.   In hot gas or hot air welding, filler material is fed into a prepared joint and a stream of hot gas (generally air) is used to heat filler and parent material. The equipment can be manual or semiautomatic. Normally, the filler material is a solid rod which is forced under light pressure into the joint. The filler rod is not melted in the gas stream but is sufficiently softened to allow it to fuse to the parent material.[147–150]

The filler rod can be circular in section, but, triangular section rods have recently been used as they allow multirun welds and fillet welds to be made more easily. For best results the filler rod should have the same composition as the substrate. Usually, only the surface is fused. Gas temperatures depend on the polymer but typically fall in the range 200–300°C, and gas flow rates range from 15 to 60 L/min.

Gases used include nitrogen, air, carbon dioxide, hydrogen, and oxygen. Compressed air is popular because it is satisfactory for many purposes and is also economical. For smaller, manual welding operations, results depend on the skill of the operator.[146–151]

Hot gas welding is used mostly for butt and fillet welding but also for lap welding of thin sheets. In lap and fillet welding, no joint preparation is needed. In fillet welding the filler is fed directly into the 90° angle of the fillet, and for lap welding no filler is used.[151]

Hauber and Pasanen[152] described the technology they employed and their test results using an advanced hot gas torch and a thermoplastic weld head with a robotic winding system. They were successful in employing the hot gas jet method to bond TPCs to 90% or more of theoretical density at surface speeds of 0.038 m/s for APC-2 and greater than 0.075 m/s for graphite-J2. An IR pyrometer was shown to be effective for closed-loop temperature control of this process provided that the fibers were fully wetted by the matrix in the prepreg tow. Interlaminar shear strength tests were shown to provide a sensitive measure of bond quality, and void content due to lack of intimate contact between bond surfaces was identified as the primary limiting factor for interlaminar shear strength.[152]

2.4.3.2 Friction welding.   The various types of friction welding include spin, rotational, and vibration welding.

Heat is produced in friction welding through mechanical rubbing of two surfaces in contact under an applied axial load. The technique is ideally suited for joining thermoplastics because the frictional heat developed at the joint line is sufficient to cause rapid surface melting without significantly raising the temperature in regions away from the rubbing surfaces. Because plastics have a relatively low thermal conductivity and melting point (compared with most common metals) short welding times (generally 0.5–3.0 s) can be used to provide joints with properties that approach those of the parent material.

Spin Welding.   Spin welding (also known as *spin bonding*) is a process that uses the heat from friction to melt substrates together. Circular thermoplastic parts are bonded by this method, and part joints or filler rods may be spun to create the weld joint. Once the melt cools, the bond is completed.

Major welding parameters include rotational speed, friction pressure, forge pressure, weld time, and melting length. Rotational speeds range from 1 to 20 m/s, friction pressures from 80 to 150 MPa, forge pressures from 100 to 300 kPa, and weld times from 1 to 20 s. Because the heating effect depends on relative surface velocity, maximum heat is generated at the outer edge in solid components. The differential generation of heat can result in weld zone stresses. Hollow sections with thin walls are more satisfactory for this process.[16]

Spin welding has been performed with a drill press, but better-quality welds can be obtained with units that impart a controlled amount of energy to the spinning part. Usually this is done by fixturing one part in a flywheel which is spun to a preset energy level. On contact with the fixed part, the energy is dissipated at the interface. Other designs use driven fixtures to impart fixed amounts of energy.[151]

The advantages of spin welding are high weld quality, simplicity, speed, and reproducibility. In most cases, little surface preparation is necessary. The main disadvantage is that in its simplest form spin welding is suitable only for applications where at least one of the components is circular and angular alignment is not required. Orbital welding can be used in the latter case, but the process is more complex. Most TPCs can be bonded, including dissimilar polymers with compatible melt temperatures.[153]

Vibration Welding.   Vibration (linear friction) welding involves the rubbing of two thermoplastics together under pressure at a suitable frequency and amplitude until enough energy is expended to melt the polymer. The vibration is stopped at that point, the parts are aligned, and the molten polymer is allowed to solidify, creating the weld.[151]

The process is rapid (weld times of 1–5 s are generally used), the vibration is typically 100–240 Hz at 1–5 mm amplitude, and pressures are similar to those used in rotational and ultrasonic welding (1–4 MPa). It is similar to spin welding except that the motion is linear rather than rotational.[151,154] The main advantage of vibration welding is its ability to weld large, complex linear joints at high production rates. Other advantages include simplicity of tooling, the ability to weld a number of components simultaneously, and the ability to weld almost all thermoplastic materials.

Vibration welding is more difficult to apply to continuous-fiber materials because of the possibility of fiber displacement. In some tests on APC-2 material, the effect of weld times between 1 and 6 s was investigated with weld pressures in the range 1–2.5 N/mm^2 and a vibration amplitude between 2 and 3 mm. The direction of the vibration amplitude was parallel to the length of the specimen. Using low pressure and vibration amplitude resulted in little or no bonding of the surfaces. Increasing either parameter resulted in higher-strength welds with a maximum of 17 N/mm^2 being attained (35% of parent material). Under all welding conditions, weld strength tended to increase with weld time. However, the degree of flash and fiber displacement also increased, and this distortion and displacement are major disadvantages of the process when applied to APC-2.[146,151,155–157]

Vibration welding has joined PEEK laminates, yielding lap shear strengths of 39.9 MPa. Linear friction welding is well suited to and has been developed for nonreinforced thermoplastic materials and works well for unidirectional composites. Results indicate that it may not be suitable for large parts or for cross- or angled-ply laminates because fibers in adjacent plies may cut each other, decreasing strength at the joint.[16,154,155,157]

Designers of injection molded automobile air intake manifolds now favor two-step molding and welding processes over molding around fusible cores. The trend is particularly strong in Europe where both techniques originated.

In the new process, the manifold is molded in two halves, a top and a bottom, with the split line along the intake runners. The two halves are subsequently vibration-welded together. This contrasts with the fusible core method in which the part is molded in a single shot around a core made from a low-melting-point metal.

Engineers have determined that most weld strength problems can be overcome by welding immediately after molding, while the part is still about 80–100°C, has not begun to distort, and is still fairly flexible. Welding tools have been produced using the same CAD data as the injection molds and then further modified by hand to accommodate minor distortions in the parts.

A technique for preventing flash from the weld from spilling out into the manifold runners and disrupting airflow has also been developed. The cross section of the weld has channels on either side of the principal welding surfaces which act as flash "traps."

Rotational Welding.   Rotational welding is suitable for bonding rotationally symmetric formed parts. One part rotates, producing heat, while the second part is held in place. After rotation is stopped, the weld "freezes." The rotational speed and friction time have an important effect. A conical surface weld is chosen to maintain the favorable properties of continuous reinforcement.

Ultrasonic Welding.   Ultrasonic welding is a bonding process that uses high-frequency mechanical vibrations, that is, ultrasonic vibrations, as a source of energy. Heat is generated by a combination of surface and intermolecular friction.

Joint design influences the quality of ultrasonic welds, where the most important factors are a loose fit and providing an energy director. A slip fit is essential because the welding process depends on movement between the two parts as well as friction and high

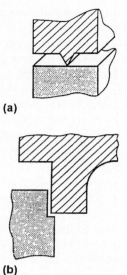

(a)

(b)

Figure 2.10 Forms of energy director. (*a*) Projection. (*b*) shear.[158]

pressure. The energy director is a small, triangular ridge, typically 0.25–0.8 mm high depending on the material. The energy director concentrates the applied power to provide rapid melting of the material contained in the director. Molten material from the energy director flows across the joint interface and fuses with the two components to form a weld. Figure 2.10*a* illustrates a simple butt joint modified with an energy director. A shear joint (Figure 2.10*b*) provides a small initial contact area and then a controlled interference along the joint as the parts collapse together.[146,158] Pressure is applied to the parts being joined, and a welding horn is applied to the area to be bonded. The horn delivers high-frequency (20- to 40-kHz), low-amplitude (20- to 60-Mm) vibrations which are concentrated by the energy directors, localizing heating and joining the thermoplastic parts. Welding time is normally 3–4 s, with parts held in place to cool for a total cycle time of 10–15 s. This has been done successfully with Gr/PEEK composites, and the lap shear strengths of the APC-2 material were virtually equal to the strength exhibited by compression-molded laminates (78.4 MPa).

The ability to weld a component using ultrasonics is governed by the design of the welding equipment, the physical properties of the material to be welded, the design of the components, and the welding parameters. For continuous-fiber composite materials, the main barrier to the use of ultrasonic welding is the difficulty of providing ultrasonic energy directors on sheet components and the consequent risk of fiber disruption at the interface under the high deformation necessary to obtain a satisfactory bond.[155,158–160]

The size and power of the welder limit the area that can be bonded in one operation. Large parts can be joined by sequential or scan welding. In *sequential welding,* a bond is made, the part is indexed, and adjoining sections are welded until the part is fully joined. In *scan welding,* parts travel under the welder at a constant speed until bonding is complete. In both cases, lap shear strengths have been realized that are equal to 80% of those of compression-molded control laminates.[161]

2.4.3.3 Electromagnetic welding.   Electromagnetic welding uses micrometer-sized particles of iron oxide, stainless steel, ceramic, ferrite, and graphite that respond to the radiofrequency (3- to 40-MHz) magnetic field. These opaque powders or inserts must be molded in the polymer matrix, and they remain in the final weld. As the high-frequency field passes through the magnetic materials, they are induced to become hot. The surrounding polymer melts, forming the bond. The induction coil must be located as close to the joint as possible if rapid bonds are to be made, and nonmetallic tooling must be used for alignment. Sandwiches, non-thermoplastic-coated composites, and reinforced, coextruded layers can be bonded. Some are bonded by incorporating magnetic particles in hot melts, liquid adhesives, or films at the joint area. Lap shear values of 26 MPa have been obtained.[162] Electromagnetic welding is quick and efficient, and industry is becoming increasingly aware of its potential, especially in joining TPCs.[16]

Induction Welding.   Induction welding is based on the principle that a magnetic field passing through a conductor generates eddy currents that dissipate in the form of heat. If the conductor is also magnetic (graphite fibers, for instance, are conductive but not magnetic), hysteresis losses also dissipate as heat. In practice, a thin layer of metal-filled adhesive or a metal screen laminated with a thermoplastic film is placed at the joint line. This metal susceptor localizes heat in the joint area once the magnetic field is applied. Lap shear strengths of 42–49 MPa have been reported for Gr/PEEK laminates welded in this fashion.

For application to plastics that have neither high conductivity nor high permeability, radiofrequency fields are generally applied to tapes of thermoplastic filled with iron oxide[163] or metal particles which are heated by the field. These tapes are placed at the interface, and upon application of a suitable radiofrequency field the particles cause the thermoplastic to fuse and create a weld. Coils made from copper tubing are connected to the output from the induction generator, and most of the field is generated in the coil. Coil design is thus an important parameter for induction heating. By incorporating empirical and computer modeling techniques, it is possible to optimize the work coil design to achieve maximum eddy current generation in the vicinity of the weld.

Some PEEK, PPS, and PEI composites of graphite without metal susceptors have been successfully induction-welded with lap shear strengths of 38.5–48.3 MPa by using APC-2 laminates and PEEK film at the interface. Some Gr/PPS composites have been welded with lap shear strengths of 23.8–30.8 MPa, whereas a Gr/PEI composite has been joined with lap shear strengths of 31.5–14.3 MPa.[163]

A new 5-kW induction welder developed for joining thermoplastics is said to significantly reduce cycle times and to have the ability to simultaneously weld large parts and multiple bond lines. The unit features the process of thermoplastic welding, which uses electromagnetic induction welding and special material to produce structural, hermetic, or pressure-tight welds with a reportedly very high degree of reliability.

Induction welding can use the amorphous bonding technique,[164] which involves making a joint by melting a layer of the amorphous polymer PEI at the weld line. (The same principle is used in resistance welding, which is discussed later.) PEI has a melting temperature of approximately 270°C, whereas PEEK has a melting temperature of 345°C, and this difference allows a joint to be made by melting the PEI while leaving the PEEK composite still fully consolidated. Induction welds using the amorphous bonding tech-

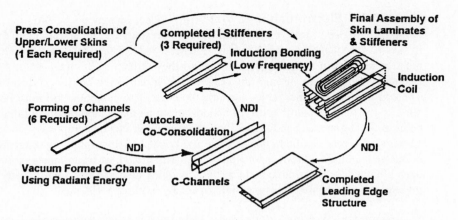

Figure 2.11   Despite requiring 14 separate manufacturing steps, Grumman Aircraft systems demonstrated that an aircraft stabilizer could be produced more economically from thermoplastic prepregs with electromagnetic induction bonding than with one-step autoclave cocuring of a thermoset.

nique have achieved lap shear strengths of 40 MPa. A refinement of this technique uses ferromagnetic particles (usually iron oxide) dispersed in a thermoplastic matrix preplaced along the joint in the form of a paste or tape. Welding times are short, up to 20 s for the largest components. The process can be used to join almost all thermoplastics, but the insert limits the joint strength.

An interesting case study demonstrated the advantages of electromagnetic induction bonding over autoclaving in joining TPC sections. Grumman fabricated an all-thermoplastic horizontal stabilizer out of prepregs of carbon fiber and PAS-2 polyarylene sulfide resin (known commercially as Ryton S). The thermoplastic was compared in two ways, one with conventional autoclave bonding of the outer skins to the I-beam stiffeners and the other with induction bonding of the two using 3M's AF-191 amorphous thermoplastic film adhesive and a vacuum bag to apply pressure. Induction heating acted directly on the carbon fiber reinforcements, eliminating the need for any metal susceptor material.

Grumman was quite encouraged that the induction-bonded thermoplastic component showed slightly greater bond strength and 10% greater stiffness than its autoclave-bonded counterpart (Figure 2.11).

Resistance Welding (Resistive Implant Welding).   *Resistance welding* (RW), also called *resistance implant welding,* is based on the principle of trapping a metal insert between two parts to be joined and then heating the insert by induction or resistance heating so that the plastic material around the implant fuses to form a joint. The heat causes the surrounding plastic material to melt, and the weld is effected by subsequent cooling under pressure. The conductive implant remains within the joint and as such affects the final strength of the weld.

Prepreg carbon fiber tape, specifically unidirectional fibers set in PEEK, can be used as a compatible implant material; a metallic implant would introduce foreign material into the thermoplastic joint. Welds of up to 127 × 25 mm can be achieved with a lap shear strength of up to 50 MPa. Because welding times are short, up to 20 sec for the largest components, resistance heating offers other advantages as a bonding technique for TPCs. It requires little surface preparation (sanding of surfaces), costs are low, and the process is relatively simple and fast. Joints can be consolidated in minutes at room temperature, compared to hours at 177°C for many epoxy adhesives.[125,151,154,165–168]

Fusion bonding of APC-2 by RW and compression molding has been investigated.[166–168] Lap shear strength from resistance-welded samples rapidly increased and asymptotically approached the compression-molded baseline in much less processing time. Resistance-welded APC-2 with PEEK film yielded a maximum lap shear strength of 44.8 MPa, with fiber motion occurring at the joint. The APC-2/PEI resistance-welded specimens achieved a maximum strength of 35.9 MPa, with no fiber motion detected. The baseline for APC-2/PEI was determined by isothermal processing (1200 s at 293°C) to be 43.6 MPa.

Recently completed studies and investigations have shown the following.

1. Within a fairly wide processing window, RW can reliably and consistently provide strengths approaching the compression-molded baseline; bond strength is observed to increase rapidly with time in the melt for a given power level; and the time required to melt (t_m) and the time in the melt required to achieve a desired strength decrease with increasing power levels.[169]

2. The joint strength of welded Gr/PEEK TPCs is extremely sensitive to the non-isothermal process history.

3. Design methodology that combines traditional stress analysis with a nonisothermal process model for joint strength allowables can predict the performance of a skin or a core structure.[170]

4. A process for RW of APC-2 composites has been developed with the aid of thermal analysis using two-dimensional finite element modeling. The numerical prediction was used as a guide in selecting the processing parameters, and single-lap shear testing and microstructure viewing were employed in bonding evaluation. Resistance-welded joints achieved a shear strength of 33.9 ± 2.3 MPa and retained this strength after hot-wet conditioning. The fiber orientation adjacent to the bond line layer was found to have an influence on the failure mode and thus the shear strength.[171]

Dielectric Welding.   Dielectric (radiofrequency) welding involves the use of radiofrequency radiation (27 MHz) to activate polymers that have a dipole and causes rapid oscillation of polymer molecules. Frictional heat from this movement causes the polymer to become molten. The process differs from induction welding in that no conductive materials are involved and that it operates at a higher frequency. Dielectric welding has been used mainly in sealing operations.[146]

Other Welding Techniques

Heated Press Joining.   Tests by Cantwell et al.[172] were undertaken to examine the possibility of joining carbon fiber-reinforced thermoplastic at temperatures below the recommended processing temperature of 380°C. The results showed that by employing a sufficiently long contact time, it is possible to obtain very high joint strengths at temperatures just above the melting temperature of the PEEK matrix. This heated press technique thermally joined carbon fiber-reinforced PEEK with lap shear strengths in excess of 50 MPa. Investigation of the variation in joint strength with welding temperature showed that with a suitable choice of contact time, it is possible to join C_f/PEEK successfully at 350°C. Below this temperature the degree of molecular interdiffusion appeared to be limited, and the resulting average lap shear strengths were low.

Polymer-Coated Material (PCM) Welding.   This method produces joints that are said to be as strong as those achievable using adhesives, but it does so in less than a minute and by an automatic, continuous process.

In order to test the PCM technique, the Welding Institute[151,162,173,174] used PEEK reinforced with 61% continuous carbon fibers supplied as unidirectional prepreg tape at a thickness of 0.125 mm. Sixteen layers of tape were used to form a laminate structure 2 mm thick. In addition, a layer of amorphous PEI was comolded onto one surface. A grade of aluminum alloy deemed suitable for aerostructure fabrication was also selected. A thin thermoplastic layer was applied to the surface of the pretreated aluminum alloy to facilitate joining to the PEEK composite.

Resistance implant welding was employed because it is relatively simple to control and guarantees maximum heat input at the joint line. The implant used was a piece of unidirectional carbon fiber prepreg tape. PEI and PES films were introduced around the implant to provide more thermoplastics in the weld. These materials are compatible with PEEK and with the thermoplastic coating on the aluminum alloy, and they melt at a lower glass-transition temperature than PEEK. During welding, the glass-transition temperatures of PES (225°C) and PEI (230°C) were both exceeded well before the PEEK began to melt (at 345°C). Thus the added PEI and PES films, the thermoplastic coating on the metal, and the comolded PEI on the surface of the PEEK all melted together to form the weld before the PEEK was damaged by heat. Welds were achieved by applying direct current through the carbon fiber prepreg implant. A three-stage welding cycle was applied, and the combination of materials produced a single lap shear specimen with a maximum strength of 40 MPa.

The main advantage of PCM welding over adhesive bonding methods is the speed with which joints can be assembled. A weld time of only 30 s was necessary to achieve the maximum lap shear strength, and this time could be reduced. Adhesive bonding techniques, on the other hand, take as long as several minutes. The bonding technology can be used with any material to which a thin layer of thermoplastic can be adhered prior to welding.

PCM welding by induction heating has been used to join a thermoplastic composite (APC-2) to alumina, L113-grade Al alloy to alumina, L113-grade Al alloy to AlN, and SiC-reinforced borosilicate glass composite to itself. The joints manufactured between APC-2 and L113-grade Al alloy were tested to a maximum lap shear strength of 30 MPa. For this lap shear strength, they required a peak current of 40 A, a pressure of 2.67 MPa, and a time of 30 s.[174]

Work coils positioned a few millimeters above the materials to be joined generate eddy currents in the metal and consequently produce a heating effect in the materials. Welding pressure is provided by a vacuum pump mounted within the machine, and throughout the weld cycle the two materials are pulled into intimate contact by this means. The weld cycle lasts about a minute during which the thermoplastic material melts and forms a perfect seal which in this case is on the Al alloy with the help of a thin interlayer.

PCM offers a number of advantages over existing joining techniques:

- Speed—joints can be made in less than 1 min and require no curing.
- Simplicity—the process is particularly easy to use and has significant potential for automation.
- Improved quality control—the process is simple to monitor and control.[146]

2.4.4 Repair

In order to make composites fully competitive with other engineering materials, field maintenance and repair materials, equipment, and procedures have been developed and are being improved.

In-flight, in-field, and in-service use of advanced composites must be accompanied by the ability to repair them. Incorporating repairability at the design stage is the ultimate goal to be achieved by relatively easy-to-use repair systems and procedures that do not require substantial equipment. Making these systems more reliable, more shelf-stable, easier to use, and able to deliver a higher percentage of the original component's properties are today's targets.[175–190]

Information on the general use and selection of adhesives for specific uses such as ship repair and in the automotive industry is available in references 191–193. Also, previous studies have related the material properties of adhesives to joint performance.[194–197] English and Armstrong[198,199] did an extensive study relating adhesive properties to lap shear performance of joints. His report as well as others also expressed the need for standardized materials and for manufacturers to provide more complete material information.[200–202]

2.4.4.1 Design decisions.   Several criteria determine how a damaged structure is to be repaired. The extent of the damage and the type of structure, primary or secondary, indicate what is to be done. In most secondary structures, either a metal plate can be riveted or bolted to the composite or a two-part epoxy can be used with fabric reinforcement. These procedures are followed for parts with minimal load requirements. Bolted repairs are the most common for battle damage repairs because of their reliability. Alternative bonding methods require more rigorous control of surface treatments, storage of heat-sensitive bonding materials, and special equipment and/or expertise.[198]

If a major structure is damaged, the air frame manufacturer or the services will generally want to bring the aircraft or part back to its facility for repair. In most cases there are no established procedures to follow, and engineering projects have to be established to determine the validity of the repair. In most cases the original composite material is used in the repair. This can be done by precuring a patch and then bonding or fastening it in place. In other cases, if the part can be removed from the aircraft, the uncured patch can be built up and the part placed in an autoclave to cure. Most parts have been made originally with a 177°C curing system, which means that if the original material is used, a 177°C cure will be required.

Though prone to many idiosyncrasies, field repair environments generally restrict users to vacuum bag cure techniques (which yield low bonding pressures), localized heat sources such as heat blankets or pads, induction curing, infrared lamps or hot air guns, and generally poor storage conditions. The materials to be used must be room-temperature-storable for 1 year, be curable at 177°C or less within 2 h, and ideally be a one-component system. Typical repair information is given in references 175–207.

2.4.4.2 Process selection.   Xiao et al.[208,209] found that for repair of TPCs, fusion bonding offered significant advantages over conventional adhesive bonding and mechanical fastening.

From the viewpoint of microstructure, the fusion bond is created when molecules from one surface diffuse into another surface. Therefore the bond, if perfect, is as strong as the resin in the composite. Furthermore, fusion bonding does not introduce an environmentally sensitive weak link into the structure as an adhesive bond does. The bonding process is also much faster when compared to the cure cycle for an adhesive. When the percentage of fiber is high in the composite, fusion bonding is often accompanied by inserting thermoplastic resin films at the bond line. Thermoplastic films may also be used to bond dissimilar materials. No special conditions are required to store thermoplastic films, another major advantage when compared with the low temperature storage required by adhesives. In addition, fusion bonding requires little surface preparation.

Nevertheless, fusion bonding also has several disadvantages. First, it usually requires high processing temperatures which may cause warpage and even deconsolidation of the composite components if they are not properly supported. The Thermabond technique utilizes an amorphous polymer film as an intermedium so that fusion bonding can be performed at temperatures lower than the T_m of the matrix resin. However, the amorphous film must have previously been consolidated with the mating surfaces of the components at the molding temperature of the resin matrix. Thus the Thermabond process does not represent a solution to the deconsolidation problem in making repairs. Second, fusion bonding repair involves localized heating which results in thermal stresses and may cause buckling and cracking of laminated structures. Therefore the pressure required for fusion bonding is usually higher than that needed to cure adhesives, and the conventional pressure systems used for repair may not be sufficient for fusion bonding repair. In addition, local melting may produce detrimental structures in crystalline thermoplastics which will cause a loss of fracture toughness.[208–210]

The above problems can be minimized by using embedded surface-heating techniques. This allows a patching-type repair to be performed at a relatively low pressure. Table 2.10 shows that high bonding strength has been achieved by resistance heating and induction methods with pressure levels that may be achieved simply by clamping.

Induction heating is fast, efficient, and precise, and it has the potential for use in healing-type repairs as well as patching repairs. Resistance heating is equally fast and relatively simple. If current density, heating time, and consolidation pressure are controlled, it can produce lap shear strengths above 20 MPa in APC-2. Resistance heating is most adaptable for patching repairs. Induction and resistance heating also allow control of the

TABLE 2.10   Fusion Bonding Methods, Variables, and Lap Shear Strengths[209]

Bonding Method	Surface Treatment	Welding Time and Pressure	Lap Shear Strength (MPa)
Induction with susceptor	Acetone	1 min, 0.25 MPa	36.5 ± 4.7
Resistance	Acetone	60s, 0.4 MPa	33.9 ± 2.3

cooling rate, which affects the crystalline structure of a semicrystalline thermoplastic and in turn its mechanical properties.

Border et al.[211] also found that induction heating was fast. Portable power units were constructed that made it well-suited for aircraft repair. When a coil was bent to fit a repair patch, it could be used on any type of repair without the need for skilled scanning of the patch. This coil represents a significant step forward in induction heating technology, as complex curves and large areas can be repaired using this method. By eliminating the need for scanning of the surface, control is also enhanced.

A study by Davies et al.[212] examined two approaches to repair processing: fusion bonding and the application of conventional epoxy adhesives to TPCs. C/PEEK composites were successfully assembled and repaired by both techniques, although both had disadvantages. Adhesive bonding required an appropriate surface treatment, but the relatively low temperatures involved (180°C) minimized the risk of damage to surrounding structure. Fusion bonding produced strong, tough joints, but high temperatures were necessary locally.

The use of other thermoplastic film adhesives allows fusion bonding to be applied at lower temperatures without the disadvantages of conventional adhesives, but careful selection of the thermoplastic is necessary if solvents are likely to be present.

2.5 FORMING AND FINISHING PMCs

2.5.1 Introduction

Thermoforming is the process of shaping a heated thermoplastic sheet by applying positive air pressure, a vacuum, mechanical drawing, or combinations of these operations. Some of the older types of thermoforming techniques developed include plug assist forming, drape forming, and matched mold forming.

Currently, processes for thermoforming thermoplastic matrix composites are in various stages of development. Some, such as matched-die flow forming and stamping, have been proven capable of high-volume commercial production. Others, such as diaphragm forming, show promise of commercial success in the laboratory but have not been proven or optimized in large-scale operations. A number of processes are being used in the aerospace industry for limited small-volume production. These are compression molding, diaphragm molding, the Therm-X (SM) process, and rubber block molding (Table 2.11).

2.5.2 Thermoforming Processes

A distinction should be made between the thermoforming of unreinforced thermoplastic sheets and continuous fiber-reinforced thermoplastic laminates. When a thermoplastic sheet is thermoformed, the melted sheet thins out and stretches to conform to the contours

TABLE 2.11   Thermoplastic Forming Methods and Equipment[213]

Method	Advantages	Disadvantages
Vacuum forming	Low pressure	Low force, impractical
Matched-metal die press	Close tolerance	Thickness mismatch
	High forming load and pressures	Friction at die interface
		Nonuniform deformation and pressure
		Long heating and cooling times
		High fabrication cost
Hydroforming	Only one solid die, undercuts and high pressure	Low temperature, no peripheral equipment
Diaphragm forming	Good fiber placement control	Long cycle time
		Temperature limitations
Rubber pad forming	Only one solid die	Lack of complex details
	High pressures	Temperature limitations

of the mold. The initial skin thickness and the depth of the draw determine the final part thickness. The surface area of the molded part is usually much greater than the initial surface area of the resin sheet, which results from the material stretching to cover the mold.

The concept of the laminate thinning out and stretching is not valid for continuous fiber-reinforced composites. Before the tool closes, the laminate is released from the clamp frame and allowed to lie on the lower tool. As the tool closes, the laminate slips into the tool from the edges to cover the contours. The thickness of the laminate does not change during the forming operation. *Drapability* and *conformability* are the preferred terms to use when describing the ability of a fabric to form to a contoured surface. As the fabric is forced to conform to the mold contours, the weave pattern becomes slightly distorted and allows the fabric to drape the surface. Movement and slippage of the warp and weft fibers relative to each other account for the ability of a fabric to conform to contours. The factors that determine a weave's ability to conform are characteristic of the particular weave design.

The thermoforming process has three key elements:

1. A laminate support frame which carries the laminate into the heat source, supports the laminate while and after the matrix melts, rapidly transfers the melted laminate from the heat source to the forming tool, and then releases the laminate onto the lower tool.

2. A heat source capable of evenly heating the laminate to its processing temperature in a reasonable period of time.

3. A forming tool capable of rapid closing speeds with sufficient clamp pressure to form the laminate.

2.5.2.1 Heating.   Heating is required in all thermoforming processes. The following methods are available.

• Infrared heating is a good choice when heating thin, well-compacted flat sheets or gentle contours in a continuous oven feeding a molding process. It is widely used in sheet thermoforming. Under the proper conditions, it is fast and effective, but it is difficult to obtain uniform heating of highly contoured parts.

- Convection heating is a more effective method than IR in penetrating plies and preventing overheating. Impingement heating is a variation of convection heating in which a multitude of high-velocity jets of heated air impinge on the surface of the heated article, greatly increasing the rate of heat transfer and reducing the length of a continuous oven.

- Contact heating results from intimate contact between the item being heated and the heating surface. It is capable of faster heating rates than convection or impingement heating but is not as convenient to use in a continuous heating line.

- Radiofrequency heating can be a very effective method of heating, but it is highly dependent upon part composition and geometry, requires specialized design and setup experience, and must be shielded to protect workers from RF radiation and to prevent interference with radio communications. Induction and dielectric heating are very similar in operation to RF heating but operate at lower electromagnetic frequencies.

2.5.3 Compression Molding (Matched-Die Press Forming)

Compressing molding thermoforming processes have had a history of commercial success in the auto industry over the past decade or more and probably will continue to be given first consideration in new applications. At least three variations of compression molding are used to thermoform fiber-reinforced thermoplastics. These processes typically employ a compression molding press and a matched pair of molds. Starting materials may be in the form of rigid, consolidated or unconsolidated sheet or stacks of rigid plies. Mold halves are aligned and mounted on parallel upper and lower platens of a compression molding press. The laminate to be formed is positioned and clamped around the edge with a clamping pressure between 138 and 276 MPa. The clamping of the edge is necessary to maintain tension on the fibers and consequently prevent wrinkling or buckling of the laminate during the forming process. The blank (the starting sheet or stack of plies) is heated and positioned between the mold halves. The press closes the mold rapidly, forming the blank to the shape of the closed mold cavity before it is cooled below its forming temperature by the mold. The mold is then opened, and the molded part removed and trimmed if necessary.[213,214] The schematic in Figure 2.12 illustrates the process with a matched mold in the pressure forming stage.

2.5.4 Stamping

Stamping is a variation of compression molding that is similar to the sheet metal forming process bearing the same name. The blank is a single sheet or a stack or plies cut to almost the exact shape of the cavity opening and to about 95–98% of the size of the cavity opening, so that virtually no flow is required to fill the cavity, and the blank and the final part are of virtually the same thickness. This variation is like most thermoforming operations in the sense that the final part is basically the original sheet formed to a different shape with very little rearrangement of the material in the sheet. A principal advantage of this method is that highly ordered unidirectional fibers tend to maintain their original

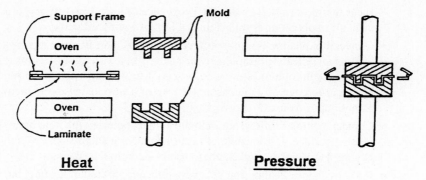

Figure 2.12   Matched mold for thermoforming.

spacing and direction in the final part. A disadvantage is that there is no built-in mechanism to prevent the heated sheet from buckling when it is draped on the mold and the mold is closed. As a result, this variation is limited with regard to the complexity of contours that can be formed. Compromises can be made by preforming the blank in some manner, adding flowable material in areas where increased thickness is required, and so on.[214]

2.5.5. Flow Molding

Flow molding is a variation of compression molding that is similar to stamping. The principal difference is that the blank is a single sheet or a stack of plies much smaller in area and thicker than the molding and is often cut into simple shapes such as rectangles. Thermoplastic composites developed for this process, such as Azdel, can typically flow up to 50%.[215] Typical molding pressures are 10,300–17,200 kPa and have been reported to be as high as 48,300 kPa.[216] Flow molding is capable of forming complexly contoured parts and can accommodate, in a single part, large variations in part thickness such as molded-in ribs and bosses as well as molded-in threaded holes, inserts, and the like.

2.5.6 Stretch Compression Molding

In unreinforced thermoforming, buckling is normally prevented by clamping the heated sheet in a frame or ring, forcing it to stretch as it changes shape. This same technique is applied to reinforced thermoplastic stamping in an effort to achieve more complex contouring without buckling. Among the several methods mentioned in the literature,[217–220] three are being studied actively.

2.5.6.1 Clamping ring with a stretchable sheet   Clamping compression molding can provide a stretching action in the sheet being formed by employing a stretchable form of reinforced thermoplastic composite sheet firmly clamped in a frame.[221] Useful forms of reinforcement include random chopped-fiber mats, swirled continuous-fiber mats, and DuPont's aligned discontinuous (or ordered staple) mats.[218]

2.5.6.2 Diaphragm Forming.

A thermoforming process called diaphragm forming is being used with reasonable success with difficult-to-form composites such as continuous carbon fiber-reinforced PEEK. It is often considered a modification of the billowing or pressure forming process. The key step in this procedure is placement of the sheet to be formed between two thermoformable or elastic diaphragms.[213,218,222–225]

This process can be carried out in an autoclave or in a hydraulic molding press fitted with the mating halves of a two-piece pressure chamber mounted on its upper and lower platens (Figure 2.13). A sheet or stack of plies is sandwiched between two somewhat larger sheets of diaphragm material (e.g., superplastic aluminum).

The diaphragms undergo a true stretching-type thermoforming action. However, the thermoplastic sheet, if reinforced with continuous fibers, cannot stretch and must slip with respect to the diaphragms. Therefore, the forming pressure is applied gradually, requiring possibly 10 min to reach final forming pressure, typically in the range 344–689 kPa.[226]

With heating, gradual pressurization, and cooling of the formed part and the mold all taking place in a closed apparatus, cycle time might range from 20 to 100 min.[213,218] A patented diaphragm molding process for thermoforming contoured parts from sheets of C_f/PEEK uses thin sheets of superplastic aluminum alloys, capable of deforming several hundred percent, as diaphragms.[218]

In ongoing investigations, Jar, Davies, Cantwell, et al.[227] studied autoclave forming of C/PEEK at 370°C and diaphragm forming at 385°C and observed that thermoplastic matrix composites allow a wide range of manufacturing methods to be considered in the production of engineering components. They also found that the potential problems associated with high forming temperatures and with the use of semicrystalline matrix materials whose structures may be sensitive to processing history do not arise, regardless of short- and long-term properties. Pantelakis, Tsahalis, et al.[228] performed similar diaphragm forming techniques and arrived at similar conclusions.

Autoclave-processed complex parts can be made from Avimid K thermoplastic composite material because of its unique tack and drape characteristics.[229] Laminate and sandwich skin panels, male and female C-channels, I-beams, and sine wave spars have been successfully fabricated using autoclave forming on low-CTE tooling. Efforts by

Autoclave Processing	Heated Pressure Chamber
1. Assemble diaphragms and laminate and clamp to mold.	1. Assemble diaphragms and laminate.
2. Load assembled mold into autoclave.	2. Load assembly into chamber and close press.
3. Heat, form, consolidate and cool.	3. Heat, form, consolidate and cool.
4. Remove mold from autoclave and remove formed part from mold.	4. Open press, and remove formed part from between diaphragms.

Figure 2.13   Diaphragm forming process.[218]

Baker and Gesell[229] will attempt in the future to concept automated fabrication in order to eliminate hand layup, which is one of the two major cost drivers associated with composite structures.

Forsberg and Koch[230] reported on the design and manufacture of thermoplastic F-16 main landing gear doors. Thermoplastic matrix composites left-hand main landing gear doors for the F-16 were fabricated using T650-42/Radel-C, a 177°C service thermoplastic matrix composite material system. These doors were full-form fit and function replacements for existing aluminum doors. They were essentially torque boxes formed by coconsolidating previously consolidated inner and outer skins. The inner skins were made using diaphragm forming. The outer skins were fabricated in an autoclave, and coconsolidation of inner and outer skins was accomplished using disposable mandrels. The finished assembly was 20% lighter than an aluminum door.

Thermoforming of glass-reinforced composites is coming. General Electric (GE) Plastics Polymer Processing Development Center is experimenting with thermoplastic sheet containing over 30% discontinuous glass fibers for an automotive application and is developing new thermoformable materials with even higher glass contents.[231] Meanwhile, Thermoforming Technologies, Inc., is developing a single-station pressure former with a profile heating capability to form reinforced thermoplastic sheet with up to 70% continuous fiber for automotive and aircraft applications.

Two government-sponsored design and manufacturing studies conducted by Lockheed Aerospace Systems Company[232] and Northrop Company[233] evaluated advanced thermoplastic matrix composite systems. The first program conducted trade studies to evaluate thermoplastics versus bismaleimides for various generic fighter components, and preliminary designs were developed followed by manufacturing verification and production demonstration tasks. These consisted of design, fabrication development, and testing of a thermoplastic F-16 main landing gear door component. The composite main landing gear door (Figure 2.14) with thermoformed details was calculated to weigh 19.14 kg. This is a weight savings of 20.2% compared to the originally designed aluminum door. The composite door was designed with manufacturability as one of the main criteria. Diaphragm forming was selected as a relatively mature process involving a short cycle time that has been successfully used in the fabrication of high-temperature thermoplastic parts and assemblies. Diaphragm forming using superplastic aluminum as the diaphragm material was used in fabricating the complex shapes from the APC-2 thermoplastic. Matched rubbermetal press forming of the inner skin was also selected, and the lower side bay bulkhead design specified diaphragm forming for the web and D-RB for the web assembly.

The goal of the second program, DMATS,[233] was to design and develop a low-cost method for manufacturing thermoplastic structures for an F-15E forward engine access door (a secondary structure) and for an advanced fighter fuselage structure (a primary structure). The autoclave diaphragm forming concept was used to produce bulkhead frames, webs, channels, upper and lower caps, and amorphously bonded stiffeners for the primary fuselage structure. Other thermoforming studies and programs are described in references 213, 218, and 234–237.

Several engineers have experimented with variations of a double-diaphragm process forming and consolidating parts in one operation, including a method similar to thermoforming that utilizes a plug assist.[238]

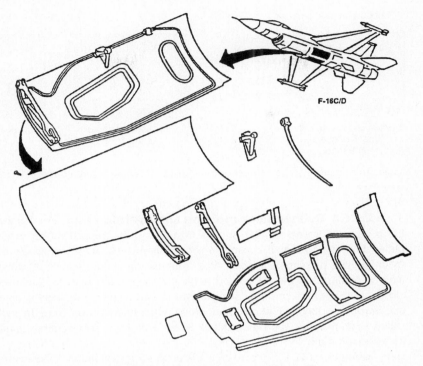

Figure 2.14   Composite main landing gear.[232]

2.5.6.3 Therm-X (SM) process.   The Therm-X (SM) process was developed and has been commercialized by United Technologies Chemical Systems Division.[239,240] It is an extension of the basic idea used in the "trapped rubber" molding process, in which molding pressure is applied to a prepreg preform on a mold half by thermal expansion of a rubber pad confined in a pressure chamber with the mold and preform. In the Therm-X (SM) process, which is carried out in a Thermoclave pressure vessel,[241] the thermally expandable process medium is a flowable elastomeric powder rather than a rubber block. This eliminates the severe pressure gradients experienced in the trapped rubber process and allows the safe generation of high pressures at high temperatures. The Thermoclave vessel is a specially designed autoclave[242] which contains a means to support the mold half, a set of tubing coils to heat and expand the powdered silicone rubber process medium, and localized electric heaters to generate the cure temperature in the immediate vicinity of the part being molded. The Therm-X (SM) process is useful primarily in the compaction and curing of difficult-to-cure, contoured thermoset advanced composites, but the method is capable of high-temperature compaction and forming of high-temperature thermoplastic advanced composites such as C_f/PEEK. For the future, process capability of 816°C and 20,700 kPa pressure has been planned, along with vessels up to 5.1 m in diameter.

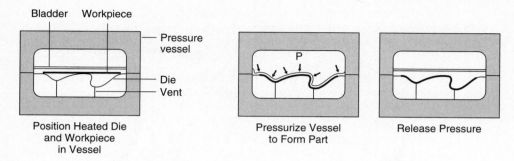

Figure 2.15   Hydroforming schematic. (Courtesy of ASEA Metallurgy.)[213]

2.5.6.4 Rubber block molding and hydroforming.[243]   Rubber block molding is another process similar in some respects to the trapped rubber process. However, rather than using thermal expansion of a rubber block to pressurize the molded part on the mold half, pressure is applied via a compression molding press. A rubber block, gradually spreading its area of contact with the sheet-mold half combination as the press closes, generates a forming action with a reduced tendency to wrinkle areas of the sheet that must be formed by lateral compression rather than by stretching. In another variation, called *hydroforming,* the rubber block is replaced by a heavy rubber diaphragm backed by hydraulic fluid[213,218,238,243,244] (Figure 2.15).

Researchers at Delft University[245] processing continuous fiber-reinforced thermoplastics have adopted a technique known as *rubber forming* in order to overcome a key drawback of more common matched-metal forming. The latter results in a high degree of detail but also is prone to nonuniform pressurization of the composite laminate, owing to the rigidity of the dies and perhaps to the nonuniform thickness of the sheet. The Delft system replaces one of the rigid metal dies with a flexible rubber die, resulting in more uniform pressure distribution. The rubber die does not have to be perfectly matched to the steel half; sometimes it is advantageous to under- or over-dimension the rubber half.

Experimentation has been accomplished with closing speeds of 0.099 m/s in a fully automated rubber forming press with 20.24×10^2 kg force and a forming area of 175.4 × 100.7 cm.

2.5.6.5 Other processes.   There are a variety of other processes in use or being studied that are more specialized or not publicized for one reason or another. Imperial Chemical Industries (ICI), for example, uses a roll forming process, very similar to a hot ingot metal roll forming process, to make long, straight, uniform cross-sectional shapes such as channels and stringers.[246] Basically, roll forming contours the sheet by simple bending and does not form compound contours.

Incremental forming[247] is a process being studied that can create large thermoplastic composite parts with relatively small equipment. It is compression molding utilizing an oven, a press, matched molds, and a transfer system. A long laminate, such as a wing skin, is molded incrementally in short lengths as the sheet being formed is indexed lengthwise through the oven and the press. The oven and the press are only the size of the section to be formed instead of the size of the entire part; thus part length is not limited by equipment size. Modular molds can change shape quickly between sections, eliminating time-consuming mold changes. Five nonidentical sections have been molded in 30 min. The potential advantages are particularly in the areas of equipment size and capital cost, labor and operating costs, and maximum part length. This technique has been proven on a prototype scale producing relatively small parts.

Vacuum forming, which is conventionally used in thermoforming unreinforced thermoplastics, is also applicable to thermoplastic matrix composites. When vacuum thermoforming, there are limitations to the types of parts that can be formed, and this technique is usually applicable to gentle curvatures with shallow draws. With either pressure application technique, plug assist or vacuum, minimizing the time from when the laminate leaves the heat source until the press is fully closed is critical. The laminate begins to cool as soon as it leaves the heat source, and if not formed quickly, it will begin to stiffen and lose its drapability and formability.[213] This transfer time is especially important with thin laminates of only one or two plies. Specific times depend on laminate thickness; however, most thermoforming equipment can make the transfer and close the tool in 5–10 s, which has been found to be adequate for PPS-based material.[248,249]

Yamaguchi, Sakatani, and Yoshida[250] examined press molding of AS4/PPS and IM7/PEEK materials and thermoforming (continuous bend forming) of HM/PEEK composite beams for future large space structures. The process forms composite beams from thin, flat sheet to hat sections continuously with multistep section rolls.

A new inner pressure molding method[251] has been developed for producing pipes with excellent workability from a carbon fiber-reinforced composite prepreg using thermoplastic matrix composite and carbon fiber while retaining the characteristics of the raw materials. Forming methods for these materials included the roll forming method and curved forming methods, but there was no forming method for producing pipes accurately. The new molding technology was developed by employing a carbon fiber-reinforced composite prepreg using PEEK as the matrix. In principle, pipes were molded using other thermoplastic resins such as PPS and nylon 66 as the matrix. Carbon fiber or glass fiber was used as the reinforcing composite, and the technology is also applicable to the production of woven and knit fabrics for use in molding.

Since the pipes are produced in the mold, they have excellent outside-diameter dimensional accuracy, surface smoothness, and virtually no voids. The pipes are remarkably strong and reliable and can have special sectional areas.

This new technology uses a thermosetting plastic that greatly improves the formability, eases forming control, and allows pipes of complicated shapes to be molded uniformly. The most important characteristics are that the pipes produced have (1) excellent toughness and strength provided by the raw materials and compatible properties, (2) excellent compressive fatigue strength and great chemical stability, (3) excellent mechanical properties at low temperatures, net shape molding, excellent dimensional accuracy, and uniform consolidation, and (4) excellent productivity in mass production, making a substantial reduction in manufacturing costs possible.

These pipes have a broad range of applications including space development (space station truss structures, manipulator arms, etc.), aeronautics (energy absorption structural materials, wheel legs, torque tubes, etc.), leisure and sporting goods, and vehicles (sports cars and linear motorcar structural materials).

Another Japanese firm[252] has developed a fiber reinforced composite material that can be melted by heating and can produce structures with complicated shapes.

Conventional composite materials usually use a thermosetting resin and are inconvenient because reshaping and forming into complicated shapes are quite difficult since the material is brittle. Asahi Chemical Industry Company, Ltd., has mixed a fibrous heat melting resin with reinforcing fibers such as carbon and glass fiber and produced a fiber-reinforced composite material that is flexible and can be bent in any direction. This material is called web interlaced prepreg (WIP) and uses carbon fiber or glass fiber as the reinforcing material. These fibers are aligned in parallel and interlaced at the individual filament level with short fiber resin, such as nylon 6, nylon 66, and PPS, PEEK, depending on the specific use. Shaping WIP at high temperature melts the resin and causes it to wrap the reinforcing material (such as carbon fiber) to provide a very strong composite material. PEEK and other engineering plastics are ideal for large products such as aircraft structural components. Finally, thermofolding is a technology that involves the heating of a laminate along a line along which the product is subsequently folded.[253] The thermofolding technique can be applied to both solid laminates and sandwiches. The solid laminate folded part has been used in a prototype electronics housing box where the consolidated laminate consisted of glass fabric layers with a PEI matrix.

With thermoplastics, a fold can be obtained without milling or bonding. Heat is applied to one of the skins, and after sufficient heating the panel can be folded. The forming operation is very quick, limited only by the heating and cooling rate of the material. The strength of a thermal fold is sufficient for nonstructural applications such as aircraft interior panels and nonstructural aircraft exterior parts.

Prototype aircraft interior panels have been made by thermofolding the sandwich along all four edges, the maximum length being 1 m. For aircraft interior panels, fire safety is of prime importance. Thermoplastics with continuous-fiber reinforcements such as glass-PEI laminate or sandwiches with this material comply with federal requirements for heat release in postcrash cabin fires.

Fokker Special Products[254] has produced Kevlar-PEI ice protection plates for the Dornier 328 turboprop since 1991 using thermofolding. A special heating system is applied to the part, heating and folding a 6-mm-wide flange over a length of 2 mm in a matter of seconds.

Research at Stanford University on dieless forming, as reported in *Plastics Technology* in March 1991,[231] has shown that a small machine can form large components from continuous-fiber materials. The process is similar to roll forming of metals.

2.5.7 Finishing PMCs

Exterior finishes on composite structures for external application on high-speed, modern jet aircraft may range from a simple primer application, which is top-coated after installation on the aircraft, to multilayered, specially prepared finishes.

As an example, consider an exterior component consisting of fiberglass prepregs and Nomex honeycomb which is required to be completely electrically conductive. Since fiberglass is not a good electric conductor, the finish applied to the product must provide the conductivity required.

The most important operation in obtaining a quality finish is to ensure a completely clean surface prior to finishing. Finishes on assemblies are usually more than cosmetic, and most play a role in the structural and functional performance of the assembly.

2.5.7.1 Coatings removal.   Paint coatings are used on military aircraft for many reasons. They act as protective layers for the substrates to which they are applied and also serve as a means for visually camouflaging aircraft. For composite materials, paints act as barriers protecting against environmental conditions and ultraviolet radiation.

Paint removal is a necessary part of aircraft maintenance and is required in order to check for corrosion on metal and to repair composite structures. In many cases, paint is removed for purely cosmetic reasons or during a change in camouflage pattern. In the past, paint removal was primarily accomplished using chemical stripping agents, which resulted in health hazards, waste disposal problems, and incomplete stripping jobs. Furthermore, many chemicals were found to degrade the organic matrix present in Gr/Ep composites and weaken the material.

Plastic media blasting (PMB) for use on Gr/Ep composites was investigated and compared with chemical and hand sanding removal to determine the potential for damage when removing paint from these materials. The research was conducted by the U.S. Air Force. The potential for streamlining stripping operations (i.e., using PMB on metal and composite airframe surfaces), for reducing health hazards, and for eliminating the potential for damage to composite materials caused by chemical strippers are key reasons behind the interest in a more complete understanding of the effects of PMB on composite materials. PMB was found to be a viable and safe paint removal method for use on Gr/Ep composites provided that special precautions are taken.

The following results from the Butkus, Behme, and Meuer[255] study show that PMB is a less damaging method of paint removal for Gr/Ep composites than is hand sanding. The overall effect of PMB paint removal on the mechanical and physical properties of Gr/Ep composites is negligible if the procedure is carried out using the "primer as a flag" criterion.

Using the Primer as a Flag.   The most important aspect of the approach taken in the PMB stripping tests was the use of the epoxy primer coat as a stopping point. Unless otherwise noted, the PMB operator used the epoxy primer coat as a visual cue to aim the blast nozzle away from an area that had already been stripped and onto another area that required stripping. This resulted in a mottled appearance of the remaining primer. However, it also minimized the amount of time the PMB blast was aimed at a single area of a surface as well as the possibility of degrading the mechanical properties of the composite by preventing damage to its surface. This technique was described as "using the primer as a flag." No internal damage to the Gr/Ep test panels was caused by either the hand sanding or the PMB method. However, considerable surface erosion and surface fiber damage were caused by the sanding methods investigated and can be caused by PMB if the primer as a flag criterion is not used. Surface damage results in a loss of mechanical properties of stripped composite materials. Using the primer as a flag criterion significantly reduces the potential for causing surface damage and subsequent degradation in the mechanical properties of composites during PMB stripping operations. No direct relationships were discovered between the amount of mechanical property degradation and specific PMB parameter combinations or between mechanical property degradation and the specific PMB media types used in the stripping operations. The results and conclusions concerning the effect of hand sanding and PMB on Gr/Ep composites should be applicable to other types of organic matrix composites.[255]

2.6 MMC MACHINING

2.6.1 Introduction

The machining of MMCs is not difficult but has different machining requirements than that for other materials. HSS tooling cannot be used for MMCs, and carbide tooling, either plain or coated, has limited capabilities. In general, carbide tooling will suffice if PCD or CVD tooling is not available, but only for very limited usage.

PCD or CVD is the most cost-effective and by far the best choice for machining MMCs. The larger the grain size, the greater the resistance to wear. Experimentation with 5-μm, 17-μm, and 25-μm grain sizes has proved that the cut edge of the tool exhibits less wear on the larger grain size. GE 1500 was the PCD grade insert that had a 25-μm grain size, and it outperformed the smaller grain size.[256]

The machining characteristics of SiC-reinforced Al MMCs are different than those of unreinforced Al alloys. Very hard SiC ceramic particles in a relatively soft matrix are abrasive, creating higher temperatures between tool and workpiece and thus faster tool wear. Modified machining parameters and special tool materials (e.g., compacted diamond or carbide), however, produce close tolerance, intricate parts with a very good finish, and surface integrity.

2.6.2 MMC Machining Behavior

Numerous studies, analyses, investigations, tests, and examinations have been conducted and reports presented on the costs, economics, benefits, and so on, of machining fiber, particulate, and whisker MMCs on a prototype or production basis and comparisons to the conventional machining of steels, aluminum and magnesium alloys, ceramics and titanium.[256–274] MMCs are considerably more difficult to machine than most conventional metals and alloys.

In an economic study on machining MMC parts, Schmenk and Zdeblick[257] found that laboratory turning and milling data for MMCs containing up to 15% volume of dispersed Al_2O_{3p} or up to 20% volume of dispersed SiC strongly suggested that the machinability of these MMC materials was quite similar to that of A390 Al-Si alloy. A390 Al and the MMC material tested contain comparable quantities of very hard, severely abrasive, dispersed particles. The oxide and carbide MMC particles and the somewhat larger silicon particles in hypereutectic A390 rapidly abrade HSS and even carbide cutting tools. Rapid tool wear, however, is much less of a problem if PCD tools can be used to machine these materials. Comparisons of machining data for these MMC materials from in-house and outside sources with the extensive data available on A390[258,260,261] indicated that the machining behavior of the MMC materials was quite comparable to that of A390. This finding is important because there are a number of well-documented cases where

A390 is economically viable for the production of high-volume components such as engine blocks and cylinder heads.

MMC disk brake rotors offer unique possibilities compared to conventional cast iron disk brake rotors. From the viewpoint of product cost, MMC rotors may well prove to be less expensive to produce than cast iron rotors for the following reasons.

- The estimated machining cost for MMC rotors is $1.57 per rotor compared to $1.72 for a cast iron rotor. This conservatively estimated savings of $0.15 per rotor (9%) does not include the expected casting cost savings for MMC materials.
- The estimated machining production rate for MMC rotors is 3131/day compared to 1900/day for cast iron rotors, an increase of ~65%.
- Tooling expenses dominate the machining cost issue for MMC materials. Expected improvements in the performance/cost ratio of PCD tooling would result in additional machining cost savings for MMC.

2.6.3 MMC Turning

Norton Diamond Film Corporation[274] performed turning tests on 6061-T6 Al composites reinforced with 15 vol% Al_2O_{3p}, comparing the wear resistance of two PCD grades and CVD thick-film diamond. The cutting speed was 680 m/min, feed rate 0.12 mm/r, depth of cut 1 mm, tool entrance angle 60°, and rake angle 6° positive. The tools were inspected at 5-min intervals during machining, and it was found that crater wear on the rake face was considerably less for the CVD tool, which may be because of the higher thermal stability and hardness of the CVD diamond. In turning, a negative and a positive rake on the PCD insert seem similar in cutting ability. The positive rake, however, produces a smaller chip and a better surface finish and requires less horsepower. As a rule in turning, rough machining generally occurs at a 2.2-mm depth of cut and a 5.1-m/s feed rate, and finishing at a depth of cut of 0.305–0.76 mm and a feed rate of 0.09–0.13-mm depending on the microfinish requirement. Microfinishes of 16–30 are routinely maintained. It should be noted that the finish appears much better to the naked eye than in readings taken by instruments.

Finally, it should also be noted that it is sometimes virtually impossible to run the turning equipment at the recommended surface finish per minute. This is especially true turning smaller diameters because conventional toolroom lathes cannot obtain the higher RPM that is needed. In roughening operations, one should avoid exceeding one-half of the PCD length or fracturing of the cutting edge may occur. It is crucial that adequate shielding and chip guards be in place while machining because of the higher chuck revolutions that are required.[256]

A machining cost comparison for 15 vol % SiC_w/SXA-2124, which included turning, was conducted by Berlin.[262] The study produced a cost model indicating that SXA required a little more than half the cost to machine than titanium and slightly more than twice that for aluminum. The only major difference between the machining cost of SXA and that of aluminum is the raw material cost.

PCD tooling has been effectively used to drill, bore, and turn 2080/15 vol% SiC_p material used for disk-brake rotors for automobiles. It appears that machining discontinuous reinforced aluminum (DRA) composite rotors is cost-competitive with cast iron rotors. In part, this stems from the high productivity rates and the fact that balancing and gauging are not required for the lighter-weight DRA rotors. These rotors can also be cast closer to net shape, thus reducing the total amount of material to be machined.[268]

Looney, Monaghan, Reilly, et al.[272] conducted a series of turning tests in which a number of different cutting tool materials were used to machine an Al/25 vol% SiC MMC. The influence of the cutting speed on tool wear, surface finish, and cutting forces was established for each tool material. It was found that carbide tools, both coated and uncoated, sustained significant levels of tool wear after a very short period of machining. The best overall performance was achieved using a solid cubic boron nitride (CBN) insert, while the worst was encountered using a solid silicon nitride (SiAlON) tool.[272]

New whiskered ceramic cutting tools are now demonstrating practical performance benefits in widespread production use on a range of difficult-to-machine materials.

One successful grade now on the market, GTE Valenite's Quantum 10 (Q-10), uses reinforcing SiC_w which, add high edge strength, increased fracture toughness, good thermal shock resistance, and excellent hot hardness to an abrasion-resistant oxide base.

In an effort to reduce machining time, a die manufacturer substituted Q-10, the SiC_w cutter, for carbide in rough and finish turning operations on a heavily scaled flow turn die forging of D-2 tool steel with a hardness of 60 R_c. When eight dies were completed, 3 weeks of machining time had been saved. Other rough and finish turning applications involving landing gears, artillery shells, and print rollers including materials, their forms, hardness, and other properties are described in reference 275.

Weinert and Biermann[276] at the University of Dortmund (Germany) feel that the main problem for conventional machining, such as turning, milling, drilling, and so on, of MMCs is tool wear caused by hard and abrasive reinforcements. They evaluated these conditions in several Al MMCs using cemented carbides, coated cemented carbides, PCD, and a CVD diamond coating on cemented carbide and on Si_3N_4 ceramic.

They concluded that different cutting tool materials should be used for turning MMCs depending on the hardness of the reinforcement. When machining SiC_p/Al or B_4C_p/Al with cemented carbides, tool wear is produced by microcutting because the reinforcements are harder than the hard phases of the cemented carbide. To reduce tool wear, the grain size of the hard phases should be larger than that of the reinforcement. Tool wear is caused by microcracking and fatigue when turning δ-Al_2O_3-reinforced aluminum. Therefore, tool wear decreases with an increase in the toughness of the cemented carbide.

When a TiAlN coating is used on cemented carbide, tool wear can be reduced in the cutting of δ-Al_2O_3-reinforced aluminum. At cutting speeds up to $v_c = 250$ m/min, wear rate is relatively low. Although the tool life of TiAlN-coated cemented carbide is lower than that of PCD, the use of this coating can be feasible because of the cost of PCD tools, particularly geometrically complex tools like drilling tools.

For the application of PCD in machining SiC- or B_4C-reinforced aluminum they found that the tool wear rate decreased with an increase in the grain size of PCD. Also, a positive rake angle and polishing of the PCD insert reduced the wear rate. The lowest wear rate can be achieved with PCD inserts that include all three modifications.

The results for turning MMCs with diamond-coated tools were very promising. Compared to the machining of δ-Al_2O_3- or SiC-reinforced Al with uncoated cemented carbide, tool wear was reduced considerably by using a CVD diamond coating. In particular,

TABLE 2.12   Guideline Parameters[a] for Machining MMCs with Syndite PCD[277]

Machining Operation	Cutting Speed		Feed Rate		Depth of Cut	
	m/min	ft²/min	mm	in.	mm	in.
Turning						
Roughing	400–600	1300–2000	>0.2	>0.008	1–2	0.04–0.08
Finishing	500–800	1600–2600	0.1–0.2	0.004–0.008	<1	<0.04
Boring						
Roughing	400–600	1300–2000	>0.2	>0.008	1–2	0.04–0.08
Finishing	500–800	1600–2600	0.1–0.2	0.004–0.008	<1	<0.004
Milling						
Roughing	250–1000	800–3200	<0.4[b]	<0.016	<3	<0.12
Finishing	500–1000	1600–3200	0.1–0.2[b]	0.004–0.008	<1	<0.04
Drilling						
Roughing; 4–12 mm diameter, 0.2–0.5 in.	25–125	80–400	0.05–0.2	0.002–0.008		
Reaming (single blade)	30–80	100–260	0.05–0.12	0.002–0.005	< 0.25 (on radius)	<0.01

[a] Imperial units are approximate equivalents.
[b] mm/tooth.

the diamond coating on a Si_3N_4-ceramic substrate led to very low tool wear when used in the machining of SiC-reinforced Al.

Clark[277] conducted a series of machining trials to evaluate PCD as a satisfactory solution for turning, boring, milling, reaming, and drilling. The base material Clark studied was $Al/SiC/15-20_p$, and the SiC_p was 15 μm diameter The use of a standard cutting fluid or lubricant (5% water-soluble oil) was the norm, particularly in operations where swarf removal was restricted, such as in drilling and reaming. Otherwise, machining dry, particularly for roughening operations, was an option.

Table 2.12 gives guideline parameters for machining with SYNDITE. The lower speeds and feeds in each range should be used to establish satisfactory machining results.

Hung, Boey, Khor, et al.[278] validated several tool wear models and studied the effect of tool materials, particle distribution, and subsurface damage of SiC_p-reinforced MMCs and concluded that roughening with uncoated WC inserts and then finishing with PCD tools was the most economical route in machining SiC-reinforced MMCs. Cast MMCs exhibited higher machinability than powder-formed MMCs mainly because of the favorable shape and distribution of the particles but also because of the fabricating processes. Regardless of the cutting tool material used for machining, cracked SiC_p and a debonded matrix-reinforcement interface were found underneath a machined surface. Such machine-induced defects could be a concern when using MMCs in a critical application

In two studies of end milling with PCD inserts Lane [26,279] found that although the machinability of aluminum composites was primarily controlled by their ceramic reinforcements, the effect of the matrix alloy was also significant. The correlation between the hardness of the matrix alloy and the flank wear of PCD cutting tools is due to strengthening of the composite matrix rather than direct abrasion of the cutting tool by second-phase precipitates. Selecting the proper heat treatment condition for a MMC component can improve tool life by 100–350%.

Lane also found that $6061/Al_2O_3/15_p$ possessed significantly different tribological properties than other foundry composites, $A359/SiC/20_p$, $A383/SiC/20_p$, and $6061/Al_2O_3/$

15_p. The polishing action of this composite on the PCD tool edge extended tool life considerably and maintained a good surface finish.

The mechanisms resulting in dramatically reduced tool wear for the 6061/Al_2O_3/15p composite as compared to the cast composites are not fully understood. Certainly the lower particle V_f, uniform microstructure, and lack of free silicon improved the machinability of the wrought composite.

The second study showed, as expected, that the V_f of the reinforcing ceramic particles had a significant effect on reducing PCD cutting tool life. However, it must be recognized that V_f is a combination of particle population and particle size. Particle size is predicted to be a substantially more important factor owing to the geometric relation between particle diameter and kinetic energy transferred to the tool edge. Either of these factors can be substantially masked by variations in the composite microstructure caused by differences in casting techniques.

Furthermore, particle composition and morphology have a strong effect on surface finish as well as tool life. Al_2O_{3p} polish the PCD tool edge, unlike SiC_p which score it. This polishing effect is somewhat reduced by the amount and morphology of the silicon phase present in the foundry alloy matrix.

Finally, these composites cause rapid initial flank wear on PCD cutting tools. In order to maintain a consistent wear rate with the attendant dimensional tolerances and surface finish, it may be necessary to slightly hone or otherwise modify the standard "upsharp" edge preparation of the PCD cutting tool.[280]

End milling is discussed further in Section 2.6.5.

Lin, Bhattacharyya, and Lane[281] continued the above studies on Duralcan composite (A359/SiC/20_p) and found the following.

1. The main types of tool wear occurring during the machining of Duralcan aluminum composite take place on flank and rake surfaces, with flank wear being dominant in the various speed-feed ranges. The primary mechanism of wear formation is believed to be abrasion between reinforcement particles and cutting tool material.

2. As expected, cutting time decreases with increasing cutting speeds and feed rates. However, the total volume of workpiece material removed increases with increasing feed rates, presenting an anomalous picture to the user. This interesting phenomenon has been explained by modifying the Taylor equation to incorporate workpiece volume removal. Proper operating conditions should be chosen according to the requirements of practical applications.

3. The surface finish of machined samples deteriorates with increasing feed rate at a constant cutting speed but does not change significantly with a change in cutting speed. The best surface finish is achieved when the tool is slightly worn rather than when it is fully sharp. However, within the range of the work perfomed by Lin et al. the surface finish eventually approached a constant value as a result of stabilization of tool nose and cutting edge radii.

4. The type of short chip produced without a chip breaker in machining Duralcan Al MMCs renders this material well suited for continuous operation.

Chambers and Stephens,[282] machining Al/5Mg reinforced with 5 vol% Saffil and 15 vol% SiC, obtained similar results:

1. MMCs reinforced with a V_f of a ceramic such as SiC can be machined at low metal removal rates with cemented carbide cutting tools. A combination of low cutting speed and high feed rate resulted in the longest tool life.

2. A significantly better performance was achieved with PCD. It is expected that with this type of reinforcement the high cost of PCD would be justified by an increased production rate.

3. Surface quality is dependent upon both the tool material and the cutting parameters. Overall, PCD with its ability to cut the reinforcement is considered to give a better-quality finish.

2.6.4 MMC EDM.   MMCs are ideal candidates for EDM, especially where complicated shapes and high accuracy are required. Only a few CMCs that are electrically conductive can be shaped by EDM. However, recent improvements in the mechanical properties of CMCs—especially the fracture toughness and strength of whisker-reinforced ceramics through better processing technology and starting materials—make them ideally suited for high-temperature and fatigue-resistant applications.

The machinability of SiC_w/2124 Al and TiB_2/SiC by EDM was investigated by Ramulu and Taya.[283] The machined surfaces of these two high-temperature composites were examined by SEM and surface profilometry to determine the surface finish. Machinability was evaluated in terms of material removal rates, tool wear, and surface finish. They found that EDM was one of the most versatile and useful technological processes for machining intricate and complex shapes in various materials, including high-strength, temperature-resistant alloys.[284] The MMCs used in this study were 15 and 25 vol% SiC_w/2124 aluminum matrix composites, in flat form with a thickness of 6.3 mm. Brass and copper were used in the EDM process as electrodes and cut the MMC and CMC under coarse, medium, and fine cutting conditions. Based on preliminary investigations, the following conclusions were made.

- The material removal rate in EDM processing increased with power to the electrode in both the TiB_2/SiC and SiC_w/Al composite materials.
- The electrode wear rate increased with an increase in conditions in the EDM process regardless of workpiece material. Electrode wear rate in the machining of TiB_2/SiC was greater than in the SiC_w/Al composite material.
- The surface quality was affected by the choice of machining process, and the brass electrode in the EDM process appeared to cause minimal surface damage. At higher cutting speeds, EDM caused severe microdamage to the surface. Surface roughness in TiB_2/SiC was found to be finer than in SiC_w/Al composites.
- Microhardness tests on 25% SiC_w/Al composites revealed that the EDM process appeared to cause surface softening at slower cutting speeds.

In a recent EDM study on an Al alloy with 9 wt% Si, 0.19 wt% Fe, 0.07 st% Cu, 0.55 wt% Mg, and 20 vol% SiC_p by Hung, Yang, and Leong,[285] statistical models of the process were developed to predict the effect of process parameters on metal removal rate, recast layer (RCL), and surface finish. It was found that the SiC_p shielded and protected the Al matrix from being vaporized, thus reducing the metal removal rate. The unmelted SiC_p dropped out from the MMC together with surrounding molten Al droplets. While some Al droplets were flushed away by the dielectric, others trapped the loosened SiC$_p$ and then resolidified on the surface to form a RCL. No crack was found in the RCL or the softened heat-affected zone below it. The input power controls the metal removal rate and the RCL depth, but the current alone dominates the surface finish of an electric discharge-machined surface.

2.6.5 Milling

In milling MMCs, CVD thick-film diamond end mills were compared to micrograin WC mills by the Norton Diamond Film Corporation. The end mills, 9.52 mm in diameter and 63.5 mm long, had three straight flutes. Cutting speed was 359 m/min, feed rate 0.038 mm per tooth, axial Depth-of-Cut (DOC) 6.35 mm, and radial DOC 1 mm. With the carbide, there was 0.79 mm of wear after 2 m of travel and removal of 13.1 cm^3 of material; with the diamond film, researchers recorded a wear land of 0.19 mm after 157.5 m of travel and removal of 655.5 cm^3. By extrapolating it was determined that the CVD diamond tool would travel 327.5 m and remove 2112.8 cm^3 of material before the tool expired (0.38 mm of flank wear). This represents more than 300 times better tool life for the CVD diamond end mill.

In another milling test,[274] micrograin carbide was compared to a Si$_3$N$_4$ end mill coated with thin-film CVD diamond. Cutting speed was 762 m/min, feed rate 0.076 mm, axial DOC 6.35 mm, and radial DOC 2.54 mm. Geometry of the end mills was identical: three flute, 60° spiral, 15.88-mm diameter, right-hand cut, right-hand spiral, 44.5-mm flute length, and 101.6-mm overall length. In this test, the micrograin carbide end mill showed 0.14 mm of flank wear after machining 104.9 cm^3; the diamond film had no measurable wear and no film chipping or spalling after 209.5 cm^3.

As tests have confirmed, HSS cannot do the job of milling MMCs. Carbide inserts in face milling cutters have marginal success. Uncoated inserts are extremely limited, with great wear apparent after removing very small amounts of stock. The TiC-coated carbide inserts show removal of three to four times more material than the uncoated inserts but are still not acceptable because of the small amount of stock removed. Once the coating fails, the ability to machine deteriorates very rapidly. As is the case for turning, PCD inserts are far superior to carbide for milling.

More research on end milling indicated that the wear of the solid carbide mills was significant, although they removed 10 times as much material as HSS end mills. PCD-inserted end mills remove material quite easily, but deep milling, as for pockets and slots, must be done in increments of ~3.1-mm depth of cut. This generally leaves slight steps in the side walls of the areas being milled. A CVD-coated end mill could work well as an alternate in this application.

2.6.6 Tapping

The most challenging operation in machining MMCs is tapping. Therefore, if machining operations were ranked on a scale of 1–10 with the easiest being a 1 then turning would be a 1, drilling a 2, face milling a 3, and end milling a 4, but tapping would top the scale and be a 12.[256]

Carbide plug taps with as many flutes as possible have been proven to work the best. The greater number of flutes is advantageous because the land area is smaller and there is less resistance to assist in the removal of the tap after threading. If a two- or three-flute tap must be used, then the backs of the flutes must be backed off manually to reduce the land. Breakage of taps larger than a no. 10 size is minimal. Number 8 and smaller break quite readily, although this can be minimized by the use of a tapping head wherein the clutch has been properly set. Synthetic materials have been successfully used for tap lubricants, allowing ease of removal (Trefolen).[256]

The tapping of MMCs requires a constant feed, and this is not possible when hand tapping. Each stoppage of the tap constitutes a dwelling in the hole allowing the tap to back off slightly because the torque has been removed. This causes rapid breakdown of the cutting edges of the flute. Less than 10 holes can be tapped by hand, whereas 100 holes have been tapped employing a properly adjusted tapping head. The speeds used should be identical to those for tapping any heat-treated aluminum. Finally, roll form taps do not work very well for MMCs because they gall and seize up very quickly. Diamond-coated taps are in the development stage and should be very effective and give excellent results.

2.6.7 Deburring

Conventional methods for removing burrs work equally well for MMCs. Carbide rotary deburring tools in an air grinder can be used to remove heavy burrs resulting from milling and drilling operations. Holes can be chamfered with standard carbide countersink tooling. Vibratory deburring is highly recommended after the removal of heavy burrs to obtain a constant radius on all outside corners and a uniform texture over the entire part. Medium selection does not seem to be a factor in obtaining good results. A plastic-impregnated medium, Al_2O_3, and a ceramic have been used with comparable results. The time needed to obtain the desirable finish is a matter of personal preference, but 15–20 min is generally sufficient.

2.6.8 MMC Drilling

Ricci[286] evaluated the following MMCs.
- SiC_f/6061-T4Al
- SiC_p/7079-T6Al
- B_4C_p/AZ61 Al-Mg.

And he used the following drills and processes.

- PCD-tipped twist and spade
- Diamond-plated twist and core
- Abrasive water jet hole cutting.

The diamond-tipped twist drills outperformed all the other drills. Core drills were found to be viable alternatives for the production of larger holes in high-V_f composites. Plated twist drills were viable alternatives for low-V_f particulate composites. Spade drills failed because of low edge strength. Abrasive water jet hole cutting was successful for rough, large-diameter holes.[286,287]

In their study of opportunities and challenges for MMCs in the automotive industry, Allison and Cole[268] compared the hole drilling machining rates of 2080/SiC/15$_p$ (2080 Al alloy reinforced with 15 vol% SiC$_p$), a conventional P/M steel connecting rod (0.5C, 2Cu), and Ti-6Al-4V. Diamond tooling is more expensive than traditional HSS or carbide tooling, but tool life can be quite long. Therefore these high tooling costs can be amortized over many parts. Because of the higher productivity rates, the capital required for machining lines can be reduced. Moreover, machining tolerances are greatly improved using diamond tooling, and so subsequent finishing operations can be reduced or even eliminated. Again, for hole drilling in discontinuous reinforced Al composites, holes are finished with such high tolerances that subsequent reaming operations can be eliminated.

Monaghan and O'Reilly[288] confirmed the conclusions of Allison and Cole and others. They found that on the 1050 series Al alloy (12%Si-balance Al)/SiC/25$_p$ tool hardness played a significant role in the efficient and cost-effective drilling of MMCs. It was observed that the PCD-tipped tools, while they lasted, were superior to the carbide tools, and that the carbide tools in turn were much better than the coated and uncoated HSS drills. This ranking applied to all the recorded values for flank and rake face wear, torque, thrust force, and surface finish. The effect of the ceramic TiN coating on the HSS drills did not improve their performance significantly compared to that of the uncoated high-cobalt HSS drills. It was found that in each case there was a similarity between the trend of the flank and rake wear results for speed and feed rate and the variation of the drill thrust force and torque, respectively.

The four vertical tail stabilizers mentioned earlier as one of the largest MMC primary structures built confirmed that existing manufacturing technology for mass production (such as cutting and drilling) was possible with MMC components.[1,274] The program proved the viability of using conventional shop procedures for drilling and fastening these MMC components. It also tested innovative manufacturing equipment such as PCD tooling in making all the drilled and countersunk holes, as well as AWJ cutting. Diamond-tipped drills were used because the MMC was considerably more abrasive than Al because of the SiC content. The tips were refurbished and reused after making 75 holes. The AWJ cutter used a thin stream of high-pressure (310 MPa) water containing powdered garnet.

Berlin[262] made a machining cost comparison of SiC discontinuously reinforced Al, unreinforced Al, and Ti and included drilling. He found that drilling was the least economical operation with all materials. While representing relatively small portions of machining time, the relative tool costs were quite high. SXA (2124/SiC/15$_w$), in particular, was extremely expensive to drill. Representing only 6.3% of total machining time, it required 39.0% of total machining cost. Therefore the average machining cost per minute was extremely high (over $2.02/min). Al and Ti follow similarly but not nearly to the extremes of MMC. This high cost was due to the relatively short lives drills typically have, especially in small diameters, and the cost of PCD drills which averaged approximately $285 each.[262]

Finally, in a prototype MMC machining study, Dwenger[256] concluded that HSS drills, both coated and uncoated, absolutely would not work with MMCs. All the drilling in his work was done with carbide drills. Hole sizes ranged from 2.3 mm through 25.4

mm in diameter. High helix drills were also used, but for normal depths of three to four times the diameter there was no proven advantage. For larger hole sizes, carbide-tipped drills were generally used. They worked equally well and had a lower initial cost than solid carbide drills. Speed and feed rates were the same as if drilling standard Al. The use of a coolant during drilling was mandatory, preferably a water-soluble oil. Coolant-fed drills worked quite well for 9.5-mm diameters and larger, especially when the hole was quite deep, for clearing chips from the flutes.

Note that a drill should never be allowed to dwell in the hole.[289] This dulls the cutting edge immediately, and drill failure will occur, leading to out-of-tolerance hole sizes or breakage of the drill.

With carbide drills, upward of 200 holes were drilled between resharpenings. However, if diamond-tipped drills were substituted it is believed that results would range from 2000 to 3000 holes between resharpenings.

Other studies conducted on 2014/SiC/15$_p$ with a WC drill in various laboratories have concluded the following.

- Tool wear was accelerated because of the presence of SiC particles.
- The use of a coolant was essential.
- Chisel edge, margin, lip crater, and lip flank wear were observed under most cutting conditions. Wear of the outer corner of the lip flank was found to have the most significant effect on performance.
- The generation of burrs on the exit side of the holes related to the rounding of the outer corner of the lip flank. Hence higher speeds and feeds resulted in the earlier appearance of such burrs.
- The most evident mechanism of wear observed on the drills was abrasion by the SiC particles present in the Al.

Yue, Lau, and Jiang[290] studied the behavior of laser microdrilling of an Al-Li (8090)/SiC/17$_p$ (average size 3 μm) MMC using a pulsed Nd:YAG laser. The concept of critical intensity was employed to predict the profile and the diameter of the drilled hole. The dimension of the hole was found to be related to the position of the laser focal spot relative to that of the workpiece. In the case of through hole drilling of thin sheet material, the diameter of the hole can be best controlled by beam intensity. Moreover, a moving laser source, by the nature of its tunneling action, can improve on the uniformity of the dimension of the processed hole; this is particularly valid for thick materials.

Jawaid, Barnes, and Ghadimzadeh[29,291] investigated a 2014 Al alloy with a 0.15 V_f of SiC$_p$, performing drilling tests with recommended grades of WC (H10F) and HSS (M2-grade) tools to determine the effect of variations in cutting speed and feed rate on tool life, forces generated, tool failure modes, and wear mechanisms.

They reported the following.

1. The results of the comparative tests carried out confirmed that tool wear was accelerated because of the presence of SiC$_p$.
2. The application of a coolant was found to be essential when drilling MMCs in order to prevent seizure and drill breakage.

3. Chisel edge, margin, lip crater, and lip flank wear were observed under most cutting conditions. Wear of the outer corner of the lip flank was found to have the most significant effect on performance.

4. The generation of burrs on the exit side of the holes was related to the rounding of the outer corner of the lip flank. Hence higher speeds and feeds resulted in the earlier appearance of such burrs.

5. The most evident mechanism of wear observed on the drills was abrasion by the SiC_p present in the Al.

2.6.9 MMC Grinding

Chandrasekaran and Johansson[292] of the Swedish Institute for Metals Research reported the results of a grinding study undertaken on an Al_2O_3 short fiber-reinforced Al alloy MMC. Variants of MMC material were tested a range of grinding conditions and abrasive wheel types, and grinding forces and surface finish were measured.

Their findings for this class of materials disclosed some aspects of the surface integrity of ground MMC materials.

1. Superabrasive wheels are to be preferred in the grinding of Al_2O_{3f}-reinforced Al alloys. In general grinding wheels using synthetic diamond (SD) grits results in superior surfaces irrespective of the fiber volume content and the type of Al alloy used. CBN-based wheels could also be effective but are associated with more exacting wheel preparation.

2. An increase in the Al_2O_{3f} content of the MMC material results in an increase in grinding forces when the abrasives used are hard and sharp enough, as is the case with SiC and superabrasives.

3. When abrasive grits of sufficient hardness and sharpness are used (as is the case with SD wheels), the role of the grinding parameters (wheel depth of cut and table speed) have only a marginal influence upon the measured surface finish parameters.

4. Both considerations of mechanical loading of the fibers under grinding and thermal considerations indicate that even under successful grinding conditions the fracture of individual Al_2O_{3f} in the MMC material into smaller bits is highly probable. Diamond grits, however, often preferentially decapitate the hard fibers.

5. Surface studies also indicate that grinding of these fiber-reinforced MMC materials is characterized by the presence of a large number of concomitant microcavities corresponding to the projected contours of individual fibers in the plane of grinding.

2.6.10 Cutting of MMCs

2.6.10.1 Mechanical cutting. Conventional cutting, turning, milling, and grinding operations can usually be applied to MMCs, but there is often a problem of excessive tool wear. In general, such problems become more significant with increasing reinforcement V_f and size. The V_f effect is evidently attributable to the tool encountering more hard material, while larger particles or fibers resist excavation more strongly and

hence stress the tooling more. Diamond-tipped or -impregnated tools are therefore usually necessary for monofilament-reinforced MMCs, while WC or even HSS tooling may be adequate for short-fiber and particulate material. The strength of the reinforcement is, however, also relevant; Al reinforced with SiC_w, for example, has been found[292] to be more difficult to machine than other composites containing weaker fibers. For most MMCs, the best results are obtained with sharp tooling, an appropriate cutting speed, copious cooling or lubrication, and a high material feed rate. These factors have, for example, been emphasized[293] with regard to Al/SiC_p composites. Diamond tooling has been found[293,294] to give a better performance than cemented carbides and ceramics and to allow relatively high tool speeds. In contrast, if carbide tooling is used, then a low cutting speed will increase tool life.[293,294] Further general features of MMC machining and cutting have also been identified.[295–297] Wire saws (in which a diamond-impregnated wire is translated along its axis while in abrasive contact with the workpiece) can be useful in cutting MMCs, although cutting rates are usually relatively slow and only straight cuts can normally be produced.

2.6.10.2 Electrical cutting.   Several cutting processes involving an electric field between the tool and the workpiece are potentially useful for MMCs. Electrochemical machining involves removal of material by anodic dissolution using a shaped cathode to determine the geometry of the cut. An ionic electrolyte is flushed through the cathode-workpiece gap, and this serves to carry away debris such as undissolved fibers.

There are difficulties in cutting and removing long fibers although this can be done for some cutting geometries by combining electrochemical action and mechanical grinding using an abrasive moving cathode. Normal ECM, however, involves no mechanical contact between electrode and workpiece, and so very little damage is induced in the remaining material.

Conventional EDM (spark erosion) involves cutting with a moving wire electrode bathed in a stream of dielectric fluid. Material removal results from the high local temperatures and liquid pressure pulses generated on the workpiece surface, which does not come into mechanical contact with the wire. As cutting is a thermal and mechanical rather than a chemical process, ceramic fibers can be cut quite readily. The process can, however, be rather slow, and damage can be relatively severe.

Lau, Yue, Lee, et al.[298] examined some of the unconventional cutting methods because they found that machining operations such as cutting and drilling were difficult to perform on composite materials with conventional tools and techniques because of their peculiar properties, including anisotropy, low thermal conductivity, and the abrasive nature of the reinforcing phases. Among the many unconventional processing techniques, laser and EDM were selected using a JK pulsed Nd:YAG laser model with a maximum average output of 120 W. Additionally, they used an excimer laser operating on a KrF gas mixture and having a maximum pulse energy of 1 J. The EDM unit was a Mitsubishi Electric DWC9Og wire cutting machine with a copper wire 0.25 mm in diameter as the electrode.

Their best results with the four composite materials [PMCs—65% laminating C_f polyacrylonitrile composite and 30 vol% Gl_f-reinforced liquid crystal polymer; MMC—20 wt% SiC/Al-Li; and CMC-50/40 wt% Al_2O_3/TiC ceramic] showed that for all the composite materials tested the material removal rate was faster in laser cut EDM than in wire cut EDM. Material damage such as excessive melting, thermal degradation of the matrix, disorientation and distortion of fibers, debonding, and delamination are commonly found in both EDM wire-cut and EDM laser-cut composite materials. This damage undoubtedly affects the reliability of the composite material in service. It is therefore important to decide between the optimum cutting rate and the extent of material damage that can be tolerated in the finished component. In addition the optimum operation window of each of these processes should also be established in order to keep damage to a minimum. Although the excimer laser can produce a cleaner cut surface with less material damage, the cutting rate still must be improved because high-power machines are currently unavailable.

A theoretical model was successfully developed to predict the maximum depth of cut in YAG laser cutting. With this equation, it was possible to design the parameters of the laser machine so as to produce the maximum depth of cut in the LBM of composite materials. However, the effects of the various laser parameters on the quality of the cut edges should be considered at the same time.

2.6.10.3 High-energy beam and fluid jet cutting.

For cutting along the fiber axis or for discontinuous composites, the fibers do not need to be fractured, and cutting, melting, or volatization of the matrix is all that is required. This can be achieved using lasers, electron beams, and so on, as in the case of unreinforced material, although the effect of the reinforcement on the thermal conductivity may influence the response of the material. At sufficiently high beam power,[298] fibers can be melted or volatilized or, more probably, fractured under the various mechanical and thermal stresses induced by the beam. However, a high level of microstructural damage tends to be induced during the cutting process in the form of cracking, interfacial debonding, and heat-affected microstructures.

Low damage levels can be combined with high cutting rates by using a concentrated jet of high velocity fluid, usually water, containing abrasive particles in suspension. This technique has been found to be applicable to both continuous[299] and discontinuous[300] MMCs. As there is no significant temperature rise, damage is only of mechanical origin and tends to be localized close to the machined surface.[301] The fluid jet technique is considered[299] to offer the best combination of cleanliness and maximum cutting speed. A related process is abrasive flow machining (AFM), which is used to generate a good surface finish. A gel containing abrasive particles flows over the workpiece surfaces under pressure. This method can be useful when polished surfaces are needed, particularly on components of complex shape.

2.6.10.3.1 Laser Beam Cutting (LBC).

LBC has been applied to MMCs. However, as with PMCs, there are problems in removing the matrix and leaving the fiber materials protruding for contact techniques (tool wear) and noncontact techniques [e.g., LBC and Water Jet Cutting (WJC)].[86]

Using a 1.5-kW CO_2 laser, scientists at TWI[86] laser-cut 8090 $Al/SiC/20_p$. A 20-mm-thick sample showed, at the cut edge, only a small amount of dross and a cut surface where no fibers were protruding from the matrix. The cuts were similar in appearance to those achievable on a monolithic Al alloy.

A Ti-6Al-4V alloy with SiC reinforcement unidirectionally aligned revealed the cut edges of fibers, a 1-mm-thick sample, protruding from the matrix. Two distinct modes of behavior in laser cutting of MMCs were identified. One was where the composites behaved as metals, and the other was where fiber and resin separation occurred in a way similar to the way it occurs in carbon fiber-reinforced polymer composites. However, at present, only limited information is available on laser cutting parameters for MMCs with different types of reinforcement.

Lau, Yue, Jiang, et al.[302] used a pulsed Nd:YAG laser to cut 8090 Al/SiC/20$_p$ to study the cutting behavior of the MMC. They found that it was possible to minimize the HAZ, to improve the quality of the machined surface by varying the laser operating parameters, and at the same time to achieve the best possible machined surface quality.

In pulsed laser cutting of SiC/Al-Li MMC, the thermal effect causes microstructural changes in the adjacent machined area. In practice, the HAZ width can be minimized mainly by using a smaller pulse energy and a shorter pulse duration. It is obvious that the higher feed rate increases the speed of cutting, however, the pulse overlap condition must be fulfilled. In determining the cutting parameters, the depth of cut, the cutting efficiency, and the HAZ width should be considered at the same time.[302]

Meinert, Martukanitz, and Bhagat[303] studied laser cutting of MMCs with a 1.5-kW CO_2 laser. They demonstrated that the absorptivity of the composites was higher than that of the unreinforced alloy 6061-T6/Al$_2$O$_3$/20$_p$ composite compared to that of the matrix alloy. This is the opposite of what occurs in conventional machining practices where the increased reinforcement content leads to decreased processing efficiency. The use of a filler alloy containing titanium, to form TiC in preference to Al$_4$C$_3$, eliminated the formation of Al$_4$C$_3$, during the welding of 6061/SiC/20$_w$ composites. Titanium additions result in the formation of an in situ TiAl$_3$ matrix having TiC reinforcement within the fusion zone of the weld. Further development of filler alloy additions and processing techniques are necessary to obtain a fusion zone having an Al alloy matrix with fine, dispersed TiC dendrites.[303]

Beryllium (Be) has all the attributes engineers want in high-performance computer disk drive arms and platters. It combines vey low density with very high stiffness and displays excellent mechanical damping and high-frequency resonance. Beryllia (BeO$_2$)/Be composite parts could make possible smaller, lighter, faster disk drives with an improved capacity. Thus far, however, the fabrication of Be has cost too much to allow widespread commercialization.

Laser fabrication resolves many of these problems. Lawrence Livermore National Laboratory (LLNL's) Hanafee and Ramos[304] successfully machined the metal using both a 400-W pulsed YAG laser and a 1000-W continuous-wave CO_2 system. They produced excellent surface finishes on an AlBeMet alloy and BeO$_2$/Be composite sheets. Cutting speeds reached 2.54 m/min on 0.5-mm-thick samples and 0.5–0.8 m/min on 1.8–2.0-mm-thick parts.

Laser cutting was fast and produced holes with very tight diameters and fine enough finishes to avoid secondary machining. It made efficient use of Be sheet. Equally important, they developed an enclosed, filtered, clean room and reduced worker exposure to Be$_p$.

Under a U.S. Air Force contract, Textron Specialty Materials (TSM) has established a TMC production facility to produce structural components for the NASP program and to support other initiatives requiring high-temperature, lightweight composite structures. In an effort to automate the cutting of composite preforms (Ti foils and SiC_f woven mat), a CNC laser cutter with compatible nesting software was implemented. Decreased material handling and cutting times in addition to increased material utilization have resulted in reduced costs and improved material through-put according to Kraus.[305]

Plasma Cutting.   Hoult, Pashby, and Chan[306] used high-tolerance plasma arc cut (HTPAC) on an MMC form (Al base-SiC) and a monolithic 2124 Al-Cu alloy and reported the results. They showed the following.

1. There are no significant differences between the optimum cut speeds of these two materials during HTPAC cutting.
2. There is a range of cut speed over which the different measures of cut quality are optimized, and this is similar for both materials.
3. For the HTPAC the cut speed appears to have a very important effect on cut surface quality.
4. The approximately optimized cut speed using this system is 55–60 mm/s when cutting these materials.
5. The kerf widths and HAZs associated with the HTPAC system are relatively small at optimum cut speeds on these materials.

Water Jet Cutting.   Lavender and Smith[307] reported their evaluation of WJ machining ZK60A $Mg/B_4C/15_p$ MMC. Their objective was to determine the effects of the WJ material removal process on the composite material surface structure and properties. These results were then compared with data generated from material that had been conventionally machined (using PCD tooling) and that exhibited the same matrix and reinforcement.

The conclusions derived from their WJ machining study indicated the following.

- The WJ-machined tensile specimens did not meet design drawing specifications for dimensional tolerances and surface finish.
- Surface profile measurements of the WJ-machined specimens indicated that surface roughness was approximately an order of magnitude greater than for diamond-machined surfaces.
- The WJ did not create a smooth, uninterrupted cut but instead prouced a "raised-ridge" surface texture.
- The WJ did not cut the ceramic particle but eroded the Mg matrix.
- The WJ did not appear to cause any microstructural damage to the material near the surface.
- The WJ did not work-harden the surface, as evidenced by the microhardness values.
- Despite the dimensional and surface finish limitations displayed by the tensile specimens, the WJ machining process appeared to be suitable for sectioning and rough machining of MMCs.

Abrasive Water Jet (AWJ) Cutting.   AWJ cutting seems to be a viable choice for forming parts out of MMCs. The abrasive particles entrained in the continuous high-velocity water jet do the machining, and there is no need to stop to replace a dull tool. Unlike the situation in laser cutting, no HAZ is produced when parts are cut by WJs. This is especially important for composites where deleterious reactions due to heat pulses can occur at the interfaces. Airborne dust is virtually eliminated. The jet can pierce the stock and thereby allows cuts to be started away from the edges. The quality of cut surfaces produced by AWJs varies significantly with the choice of cutting parameters. A small-diameter jet has inherent omnidirectional cutting potential, and this makes it a logical candidate for integration with a robotic system.

The micromechanisms of the AWJ cutting characteristics of cast Al 2014/SiC$_p$ composite and Dural Al 357/SiC$_p$ composite were examined by Rohatgi, Dahotre, Gopinathan, et al.[308] They found the following.

1. The AWJ-cut surfaces produced two distinct zones of roughness on the cut surface in the matrix as well as in the composite material. The top region was considerably smoother compared to the bottom region which was much rougher. The ratio of the height of the smooth region to that of the rough region decreased as the traverse speed increased for both the matrix and the composite materials.

2. The depth of cut increased with decreasing traverse speed in both the matrix alloys and the composites. For a given traverse speed, the depth of cut is greater in the base alloy than in the composites studied.

3. The presence of iron on the WJ-cut surfaces, as reflected by SEM, was due to the presence of garnet particles that were lodged in the AWJ-cut surface at the end of wear tracks.

Hamatani and Ramulu[309] conducted an experimental investigation on the machinability of particulate-reinforced ceramic TiB$_2$/SiC and metal SiC/Al matrix composites by AWJs. Both piercing and slot cutting experiments were conducted to determine the influence AWJ machining has on the material.

Based on this preliminary investigation of the machinability of SiC/Al and TiB$_2$/SiC, the following conclusions were made.

1. SiC/Al composite was easily machinable by the AWJ method, and a good surface finish was produced. The degree of orthogonal accuracy in the cut surface seems to be better under slow cutting conditions.

2. AWJ machining of the CMC TiB$_2$/SiC also was feasible. Better holes with minimal damage could be produced in CMCs by an AWJ at a low standoff distance.

3. Slotted edge damage in the MMC was found to be greater than in the CMC.

2.7 JOINING MMCs

In assessing the use of MMCs for specific applications, joining has been a major concern. The underlying issue is that the heat typically used to fabricate metal-to-metal joints has the effect of promoting reactions between the matrix and the reinforcement of MMCs. This can cause both structural and metallurgical problems in the product. Structural problems include cracking at matrix-reinforcement interfaces, and for continuous fiber-reinforced materials, cracking of the fibers themselves. Metallurgical problems include loss of protective coating on the fibers (for continuous fiber-reinforced materials), partial or complete dissolution of the reinforcement, and potential detrimental phase formation caused by the matrix alloying with the reinforcement material.[310,311]

2.7.1 Joining Issues

Joining issues involve assessing microstructural changes in welds, examining alternate welding and joining processes, and developing new processes tailored to these materials. Additionally, the application of rapid thermal cycle processes to minimize particle-matrix interactions in particulate-reinforced Al composites is continually being investigated. The concern has been that during fusion welding (particularly with Al/SiC composites), the SiC_p break down in the superheated molten Al, allowing the subsequent precipitation of detrimental phases on cooling. Large Al_3C_4 plates dominate the structure. Scientists and welding engineers have postulated which mechanisms create these changes and have attempted to predict which welding processes would be most applicable.

The application of solid-state processes has been considered advantageous because there is no melting to allow dissolution of the reinforcing particles. The resulting structure shows essentially a base metal distribution of MMC particles all the way to the bond line. As for continuous fiber-reinforced Ti materials, considerable work has been done in developing solid-state variations of existing rapid thermal cycle processes, including resistance spot diffusion bonding and resistance brazing. These approaches have the advantage of achieving a bond at the contacting surface but avoiding the high heat that would degrade the reinforcing fibers. The application of other processes, including ultrasonic welding (USW) and resistance implant welding, has also been considered.

The particles, whiskers, or continuous fibers used for reinforcement have a V_f typically between 10 and 60%. Because these materials usually have high strength and high moduli and retain their strength up to high temperatures, operations such as forging and machining are difficult. As a result, joining simple parts provides an attractive method for making more complex products.

There are a number of problems peculiar to MMCs. Particulate reinforcements may have densities different from that of the matrix, and this can lead to severe segregation effects. Second, undesirable chemical reactions can take place between the particle and the matrix, especially when the matrix is molten. This has been shown to be a particular problem in Al alloys reinforced with SiC and can lead to formation of undesirable aluminum carbides. Third, the ductility of the materials is very low compared with that of the matrix material, and thus the risk of cracking in the weldment, particularly during fusion processes, is higher.

These problems can be overcome by careful optimization of the process and by applying a detailed understanding of the metallurgical phenomena peculiar to these materials to development of the joining procedure. One aspect of the problem, for example, has been shown in the joint line of a resistance weld between a 6082 Al alloy and a 6082 alloy reinforced with SiC_p. Segregation of the particulates in micrographs can be clearly seen, but no evidence has been found of reactions between the matrix and SiC_p. Correct manipulation of the welding parameters can minimize segregation.

2.7.2 Solid-State Processes

2.7.2.1 Friction welding (FRW).   This process has proved very successful in making sound joints. Two components are moved against each other under pressure, developing frictional heat. The relative motions can be of several types (e.g., rotary, orbital, or linear). TWI[312] has examined rotary and linear motions, however, rotary motions have been the most widely used. A soft layer is formed at the interface and undergoes intense plastic deformation, much of the material being extruded from the joint. Once this layer has formed, a solid-state bond exists between the two components, and the bond is normally allowed to cool under pressure.

From a metallurgical point of view, examination of the interface (which is often hard to find exactly) shows a continuous microstructure with no evidence of microstructural discontinuities across the interface. Sometimes there is evidence of fragmentation of the reinforcement caused by repeated impacts during the friction process, but this phenomenon is not always observed.

So far, only limited mechanical property data have been obtained,[312] but the results appear encouraging. Yield values in cross-weld tensiles of up to 90% of the parent material value have been obtained. It is probable that suitable postweld heat treatment would give even better results because the age hardening matrix would have been subjected to solution treatment or averaging, depending on location and thermal cycle.

Other investigations have found that the continuous-drive FRW process is a useful method for joining SiC_p/Al MMCs. Unlike conventional welding methods, FRW of Al-based composites produces no pores or clusters of SiC_p in the HAZ. In general, the plastic deformation occurring during the welding operation has no significant effect on the overall distribution of SiC_p in the material. This in turn implies that the stiffness (E modulus) is maintained across the joint. It therefore follows that the resulting HAZ strength level is mainly controlled by reactions taking place within the Al matrix during the weld thermal cycle. As a result, full recovery of HAZ strength can be achieved by using an appropriate postweld heat treatment involving solution heat treatment at 535°C followed by artificial aging at 160°C for 10 h.

2.7.2.2 Inertia welding (IW).   A process that is somewhat similar to FRW, has been investigated in several studies as a joining medium for SiC/6061 Al and Al_2O_3/6061 Al.[313–315] The former material was studied for possible incorporation into the structural components of lightweight torpedoes. The MMC under consideration (23 wt% SiC) offers a very favorable strength-to-weight ratio, has near-isotropic properties, and is very heat-resistant (about 65% more tensile strength than that of unreinforced 6061 Al at 260°C).

One naval study has involved the pistons of the MK 46 torpedo engine which are made of 6061 Al alloy. Operation of the engine subjects the piston tips to intense heating and causes melting and erosion in the upper portion. Current practice is to replace the pistons after five torpedo runs. To reduce maintenance and associated costs resulting from the relatively short operational life of these pistons, the feasibility of replacing the top portion of the piston with a SiC/Al MMC using IW as the joining method was investigated. IW was selected because it is a low-cost procedure that permits the joining of dissimilar metals and requires no special environment. The cylindrical geometry of the pieces to be mated allows use of this technique with very little preparation and no special tooling.

The inertia-friction welding (IFRW) process has been widely used for joining dissimilar materials and exhibits several characteristics that make it a viable candidate for welding Al MMCs and dissimilar materials to conventional monolithic Al alloys.[313] These characteristics include an absence of melting at the weld surface, rapid heating and cooling rates, and the expulsion of severely heat- and deformation-affected material during the final forging and upset stage of the process.

A major problem with fusion welding of Al/Al$_2$O$_3$ materials is particulate coalescence and a nonuniform particulate distribution in the fusion zone. Although somewhat controversial, it is generally accepted that melting does not occur at the interface during IFRW.[313] In addition, the production of high-joint-efficiency welds in Al$_2$O$_{3p}$-reinforced composites requires retention of the particulates and grain structures across the weld interface. This microstructure retention is promoted in IFRW by the extremely high heating and cooling rates. These high heating rates and short periods at peak temperature tend to reduce or eliminate particulate-matrix reactions as well as promote a uniform particulate distribution and reduce grain coarsening near the weld interface.

In one investigation,[313] IFRW was explored as an alternative to fusion welding of particulate-reinforced Al MMCs because of its solid-state nature. The purpose was to identify the procedural variables that influence the weldability of 6061 Al reinforced with Al$_2$O$_{3p}$ during IFRW. Extruded rods of 6061 Al/Al$_2$O$_3$/15$_p$ underwent IFRW. The findings were as follows.

- Tensile tests conducted on similar and dissimilar materials revealed joint properties that were approximately 40 and 50% of those of the parent materials, respectively.
- Reduced-radius bond tests showed limited ductility of the similar and dissimilar MMC joints.
- Metallographic analysis of the weld region showed good interfacial mixing at the bond-line. Al$_2$O$_{3p}$ were continuous along the bond interface, and the bond region appeared to be sound and free of any defects.
- The process variables, such as the welding pressure and the speed required to achieve desirable amounts of flash and near-base-metal properties, were lower than those needed for joining unreinforced Al alloy 6061-T6.
- The welds between composite and monolithic Al exhibited room-temperature strengths of 83% of those of the composite base material.
- The tensile strength of the welds exhibited high joint efficiencies when high axial pressure and flywheel speed were used.

2.7.2.3 Diffusion welding (DFW).   At TWI,[316] preliminary trials of DFW involving pure Al/SiC/20$_p$ and 6061 Al/SiC/20$_p$ produced successful bonds using Cu, Ag, or no interlayers. Metallographic examination confirmed the need for precise control of the process variables to ensure that no particulate-rich zones or particulate-free zones formed at the interface.

Mechanical data on the process is limited, but it is known that a bond that appears sound when examined microstructurally does not necessarily produce good mechanical properties. DFW of Al is not easy because of the tenacious and stable oxide, and so it is

TABLE 2.13   Material Combinations Investigated During Program[317]

Aluminum matrix combinations

    6061 Al to 6061 Al

    P100 Gr_f-reinforced 6061 Al to B_4C_p-reinforced 6061 Al

    P100 Gr_f-reinforced 6061 Al to 6061 Al

    B_4C_p-reinforced 6061 Al to 6061 Al

Magnesium matrix combinations

    AZ31B Mg to AZ31B Mg

    B_4C_p-reinforced AZ91 Mg to 6061 Al

    P100 Gr_f-reinforced AZ91 Mg with AZ61 cover sheets to itself

    B_4C_p-reinforced AZ91 Mg to itself

    P100 Gr_f-reinforced AZ91 Mg with AZ61 cover sheets to B_4C_p-reinforced AZ91 Mg

    P100 Gr_f-reinforced AZ91 Mg with AZ61 cover sheets to SiC_p-reinforced ZK60A Mg

difficult to achieve reproducibly satisfactory bonds in the matrix alloys. Addition of the reinforcing particulate imposes further restrictions on process variables.

High-strength, long-life Gr-metal bonds have been developed by the Jet Propulsion Laboratory in Pasedena, California. The highly reliable technique has applications in aircraft and space components, brakes, lightweight weapons, and other devices that use Gr as a low-weight, high-strength structural member. The method requires the use of sputtered metal bonding agents to provide a graded diffusion bond to which both Gr and metal adhere. Bond strengths of more than 20.6 MPa are possible. The mating withstands high temperatures and forms an efficient electric conduction path.

A recently completed development study conducted by Rosenwasser, Auvil, and Stevenson[317] documented the results of a low-temperature, solid-state bonding (LTSSB) technique for producing high-quality joints between MMC parts. These materials (Table 2.13), are shown in Figure 2.16. Composite materials could be bonded at temperatures and pressures low enough to prevent dimensional change (deformation) and/or damage to the composite materials. Thus net shape bonding could be accomplished without the need for subsequent machining. The bonding could be performed at locations (such as in space) where subsequent machining would be impractical or impossible. The room-temperature shear strengths of the resultant bonds ranged from about 13.8 to over 1380 MPa. Except for joints between Al or Mg alloys bonded to themselves, the LTSSB joints were always stronger than the shear strengths of the composites themselves. Measured shear strengths were much higher than those reported for adhesively bonded or brazed joints, permitting greater design freedom and a larger margin of safety for space structure designers. The shear failures of composite specimens bonded using the LTSSB approach occurred in the composite materials even when tested at −157°C or 121°C (nominal space conditions). The LTSSB joints exhibited good low-cycle fatigue properties and their thermal conductivity was very high, as much as a factor of 10 greater than that of adhesively bonded joints. This higher thermal conductivity is important because it reduces the thermal gradient across the joint and increases heat transfer. This is important for thermal management applications such as space radiator components and for threat survivability requirements.[317]

Min, Fisk, Hartle, et al.[318] also reported on the diffusion bonding of Gr/Al tubes,

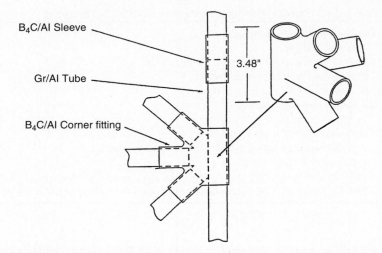

B₄C/Al Sleeve

Gr/Al Tube

B₄C/Al Corner fitting

3.48"

Figure 2.16   Typical P100/6061 tube, B_4C_p/6061 fitting truss structure design.[317] 1 in. = 25.4 mm.

demonstrating their flight-worthiness and testing three-bay truss towers made of these tubes.

The flight technology experiment concept shown in Figure 2.17 employed Gr/Al consolidated by either diffusion bonding or pultrusion (Figure 2.18). Both approaches started with layups of liquid metal-infiltrated Gr/Al precursor wires and 6061 Al foils. In the diffusion bonding process, consolidation took place by internal pressurization of the tube layup against an external die.

A truss was successfully assembled[319] using the diffusion-bonded Gr/Al tubes and B₄C/Al fittings designed for this truss. Adhesive bonding was used for the assembly. The truss was tested under static tension, compression, and torsion loading to measure the load deflection response, and under dynamic loading to identify the fundamental frequencies and mode shapes. With acceptable experimental and predictive errors, the truss behaved well under both types of loading.

Dunford and associates[320] examined Al-Li alloy MMC sheet material containing up to 20 wt% SiC and found that adequate formability of 8090 MMC was clearly possible at 530–540°C as well as subsequent diffusion bonding. Their tests indicated that the 20 wt% SiC_p/Al-Li 8090 alloy MMC could be diffusion-bonded with the possibility of combining SPF with DB in the fabrication of multisheet structures. This MMC has the advantage of a lower-density matrix compared with MMCs based upon 2XXX or 7XXX alloys.

In another series of tests performed by Dunford and Partridge[321–323] a particulate-reinforced MMC containing 17 vol% SiC_p in an Al-Li 8090 (Al-2.4 wt% Li-1.2 wt% Cu-0.6 wt% Mg) alloy matrix in sheet form was manufactured using a proprietary thermo-mechanical processing route. They found that solid-state DB of Al-Li 8090 alloy without additional interlayers produced joints with shear strengths and microstructures similar to those of the base metal. For SPF/DB structures, it was predicted that sheet thicknesses up to 2 mm could produce strain without peel fracturing.[321]

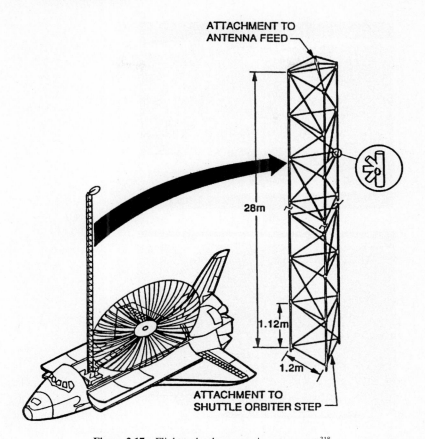

ATTACHMENT TO
ANTENNA FEED

28m

1.12m

1.2m

ATTACHMENT TO
SHUTTLE ORBITER STEP

Figure 2.17   Flight technology experiment concept.[318]

The NASP materials and structures program involves a series of titanium matrix composite development activities that entail the development of fiber-matrix materials, SCS-6, SCS-9, and Sigma. In order to demonstrate the material and structural performance of TMCs, several subcomponents have been fabricated and tested. These subcomponents are representative of X-30 structural types including fuselage, wing, and nozzle. An example of a lightly loaded hat-stiffened structure such as a fuselage is made of both SCS-6 materials, SCS-9$_f$, and β21-S matrix and is assembled by DB and mechanical fastening. It includes a complex splice joint involving manufacturing details such as joggles, buildups, and single-sided assembly.[324]

Many different approaches have been used to incorporate reinforcing fibers into metal matrices, some more successful than others. One of the most challenging scenarios is the synthesis of MMCs for high-temperature structural applications. These MMCs are generally composed of brittle ceramic fibers in high-strength, creep-resistant alloy matrices. Bampton, Cunningham, and Everett[325] have developed a method for synthesizing this type of MMC. It utilizes a thin, transient liquid layer at a critical stage of composite consolidation. The so-called foil-fiber-foil (F/F/F) MMC consolidation approach has proven

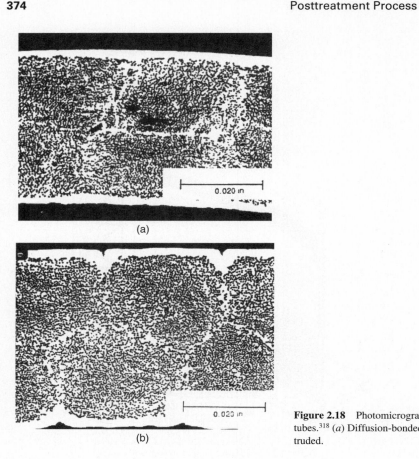

(a)

(b)

Figure 2.18 Photomicrograph of two-ply tubes.[318] (*a*) Diffusion-bonded. (*b*) Pultruded.

very effective in several MMC systems, most recently in the Ti alloy-SiC$_f$ system β21-S/SCS-6.[325] These MMC systems all have the prerequisites for successful F/F/F processing, namely, ductile matrix alloys and large-diameter monofilament fibers. This transient liquid consolidation (TLC) is a fabrication method that could be employed in joining MMCs as well as CMCs, aluminides (γ-TiAl/Al$_2$O$_3$), and superalloy-Al$_2$O$_3$).

Fukumoto, Hirose, and Kobayashi[326] successfully diffusion-bonded continuous SiC$_f$-reinforced Ti-6Al-4V composites to Ti-6Al-4V alloy. They found that the joint strength increased with bonding time and reached a maximum at 850 MN/m^2 for V_f = 30% composite and 650 MN/m^2 for V_f = 45% composite.

In V_f (30) composite the maximum joint strength was obtained for a bonding pressure of 7 MPa and a bonding time of 7.2 ks, and for a bonding pressure of 12 MPa and a bonding time of 3.6 ks. The maximum joint strength was 85% of the Ti-6Al-4V base metal tensile strength.

Second, the void ratio at the interface between the composite matrix and Ti-6Al-4V decreased with increasing bonding time. The bond strength between fiber and Ti-6Al-4V plate may not contribute to the joint strength under these conditions. The joint strength was controlled by bonding of the composite matrix and the Ti-6Al-4V plate.

The bondability of SiC/Ti-6Al-4V composite and Ti-6Al-4V alloy was better for V_f (30) than for V_f (45) composite. This is because the bonding pressure for each fiber is higher and the distance from one fiber to another is greater in V_f (30) than in V_f (45) composite, and so plastic deformation of Ti-6Al-4V alloy occurs easily in V_f (30) composite.

2.7.3 Fusion Processes

Fusion processes are economical and flexible, and they generally require only modest capital investment. They are well established for Al alloys and are attractive for joining MMCs. However, experience has highlighted a number of problems that must be considered.

The welding of Al MMCs differs from the welding of monolithic Al alloys in that the weld pool of the composite contains a solid phase because the reinforcement does not necessarily melt at welding temperatures. Thus, in the strictest sense, the composite weld pool is actually a partially melted zone. The most obvious effect of the presence of the reinforcement is the increased viscosity of the Al MMC weld pool. This weld pool does not flow and wet as readily as that of monolithic Al alloys, and mass transfer is limited because of its viscous nature.

Therefore heat flow by convection in the weld pool is believed to be less effective than in Al alloys, and conductive heat flow through the Al alloy matrix is thought to play a larger-than-normal role in determining the temperature distributions and cooling rates in and around the weld pool of Al MMCs. The differences in heat flow in the MMC, relative to that in monolithic Al alloys, can affect the resulting microstructures and the stress distributions in the MMC weld. The use of other reinforcements, such as Al_2O_3, which is stable in molten Al, can solve the problem.[327]

A second problem is rejection of the particulate by the solidifying interface. Most MMCs are reasonably homogeneous, but following fusion welding, ceramic particles agglomerate in the interdendritic regions, reducing their efficiency as a reinforcement. The magnitude of this effect and its significance are clearly influenced by the particle size and the cooling rate, which determines the dendritic cell size.

With certain particle-matrix combinations, there is a significant risk of particle-matrix interactions. However, the reaction can be suppressed by using high-silicon alloy matrices or by avoiding high-energy density processes that raise the peak weld pool temperature.

A third problem may occur with some MMCs made by P/M techniques. If the occluded gas content is too high, gas will come out of solution at high temperatures, causing extensive cracking in the HAZ. However, this does not appear to be a problem with current powder route MMCs.[312]

2.7.3.1 Fusion welding

Gas Tungsten Arc Welding (GTAW).   Bhattacharyya et al.[328] initiated a study to determine the influence of Al_2O_3 microsphere reinforcement on the mechanical behavior and wettability of a 6061 Al MMC.

Although the composite has superior properties compared to those of the matrix, the fracture toughness appears to be much lower. It is evident that the structure of the reinforcement particles greatly influences the material behavior. Tests showed that the composite material had good weldability using both GTAW and gas metal arc welding (GMAW) processes. However, larger heat-affected zones and increased porosity in the weldments were common in the composite, and the Al_2O_3 microspheres appeared to contribute largely to these phenomena. Comral-85, a standard heat-treatable 6061 Al matrix reinforced with Al_2O_3 microspheres was selected. The reinforcing particles had an average diameter of 20 μm, and a nominal 25% V_f. Tensile strength and stiffness of Comral-85 weldments were found to be greater than those of 6061 weldments. The fatigue strength was also superior for Comral-85. Compared to 6061 Al alloy, Comral-85 MMC is about 10% stronger in yield at 25°C and more than 100% stronger at 200°C. It is also 25% stiffer, but its toughness is significantly lower than that of 6061. It can be readily welded by both GTAW and GMAW. However, it displays a large heat-affected zone because of its lower thermal conductivity. Additionally, welded specimens, compared to 6061 welded specimens, show higher levels of tensile, bending, and fatigue strength but very low toughness. Finally, weldments display extensive porosity probably as a result of gas bubble attachment to reinforcement particles via surface tension forces.

In studies conducted by Gittos and Threadgill of TWI,[312] it was found that molten Al MMC materials were very viscous, even at temperatures considerably in excess of the melting point. This means that it is difficult to use conventional filler alloys. The two liquids do not mix satisfactorily in the few seconds that the pool remains molten, and care must be taken to minimize the melting of the MMC material. This effect may not be a problem if low-strength welds are required, but it has significant implications for the design of welded structures that incorporate MMCs. However, a prime Al producer, Duralcan, has demonstrated that Al MMCs can undergo GTAW or GMAW with procedures chosen to minimize dilution, thus overcoming the poor miscibility of the molten MMC. A second feature is rejection of the particulate by the solidifying interface. Most MMCs are reasonably homogeneous, but following fusion welding, ceramic particles agglomerate in the interdendritic regions, reducing their efficiency as a reinforcement. The magnitude of this effect and its significance are clearly influenced by the particle size and the cooling rate, which determines the dendritic cell size. In large weld pools, gravitational effects may also lead to enhanced macrosegregation, although there are no known investigations of this.

Duralcan[327] has a major research program under way involving the welding of 101.6-mm-diameter composite driveshafts to Al yokes for light trucks. Individual welds are completed in 11–12 s at speeds of 152.4 cm/min. The seamless tubes that make up part of the driveshafts consist of 6061 Al reinforced by Al_2O_3 particulates. The choice of Al_2O_3 over SiC, the material normally used to reinforce Al, is based on the former's superior weldability. To keep the Al_2O_{3p} from adhering to each other while the matrix material is in the molten state, at least 3.5% Mg should be present. A filler metal often used in such situations is 5356, which contains 5% Mg. The Al_2O_3 in the weld bead is introduced by dilution from the base metal as a result of fusion during welding. SiC-reinforced Al can be welded, however, rapid weld passes generated by low-heat input processes should be used. The SiC in the weld pool must be kept cold in order to prevent the material from decomposing and forming Al_4C_3. Resultant mechanical properties are just as good with welded 6061 Al matrix composites reinforced with Al_2O_{3p} as they are with welded 6061 Al. Welded MMC structures have achieved a 96.133×10^3-MPa modulus.[329]

The University of Wisconsin at Milwaukee used manual GTAW and Al filler metal to carry out some preliminary work on the weldability of a 6061 Al/SiC/l5$_p$ MMC.[330]

Gas Metal Arc Welding (GMAW).   Reynolds[331] examined welding procedures for 6061/Al$_2$O$_{3p}$ composite base plates using primarily the GMAW process. Continuous composite welding electrodes (1.14 mm in diameter) were successfully produced by extrusion, rolling, and drawing. Welds were prepared using short-circuit transfer and spray transfer GMAW procedures employing conventional 4043 and composite 4043/10% Al$_2$O$_{3p}$ welding electrodes. The welding performance of both the conventional and composite 4043 welding electrodes on the composite base plate was unsatisfactory. There was little retention of Al$_2$O$_3$ in both the conventional and composite electrode welds and extensive weld metal porosity and cracking in the composite electrode welds made by either the short-arc or spray transfer welding process. This was largely due to poor wettability of the base plate and the dispersed Al$_2$O$_{3p}$ by the 4043 filler metal composition. A Mg-containing filler metal composition provided significant improvements in both weld metal soundness and Al$_2$O$_{3p}$ retention. The as-welded mechanical properties of welds produced with a conventional Mg-containing 5356 electrode were found to be superior to those achieved with a conventional 4043 electrode. Mg additions to welds produced with the 4043/Al$_2$O$_{3p}$ composite electrode improved both soundness and Al$_2$O$_3$ retention but not to a sufficient degree to make these welds suitable for mechanical testing. Mg-containing filler metal compositions such as 5356 appear to be promising starting points for the construction of composite electrodes for GMAW of Al/Al$_2$O$_{3p}$ composite base plates and future work. Other compositions that might be suitable on the basis of their wetting characteristics toward Al$_2$O$_{3p}$ are Cu, Si-containing compositions such as A390.

Resistance Welding.   There is a phenomenon that so far defies rational explanation: a mass segregation effect found in some resistance welds in which the SiC$_p$ segregate preferentially to the periphery of the molten nugget. It is not clear whether this is caused by a hydrodynamic effect generating a centrifugal force or whether it is a surface tension effect. At any rate, the explanation is of academic interest only, as it can be suppressed by careful control of welding parameters.[312]

Both RW and forge welding are quite adaptable to a range of DRA composites. Problems associated with welding these materials typically have two causes. For SiC-reinforced Al MMCs, the high temperatures and relatively long times at temperature permit dissolution of the carbides and subsequent precipitation of detrimental phases (Al$_4$C$_3$). For Al$_2$O$_3$-reinforced Al MMCs, similar conditions allow the individual reinforcing particles to agglomerate, which reduces material performance.

Al MMCs are generally spot-welded employing conditions similar to those used for conventional Al alloys. However, these materials typically require substantially less current (about half) for welding. This is probably because of the increase in bulk resistivity associated with addition of the particles.

It is believed that the minimal reaction of the SiC to form Al$_4$C$_3$ is due to two factors. First, in resistance welds, the temperature of the liquid nugget only barely exceeds the melting point of the material,[332] providing little if any driving force for the reaction. Second, temperature excursions above the liquidus are relatively short, typically less than 0.1 s, which minimizes any potential reaction.

Spot welds on DRA composites do show some redistribution of the reinforcement in the fusion zone. Typically, this zone is relatively free of reinforcement near the fusion line, with an increased density of reinforcement near the center of the weld. It is possible that this redistribution of the composite is the result of macroscopic liquid flow during solidification, but the effect is still under study. The influence of this reinforcement-free zone on the mechanical performance of the weld is not known.

In the fuselage section designed for the national aerospace plane, there are hundreds of "hat-type" structural shapes of SiC_f/Ti that are spot-welded to 1.2-m^2 composite panels. The MMC is expected to withstand temperatures up to 760°C and is considered critical to the primary fuselage structure of the national aerospace plane.

Laser Beam Welding (LBW).   To make practical complex engineering components from MMCs, a method for joining MMCs to each other and to monolithic materials is a key technology. In the welding of MMCs, the fusion zone and the heat-affected zone should be minimized to avoid degradation of mechanical properties as a result of fiber-matrix reaction at elevated temperatures. From this point of view, LBW seems to be the most feasible fusion welding process because the laser beam can provide a controllable heat source with high energy density.

Hirose and his associates[333] at Osaka University joined 3- and 10-ply SiC_f-reinforced Ti-6Al-4V composites using a laser beam. With a 300-μm-thick Ti-6Al-4V filler metal, fully penetrated welds without apparent fiber damage were obtained in welding directions both parallel and transverse to the fiber direction by controlling the welding heat input. Excess heat input resulted in the decomposition of SiC and subsequent TiC formation, and as a result a decrease in joint strength. The welding of the 3-ply composite, in which full penetration was achieved at lower laser power, exhibited higher flexibility in heat input than that of the 10-ply composite. Heat treatment at 900°C after welding improved joint strength because of homogenization of the weld metal and decomposition of TiC. The strengths of the transverse weld joints after heat treatment, were approximately 650 and 550 MPa for the 3- and 10-ply composites, respectively. With the welding direction parallel to the fiber direction, the strengths both parallel and transverse to the weld joint were equivalent to those of the base plate.

Pulsed laser welding of A356 Al/SiC/20$_p$ and A356 Al/SiC/20$_p$ composites was used by University of Tennessee Space Institute scientists[334] to produce diverse microstructures in and around the weld. Nonequilibrium and unconventional phases such as Al_4C_3, $Al_4Si_2C_5$, and several unidentfied phases evolved during the processing. The surface of the SiC_p was physically and chemically modified. The extent of microstructural changes was at a minimum in the low-duty-cycle welds (50 and 57%) and at a maximum in the high-duty-cycle welds (80 and 91%). Control of the heat input, and thereby control of the microstructure in and around the joint region, was achieved by operating the laser in the pulsed mode. The composite processed with intermediate-duty cycles underwent the optimum changes in microstructure and possessed optimum strengths. Tensile tests of the longitudinal welds showed enhancement of mechanical properties (tensile strength and percent elongation), whereas tests of transverse welds showed degradation of the strength and ductility of both composites.

Kawall and Vlegelahn[335] investigated the feasibility of welding an Al_2O_3-reinforced 6061 Al alloy. They found that in joining this composite material by conventional welding and using a filler metal, the joint was left with a fewer number of Al_2O_{3p}. Therefore, laser welding was tried as an alternative, and the following results were obtained.

- Al_2O_3-reinforced 6061 Al alloy coupled better than the plain 6061 Al alloy at low power. However, above 3.5 kW, coupling deteriorated and a considerable portion of the beam was lost to reflection.

- Plasma suppression was a major problem in composite laser welding. Gas flow over the weld had to be very closely controlled to achieve plasma suppression. For best results, the gas flow must be turbulence-free and the welding direction should be such that the newly formed weld moves with the flow of the shield gas.

- Loss of reinforcing Al_2O_{3p} in the weld zone was another problem in the laser welds. This loss of particles was more pronounced in the keyhole region and near the top of the weld zone.

- Mechanically sound welds were not obtained in this preliminary study. However, welds obtained without the use of filler materials showed further promise in laser welding of the composite material.

In more recent development work with MMCs, Dahotre, McCay, and McCay[336,337] employed a laser-induced liquid-phase reaction synthesis-assisted joining technique for SiC_p/Al alloy composite materials. They were produced by synthesis of a suitable material product as a result of interaction between composite and Ti or Ti alloy reactive filler material induced by laser energy in the joint region. Such a reaction product minimized or eliminated the formation of a deleterious Al_4C_3 phase in the joint region depending upon the type and nature of the interfacial reactive filler material and also upon the laser processing parameters. A laser beam was utilized both to synthesize the interfacial reactant mixture and to heat the base material adjacent to the joint region to minimize thermal stresses. The technique, along with suitable filler material, can be extended to a variety of MMC systems including combinations of Gr/Al, B/Al, B_4C/Mg, steel/Al, W/Al, Al_2O_3/Al, and Gr/Cu.

In their laser cutting studies Meinert, Martukanitz, and Bhagat[303] also examined LBW of 6061/SiC/20$_w$ and especially suppression of the formation of Al_4C_3 in SiC-reinforced materials. In their work filler alloy additions of Ti were supplied to the weld to supersede the formation of Al_4C_3.

Their findings showed that laser welds made with Ti filler additions to the 6061/SiC/20$_w$ composite indicated that the formation of Al_4C_3 had been suppressed. Instead of large plates of Al_4C_3, a very fine dendritic structure was found in the fusion zone. These dendrites were extremely small (1–5 µm) and evenly dispersed throughout the fusion zone. Microprobe analysis revealed that the dendrites consisted only of Ti and C and that the matrix was composed primarily of Ti, Al, and Si.

The Ti additions resulted in the formation of an in situ titanium aluminide matrix having TiC reinforcement within the fusion zone of the weld. Further development of filler alloy additions and processing techniques is necessary to obtain a fusion zone having an Al alloy matrix with fine, dispersed TiC dendrites.

Fukumoto, Hirose, and Kobayashi[338] applied LBW to the joining of SiC_f-reinforced Ti-6Al-4V composite to Ti-6Al-4V alloy. The weldability obtained for a wide bead (900 μm wide) was superior to that for a narrow bead (400 μm wide), the maximum joint strength of 991 MN/m^2 being obtained at the optimum laser beam position for the wide bead. When the beam position was closer to the composite than the optimum range, the SiC_f were damaged and segregation of C and Si occurred near the damaged fibers. This caused a deterioration of joint strength. When the laser beam position was further from the composite than the optimum range, the joint strength was reduced by incomplete welding and/or formation of grain boundaries between the composite matrix and the Ti-6Al-4V plate.

Finally, the joint strength was improved by heat treatment at 900°C for approximately 3.6 ks. It was therefore possible to apply LBW to the joining of MMCs to metal. The laser beam position should be controlled within the range $190 \leq x \leq 320$ μm to obtain sufficient joint strength (i.e., above 850 MN/m^2), where x is defined as the distance between the center of the bead and the edge of the composite. Such precise control of the heat source is almost impossible in other fusion welding processes. Thus LBW must be one of the most favorable fusion welding processes for continuous fiber-reinforced composites.

Electron Beam Welding (EBW).   This process appears to have potential for joining at least some SiC-reinforced Al MMCs. Like LBW, the EBW process is capable of producing rapid thermal cycles. However, the physics of beam-material interaction differs for the two processes. Heating during LBW results from the absorption of photons by the substrate, whereas heating during EBW results from the transfer of kinetic energy to the atoms of the substrate via collisions with the high-energy electrons of the beam.

Lienert, Brandon, and Lippold[339] determined that EBW could be used successfully to join the same cast DRA (A356 Al/SiC/15$_p$) without the formation of large amounts of Al_4C_3. Electron beam welds and CO_2 laser beam welds were produced at identical energy inputs, travel speeds, and focused beam diameters. The use of EBW resulted in sound welds with a uniform distribution of reinforcement and a fine grain size. The electron beam welds exhibited an order of magnitude less Al_4C_3 than the laser beam welds made using the same parameters. The electron beam welds made at 85 mm/s with sharp beam focus contained almost no Al_4C_3. Electron beam welds made at slower speeds and/or with defocused beams contained somewhat more Al_4C_3.

The different mechanisms of energy transfer in EBW and LBW appear to affect the final weld microstructures of SiC-reinforced Al MMCs and thus the weld properties. This occurs despite the fact that both welding processes allow for rapid thermal cycles with low overall heat input. Unfortunately, EBW cannot be successfully applied to all SiC-reinforced Al MMCs. For example, EBW on a wrought 2014 Al/SiC/15$_p$ MMC cut the test part. It is not known whether the problem resulted from the manufacturing process (P/M) or was inherent to the material.

During EBW of 356 Al/SiC/15$_p$, neither constituent preferentially absorbs the energy of the electrons, and heating is more uniform than with LBW.

A second important difference exists between LBW and EBW that may affect the retention of SiC. The vacuum used during EBW promotes greater evaporation of alloy additions (relative to LBW), which results in lower peak temperatures in the weld pool and greater stability of the SiC.

Finally the results of this follow-up study[340] can be more broadly interpreted to suggest that other high-melting-efficiency welding processes that allow rapid welding cycles and minimal superheating of the molten pool may also be used to join SiC-reinforced Al/SiC MMCs without the formation of potentially damaging levels of Al_4C_3.

Cast Welding.   A new method, cast welding, has been developed for metallurgically bonding inserts into a cast component during a normal casting operation. Recently, there has been interest in making pistons or other cast components that are strengthened in critical areas by ceramic reinforcement (e.g., SiC_p or Al_2O_3 short fibers). This method consists of making a preform of the ceramic reinforcement, placing it in the die, and squeeze casting or applying high pressure in order to infiltrate the ceramic preforms and form a composite component.[341,342]

This is often unsatisfactory owing to preform deformation, positioning, component size limitations, and expense. The only other method currently used for producing a component made from two different materials is to fabricate the cast and composite portions separately and perform a subsequent welding or brazing operation. However, it is very difficult to weld MMCs, and the long times involved in brazing can cause severe degradation of the properties of both the composite and the cast structure. Also, it may be impossible to access the joint location in order to perform the weld.

Cast welding, has been used to obtain the following successful bonds.

Between cast Al-12 wt% Si and inserts of

Al-12 wt% Si/SiC/55$_p$
Al-12 wt% Si/SiC/15$_p$
2014 Al/Al$_2$O$_3$/15$_p$
6061 Al/Al$_2$O$_3$/15$_p$
A356 Al/SiC/15$_p$
ZA-12
ZA-12/SiC/55$_p$
ZA-12/SiC/15$_p$

Between cast ZA-12 and inserts of

Al-12 wt% Si
A1-12 wt% Si/SiC/55$_p$
ZA-12
ZA-12/SiC/55$_p$
ZA-12/SiC/15$_p$

Between cast A1-12 wt% Si-1.5 wt% Mg-1 wt% Ni-1 wt% Cu and inserts of

A1-12 wt% Si/SiC/55$_p$
ZA-12/SiC/55$_p$

Capacitance-Discharge (C-D) Resistance Spot Welding. This process has much potential for the joining of Al, Mg, and Ti MMCs. It uses the electric energy discharged from a bank of capacitors to rapidly heat the weld zone via I^2R heating.[*] Compared with conventional RW processes, C-D RW is characterized by an extremely short weld thermal cycle, typically less than 10–20 ms in duration. This short cycle promotes concentration of the weld in a small zone near the weld interface rather than allowing it to be conducted away from this region into the surrounding base material. The process is also characterized by an extremely rapid cooling rate, which further minimizes the effect of heat on the surrounding material. The C-D welding process has been successfully applied to the following Al and Mg MMCs.

> 6061 Al/SiC/ up to 40_p
> 6061 Al/ $B_4C/30_p$
> AZ61 Mg/$B_4C/40_p$
> 6061 Al/SiC/48_f
> 2024 Al/$B_4C/30_p$

The above weld microstructures exhibit no porosity and no observable alteration of the SiC and B_4C reinforcement.

A recently completed study[343] showed that solid-state and fusion welds could be produced between sheets of monolithic and SiC_f-reinforced Ti-6Al-4V using C-D resistance spot welding. The following conclusions were reported.

- Capacitor voltage (and corresponding current) and electrode force could be controlled to reproducibly generate solid-state and fusion spot welds in both monolithic and fiber-reinforced Ti-6Al-4V sheet.

- Solid-state and fusion welds in Ti-6Al-4V sheet exhibited tensile shear strengths above the minimum requirements set forth in military specification MIL-W6858D. Shear fracture in the solid-state welds occurred through the weld interface region, whereas fracture in the fusion welds was caused by nugget pullout.

- Some of the solid-state welds produced in the fiber-reinforced sheet exhibited complete welding across the weld interface with no evidence of fiber displacement or degradation. Increased voltage promoted the initiation of melting at the fiber-matrix interface and, at sufficiently high voltage, across the entire weld interface. Excessive interface melting promoted appreciable fiber dissolution and displacement.

- Optimized solid-state welds in the fiber-reinforced material exhibited tensile strength and shear stress levels 60 and 80%, respectively, of those of similar welds produced in the monolithic material. This reduction in fracture strength was attributed to fracture along a layer of fibers in the weld HAZ adjacent and parallel to but remote from the weld interface.

---

[*] It should be noted that during resistance spot welding of Ti, I^2R heating (I, current; R, resistance) occurs principally within the Ti sheets because of the high resistivity of Ti and not at the faying surfaces because of high interface resistance. This behavior is in contrast to that of steel and Al, for which heating at the faying surfaces is much more important.

- Dissimilar solid-state and fusion welds were produced between monolithic and fiber-reinforced sheets. The average tensile shear strength and stress of the welds produced at 160 V were 70 and 80%, respectively, of that of similar welds produced in the monolithic material. The fracture of all welds occurred remote from the weld interface along an adjacent layer of fibers.

Work on C-D welding of Al MMCs has also been reported.[344] The capacitive system used was basically a forge butt welding process. Specifically, the system used oriented parts in a butt configuration, with the ends separated by a projection on one of the surfaces. A force was then applied to the workpieces, and energy from the discharging capacitors was introduced. The result was surface melting followed by impact and forging of the components. This work has shown that Al_4C_3 formation can be precluded for several types of SiC-reinforced Al MMCs. Consistent with the other RW processes, the short exposure times and rapid cooling rates (approximately $10^6/°C·s$) were responsible for retention of the SiC.

2.7.4 Brazing, Soldering, and Transient Liquid-Phase Bonding (TLP)

Brazing is an attractive alternative to other methods for attaching consolidated MMC preforms to superplastically formed, diffusion bonded Ti structures. Brazing requires minimal bonding force and temperature exposure, eliminating or minimizing mechanical damage to fibers or fiber-matrix interactions.

Suganuma, Okamoto, and Suzuki[344] described the brazing of 6061 Al reinforced with 5, 10, and 15 vol% short Al_2O_{3f} using Al-Si and Al-Mg interlayers. However, they reported bonding temperature range of 580–610°C suggests that TLP bonding may have actually occurred. A 150-μm-thick sheet of Al-10Si was used for joining, as well as a 140-μm-thick sheet of Al-1.5 Mn clad with 20 μm of Al-10Si on each side. The bonding time for each test was 10 min at a vacuum of 6.7 MPa. Bonding the composite to itself proved easier than joining the composite to monolithic 6061 Al. The tensile strengths for the similar material bonds of 5% Al_2O_{3f} material exceeded 230 MPa.

Blue, Lin, Lei, et al.[345] evaluated the infrared joining technique as a method of joining TMC to Hastelloy-X. Joining temperatures at about 1000°C were used with little or no degradation of the TMC. The process was very rapid, lasting a few seconds to a few minutes, and provided void-free joints and bond shear strengths on the order of those obtained with conventional joining processes. Shear strengths ranged from 55 MPa for 3T overlap to 128.6 MPa for 1.5T and 198.4 MPa for 1T. The three-point strength of the TMC was 1644–1680 GPa and 81–95 GPa for the modulus. The percent total strain to failure was also improved, and the brazing affected zone was reduced. With further refinement of the process and joining materials, these properties can be optimized.

The TMC was SCS-6/β21, and the Metglas Ti-based brazing foil with a 0.065-mm thickness, MBF 5012, was used as the joining material. The composition of the foil was 60 wt% Ti, 20 wt% Cu, and 20 wt% Ni and had a liquidus temperature of 936°C.

Khorunov and his associates[346] successfully soldered Al matrix composites with stainless steel and B_f as reinforcement materials.

The flux-free joining was conducted in two steps: first, the composite materials were covered with filler metal using the method of abrasive tinning, and then they were joined by passing heating current under pressure using resistance spot welding.

In the evaluation Zn-, Al- and Sn-based filler materials were examined. It was found that Sn-based filler metals did not ensure sufficient joint strength. Al-based filler metals were strong, but their liquidus temperatures were close to the fusion temperature of the Al alloy's matrix. Zn-based filler metal met the above requirement best, and as a result a solder was selected with the following composition: 65 Zn-20 Al-15 Cu. The alloy fusion temperature was 450°C, spread well over the Al matrix, and possessed rather high strength—over 300 MPa. The optimal tinning temperature was found to be 450–470°C.

It was determined that to produce reliable joints the joining temperature had to be 480–500°C for 4–6 s. Application of pressure to the specimens during the joining process helped to remove excess solder from the joint and drew together the outer layers of the reinforcing fibers.

Transient liquid-phase bonding has proven fairly successful in joining Al MMCs. The process employs a filler material or interlayer to produce a transient liquid layer at the interface between the components to be joined. Solidification of the joint occurs isothermally by diffusion. The advantage of TLP bonding of Al MMCs is that it involves lower peak processing temperatures than those used in other fusion welding processes. The lower processing temperatures result in less damage to the engineering properties of the composite in and around the joint. However, there is concern about damage to the properties of the composite base material during processing because the thermal cycles for TLP bonding hold the entire assembly at temperatures near the solidus for a longer time than in the other fusion welding processes.

TLP bonding processes for Al MMCs are controlled by several parameters, including surface finish, type and thickness of the interlayer or filler metal, processing time and temperature, and clamping pressure on the joint. The surface of Al_2O_3 must be removed to facilitate wetting of the substrate by the interlayer. The choice of the correct combination of processing parameters is critical in avoiding such problems as reinforcement-enriched or -depleted zones and Kirkendall porosity. In general, the thinnest interlayer possible should be used to allow for the shortest processing times and a narrow liquid zone. Additionally, the lowest possible temperatures should be used to limit damage to the material properties of the composite. Although the application of pressure can aid in improving interlayer-substrate contact, excess pressure can lead to expulsion of the liquid zone, which results in reinforcement-rich areas at the joint line.

2.7.4.1 SiC-reinforced Al MMC.

TLP bonding with gold and Al-Si-Mg interlayers has been used to join two 6061 Al/SiC/25$_p$ composite sheets.[347] Based on thermomechanical simulation studies, a process window of time-temperature combinations was developed that would not produce damage to the properties of the base material. The window ranged from 30 min at 565°C to 10 min at 580°C. A 0.025-mm gold interlayer was shown to produce better results than the Al-Si-Mg material. Optimum conditions of 30 min at 567–580°C were reported for the gold interlayer. Three distinct zones were observed in the vicinity of the bond interface: a particle-enriched zone, a diffuse flow zone that showed a pattern of material flow, and the undisturbed base material. No mention was made of the formation of Al_4C_3 under these conditions. Low tensile joint efficiencies (<30%) were attributed to voids at the joint interface.

2.7.4.2 Al_2O_3-reinforced Al-MMC.

The joining of 6061 $Al/Al_2O_3/15_p$ using TLP bonding has been reported.[348] Three types of interlayers demonstrated an ability to join the material: gold, copper, and Al-12Si. The bonding tests were performed in a vacuum (1.3×10^{-8} Pa) on mechanically cleaned parts using a clamping pressure of less than 70 kPa. Joining with a 25-μm-thick interlayer of gold at 580°C for 130 min produced the strongest bonds. The joints had a 323-MPa yield strength and a 341-MPa UTS with a joint efficiency of 95%. Joining with a 125-μm-thick layer of Al-12Si at 585°C for 20 min yielded bonds with nearly identical properties, however, the bond line contained residual filler material as well as voids. A 25-μm-thick layer of copper produced bonds with much lower strengths. An increase in the V_f of the reinforcement was observed near the bond lines of joints made with gold and copper, however, increased loading of reinforcement was not observed for bonds made with Al-12Si.

2.7.5 Adhesive Bonding of MMCs

Bergquist et al.[349] found that the application of conventional surface pretreatments to Al alloys reinforced with SiC_p produced surfaces that are quite different from those produced on bulk matrix metals. Grit blasting smeared the surfaces with matrix metal, leaving them Al-rich. Standard etching treatments aggressively removed Al, produced surfaces enriched with loosely bound SiC_p, and changed overall surface toughness. Examples were the FPL etch (Forest Products Laboratory, sodium dichromate-H_2SO_4) and the P2 etch (ferric sulfate plus H_2SO_4). Anodization limited SiC_p buildup and did not contribute to SiC enrichment. It was further found that the above changes depended on the time and type of the treatment. Three treated surfaces were selected for bonding at 121°C with a commercial rubber-toughened epoxy film adhesive. Thick adherend lap shear bond strengths were found to be statistically independent of surface treatment and SiC_p reinforcement.

Very strong bonds, fully equivalent to if not better than those in bulk Al, can be formed in MMCs using standard bonding procedures and conventional film adhesives.

Davies and Ritchie[350] were concerned with identifying reasons for the environmental failure of adhesive joints using Al alloy substrates and the potential for improvement in performance by the development of modified anodizing treatments. In a series of tests they demonstrated that adhesive penetration occurred in the porous surface oxides created by phosphoric and chromic acid anodizing of Al alloys. The extent of this penetration was determined by the oxide morphology produced. In the case of the phosphoric acid anodize (PAA) oxide extensive penetration resulted, but only partial penetration was evident within the chromic acid anodize (CAA) oxide. However, the CAA oxide morphology can be modified to allow extensive adhesive penetration to occur. From the data obtained from the durability tests there seems to be a correlation between the extent of adhesive penetration into the surface oxide and resulting durability performance. All this evidence supports the "microcomposite" hypothesis and its positive influence on the mechanical properties of the critical interphase region of an adhesive joint.

2.8 Forming MMCs

Designers desirous of using MMCs in future applications continually encounter one of the main drawbacks of MMCs—their poor formability.

Extensive studies have been conducted and continue to explore the superplasticity and superplastic forming of MMCs (microstructure, manufacturing methods, and thermomechanical processing).

2.8.1 Isothermal, Conventional, and High-Strain-Rate Superplasticity (HSRS)

In the case of isothermal superplasticity, the superplasticity occurs according to the strain rates. For example, the strain rate at which superplasticity takes place in 2124/SiC/20$_w$ composite is about three orders of magnitude higher than that observed in other Al/SiC composites, which is about 10^{-4}–10^{-5}/s. Again, the latter strain rates are those usually observed in conventional superplastic Al alloys.

Huang, Liu, Yao, et al.,[351] reported their work covering the superplastic behavior of a 20% SiC$_w$/6061 Al composite manufactured by the squeeze casting method under the extruded state and tensile test conditions. Their results showed that the material can exhibit extended ductility (300%) during isothermal (550°C) tensile deformation at a relatively high strain rate of 1.7×10^{-1}/s.

Mahoney and Ghosh[352] demonstrated that after thermomechanical processing, 7475/SiC/15$_w$, behaved superplastically (350%) at conventional superplastic strain rates of 2×10^{-4}/s; this is also the strain rate at which superplasticity is observed in the monolithic matrix alloy 7475. These results are similar to those observed in the P/M 64/SiC/10$_p$ composite system.[352] (Alloy P/M 64 is essentially a powder metallurgy version of the 7064 ingot alloy.) The above results suggest that superplastic deformation in P/M 64/SiC/10$_p$ and 7475/SiC/15$_w$ composites is dominated by the behavior of the matrix. This is a result quite different from that observed in the reinforced 2124 and 2024 alloys. Specifically, whereas both the 7475 and the P/M 64 unreinforced matrix alloys are superplastic, the unreinforced 2124 and 2024 matrix alloys are not. Despite this fact, both 2124/SiC/20$_w$ and 2024/SiC/20$_w$ composites behave superplastically and, furthermore, 2124/SiC/20$_w$ exhibits superplasticity at exceptionally high strain rates.

In HSRS studies[352–354] it is particularly noted that 2124/SiC/20$_w$ exhibited superplastic properties at exceptionally high strain rates. This HSRS has been observed not only in 2124/βSiC$_w$ but also in 2124/β-Si$_3$N$_{4w}$,[353–354] 7064/α-Si$_3$N$_{4w}$,[354] 6061/β-Si$_3$N$_{4w}$,[355] and mechanically alloyed IN 9021/SiC$_p$.[356] These HSRS results represent an important breakthrough, because a major limitation of current SPF technology is the slow forming rate.[357,358]

2.8.2 Applications of Superplasticity

In forming wedge-shaped, hollow MMC blades, configurations have been prepared with SCS-6$_f$-reinforced Ti sheets. The above choice of fiber and matrix has demonstrated high strength and high stiffness of continuous SiC$_f$-reinforced Ti alloy with the retention of usable properties up to 982°C, making it a very attractive material for use in high-

TABLE 2.14   Ti Matrices Used in Production of SiC/Ti Alloy Composites

CP Ti

Ti-6Al-4V

Ti-3Al-2.5V

Ti-15V-3Cr-3Sn-3Al

Beta C Ti 3Al-8V-6Cr-4Mo-4Zr

IMI 829 Ti-5.5Al-3.5Sn-3Zr-1Nb

Titanium aluminides (alpha 2 type)

temperature applications.[359] Ti alloy matrices that have been successfully composited with continuous SiC are shown in Table 2.14.

A prime example of the application of the SPF/DB approach involves a high-speed missile inlet. The reinforcement material is applied to the cowl lip and the main air duct, locally stiffening these areas with minimal thickness buildup and eliminating the requirement for splitters.

2.8.3 Materials

Like Al/SiC composites, not all Al/Si_3N_4 composites exhibit the superplasticity phenomenon. Table 2.15 lists the Si_3N_{4w} reinforced composite combinations that are superplastic and nonsuperplastic. Results from $2124/SiC_w$ and $6061/SiC_{w'}$ are included for comparison. As shown in Table 2.15, specific combinations of matrix and whiskers in an MMC are critical to the development of HSRS. A simple criterion for superplasticity based on the individual type of whisker or alloy matrix appears to be impossible. This viewpoint is supported by the observation that $2124/\beta\text{-}Si_3N_{4w}$ and $6061/\beta\text{-}Si_3N_4$ are superplastic but $2124/\alpha\text{-}Si_3N_{4w}$ and $6061/\beta\text{-}SiC$ are not.

2.8.4 Applications of SPF MMCs

The effort and work program initiated by the U.S. government and General Electric[360] to design, develop, fabricate, and demonstrate the structural and environmental suitability of reinforced Ti fan and compressor blades was successful. With $SCS\text{-}6_f$ as the reinforcing material and the design of a SPF/DB-reinforced Ti MMC advanced engine fan blade that meets the mechanical requirements, the viability for advanced blading for military engines was shown. An $8090/SiC/16.6_p$ hatch cover was fabricated by SPF for a British

TABLE 2.15   Superplasticity of Al/SiC_w and Al/Si_3N_{4w} Composites[358]

Matrix	Whisker	Superplastic
2124 Al	$\beta\text{-}Si_3N_4$	Yes
6061 Al	$\beta\text{-}Si_3N_4$	Yes
7064 Al	$\alpha\text{-}Si_3N_4$	Yes
2124 Al	$\alpha\text{-}Si_3N_4$	No
2124 Al	$\beta\text{-}SiC$	Yes
6061 Al	$\beta\text{-}SiC$	No

Aerospace aircraft and is now in production.[361] Finally, subcomponents that are representative of NASP structural types, including the fuselage and wing, have been fabricated by several methods including SPF and SPF/DB in order to demonstrate the material and structural performance of TMC. One of the most complex subcomponents made was the lightly loaded splice subcomponent. It was representative of a lightly loaded, hat-stiffened structure such as a fuselage. Both SCS-6 and SCS-9$_f$ and β21S matrix was used. It included a complex splice joint with manufacturing details such as joggles, buildups, and single-sided assembly. Two of the largest remaining components were a stabilator torque box and a fuselage section. The stabilator was 1.8 m × 2.4 m overall and consisted of large, flat skins, I-beam spars, and intercostals, some of which had been joined by SPF/DB.

2.9 FINISHING AND CORROSION PROTECTION

MMCs fabricated with high-modulus graphite or SiC reinforcements have excellent structural properties. However, serious corrosion problems can occur, especially for Al/Gr MMCs, which contain some of the most powerful couples in the galvanic series. For Al/Gr, accelerated corrosion is likely to occur when the metal foils that cover the surface of the MMCs are penetrated by pitting attack, which leads to the establishment of a Gr-metal couple.[362–365] For Al/SiC MMCs, this galvanic problem might be less severe because of the insulating nature of SiC.

Anodizing of Al alloys is an electrochemical method of converting Al metal into Al_2O_3 by adding an external current in an acid electrolyte. The most widely used electrolyte is H_2SO_4. There are two types of sulfuric anodizing: conventional anodizing, which is performed at room temperature and provides about 7–15 μm of oxide thickness and a fairly hard surface, and hard coat anodizing, which is performed at around 0°C and provides about 50 μm of oxide thickness with extreme hardness. The oxide film consists of a thin, continuous barrier layer below a thick, porous layer. The structure of the porous layer was characterized by Keller et al.[366] as a close-packed array of columnar hexagonal cells that contain a central pore normal to the substrate surface. The porous layer can be sealed in hot water or a dichromate solution to close these pores.

In a program developed by Lin, Greene, Shih, et al.,[367] 6061/SiC/25$_p$ with 10-μm particulate powder was studied. Anodic coatings were produced by anodic oxidation in an acid bath to form an oxide layer. The procedures used in the study consisted of the following steps.

CONVENTIONAL ANODIZED COATINGS

1. The part was wiped with hexane.
2. The part was immersed in hexane at 50°C for 15 min and then immersed in an alkaline solution at 66°C for 8 min.
3. The part was immersed in a deoxidizer at room temperature for 10 min.
4. The part was immersed in 10 vol% H_2SO_4 at room temperature and constant-current density applied. The coating time was about 30 min. Sulfuric acid anodize (SAA)

5. The anodized test part was immersed in hot water at 90–100°C for 20 min. Hard coat with sulfuric and oxalic acids hot water and sulfuric and oxalic acids (HWS)[368–369]

HARD ANODIZING COATINGS

1. The same procedure as for conventional anodizing coatings was used for degreasing and deoxidizing.
2. The test part was immersed in 15 vol% H_2SO_4 at 0°C, and a constant current density was applied. The coating time was about 1 h.
3. The anodized test part was immersed in hot water at 90–100°C for 30 min.

The results showed that, for 6061 Al, conventional and hard-anodized coatings (SAA + HWS) provide excellent corrosion resistance to 0.5 N NaCl (open to air). Anodized coatings (SAA + HWS) on Al/SiC provide satisfactory corrosion protection, but they are not as effective as for 6061 because the structure of the anodized layer is affected by the SiC_p. The corrosion resistance of hard-anodized Al/SiC is less than that of conventionally anodized Al/SiC because the area fraction of the continuous barrier layer for hard-anodized SiC/Al is less than that for conventionally anodized SiC/Al. A mechanism for the formation of anodized layers on Al/SiC has been proposed by Lin et al.[367]

Toyota Technological Institute and Toyoda Gousei Company, Ltd., have jointly developed a new composite jet electroplating technique for MMC coating.

With conventional electroplating methods, increasing the electric current density generates an ion-deficient diffusion layer on the plated object surface, resulting in faulty plating, and so plating at a high current density had been impossible. The new jet electroplating technique is based on jet flow, with the ions supplied forcibly, and so the diffusion layer on the object surface is very thin, allowing plating at a high current density and at a high deposition rate.

For good adherence between the substrate and the deposit, a lower reinforcing particle V_f in deposit near the substrate is better, but for good wear resistance, a higher reinforcing particle V_f in the deposit surface is better. Therefore, a gradient is necessary in the reinforcing particles V_f inside the deposit.

An Al_2O_3/Ni_p system was used as the model composite initially, however, the technique is also applicable to other composite systems. The moves to lighten automobiles in recent years have triggered the use of Al alloys in place of steel materials, but the inadequate hardness of sliding parts is a problem. The new technique resolves this problem, and commercialization enables Al composite alloys to be used for electroplating even for the sliding parts of automobile engines and compressors.[368,369]

2.10 CMCs

Although a number of promising fiber-reinforced ceramics have been produced and characterized by mechanical testing, several other considerations must be taken into account if these materials are to be used in practical applications.

One major consideration is machining of CMC materials. The term *machining* is used here to include EDM, grinding, drilling and cutting of holes and other types of material removal operations. Holes can be required for mechanical joining as well as for other purposes. Other types of cuts are required for specific applications.

Another major consideration is joining. In general, a CMC needs to be attached to other substructures of the same material or of other materials such as metals. In some cases, such as in the use of whisker-reinforced material as a cutting tool bit, a simple mechanical joint identical to that used for monolithic ceramics can be utilized. However, for many of the applications envisioned for fiber- or whisker-reinforced ceramics, more complicated joining systems need to be developed.

2.10.1 Machining of CMCs

Conventional machining of high-performance ceramic materials is relatively well developed and can be adapted for CMCs. Because of the hardness of many of these materials, diamond is usually required as the abrading material. As material hardness is increased, conventional cutting becomes increasingly more time-consuming and expensive. Consequently, there has been increasing emphasis on newer machining techniques.

Perhaps the most promising method for machining CMCs is laser machining. Ridealgh et al.[370] have provided some fundamental information on laser cutting of a glass-ceramic matrix composite and compared laser cutting with conventional cutting using a diamond abrasive. They used a material containing about 50 vol% of uniaxially aligned Nicalon fibers. A 2-kW continuous-wave CO_2 laser, along with a moving table, was used. Beam diameters were 1–3 mm, and power and table speed were varied to give a range of conditions. The power level was varied from 0.9 kW to 1.6 kW, and speed from 22 mm/s to 505 mm/s. Cuts were made both parallel and perpendicular to the fiber orientation direction. After laser processing, cut widths were microscopically determined.

Conventional cutting (sawing) was carried out using two speeds, 250 and 4000 r/min, with a diamond abrasive wheel. Wheels having thicknesses of 0.3 and 1 mm were used.

The laser cutting did not cause any microcracking, and so there was no difficulty in measuring the dimensions of the cuts. With 1 kW power and a 1 mm beam diameter, full-section (9-mm) cut depths were attained either perpendicular or parallel to the fiber alignment direction at table speeds of less than about 40 mm/s. Cut depth decreased to less than 0.1 mm at a speed of 505 mm/s. Cut width was decreased from about 0.15 mm to 0.05 mm over the same range of table speed. In contrast, a full-depth cut was not attained under any of the conditions used with a 3-mm beam diameter.

The use of a diamond abrasive wheel resulted in cutting of the composite by continuous multifracturing of the constituents, causing considerable damage to the glass-ceramic matrix at the cut face. At the lower sawing speed, damage varied along the length of the cut. Damage was greater, but more uniform, at the higher sawing speed. Local multiple fracture of the matrix and debonding at the fiber-matrix interface resulted in removal of matrix material, leaving protruding fibers. Similar features were observed, to a lesser extent, in saw cuts made parallel to the fibers.

Fibers also protruded from laser-cut surfaces but to a lesser extent than from the diamond-sawn surface cut at the higher speed. No debonding of fibers was observed with laser cutting. A glassy appearance of the matrix indicates that laser cutting involves melting and some resolidification. Protrusion of the fibers might be attributable to the higher melting point and thermal conductivity of SiC relative to those of the matrix material. However, the rounded ends of the fibers, in contrast to the jagged ends observed with diamond cutting, indicated that some melting of fibers had occurred. Hence it was concluded that the main mechanism of separation of fibers was melting and not fracture under stress as the matrix melted.

In general, the work indicated that laser cutting can be an economical process for CMCs and that the cut surface suffers less damage than with conventional cutting using a diamond abrasive wheel.

Other newer techniques for machining advanced ceramic materials can be adapted for CMCs, although much development work needs to be done. Machining with high-pressure water jets is a well-known technique that is beginning to be applied to ceramics.[371] WJ machining makes use of the common phenomenon of erosion produced by water streams. Although the natural process is very slow, compression of water under pressures as high as 370 MPa and the use of a nozzle having a diameter of about 10–20 μm result in water velocities up to three times the velocity of sound and a great enhancement of cutting speed.

The cutting action of a high-pressure WJ involves compressive shearing. Although compressive shearing of many advanced ceramic materials is made difficult because of the high hardness of these materials, the WJ process can be made more practical by the addition of solid particles to the water stream. Particles suspended in the water have the same velocity as the water and create a higher level of kinetic energy because of their higher mass. The introduction of particles also creates a very effective pulsed erosion process because of the mass difference between the particles and the water. Since the water pressurizing system requires very pure water for minimization of corrosion and erosion of the system, the particles are introduced at the nozzle exit.

Another technique suitable for the machining of advanced ceramics, including CMCs, is EDM.[372] The workpiece and the shaping tool are the electrodes in EDM, and a liquid dielectric separates them so that there is no direct physical contact and, consequently, no mechanical stress on the workpiece. Machining advanced ceramics into intricate shapes with close tolerances and a mirror finish is possible.

The last method of machining to be described here is USM.[373] USM is nonthermal, nonchemical, and nonelectrical. Materials that have been machined by this process include Al_2O_3, SiC, Si_3N_4, and borosilicate glass, all of which have been used as matrices in CMCs. Typical linear oscillation frequency is 20,000/s. When used with an abrasive slurry flowing around the cutting tool, the workpiece is microscopically ground. The metal toolholder and tool must be designed to transmit the acoustic energy properly and to resonate within the bandwidth of the transducer used. The maximum stroke usually required for grinding operations is 0.064 mm.

The abrasive is a key aspect of USM. B_4C, SiC, and Al_2O_3 are typical abrasives. The abrasive is suspended in water, usually at a concentration of 20–50%. The slurry is circulated at a high rate to cool the tool and the workpiece, supply fresh abrasive to the cutting location, and remove abraded particles. Abrasive size depends upon the desired speed of cut and surface finish.

It seems likely that all these methods, and perhaps others, will have a place in CMC technology.

2.10.1.1 EDM CMCs. One of the recent, and successfully applied nontraditional machining techniques that utilizes electrothermal energy is EDM. Unlike traditional methods, this procedure does not depend on material hardness but requires that the resistivity of the workpiece be lower than 100 Ω·cm.[374] The tool does not touch the workpiece but is connected to electrodes separated by a dielectric fluid. There are two principal types of machines, the ram-type, also known as die sinking, and the wire type. For ram-type machines, the workpiece is generally the cathode and the shaping tool is the anode. However, in wire-cut EDM, the reverse is true.[374] The addition of computer numerical control provides nearly rectangular-shaped current pulses where the rise in voltage is held constant for a period of time known as *on-time*. During such on-times, sparks are generated, and electric energy is converted to thermal energy. This thermal energy can remove material from the surface of the workpiece by two erosion mechanisms: melting and spalling. With the melt mechanism the material is removed when the melted region is swept by the dielectric fluid.[375] With the spalling mechanism, a steep temperature gradient exists and creates internal stresses high enough to overcome the bond strength, causing cleavage of crystals on the surface.

Gadalla and Bedi initiated a study[30] on TiB_2/BN composites, materials used in the vacuum metallizing industry. Although TiB_2/BN can be readily machined with traditional tools, EDM is more attractive because it allows intricate shaping and is expected to yield higher material removal rates since the composite hardness does not affect machining in the latter case.

TiB_2 conducts current and forms a liquid phase at the interface with BN. Neighboring crystals of BN and some TiB_2 spall as a result of thermal shock. Composites rich in TiB_2 or with fine TiB_2 grains gave high material removal rates. Increasing the amount of the conducting phase by 10 times was as effective as decreasing the grain size from 11 μm to 7 μm. They found that coarse TiB_2 could withstand high pulse durations before the wire broke. The material removal rate increased with pulse duration, frequency, and current. For the same composition and grain size, increasing the pulse duration or current increased the crater depth (the roughness) up to a certain value, beyond which increasing these parameters yielded a smoother surface. The conductivity of the dielectric was effective only for compositions rich in TiB_2. In such cases, higher water conductivity lowered the energy required for material removal.[376]

Generally, most of the properties of TiB_2 are similar to those of SiC. TiB_2 (16 vol%) reinforcement of SiC exhibits a 30% higher flexural strength (478 MPa) than SiC at 20°C. This strength is unaffected by temperatures up to at least 1200°C.[37]

The addition of TiB_2 to a SiC matrix significantly reduces the grain size of the matrix in the sintered microstructure. TiB_{2p} reportedly pin grain boundaries and inhibit grain growth.

Electric resistivity is lower for the SiC/TiB_2 composite because of the good conductance of TiB_2.

With appropriate composite ratios, SiC/TiB_2 can be electric discharge-machined. For example, an EDM rocker arm insert made from a TiB_{2p}-toughened SiC composite is 50–75% tougher than direct sintered SiC.[31,377,378]

In laboratory tests, Cales, Martin, and Vivier[43] found that new conductive Si_3N_4- and Al_2O_3-based composites with high mechanical properties could be developed for this purpose by adding amounts of TiC and/or TiN_p in the ceramic matrix. They concluded that they had developed special grades of conductive ceramic composites with either a Si_3N_4 or an Al_2O_3 matrix for EDM. These composites exhibit high fracture strength, at least equal to that of the matrix, and high electric discharge machinability. They are more easily machined than WC, and tool consumption is considerably lower. Machined surfaces with low roughness and high material removal rate have been obtained after optimization of the EDM conditions. In the future, they are expected to replace conventional Si_3N_4- and whisker-reinforced Al_2O_3 ceramics in a number of applications where complex machined parts are required.

The fracture toughness of SiC whisker-reinforced Al_2O_3 is nearly double that of the material without the fibers. The same is true with Si_3N_4-based composites. The particle size and percentage of TiC or TiN to be added to the matrix can be adjusted to make it electrically conductive enough to carry out the EDM process without significantly compromising the ultimate properties and performance requirements of the material. Tests have shown that for Si_3N_4-based composites with TiN added, EDM can be performed at a conductivity higher than $2 \times 10^{-2}/\Omega \cdot cm$ (and preferably $5 \times 10^3/\Omega \cdot cm$). In contrast, for wire EDM of SiC_w-reinforced ZrO_2-toughened Al_2O_3 with TiC added, the minimum conductivity value was found to be $1/\Omega \cdot cm$. With a 30 vol% TiC addition to the Al_2O_3-based composite, the bend strength was reported to be 860 MPa, while up to 50 vol% of TiN_p could be added to the Si_3N_4 matrix without reducing its fracture toughness. This is one example where difficult-to-machine materials such as ceramic composites can be tamed by making them electrically conductive and able to be processed by EDM.

The machinability of this composition is twice that of WC and approaches that of steel. The strengthening feature is not evident when coarser-grade TiN_p are used, suggesting that there is a critical size effect as well. The major drawback with such compositions is the temperature capability, which is realistically limited to about 1000°C because of the onset of oxidation of the metallic phases beyond that point.[379]

Such findings suggest that there is a reason to expect further development of EDM as a potential finish machining technique with reasonable material removal rates.[380–384]

An electroconductive ceramic composite has been developed and tested for high-temperature extrusion dies, cutting tools, and wear parts. By providing an electroconductive ceramic, the ceramic can be electric discharge-machined to virtually exact shape subsequent to furnace sintering, hot pressing, or hot isostatic pressing to the final density and hardness of the ceramic. Thus complex geometries and features such as holes, chamfers, slots, angles, changing radii, and so on, can be electric discharge-machined into the composite that could not previously be either provided or produced economically. After the EDM, minimal diamond grinding for final dimensionality or surface condition may be provided, if desired, for the particular application. Although the ceramic composite is specifically directed toward die construction, tooling, and wear parts, the properties of the ceramic allow other applications, in particular where intricate shapes are required.

The new ceramic composite includes a base nonconductive ceramic component such as Al_2O_3 to which is added sufficient amounts of an electroconductive ceramic component such as TiC, TiB_2, ZrB_2, or TiN to achieve an electric resistance of less than about 10 $\Omega \cdot cm$ specific resistance. To overcome the decrease in the toughness and/or strength of the ceramic composite resulting from the addition of electroconductive ceramic, SiC_w are added to the ceramic composite in sufficient amount to negate the detrimental effect on mechanical properties, wear properties, and the hardness of the electroconductive ceramics added. This results in an improvement in fracture toughness two- to three-fold over that of the individual ceramic components. Strength and hardness are also increased.

The surface quality obtainable with EDM of the new composite is comparable with that achieved by the best conventional grinding methods used to obtain a high-quality surface finish. For example, a billet of the new electric discharge-machinable whisker ceramic composite was cut by a traveling wire electrodischarge machine, and upon examination minimal removal of material, on the order of 0.01 mm, produced a surface finish of 16 μ in. This surface finish is comparable to surface finishes obtained by costly diamond grinding processes.

The electrodischarge cutting speeds of the new ceramic composite compare with the electrodischarge cutting speeds of typical tool steels and exceed the cutting speeds typically used to cut carbide materials. Moreover, the new ceramic composite does not exhibit edge chipping or a limitation with respect to minimum EDM section thickness, which is indicative of a minimal reduction in strength associated with the thermally affected surface material.[380–384]

2.10.1.2 Grinding.

Rezaei, Suto, Waida, et al.[46] conducted an experimental investigation into the creep feed grinding of CMCs. They believe that creep feed grinding with a diamond wheel has been shown to be the most efficient method of stock removal.[47] The main problem with grinding of ceramic is the rapid wear of the diamond abrasives which makes the process very costly. To increase the efficiency of the grinding process, novel grinding wheels have been developed that have superior cooling and lubrication ability.[48] Dressing diamond wheels has also been the subject of other research, and unique dressing methods have been reported.[49] Their investigation compared Si_3N_4, Al_2O_3, SiC, ZrO_2, Al-reinforced Al_2O_3, SiC_w-reinforced Si_3N_4 and C_f-reinforced Si_3N_4. The last-mentioned material was ground in three directions: longitudinal, transverse, and normal to the fiber direction. All specimens were carefully prepared by grinding all sides. They were all 100 mm long except for the fiber-reinforced ceramics, which were 70 mm long. The grinding depth of cut was set at 2 mm, and the table speed at 30 mm/min. For the case of whisker-reinforced Si_3N_4, a table speed of 18 mm/min was selected as this composite was found to be more difficult to grind. They concluded the following.

- Among the ceramic materials, SiC requires the highest grinding power, but Si_3N_4 induces the highest normal grinding force. The latter was found to be the most difficult ceramic to grind.
- Al-reinforced Al_2O_3 is slightly easier to grind than Al_2O_3, and consequently the surface finish produced is rougher than that of Al_2O_3.
- C_f-reinforced Si_3N_4 was ground with a tenth of the power required for Si_3N_4 when the direction of grinding was along the length of the fiber. When the grinding direction was normal to the fiber, the grinding power was half that required for Si_3N_4.
- In contrast to other ceramic composites investigated, the SiC_w-reinforced composite was more difficult to grind than the pure Si_3N_4. Consequently, the surface roughness of the ground composite decreased.

- The surface roughnesses of the fiber-reinforced Si_3N_4 specimen and the Al-reinforced Al_2O_3 generally increased by a factor of 1.4 as compared with those for the pure ceramics.

2.10.1.3 Drilling.   High-power electron beams can be used to machine vias and interconnecting structures in CMC green sheets. For example, unfired Al_2O_3 ceramic sheets (composed of 90% Al_2O_3, 6.7% polyvinyl butyral, and 3.3% dioctyl phthalate) machined using 100-keV electrons and a 1-mA beam current. The advantages of this technology are direct maskless metallization, noncontact machining of high-density vias, and interconnecting structures of fine dimensions. These features translate into the design and development of state-of-the-art multilayer ceramic (MLC) packaging on a quick turnaround basis from designing to building. The electron beam technology reported by Sarfaraz, Yau, and Sandhu[385] offers accurate machining of three-dimensional interconnecting line structures. Vias of aspect ratio >3 and diameters as small as 30 μm have been fabricated.

Laser drilling of unfired ceramic sheets has been reported previously. However, it has been limited to generating only via holes.[386] Although small via holes can be produced using laser drilling technology, the fabrication of separate screening masks is still required for full metallization of the interconnecting wiring patterns. Thus the major advantage of the described electron beam machining process is its maskless ability to generate fully metallized ceramic sheets.

The amount of material removed is in general proportional to the power transferred from the electron beam to the target material. A different beam energy and beam current may be needed to optimally process different material sets.

At this time, the electron beam process throughput is not comparable to that of the conventional technique. The technology, however, offers a quick turnaround in the building of a prototype product, which is key for today's competitive marketplace for advanced packaging technology. In addition, with the high-resolution capability and the noncontact machining process, electron beams can deliver smaller features and higher-density wiring patterns with superb accuracy.

2.10.1.4 Cutting

AWJ.[387–392]   Tuersley et al.[380] examined the material removal rates and near-surface damage of CMCs (7.5% SiC_w/Al_2O_3). Savrun and Taya[300] commented on the surface of the 7.5% SiC_w/Al_2O_3 CMC after machining over a range of traverse speeds. For cutting speeds of 12.7, 25.4, and 50.6 mm/min, they recorded surface finishes of 3.18, 3.56, and 3.81 μm, respectively. In this instance, a 70-mesh SiC abrasive at a flow rate of 4.95 kg/min and a water jet pressure of 275 MPa was used to cut plates 6.3 mm thick with a standoff distance of 2.54 mm. It was noted that surfaces became rougher and striations became more visible as the cutting speed increased, and there was evidence of limited plastic deformation occurring at higher speeds. A similar investigation made on a 20% vol TiB_2/SiC_p-reinforced CMC by Hamatani and Ramulu[309] found similar results, bearing in mind the differences in machining parameters. In this case, the much softer garnet abrasive was used (80-mesh) and at the more generally used flow rate of 0.54 kg/min. The water jet pressure was a comparable 240 MPa, and the stand-off distance was varied between 1.5 and 8.0 mm.

A comparative study[300] of the merits of AWJ techniques versus diamond saw cutting performed on SiC/Al$_2$O$_3$ composite found that while the material removal rates using the AWJ method were up to 20 times greater, there was an order of magnitude difference in the surface finish obtained, the diamond saw producing a 0.3- to 0.4-μm finish compared to the 3- to 4-μm finish produced by the AWJ procedure. It was also concluded that the AWJ technique is a lot more dependent on the matrix material than on the reinforcement particulate or whisker material. This suggests that there could be advantages in adoption of the AWJ procedure for rough machining.

Kahlman, Karlsson, Carlsson, et al.[389] examined the AWJ machining and wear of a variety of ceramic and CMC material systems.

The materials included in their tests are listed in Table 2.16. The results from the wear tests showed very drastic wear behavior for several of the ceramic materials. The LPS-SiC and the TiB$_2$/B$_4$C composite, for example, which have shown very promising results as erosion-resistant materials, had rapid wear in this AWJ test. The measured wear lengths and wear volumes are shown in Table 2.16.

Electric discharge machining of ceramics, for instance, removes material by a similar type of thermal spalling as AWJ machining, but ceramics often need secondary machining to remove surface debris before being used in machine applications. This can be avoided when WJs are used because the loose fragments are washed away, leaving a smooth surface with a wavelike appearance, when cutting thicker billets.[390–391]

To machine CMC materials it is necessary to use other abrasives (such as Al$_2$O$_3$ in lieu of garnet) with higher melting points, but the nozzle materials currently in use limit this approach because the life of a standard nozzle decreases from about 2 h to 15 min when higher-melting-point materials are used.

Schwetz, Sigl, Greim, et al.[392] investigated fine-grained B$_4$C and new B$_4$C/TiB$_2$ composites with improved mechanical properties that were developed to be used as nozzle materials for AWJ machining.

The rate of erosion caused by abrasive particle impact on dense plates of B$_4$C, B$_4$C/TiB$_2$, and SiC ceramics was measured with an AWJ at different impingement angles. Finally, the wear resistance of B$_4$C ceramics was compared with that of hard metals by determining the life of AWJ nozzles made from these materials.

B$_4$C outperforms hard metals as AWJ nozzle material especially under conditions where very hard abrasives such as Al$_2$O$_3$ are used. The combination B$_4$C nozzle-Al$_2$O$_3$ abrasive is economical for difficult cutting jobs, especially for the cutting of very hard, tough workpiece materials such as modern ceramics, cermets, and CMCs.

Laser Beam.   CO$_2$ laser scribing and cutting were studied on a C$_f$/SiC matrix composite nominally containing 45 vol% of carbon fibers by Trubelja, Ramanathan, Modest, et al.[393] The scribing was performed in a continuous-wave (CW) mode using laser powers between 750 and 1500 W and specimen translation velocities between 0.5 and 4 cm/s. The laser spot size was 300 μm in diameter. Reasonably good agreement between theory and experiment was found. The microstructures of the laser-cut surfaces indicated the formation of redeposit by condensation from the vapor phase. The four-point bending strength of the laser-cut composite was found to be approximately 20% lower than the corresponding strength of the diamond-cut composite. Strength was fully recovered after 180 ± 10 μm of the material was removed from the lased surface by grinding.

TABLE 2.16   Ceramic and CMC Materials Evaluated in AWJ Tests[389]

Material[a]	Fracture Strength, σ_B (MPa)	Poisson's Ratio, ν	Modulus of Elasticity, E (GPa)	Thermal Expansion Coefficient, α ($\times 10^{-6}/°C$)	Thermal Conductivity, λ (W/m-K)	$\Delta T_{max} = \dfrac{\sigma_B(1-\nu)}{E\alpha}$ °C	$R' = \Delta T_{max}\,\lambda$ (W/mm)	Wear Length, l (mm)	Wear Volume, V (mm³)
Al₂O₃ (S)	350	0.22	350	8.1	27	95	2.6	>46	180
TiB₂ + 30 vol% B₄C (HIP)	460	0.20	480	6.7	60	115	6.9	31.8	44.8
TiB₂ (HIP)									
No. 1	550	0.20	490	7.4	70	120	8.5	31.7	45.1
No. 2	550	0.20	490	7.4	70	120	8.5	31.5	44.1
Al₂O₃ + 25 vol% SiC$_w$ (HP)									
No. 1	900	0.20	390	6.0	36	310	11.1	14.6	—
No. 2	900	0.20	390	6.0	36	310	11.1	13.6	—
SiC (LPS)									
No. 1	515	0.20	420	5.1	77	190	15	29.7	39.8
No. 2	515	0.20	420	5.1	77	190	15	28.4	29.9
Si₃N₄ (HIP)	670	0.20	310	3.7	34	465	16	30.4	32.2
Si₃N₄ + 30 vol% SiC (HIP)									
No. 1	750	0.20	350	3.8	40	450	18	20.6	13.8
No. 2	750	0.20	350	3.8	41	450	18	16.6	7.8
SiC (SSS)	410	0.17	410	4.0	102	210	21	17.7	8.0
WC + 7 vol% Co (S)	—	—	—	—	90[b]	775	~70	9.6	—
WC + 11 vol% MO/MoC (S)	—	—	—	—	>100[c]	530	~100	4.3	—

[a] HIP, Hot isostatically pressed; HP, hot pressed; S, sintered; SSS, solid-state-sintered; LPS, liquid-phase-sintered.
[b] $\lambda_{Co} = 70$.
[c] $\lambda_{Mo} = 140$.

The oxidation resistance of the laser- and diamond-cut composites was studied at 1103, 1304, and 1402°C, as well as the oxidation behavior at these temperatures. They found that for both materials there was an initial mass loss attributable to the oxidation of carbon and possible active oxidation of SiC, followed by a slow mass gain due to passive oxidation of SiC. At 1304°C the rate of passive oxidation of SiC in the laser-cut material was somewhat higher than in the diamond-cut material. At 1402°C, the diamond-cut surface oxidized more rapidly than the laser-cut surface. The differences in oxidation rates were attributed to differences in microstructure.

2.10.1.5 Cutters

Composites and Cutting Tools. Cutting tools exemplify the application of materials in several high-stress and high-temperature environments. In metal cutting, the cutting tool, acting as a blunt wedge, removes material from the workpiece by deforming a thin layer of the workpiece and separating the material near the cutting edge of the tool by a high-strain-rate plastic shear mechanism. A significant part of the mechanical energy supplied in the work of cutting is converted to thermal energy and high temperatures are consequently generated. Heat produced by the primary shearing process flows into both the workpiece and the chip, with the majority going into the chip, while friction developed at the tool-chip interface and rubbing along the tool flank heats both the tool and the workpiece.

The primary mode of mechanical wear for ceramic cutting tools is abrasion, the removal of tool material by a scoring action of protruding asperities and by hard phase inclusions in the workpiece and chip. The two-body abrasive wear resistance of metals that can accommodate large strains prior to fracture has been the subject of numerous investigations, and it has been shown that the wear resistance is determined by the hardness of the material.

There have been no clear-cut scientific selection criteria developed to match the specific workpiece with the optimum and most cost-effective type of cutting tool. Most often users follow the recommendations of a vendor or the guidelines in handbooks and use their own trial-and-error experience to arrive at the proper choice of cutting tool.

The scientific approach would involve examining all the measured data and the criteria developed as the fallout of all test results for the user to be able to systematically arrive at the best choice for a specific machining application.

Cutting Tool Materials. Because hard materials are used to cut softer materials and because ceramics are inherently hard, an interest in the use of ceramics for cutting tools developed in Germany around the beginning of the century (1905).[394] Al_2O_3 was the material considered for cutting tools. Patents were issued in 1912 and 1913 in the United Kingdom and in the Federal Republic of Germany, respectively. Cemented carbides were introduced in Europe in 1926, and during World War II, carbides were used to increase the cutting capability of the earlier HSS tools. This was followed by the introduction of Al_2O_3-coated WC. In the 1960s, ceramic cutting tools were based on sintered or hot-pressed polycrystalline Al_2O_3. Two factors limited the widespread adoption of these Al_2O_3 tools by the metal cutting industries. The inherent low fracture toughness of the material limited the range of application, and the low thermal conductivity increased the susceptibility of the tool to damage by thermal shock.

It is apparent that the strengths of ceramic tools have been improved since the end of the 1960s through better control of the microstructure and by the use of additives. A better understanding was developed, and as a result, mechanical, thermal, and chemical properties of ceramics were obtained and adapted to ceramic cutting tools. In addition, composition variables gave rise to the cermets, and additions of Ti, TiC, and WC were made to improve the strength of Al_2O_3 cutting tools.

Harder materials such as PCD and cubic boron nitride were developed for use during 1972–1974. CBN, Al_2O_3-TiC ceramics, and the newer CMCs based on Si_3N_4, SiC_w/Al_2O_3, and Al_2O_3/ZrO_2 are very promising because of their wear resistance and effective application to the machining of Ni- and Co-based superalloys in the jet engine industry. PCDs find excellent application in machining abrasive materials such as Si-Al automotive alloys and Gl- and C_f-reinforced composite materials.

The demand for CMCs will strengthen as the ceramic industry scales up manufacturing processes, costs become more competitive with those of conventional materials, and users gain confidence in the composites' properties and reliability. Major market segments are the cutting tool inserts TiC_p/Si_3N_4, TiC_p/Al_2O_3, and SiC_w/Al_2O_3.

Alumina.   Two innovations in Al_2O_3 material technology aimed at increasing the fracture toughness of Al_2O_3 have found commercial utilization as ceramic cutting tools. The first, transformation-toughened Al_2O_3, was introduced into the marketplace about 1980 and was targeted to expand the application range of ceramics for steel machining. This family of ceramics consists of particles dispersed in an Al_2O_3 matrix. When a load is applied, the tensile stress field surrounding a crack tip causes the tetragonal ZrO_2 to transform to the stable monoclinic polymorph. This transformation, with an accompanying volume expansion of the dispersoid, leads to an increase in fracture toughness and retards crack propagation.

The newest Al_2O_3-based material, SiC_w-reinforced Al_2O_3, was added to the cutting tool arsenal only within the past 5 years. The whisker reinforcement produces a twofold increase in fracture toughness of the composite relative to monolithic Al_2O_3. SiC_w-reinforced Al_2O_3 has received widespread acceptance in the aerospace industry, where it is considered the state-of-the-art cutting tool material for rough machining of superalloys.

One of the first ceramic composite cutting tools was SiC_w/Al_2O_3.[395–399] Advanced Composite Materials Corporation (ACMC) grows SiC crystals from rice hulls, which are composed mostly of relatively pure silica. The silica reacts with carbon when heated to 1800°C. Joining these SiC_w with Al_2O_3 produces ARtuff, ACMC's high-performance ceramic composite. ARtuff ceramic composites make high-performance cutting tools for superalloys, nozzle inserts for water jet looms for the textile industry, cutting tools for the furniture industry, and can-forming tools for the beverage industry. Like most whisker-reinforced ceramics, ARtuff has several advantages, including strength, fracture toughness, and resistance to wear and thermal shock.

Another firm, Greenleaf Corporation, uses ACMC's SiC_w in its machine tool inserts. WG-300 Al_2O_3 with SiC_w forms dozens of cutting tools that make cuts on Ni-based alloys without smearing.

WG-300 has high thermal and mechanical shock resistance that holds up 6 to 8 times better than that of cemented carbides. It has been tested at speeds as high as 8 times that of carbide at 1.5 times carbide feed rates and looks particularly good for machining forged high-Ni alloys (e.g., Inconels and Hastelloys), where high thermal and mechanical shock are encountered.

Whisker reinforcement more than doubles the fracture toughness of oxide ceramics up to temperatures as high as 1000°C; this makes them more resistant to catastrophic failure and reduces their chances of failure with time. The composite's thermal conductivity is also about 40% better than that of the unreinforced matrix material.

Because of WG-300's increased toughness, cutting tools made with it have sharp rather than beveled edges. Cutting tools with beveled edges cause finished cuts in the hardest alloys to look smudged. Cutting and shaping tools using WG-300 inserts have 10 times the metal removal rate of those fitted with WC. Whiskers also greatly enhance the material's thermal conductivity and thermal shock resistance, permitting use with coolants.

Greenleaf Corporation also makes another whisker-reinforced cutting tool, WG-100, which forms drills, mills, routers, and other tools for machining advanced composites such as CCC, Kv/Ep, and nonmetallic substances.

The Al_2O_3 with 5% ZrO_2 has been used to machine gray cast iron for automotive components. Toughened Al_2O_3 with 20% ZrO_2, has been used at cutting speeds of up to 5000 ft^2/min and was adopted by the steel industry because of its capability for breaking steel chips. Al_2O_3-TiC composites, containing at least 20% TiC, formed by hot pressing and sintering in an inert atmosphere have been used for milling and rough turning applications of cast iron and high-temperature Ni-based superalloys. These Al_2O_3/TiC carbide tools are cost-competitive with traditional carbide tools and can operate at significantly higher cutting speeds. A cutting tool insert of SiC/Al_2O_3, Quantum 10, is claimed to outperform other ceramic tool inserts, such as coated carbides, cermets, and CBN, both in cutting speed and tool life when cutting hardened steels and superalloys. Competitive materials are being produced in Japan where a ceramic composite ultrahard tool material based on SiC_w and ZrO_2 in an Al_2O_3 matrix has been reported to have a fracture toughness of 8 MPa m$^{1/2}$ and a three-point bend strength of 1600 MPa.

ZrO_2.   Two new cutting tools from toughened ZrO_2 ceramic have been introduced.[400] Corning claims that the material offers three times the ductility of a comparable ZrO_2 material as well as metallike strain-to-failure characteristics normally found only in fiber-reinforced ceramic materials.

The toughened ZrO_2 combines with Al_2O_3 and has successfully machined ferrous-based alloys. Initial test results show that a ceramic composite cutter offers up to six times the wear resistance of cutting tools made from coated carbides.

The Greenleaf Corporation claims that the new cutting tools have outperformed all traditional ceramic and nonceramic products at both high and low cutting speeds in most applications. The ceramic material doesn't chip or fracture at the edges, as often occurs in conventional ceramic cutting tools. Greenleaf has found that the new products have twice the strength of the company's more traditional Al_2O_3 products.[401]

Norton Company has developed Norzide zirconia (TZP). The material belongs to a relatively new class of sintered ZrO_2-based ceramics characterized by high strength, high toughness, and superior wear resistance. According to Norton,[400] one formulation of the material, Norzide TZ-110HS, has been used as a bed knife in a granulator and lasted three times longer than the submicron grain size grade C-10 WC previously used. In another knife application—a slitting operation—the same material's precision-honed cutting edge outlasted razor blade grade stainless steel by a factor of 24.

Withers[402] examined two composites, sol-gel-derived $Al_2O_3/ZrO_2/SiC_w$ and pressureless-sintered ZrO_2/TiC composites which were determined to posess the optimum properties for cutting tool applications. The ZrO_2/TiC composites, which were prepared and consolidated by novel processing, exhibited exceptional high-temperature properties. Machining tests were performed on HSS with these composites. The SiC_w-reinforced composites and the ZrO_2/TiC composites had a significantly improved volume of metal removal over that of commercial cutting tools for machining steel. The ZrO_2/TiC composites have the advantage, in addition to superior cutting performance, of being produced via a cost-effective method (pressureless sintering).

Si_3N_4.   Ford Motor Company introduced an S-8 ceramic cutting tool material consisting of Si_3N_4 with additions of Y_2O_3 and Al_2O_3. It has been used to machine production auto engines. Several General Motors divisions also use these tools for bore-cutting 2.5-L engines, turning brake drums, and face milling 4.1-2 V-8 engine blocks.[403] Norton Company has developed a Si_3N_4-reinforced with SiC_w ceramic cutting material.

Sialon (Si_3N_4/Al_2O_3).   Kennametal has developed several sialon cutting tools. Kyon 2000 has an impact resistance approaching that of most ceramic-coated carbides and is capable of cutting at the high speeds normally associated with conventional ceramics. Kyon 2000 also offers excellent thermal shock resistance; coolant can be used to improve both tool life and surface finish. The recorded productivity gains of sialon cutting tools, such as Kyon 2000, are increased metal removal rates of 100–300% on cast iron and 400–800% on high-temperature nickel-based alloys when compared to the removal rates for carbides. Recently introduced is Kyon 3000 which has a better performance than Kyon 2000.

Titanium Carbide (TiC).   General Telephone (GTE) has developed a Si_3N_4/TiC composite with a high fracture toughness and hardness better than those of sialon alloys. Originally introduced as Quantum 5000, the composite consists of a matrix phase (70%) and a dispersed phase of TiC (30%). The Si_3N_4 matrix phase also contains additions of Y_2O_3 and Al_2O_3. GTE claims higher productivity (higher feed rates and cutting speeds) than those for sialon and has introduced an improved version known as Quantum 6000 which it claims has superior performance.

Titanium Boride.   In a series of developments in Japan, several laboratories have produced composites of 70 wt% Ti(C,N)-30 wt% TiB_2 with a three-point flexural strength of over 800 MPa at room temperature, $K_{IC} > 5$ MPa m$^{1/2}$, H > 2500, and $\alpha = 8 \times 10^{-6}/°C$, as well as excellent cutting tool behavior. The Ti(C,N)/Cr_3C_2 system is also now under investigation. Both Ti(C,N)/TiB_2 and Ti(C,N)/Cr_3C_2 system cutting tools have been verified to have a longer lifetime in the high-speed (300-m/min) cutting of plain carbon steel than WC-Co and TiN cermet tools have under comparable circumstances. These new composite materials are both now under evaluation for the cutting of heat-resistant and stainless steels.

Composite Cutter Applications.   A new group of ceramic cutting tool materials, which was introduced in recent years and is now used commercially in a wide range of metal cutting applications, including turning and milling, is comprised of three main categories: metal oxides, metal nitrides, and metal carbides (Table 2.17).

TABLE 2.17   Materials Comparison Chart

Material	Strengths	Weaknesses	Typical Applications
HSS	Superior Shock Resistance Versatility	Poor Speed Capabilities Poor Wear Resistance	Screw Machine and Other Low-Speed Operations Interrupted Cuts, Low-Horsepower Machining
Carbide	Most Versatile Cutting Material; High Shock Resistance	Limited Speed Capabilities	Finishing to Heavy Roughing of Most Materials, Including Irons, Steels, Exotics, and Plastics
Coated Carbide	High Versatility; Good Performance at Moderate Speeds	Limited to Moderate Speeds	Same as Carbide, Except with Higher Speed Capabilities
Cermet	High Versatility; Good Performance at Moderate Speeds	Low Shock Resistance; Limited to Moderate Speeds	Finishing Operations on Irons, Steels, Stainless Steels, and Aluminum Alloys
Ceramic Hot/Cold Pressed	High Abrasion Resistance; High Speed Capabilities; Versatility	Low Mechanical Shock Resistance; Low Thermal Shock Resistance	Steel Mill-Roll Resurfacing, Finishing Operations on Cast Irons and Steels
Ceramic Silicon Nitride	High Shock Resistance; High Thermal Shock Resistance	Very Limited Applications	Roughing and Finishing Operations on Cast Irons
Ceramic Whisker Reinforced	High Shock Resistance; High Thermal Shock Resistance	Limited Versatility	High-Speed Roughing and Finishing of Hardened Steels. Chilled Cast Iron, High-Nickel Superalloys
Cubic Boron Nitride	High Hot Hardness; High Strength; High Thermal Shock Resistance	Limited Performance on Materials Following; 38 RC; Limited Applications; High Cost	Hardened Work Materials in 45–70 Rockwell C Range
Polycrystalline Diamond	High Abrasion Resistance; High Speed Capabilities	Limited Applications; Low Mechanical Shock Resistance	Roughing and Finishing Operations on Abrasive Non-Ferrous or Non-Metallic Materials

Si_3N_4-based ceramics, generally used for machining gray cast irons, also have sufficient toughness for milling applications. However, the low toughness of the oxides and carboxides allows for interrupted cuts only at very low feed rates. Consequently, ZrO_2 is added to Al_2O_3 to improve the toughness of this composite cutting tool.

A hot-pressed ZrO_2/Al_2O_3 ceramic cutting tool[404] insert is being used by an aircraft engine manufacturer to machine a variety of engine parts made of Inconel 718. Often many of the company's parts must be production-machined in the fully heat-treated condition, some as hard as 48 HR_C. Cutting speeds were increased from 34 m/min for cemented carbide to 244 m/min, which is nearly double the speed of conventional hot-pressed ceramics and a sevenfold improvement over the speed of cemented carbide.

Another category is the Al_2O_3/TiC material. The addition of up to 30 wt% of TiC to Al_2O_3-based cutting tools has resulted in what is often called a *composite* or *black ceramic* and has increased the transverse rupture strength (breaking strength) of the basic Al_2O_3 ceramic. These ceramic materials have been very successful in turning hard cast irons and heat-treated steels hardened to 64 to very fine surface finishes and tolerances normally obtained only by grinding. Parts machined with this category of ceramic cutting tool range from small gear parts for the automotive industry to large form rolls for steel mills. The cutting speed is 1.3 m/s, and the feed rate is 0.25 mm/r. Productivity can double with this type of material compared to cemented WC.

Ceramics that combine Al_2O_3 and TiC are known by many names in addition to hot-pressed and black ceramics. Other terms used to describe such ceramics include *cermets, composite ceramics,* and *modified ceramics.* Hot-pressed ceramics have found application as replacements for conventional carbide in many jobs. For example, Al_2O_3/TiC hot-pressed ceramic inserts have replaced carbide tooling to effect a productivity increase and an improvement in surface finish, eliminating the need for a subsequent polishing operation.[405]

When correctly applied Si_3N_4-based composite tools are used, there are multiple benefits:

- Increased machine tool capacity
- Better quality parts
- Reliable tool life without insert breakage
- Fewer tool changes and less machine downtime
- Increased tool life.

These add up to improved productivity and reduced cost.[406]

The higher toughness and thermal shock resistance of Si_3N_4-based ceramics, compared these values for Al_2O_3, permit the rough turning and milling of cast irons under severe conditions. These conditions include heavy interruptions, variations in depth of cut, rough scale, and the use of a coolant. High-temperature Ni-based alloys can also be turned quite efficiently with this type of cutting tool. However, steels are unsuitable for machining with Si_3N_4 ceramics.

Inconel 718 castings are machined using a Si_3N_4-based ceramic with a 90% reduction in cutting time and an 84% reduction in total machining cost.

In another application using Si_3N_4 tools, a gray cast iron automotive brake disk increased the feed rate over that of Al_2O_3 from 0.38 mm/r to 0.51 mm/r and provided an average tool life of 400 pieces per cutting edge versus 150 for Al_2O_3.

Another class of cutting tools are the sialons.[407] Field tests have shown them to be successful in the machining of cast iron, Ni-based superalloys, and Al-Si alloys. In these tests the sialon tools exhibited longer tool life (at both low and high cutting speeds) than the cutting tools traditionally used for these materials.

When pressureless-sintered β-SiAlON is used, materials perform better than hot isostatically pressed β-Si_3N_4 ceramics in machining Ni-based superalloys, and the longest tool life has been obtained for (α + β)-SiAlON materials.[407]

Cutting tools are the leading edge of commercial ceramics, and breakthroughs there have a way of working their way back to other applications where wear and heat resistance are important. So it is heartening to see how researchers have solved sintering problems that occur when Si_3N_4 cutting tools are strengthened with TiC.

While TiC-Si_3N_4 tools do have excellent properties, the two materials react to form nitrogen gas when sintered. The released gas forms pores which concentrate stresses and weaken the tool. The solution to the problem is to coat TiC_p with thin TiN coatings. The TiN film keeps the TiC from reacting with Si_3N_4 during sintering, and so no nitrogen gas or by-products are formed. The resulting tool is easier to sinter and performs better than TiC-Si_3N_4 tools.

The fourth and final grouping of tools are those containing SiC_w, which are now commercially available from various manufacturers as reinforcements for cutting tools. The whiskers improve toughness, strength, thermal shock resistance, and reliability and can be added to oxide, nitride, and carboxide ceramics. Commercially available grades contain fewer than 30% whiskers, resulting in higher toughness and better thermal conductivity.

SiC_w-reinforced tools have been applied mostly in the machining of high-temperature alloys, resulting in significant productivity increases. The rough turning of a turbine rotor made of Incoloy 901 (40 HR_c) was machined using round inserts 12.7 mm in diameter at a cutting speed of 4.5 m/s and a feed rate of 0.15 mm/r. Stock removal was about five times faster than with conventional cemented carbides.

SiC_w-reinforced Al_2O_3 inserts have been evaluated in the machining of Inconel 718 also.[397] Thangaraj and Weinmann used 12.7-mm-diameter round inserts at cutting speeds ranging from 6.0 m/s to 13.0 m/s, and the feed rates ranged from 0.13 mm/r to 0.51 mm/r. They found that tool failure in the cutting of the relatively soft (220 HB) Ni-based superalloy was due to excessive wear. Flank wear played a larger role at the lower speeds, but depth-of-cut notch wear was significant at the higher speeds. Abrasion, adhesion, and chipping were found to be dominant wear mechanisms.

Al_2O_3/SiC composite inserts generally contain 30–45% SiC_w. Extensive physical, mechanical, and performance data on Al_2O_3/SiC_w material are now available.[408–409] It has been reported[410] that the "characteristic strength" of hot-pressed Al_2O_3 bodies increased from about 504 MPa to 690 MPa as the content of SiC_w increased from 0 to 30 wt%. Adding SiC_w, followed by hot pressing, can more than double Al_2O_3s fracture toughness to a value of 8.7 MPa $m^{1/2}$.[411] In actual metal cutting of alloys such as Inconel 718, composite tools have performed up to three times better than conventional ceramic tools and eight times better than carbide tools.[412,413]

SiC/Al_2O_{3w} orientation, fracture behavior, and cutting performance have been studied at laboratories and universities throughout the world.[398,399,414,415]

2.10.2 Joining CMCs.

For most applications of CMCs, composite-to-composite or composite-to-metal joints are required. Methods for bonding only matrix material (without considering fiber or whisker interactions) are entirely analogous to those used for bonding monolithic ceramics. Cawley[416] has reviewed bonding techniques for ceramic composites, emphasizing the fact that the joining of ceramics can in most cases be viewed as a localized variation of a ceramic forming process. The processes listed include metal brazing, silicate brazing (solder glasses), diffusion bonding (a variation of the sintering process for forming ceramics), cementing, and fusion welding (a variation of fusion casting). Resin bonding (adhesive bonding) can also be used as well as mechanical fasteners.

The general methods described by Cawley should be applicable for CMC joining but have mostly been used or investigated for monolithic ceramic joining. Metal brazing is suitable for both ceramic-to-ceramic and ceramic-to-metal brazing. A problem with brazing is that most metals do not readily wet ceramic surfaces. One method of overcoming this problem is to metallize the ceramic surface, which can be done using several processes. The *active filler metal process* incorporates an active metal such as titanium into the filler metal. During brazing, the metal segregates to the surface of the ceramic where it reacts to form a wettable surface. Active filler metals have been used in joining both oxide and nonoxide ceramics to one another and to metals. It can be inferred that this process would also be suitable for joining oxide ceramics to nonoxide ceramics.

Brazing using a silicate glass is similar to metal brazing but is generally less complicated because wetting is generally not a problem. Both oxide and nonoxide ceramics based on Al_2O_3 and Si_3N_4 have been brazed using silicates.

In producing a ceramic material such as Si_3N_4, sintering aids are added to form grain boundary phases that become liquid during the sintering process. The interphase formed during silicate brazing is similar (or even identical) to this grain boundary phase, suggesting that TLP bonding for composites is a possibility. In this case, a low-temperature-melting material would be placed between the ceramic composite faces and the temperature raised above the melting point. The molten interfacial layer could then dissolve some of the composite matrix material, forming a variable composition alloy. After a certain amount of dissolution, the alloy would become sufficiently rich in the more refractory matrix material to solidify. The initial melting would aid in the removal of porosity, and capillary forces would draw the surfaces together. This process could enable reinforcing fibers or whiskers to be incorporated into the joint material.

A number of monolithic oxide and nonoxide ceramics have been joined using the diffusion bonding process. Temperatures have ranged from 1350°C for joining Al_2O_3 to Al_2O_3 to 1800°C for joining Si_3N_4 to Si_3N_4. In addition, it has been necessary to apply pressure to provide a sufficient driving force for bonding. This method is probably of only limited utility for CMC joining. The high temperatures required for most of the matrix systems described would be detrimental to fiber or whisker properties because of thermal degradation and to reactions with the matrix material. In addition, it is difficult to apply uniform pressure to an interface in a complex component.

Cement bonding techniques might be applicable for bonding CMCs in limited situations. For example, in the aluminum phosphate-bonded system, bonding with material having a composition similar or identical to that of the matrix is a possibility. Because of the generally high temperatures required and the disruption of microstructure resulting from melting ceramic materials it does not seem likely that fusion welding will find wide application in CMC joining.

Another type of ceramic-to-metal bonding technique that has been described[417] is friction welding. In this process, a ceramic and a metal specimen are rotated against each other under pressure, and the frictional heat causes the metal to flow. The nature of the bond is not fully understood but is believed to be part mechanical and part chemical.

Glass-to-metal bonds have also been created by electrostatic bonding. In this process, the surfaces are placed in intimate contact and a very high voltage is applied across the joint. This results in ion migration toward the joint, which produces electrostatic attraction between the two materials. The exact mechanism by which this method works is not understood.

Although all the general methods described above might be applicable to CMC joining, relatively little has yet been reported on nonmechanical composite joining. However, Rabin[418,419] has described joining work on Nicalon$_f$-SiC matrix material produced at ORNL by the thermal gradient forced-flow method.

The object of the work was to provide a method that produced a joint usable to 1000°C at a processing temperature not appreciably exceeding 1400°C, to minimize fiber degradation. Two methods were used. In the first method Ti and C powders, along with 15 wt% Ni powder, were reacted at the joint interface to form TiC. The Ni reduced the reaction temperature from about 1600°C to 1200°C and also formed a liquid phase that aided in densification. The resulting microstructure was that of a cermet of interconnected TiC grains surrounded by a Ni-rich matrix. The reaction was conducted in argon under a pressure of 20–50 MPa at a maximum temperature of 1400°C. This method produced a joint about 125 μm thick, with some residual porosity believed to have resulted from defects in slurry application of the starting materials. Small particles of SiC were present in the joint material.

The second method involved infiltration of C with molten Si. The C source was either fine C powder or woven C cloth. In both cases, Si infiltration was carried out at 1480°C for 15 min in a 4-Pa vacuum. When powdered reactants were used, a thin, uniform layer was applied to both joining surfaces using a slurry method. Prior to joining, the binder was removed by slow heating in air to 400°C.

A typical joint produced by this method was 200 μm thick and exhibited a uniform distribution of SiC crystals in a Si matrix.

When C cloth was used, the SiC crystals had the morphology of the original C$_f$ and several percent of unreacted C was typically present. There was usually a layer of free Si present at the interface, which is probably undesirable from a mechanical property standpoint. Figure 2.19 shows a joint of this type.

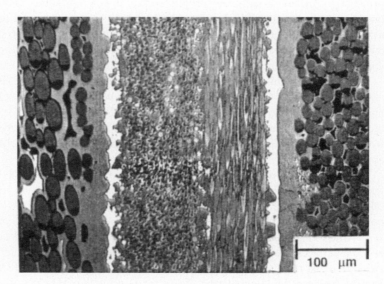

Figure 2.19   Optical micrograph showing a joint between CVI-produced Nicalon$_f$ and SiC matrix specimens that was formed by direct silicon infiltration of carbon cloth placed between the specimens.[418]

In addition to joining completed ceramic composite materials, joining can be carried out at some intermediate state as is sometimes done with monolithic ceramics. An example of this procedure, providing a lap joint with a shear-resistant interface, has been reported by DuPont[420] for CVI-produced structures. Using a method called *coinfiltration bonding,* subcomponents are brought together early in the CVI process and additional CVI is carried out. Bond shear strength at room temperature was about 29 MPa. Values at elevated temperatures were 21 MPa at 1000°C and 25 MPa at 1400°C. Shear strengths remained relatively constant when tested at room temperature after a 1- or 50-h exposure in air at about 1100 and 1400°C.

Mechanical bonds can decrease the possibility that a joined assembly will fail at the joint in a catastrophic manner, as can occur with a nonmechanical bond involving only matrix material or matrix material bonded to a metal. However, a mechanical bond of two ceramic composite materials using a metal fastener can be a problem at the high temperatures involved for many applications.

One potential problem is that the CTE values for the fastener metal and the composite can be appreciably different (with the metal typically having greater expansion), resulting in loose joints or excessive stresses, depending upon the temperature regime. A snug radial fit at room temperature results in radial stresses at elevated temperatures. To achieve a snug radial fit at some elevated temperature, the room-temperature fit must be loose. A snug axial fit at room temperature results in a loose fit at elevated temperatures. To obtain a snug axial fit at some elevated temperature, the joint must be stressed at room temperature. Another problem can be that the metal fastener is not capable of withstanding the high temperatures required for the application.

For cases where a particular metal fastener is satisfactory from a temperature standpoint, joint configurations that can provide stress-free joints with composites at elevated temperatures have been designed. For example, Sawyer et al.[421] have designed and tested elevated-temperature thermal-stress-free joints for CCC that are applicable to CMCs as well.

2.10.2.1 Hot press diffusion bonding.

The use of a SiC monofilament to reinforce a Si_3N_4 matrix has led to refractory CMCs with outstanding high-temperature properties. These composites provide many of the properties required for future generation aerospace applications and are viable candidates for structural applications where use temperatures exceed 1350°C.[422]

The fiber used as a reinforcement is a SiC monofilament produced by a high-temperature vapor deposition process using a C monofilament as a substrate. The vapor deposition process and the stoichiometric composition of the composites are responsible for the outstanding tensile properties (over 4 GPa at room temperature). Most studies were conducted with 140-µm fiber (SCS-6), and limited studies were conducted with 75-µm fiber (SCS-9). Both fibers, from Textron Specialty Materials, are products of vapor deposition of SiC on a C monofilament. The matrix is Si_3N_4 with 1.25% MgO and 5% Y_2O_3 as sintering aids.

Joining experiments with these materials have been conducted in anticipation of a need to fabricate complex shapes from an assembly of elementary geometric panels. For example, experiments evaluated a direct bonding approach wherein two densified composite panel specimens were hot-pressed together using a monolithic monolayer as a high-temperature stable bonding agent. Double-notch shear tests showed that failure oc-curred outside the monolithic bond line region at levels of 140 MPa. No change in the in-terlaminar shear strength of the bonded specimens was observed after the specimens were exposed to thermal shock with a temperature gradient of up to 1200°C.

Recent work has centered on hollow airfoil applications. To produce complex shapes such as these, it is necessary for the preform tape to conform to the curvature of the part and to achieve multidirection reinforcement. The tape contains a 75-μm-diameter fiber (SCS-9), which allows it to be more easily wrapped around the leading edge of the airfoil. Other plies would be wrapped at opposite orientation angles and along the axis of the airfoil to give balanced and multidirectional reinforcement. Fabrication was per-formed by hot pressing. However, recent studies have centered on the use of HIP for con-solidation of parts, and initial results indicate that it does not damage the fiber and pro-duces a higher-strength composite.

A number of components have been made using the tape layup and hot press fabri-cation technique. They include vanes produced for an earth-to-orbit rocket engine turbine for General Electric under an ongoing NASA program.

2.10.2.2 Adhesive bonding.

Y-Si-Al-O-N is one of the most promising ad-vanced engineering ceramics. These materials exhibit high strength and high toughness, and they show good wear and chemical resistance. At present, slip casting and injection molding are the main techniques for shaping, which limits the size and shape of the com-ponent that can be formed. The obvious alternative to these fabrication techniques is to use a joining method that can build up complex or large shapes from a series of smaller, more simply shaped components.

Techniques such as diffusion bonding, welding, soldering, and brazing cannot be directly applied to Si-Al-O-N ceramics because of their tendency to decompose rather than melt at high temperatures. However, Y-Si-Al-O-N ceramics contain a small percent-age of an intergranular glassy phase that melts at approximately 1350°C (eutectic temper-ature in the Y-Si-Al-O-N system).[423] Hence joining should be possible if this liquid form-ing component of the ceramic is used.

The present joining adhesives selected differ from the pure-glass and composite adhe-sives in that they contain more solid phase than liquid phase at the joining temperature. The main constituent of the adhesive is α-Si_3N_4 (45 wt%), and the remainder is a mixture of ox-ides (Y_2O_3, Al_2O_3, and SiO_2). The advantage of this type of adhesive is that it is closer in composition to the ceramic being joined, and so the properties of the joint are not much dif-ferent from those of the parent material. Also, at the joining temperature (1600°C), the α-Si_3N_4 reacts to form β-SiAlON, which has an acicular nature that reinforces the joint. Material joined with pure-glass forming adhesives is more likely to break at the joint be-cause the layer of glass between the adherends has a low resistance to crack propagation. Figure 2.20 is a schematic representation of adhesive joining of Y-Si-Al-O-N ceramics.

2.10.2.3 Brazing.

Understanding materials behavior during brazing of CMCs is an area of critical need. This broad class of materials serves a wide range of technolog-ically important applications, including engine components, wear-resistant parts, metal extrusion dies, heat exchangers, machine tools, and medical products. Ceramic compos-ites are often most easily introduced into an application by joining them to a metal sup-port structure.[424]

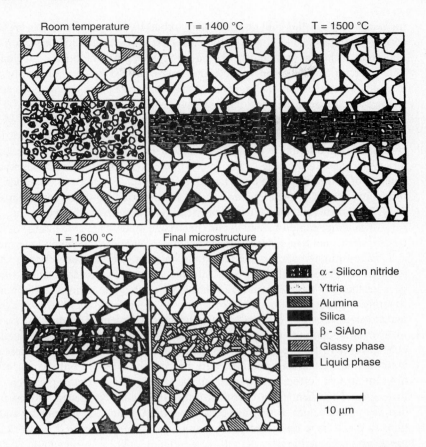

Figure 2.20   Schematic representation of the adhesive joining process in Y-Si-Al-O-N ceramics.[423]

The active filler metal process has been developed to the point where reliable joints can be produced.[425,426] Silver-copper eutectics containing a small percentage of Ti and other minor alloying additions are among the most successful. The braze occurs by initial formation of a Ti-rich metal or metal oxide layer. Further transitions to the braze metal itself occur, as expected, via gradients in both compositions of the oxidation state. Other aspects of the joining involve the role of SiC_w at the interface.

A study on the brazing of SiC_w/Al_2O_3 was performed by Lannutti et al.[424] They selected Incusil ABA as an active braze metal material which consists of approximately 59.2 wt% Ag, 27.1 wt% Cu, 12.1 wt% In, and 1.2 wt% Ti.

2.10.2.4 Phosphate bonding.

The preparation of continuous-fiber composites by hot pressing or HIP is expensive, and the high temperatures required may be detrimental to the fibers. The high temperatures may also create problems involving the fiber-matrix interface and differential thermal expansion.

One way to prepare ceramic composites is to use phosphate bonding techniques. Phosphate binder systems have been used in common refractories (mostly Al_2O_3-based) for many years in the United States and abroad,[427–432] but their use in high-performance ceramic composites, especially nonoxides, remains relatively unexplored. Aluminum phosphate has been studied as a binder for Si_3N_4. Phosphate bonding techniques can provide a means for low-temperature processing of nonoxide ceramic composites, allowing exploitation of their unique properties (such as electrical) in applications such as electromagnetic windows where very high strength is not required.

Phosphate bonding can be used to prepare CMCs by well-developed techniques employed in the PMC field. The general concept is the same: a continuous tow or cloth is combined with a liquid or paste matrix precursor, and with time and heat this matrix material solidifies. In the case of phosphate-bonded ceramics, the matrix precursor is a suspension of ceramic particles in a liquid, instead of a monomer or oligomer. Nonoxide reinforcements (SiC_w, platelets, and Si_3N_{4w}) introduced into aluminum phosphate-bonded Si_3N_4 have been found to be chemically compatible with the binder.[433]

The temperatures needed to cure phosphate binders are in the range of room temperature to 300°C, not that different from the temperatures used in the polymer field. The layup and shape-forming methods used (vacuum bagging, pressing, autoclaving, etc.) are all applicable. In conventionally processed sintered composites, problems arise with densification because the reinforcement (fibers or whiskers) inhibits the shrinkage that must accompany sintering. Processing composites using phosphate bonding avoids these problems because linear changes are near zero and practically any reinforcement loading can be used.[434,435]

For the continuous oxide fibers (silica and aluminosilicate) evaluated by Martin et al.[435] protective fiber coatings were necessary in order to prevent detrimental reactions with the matrix. Phenolic coatings were found to be effective only when composites were fired in nitrogen, effectively pyrolizing the phenolic. The resulting carbon coating on the fibers protected them from chemical interaction with the matrix. If this carbon coating had been detrimental to the dielectric properties, which are crucial for radome applications, it could have been removed by refiring the composites at 300–400°C in an oxidizing atmosphere. Although strengths of continuous fiber-reinforced composites in general were somewhat low, the highest strength observed (36 MPa for BPO_4/Si_3N_4 matrix, Astroquartz fused silica fabric) is adequate for radome applications, especially when accompanied by significant toughening of the material. The strength was very encouraging, considering that the material had 35% porosity.

The techniques for making continuous-fiber composites are very similar to those used in the polymer composites field. Filament winding, infiltration by three-dimensional braids, tape wrap, and cloth layup techniques can all be used to produce phosphate-bonded composites in net shapes. Another attractive feature of the phosphate bonding technique is negligible linear changes during firing offering net shape and net size capabilities.[434,435]

In another study by Martin et al.,[435] phosphate bonding techniques were used to prepare ceramics in the system Si_3N_4-Al_2O_3-B_2O_3-H_3PO_4 at low temperatures (700–900°C). The mechanical properties and phase composition showed a strong dependence on Al_2O_3 particle size and firing temperature. The use of 0.05-μm Al_2O_3 resulted in a stable phase composition and stable properties. The incorporation of B into the binder system allows processing flexibility of materials before firing. The combination of strength, thermal expansion, and dielectric properties of phosphate-bonded Si_3N_4 materials makes them promising candidates for applications such as advanced missile radomes and electromagnetic windows.

2.10.2.5 Mechanical fasteners.   The purpose of the study in a recently completed refractory composite fastener effort under the Refractory Composite Materials and Structures Augmentation Program (RC MASAP) was used to support the NASP airframe design requirements.[436]

This program addressed the initial characterization needed to evaluate fasteners for tension and pin shear as a function of temperature, torque, cyclic oxidation at reduced pressure and residual properties after cycling.

Nine CCCs, five CMCs, and four ceramic whisker-reinforced fastener systems were screened for pin shear, thread shear, tension, cyclic oxidation at reduced pressures, and torque. The CCC fasteners were examined for use at a highest temperature of 1649°C, while the ceramic and CMC fasteners were examined for use at up to 1371°C. Fastener requirements were based on the need for a shear fastener application in this temperature range.

The conclusions from the RC MASAP fastener test program are as follows.

1. Screening data was obtained on 21 different RC fastener systems. Three fastener systems, Refractory Composites, Inc. SiC/SiC, Kaiser Aerotech's SiC/SiC, and Cercom's SiC_w/Si_3N_4 fasteners were identified as the leading candidates and as being capable of use at 1371°C (Figure 2.21.)[436]

2. Only one fastener system that was tested at 1649°C, RCI's SiC/SiC, met the established material requirements. (Kaiser Aerotech's SiC/SiC fasteners arrived too late to test at 1649°C, and Cercom's ceramic fastener was not tested beyond 1371°C.) Others haven't been tested or arrived too late to test.

3. The reinforcement (fiber, whisker) used in the layup affected the strengths of the fasteners.

4. A through-the-thickness braided fastener manufactured by Kaiser Aerotech had the best combined properties of pin shear and tension for a CCC fastener. This fastener could be used at up to 1371°C if mechanical loads were kept sufficiently low. Coating technology has not been developed and therefore kept this fastener from higher temperature usage and application.

5. Combined temperature and reduced pressure cycling with a 871°C maximum temperature showed that mass loss was insignificant for all the fasteners tested after 20 cycles.

6. Fasteners were graded for compliance with dimensional specifications and for visual anomalies.

   • Most CCC fasteners were not within specified dimensions.
   • Ceramic fasteners were much better at meeting specified dimensions.
   • CCC fasteners had surface flaws.
   • Visual appearance was not always indicative of fastener performance.
   • Of the CMC fasteners, DuPont's C/SiC fasteners looked the best but did not perform the best mechanically.
   • Whisker reinforced ceramics were good in appearance and dimensional measurements.

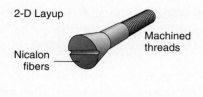

2-D Layup

Machined
threads

Nicalon
fibers

3-D Braided
preform

Machined
threads

Nicalon
fibers

DuPont Noveltex C/SiC

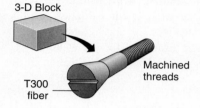

3-D Block

T300
fiber

Machined
threads

Hitco SiC/C

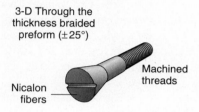

3-D Through the
thickness braided
preform (±25°)

Nicalon
fibers

Machined
threads

Cercom Whisker Reinforced Ceramics

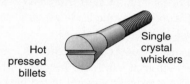

Hot
pressed
billets

Single
crystal
whiskers

LTV SiC$_w$/Si$_3$N$_4$

Hot
pressed
billets

Diamond
wheel
ground
threads

Whisker type	Matrix	Whisker loading	TPI	Thread form	DIA
SiC	Al$_2$O$_3$	39%	20	UNC-20	1/4
SiC	Si$_3$N$_4$	5%	20	UNC-20	1/4
Si$_3$N$_4$	Si$_3$N$_4$	15%	20	UNC-20	1/4

Whisker type	Matrix	Whisker loading	TPI	Thread form	DIA
American Matrix, Inc SiC	Allied Signal GN-10 Si$_3$N$_4$	30%	20	Modified V-60	1/4

Figure 2.21   Fastener designs used by several companies.[436]

7. Static mechanical results determined the fastener systems meeting requirements.

  • For the CMC fasteners, Hitco's SiC/C fasteners met the requirements for pin shear.
    DuPont's coated C/SiC fasteners did not meet pin shear or tensile requirements. All
    CMC fasteners that had SiC$_f$ were able to meet room-temperature pin shear require-
    ments. The C/SiC CMC fasteners did not meet requirements when coated. Without
    a coating, they did not meet room temperature requirements but did meet one of the
    elevated-temperature requirements. RCI's and Kaiser Aerotech's SiC/SiC fasteners
    met both room-temperature pin shear and tensile requirements.

  • For the whisker-reinforced ceramic fasteners, the results were mostly positive.
    The exception was the SiC$_w$/Al$_2$O$_3$ ceramic, which had consistently lower values.

8. Residual property measurements taken after thermal or reduced pressure cycling provided mixed results. Strength gains were noted for several fastener systems after cycling. Other fasteners showed strength losses that correlated with mass loss.

- DuPont's uncoated C/SiC, Carbon-Carbon Advanced Technologies, Inc. (C-CATs) CCC with molded threads, and B.F. Goodrich (BFG's) CCC had strength retentions in tension and pin shear above 90% after the 871°C low cycle. The rest of the fasteners showed strength losses in either pin shear or tension or in both properties after cycling.

- After the 1316°C medium cycle, RCI's SiC/SiC and DuPont's coated and uncoated C/SiC fasteners showed strength retentions above 100%. The remainder of the fasteners showed strength decreases for one or both properties.

- After the 1649°C high cycles, C-CATs CCC fasteners with molded threads showed a tensile strength increase of greater than 90%. For pin shear, C-CATs CCC fasteners were twice as high as at room temperature.

REFERENCES

1. Parker, I. 1991. Fasten or bond? *Aerosp. Compos. Mater.* 3(5):32–3.

2. Pen Associates, Inc. *A Guide to Cutting and Machining Kevlar® Aramid.* E. I. Du Pont de Nemours and Pen Associates, Wilmington, DE.

3. Beard, T. 1989. *Machining composites—New rules and tools. Mod. Mach. Shop* April:74–85.

4. Kim, K. S., D. G. Lee, Y. K. Kwak, et al. 1992. Machinability of carbon fiber-epoxy composite materials in turning. *J. Mater. Process. Technol.* 32(3):553.

5. Konig, W., C. Wulf, P. Grab, et al. 1985. Machining of fiber reinforced plastics. *Ann. CIRP* 34(2):537.

6. Santhanakrishnan, G., R. Krishnamurthy, and S. K. Malhotra. 1988. Machining of composites: Surface morphology and tool wear studies. *Met. Matrix Carbon Ceram. Matrix Compos. 1988,* ed. J. D. Buckley, pp. 411–7, January 1988, Cocoa Beach, FL. NASA CP 3018.

7. Ramulu, M., M. Faridnia, J. L. Garbini, et al. 1991. Machining of graphite/epoxy composite materials with polycrystalline diamond (PCD) tools. *ASME Trans.* 113(4):430–6.

8. Cheng, H. -Ho. An analysis of drilling of composite materials. Ph.D. Thesis, University of California at Berkeley. 89-16702.

9. Wong, T. L., S. M. Wu, and G. M. Croy. 1982. An analysis of delamination in drilling composite materials. *Proc. 14th Nat. SAMPE Tech. Conf.,* pp. 471–83, October 1982, Atlanta, GA.

10. Wu, S. M. 1982. How to reduce the drilling cost for the aerospace industry. *Proc. 14th Nat. SAMPE Tech. Conf.,* pp. 438–43, October 1982, Atlanta, GA.

11. Horng, S. Y., S. M. Wu, and T. Voss. 1982. An end effector for robotic drilling. *Proc. 14th Nat. SAMPE Tech. Conf.,* pp. 464–70, October 1982, Atlanta, GA.

12. Kinkaid, R. G. 1991. PCD drills for composite and metal matrix materials. *SME Superabrasives Conf.,* June 1991, Dearborn, MI. SME MR-91-174.

13. Vaccari, J. 1990. Machining of graphite/epoxy materials with PCD tools. *Amer. Mach.* 16:19, 1990.

14. Office of Technology Assessment, U.S. Congress. 1988. *Advanced Materials by Design.* U.S. Government Printing Office, Washington, DC.

15. Schwartz, M. M. 1983. *Composite Materials Handbook,* 1st ed. McGraw-Hill, New York.

16. Schwartz, M. M. 1992. *Composite Materials Handbook,* 2nd ed. McGraw-Hill, New York.

17. Koplev, A., A. Lystrup, and T. Vorn. 1983. The cutting process, chip and cutting forces in machining composites. *Composites* 14(4):371–6.

18. Stotler, C. L., and S. A. Yokel. 1989. PMR graphite engine duct development. NAS 3-21854, NASA CR 182228.

19. Kuhl, T., G. Devlin, and J. Bunting. 1989. The mechanics and economics of advanced tooling systems for drilling and countersinking composite materials. SME AD-89-645. FASTEC '89.

20. Sap, F. 1986. To drill good holes in composites one approach is to use a 40,000-rpm drill. *Am. Mach.* August 1986:23–5.

21. Dean, C. D., S. J. Brox, R. A. Hay, et al. 1993. The use of CVD diamond coated tools for machining advanced composite materials. Machining of Composite Materials II, October 1993, Pittsburgh, PA.

22. Colligan, K., and M. Ramulu. 1992. *Manuf. Rev.* 5.

23. Wern, C. W., M. Ramulu, and K. Colligan. 1993. A study of the surface texture of composite drilled holes. *J. Mater. Proc. Technol.* 37(14):373.

24. Malhotra, S. K. 1990. Some studies on drilling of fibrous composites. *J. Mater. Proc. Technol.* 24:291.

25. Hocheng, H., and H. Y. Puw. 1993. Machinability of fiber-reinforced thermoplastics in drilling. *J. Eng. Mater. Technol.* 115(1):146.

26. Lane, C. 1992. The effect of different reinforcement on PCD tool life for aluminum matrix composites. *Proc. Mach. Compos. Mater. Symp., ASM,* November 1992, Chicago, IL.

27. Michel, G. 1992. A new approach to machinability testing: Application to machining of metal matrix composites. *Proc. Mach. Compos. Mater. Symp., ASM,* November 1992, Chicago, IL.

28. Sullivan, S. 1992. Machining, trimming and drilling of metal matrix composites for structural applications. *Proc. Mach. Compos. Mater. Symp., ASM,* November 1992, Chicago, IL.

29. Jawaid, A., S. Barnes, and S. R. Ghadimzadeh. 1992. Drilling of particulate aluminum silicon carbide metal matrix composites. *Proc. Mach. Compos. Mater. Symp., ASM,* pp. 35–47, November 1992, Chicago, IL.

30. Gadalla, A. M., and H. S. Bedi. 1991. Effect of composition and grain size on electrical discharge machining of BN-TiB$_2$ composites. *J. Mater. Res.* 6(11):2457–62.

31. McMurtry, C. H., W. D. G. Boecker, S. G. Seshadri, et al. 1987. *Ceram. Bull.* 66(2).

32. Di Ilio, A., A. Paoletti, V. Tagliaferri, et al. 1987. Progress in drilling of composite materials. *Ceram. Bull.* 66(2):199–203.

33. Mehta, M., T. J. Reinhart, and A. H. Soni. 1992. Effect of fastener hole drilling anomalies on structural integrity of PMR-15/Gr composite laminates. *Proc. Mach. Compos. Mater. Symp.,* pp. 113–26, *ASM,* November 1992, Chicago, IL.

34. Weiss, R. A. 1989. Portable air feed peck drilling of graphite composite, titanium and other materials in dissimilar combinations. SME AD 89-642.

35. Miller, J. A. 1987. Drilling graphite/epoxy at Lockheed. *Am. Mach. Automat. Manuf.,* October:70–2.

36. Computer-controlled drill designed to meet unique B-2 fabrication needs. *Aviat. Week Space Technol.* April 17, 1989:51–2.

37. Colligan, K. 1994. New tool drills both titanium and carbon composites. *Am. Mach.* 138(11):56–8.

38. Mehta, M., and A. A. Soni. 1990. Investigation of optimum drilling conditions and hole quality in PMR-15 polyimide/graphite composite laminates. *Proc. 6th Ann. Adv. Compos. Conf., ASM/ESD,* pp. 633–5, October 1990, Detroit, MI.

39. Elber, G. 1994. Understanding alternative techniques helps solve cutting problems. *Adv. Compos.* 9(1):30–3.

40. Peterson, S. A. 1980. Cutting holes in fabric-faced panels. NASA Technical Brief, Fall 1980, p. 391.

41. Caprino, G., and V. Tagliaferri. 1995. Damage development in drilling glass-fiber-reinforced plastics. *Int. J. Mach. Tools Manuf.* 35(6):817–29.

42. Hocheng, H., and C. C. Hsu. 1995. Preliminary study of ultrasonic drilling of fiber-reinforced plastics. *J. Mater. Proc. Technol.* 48(1–4):255–66.

43. Martin, C., B. Calés, P. Vivier, et al. 1989. Electrical discharge machinable ceramic composites. *Mater. Sci. Eng.* A109:351.

44. Gunderson, G. L., and J. A. Lute. 1992. The use of preformed holes for increased strength and damage tolerance of advanced composites. *Proc. Am. Soc. Compos. 7th Tech. Conf.,* pp. 460–64, University Park, PA, October 1992.

45. Marx, W., and S. Trink. 1978. *Manufacturing methods for cutting machining and drilling composites,* vol I: *Test Results,* vol II: *Final Report,* August 1976–August 1978. Grumman Aerospace Corporation. Bethpage, NY. AFML-TR-78-103, Contract F33615-76-C-5280.

46. Rezaei, S. M., T. Suto, T. Waida, et al. 1992. Creep feed grinding of advanced composites. *Proc. Inst. Mech. Eng.* 206(2):93.

47. Suzuki, K., T. Uematsu, S. Asano, et al. 1988. Advances in grinding composites and WC. *Ind. Diamond Rev.* June:278.

48. Suto, T., T. Waida, H. Noguchi, et al. 1990. High performance creep feed grinding of difficult-to-machine materials with new-type wheels. *Bull. Japn. Soc. Precis. Eng.* 24(1):39.

49. Rezaei, M., T. Suto, T. Waida, et al. 1989. A novel dressing technique for diamond wheels. *Ind. Diamond Rev.* 6(89):258.

50. Water jet cutting. *Aircr. Eng. Aerosp. Technol. Proc.* 66(2):5–6, 1994.

51. Murray, T. P. 1989. Machining composites for medical devices. *Compos. Manuf.* January 1989. SME EM-89-104.

52. Schreiber, R. R. 1992. Composites: A mixed blessing. *Manuf. Eng.* 109(1):87–92.

53. Ramulu, M., M. Faridnia, J. L. Garbini, et al. 1991. *J. Eng. Mater. Technol.* 112:430–6.

54. More, E. R. 1976. Manufacturing methods for composite fan blades. Hamilton Standard Division, United Technologies Corporation, Windsor Locks, CT. AFML-TR-76-138, Contract F 33615-74-C-5135.

55. Boszor, S. M., and R. S. Ecklund. 1991. Manufacturing technology for advanced propulsion materials. Phase V: Composite airfoil manufacturing scale-up for fan applications. United Technologies–Pratt & Whitney. WL-TR-91-8009, FR 21372, Contract F 33615-85-C-5152.

56. Pilkington, D. J., G. Bryce, and W. Nimmo. 1988. Improvements in automated ply cutting and handling. *SAMPE J.* 24(5):9–13.

57. Knife and ultrasound slice prepreg. *Mach. Des.* November 10, 1988:70–1.

58. Kim, J. -D., and E. -S. Lee. 1994. A study of the ultrasonic-vibration cutting of carbon-fiber reinforced plastics. *J. Mater. Proc. Technol.* 43:259–77.

59. Miller, R. K. 1991. *Waterjet Cutting: Technology and Industrial Applications.* Fairmont Press, Lilburn, GA.

60. Mason, F. 1989. Water and sand cut it. *Am. Mach.* October:84–95.

61. Schroter, D. C. 1989. Machining with waterjets and hydroabrasives. *SME Int. Conf. Exposit.* May 1989, Detroit, MI. SME MS89-482.

62. Howarth, S. G., and A. B. Strong. 1990. Edge effects with waterjet and laser beam cutting of advanced composite materials. *Proc. 35th Int. SAMPE Symp.,* pp. 1684–97, April 1990, Anaheim, CA.

63. Profitable waterjet cutting for aerospace materials. *Modern Applications News* August 1995:30.

64. Bartlett, D.C. 1991. CNC water-jet machining and cutting center. Allied-Signal Aerospace, Kansas City, MO. KCP 613-4397, DE AC04-76DP00613.

65. Irving, R. R. 1989. Water jet cutting. *Metalworking News,* August 21:19.

66. CTL expansion includes water-jet cutting. *Adv. Compos.* September/October 1988:18.

67. The Paser abrasive jet enhances Rockwell's composite cutting. Flow Systems, Inc., Kent, WA, 1988. FS-055.

68. Wightman, D. F. 1990. Waterjet cutting and hydrobrasive machining of aerospace components. SME MR90-672.

69. Hashish, M. 1991. Optimization factors in abrasive-waterjet machining. *ASME Trans.* 113(1):29–37.

70. Hashish, M. 1991. Current capabilities for precision machining with abrasive-waterjets: Nontraditional machining. SME MR-91-520.

71. Hashish, M. 1992. State-of-the-art of abrasive-waterjet machining operations for composites. *Mach. Compos. Mater. Symp.,* pp. 65–73, *ASM,* November 1992, Chicago, IL.

72. Kulischenko, W. 1986. Abrasive jet machining. *Adv. Ceram. Conf.,* ed. D. Richerson, pp. 176–84, Dearborn, MI. SME MR 76-694.

73. Advanced composites for aerospace require waterjet technology. *Modern Applications News* September 1992:40–3.

74. Water-jet cutting expands opportunities for West Coast fabricator. *Weld. Des. Fabr.* April 1993:18.

75. Davis, D. C. 1987. Robotic abrasive water jet cutting of composite aerostructure components. *SME Robotic Syst. Aerosp. Manuf. Conf.,* ed. T. J. Drozda, pp. 382–96, September 1987, Dearborn, MI.

76. Stover, D. 1992. Out on the cutting edge. *Adv. Compos.* September/October:46–53.

77. Groppetti, R., and A. Cattaneo. 1993. A model for hydro and hydroabrasive jet machining of carbon fiber reinforced plastic composites. *Proc. Int. Conf. Mach. Adv. Mater.* pp. 297–404, July 1993, Gaithersburg, MD. NIST SP 847.

78. Vaccari, J. 1989. Abrasive-waterjet job shop thrives. *Am. Mach.* October 1989:80–1.

79. Arola, D., and M. Ramulu. 1994. Micro-mechanisms of material removal in abrasive waterjet machining. *Proc. Adv. Mater.* 4(1):37–47.

80. Burnham, C. 1990. Abrasive waterjets come of age. *Mach. Des.* May 10:93–7; Cutting tool links CAD to water-jet robot. *Mach. Des.* January 11:16; Lockheed-Georgia adds AWJ. *Am. Mach.* January:25.

81. Zaring, K., G. Erichsen, and C. Burnham. 1992. Procedure optimization and hardware improvements in abrasive waterjet cutting systems. *Proc. Mach. Compos. Mater. Symp., ASM/TMS,* ed. T. S. Srivatsan and D. M. Bowden, pp. 75–9, November 1992, Chicago, IL.

82. Ramulu, M., and D. Arola. 1993. Water jet and abrasive water jet cutting of unidirectional graphite/epoxy composite. *Composites* 24(4):299–308.

83. Hashish, M. 1987. Prediction of depth of cut with abrasive waterjets. *ASME Symp. Model. Mater. Process.,* ed. A. A. Tseng, vol. 3, pp. 65–82, New York.

84. Geskin, E. S., L. Tisminetski, D. Verbitsky, et al. 1992. Investigation of waterjet machining of composites. *Proc. Mach. Compos. Mater. Symp., ASM,* pp. 81–7, November 1992, Chicago, IL.

85. Hashish, M. 1993. Comparative evaluation of abrasive liquid jet machining systems. *ASME Trans.* 115(1):44–50.

86. Street, W. G., C. Ferlito, and S. Riches. 1992. Trends in laser cutting of advanced materials. *TWI Bull. 5:*108–10.

87. Doyle, D., and J. Kokosa. 1987. Laser processing of Kevlar: hazardous chemical by-products. *Proc. LAMP 87,* pp. 285–89, May 1987, Osaka, Japan.

88. Caprino, G., V. Tagliaferri, and L. Covelli. 1995. The importance of material structure in the laser cutting of glass fiber reinforced plastic composites. *J. Eng. Mater. Technol.* 7(1):133–38.

89. Caprino, G., V. Tagliaferri, and L. Covelli. 1995. Cutting glass fibre reinforced composites using CO_2 laser with multimodal-gaussian distribution. *Int. J. Mach. Tools Manuf.* 35(6): 831–40.

90. Viegelahn, G. L., S. Kawall, R. J. Scheuerman, et al. 1991. Laser cutting of fiberglass/polyester resin composites. *Proc. 7th Adv. Compos. Conf., ASM/ESD,* pp. 143–7, September/October 1991, Detroit, MI.

91. Schwartz, W. H. 1989. Fasteners hold their own. *Assem. Eng.* April:24, 25.

92. Albee, N. 1988. Aspects of mechanically fastening advanced composites. *Adv. Compos.* November/December:54–60.

93. Kim, R. Y. 1987. Fatigue strength. In *Engineered Materials Handbook,* vol. 1, *Composites,* 436–44. ASM International, Materials Park, OH.

94. Smode, F. E. 1992. In *Composite Material Fastener Reduces Lightning Dangers,* ed. D. Whittaker, 63–65. Sterling, London, U.K.

95. Fasteners for composites. *Aircr. Eng.* July 1988:2, 3.

96. McCarty, L. H. 1989. Lightweight fasteners reduce unit stresses on composites. *Des. News,* September 4:186–7.

97. Dahl, J. M., and G. N. Maniar. High strength fastener applications of Aer Met 100 and Custom Age 625 PLUS. *Symp. Fasteners, 8th Mater. Expos. ASM/TMS,* Rosemont, IL, October, 1994.

98. Dubble, M. 1989. Composite fasteners face the aerospace challenge. *Mach. Des.* 61(14): 102–4.

99. Hi-Lex, a trademark of Hi-Shear Corporation. Torrance, CA.

100. Rausch, J. 1988. Fastening fundamentals for fabricated composites. *Assem. Eng,* October:34–6.

101. Dastin, S. 1982. Machining and joining techniques. In *Handbook of Composites,* ed. G. Lubin, 552–91. Van Nostrand Reinhold, New York.

102. Composite fasteners. *Rotor Wing* May 1993:53.

103. Bowtell, M. 1991. Advances presented in plastics and composites joining. *Adhes. Age* August:42–4.

104. *Adhes. Age,* May 1993:40–1.

105. Rouse, N. E. 1985. Improved methods for thermoplastic bonding. *Mach. Des.* April 11:72–9.

106. Johnson, W. S., ed. 1988. *Adhesively Bonded Joints: Testing, Analysis, and Design.* ASTM International, Materials Park, OH.

107. Manufacturers are developing improved structural systems. *Aviat. Week Space Technol.* December 12, 1988:67–72.

108. Seidl, A. L. 1989. Repair of composite structures on commercial aircraft. *Proc. 15th Ann. Adv. Compos. Workshop, SAMPE,* January 1989, San Francisco, CA.

109. Carter, A. B., III. 1989. Preserving the structural integrity of advanced composite materials through the use of surface mounted fasteners. *SAMPE J.* 25(4):21–5.

110. Lachman, L. M., ed. 1973. *Advanced Composites Design Guide,* 3rd ed. Air Force Materials Laboratory, Dayton, OH.

111. Joining of advanced composites. *Engineering Design Handbook,* U.S. Army Material Development and Readiness Command, 1979. DARCOM-P 706316.

112. *Adhesive Bonding Handbook for Advanced Structural Materials.* European Research and Technology Centre, Noordwijk, Netherlands, 1990. ESA PSS 03 210, N91-32234.

113. Kim, K. S., W. T. Kim, D. G. Lee, et al. 1992. Optimal tubular adhesive-bonded lap joint of the carbon fiber epoxy composite shaft. *Compos. Struct.* 21(3):163.

114. Schwartz, M. M. 1994. *Joining of Composite Matrix Materials.* ASM International, Materials Park, OH.

115. Pocius, A. V., and R. P. Wenz. 1985. *SAMPE J.* 21(5):50.

116. Stevens, T. 1990. Joining advanced thermoplastic composites. *Mech. Eng.* March:41–5.

117. Osterndorf, J., R. Rosty, M. J. Bodnar. 1989. Adhesive bond strength and durability studies using three different engineering plastics and various surface preparations. *SAMPE J.* 25(4): 15–9.

118. Rose, P. W., and E. M. Liston. 1985. Treating plastic surfaces with cold gas plasmas. *Plast. Eng.* October:41–5.

119. Adhesive bonding of Vectra liquid crystal polymers. 1986. Celanese Corporation. RCH-36-49.

120. Adhesive bonding of Torlon parts, Bull. TAT 36A, Amoco Chemical Company, Chicago, IL.

121. Rogus, K., and J. Aragon. 1989. Plasma etch: A more controlled surface preparation for bonding composite structures. *Proc. 34th Int. SAMPE Symp. Exhib.,* May 1989, Reno, NV.

122. Occhiello, E., M. Morra, G. L. Guerrini, et al. 1992. Adhesion properties of plasma-treated carbon/PEEK composites. *Composites* 23(3):193–200.

123. Kinloch, T. J., and B. R. K. Blackman. 1993. Continuing challenges facing adhesive technologist. *Symp. Soc. Chem. Ind. Brit. Inst. Met.* March 1993, London.

124. Tech monitoring, materials technologies, structural adhesives. SRI International, 1991, p. 95, Menlo Park, CA.

125. Silverman, E. M., and P. A. Griese. 1989. Joining methods for graphite/PEEK thermoplastic composites. *SAMPE J.* 25 (5):34.

126. Goeders, D. C., and J. L. Perry. 1991. Adhesive bonding PEEK/IM-6 composite for cryogenic applications. *Proc. 36th Int. SAMPE Symp. Exhibit.,* ed. J. Stinson, R. Adsit, and F. Gordaninejad, p. 348, San Diego, CA.

127. Yoon, T. -H., and J. E. McGrath. 1991. Adhesion study of PEEK/graphite composites. *Proc. 36th Int. SAMPE Symp. Exhibit.,* ed. J. Stinson, R. Adsit, and F. Gordaninejad, p. 428, San Diego, CA.

128. Occhiello, E., and F. Garbassi. 1988. Surface modifications of polymers using high energy density treatments. *Polym. News,* 12:365.

129. Boenig, H. V. 1982. *Plasma Science and Technology.* Cornell University Press, Ithaca, NY.

130. Wingfield, J. R. J. 1993. Treatment of composite surfaces for adhesive bonding. *Int. J. Adhes. Adhes.* 13(3):151–6.

131. Politi, R. E. 1987. Recent developments in polyimide and bismaleimide adhesives. In *High Temperature Polymer Matrix Composites,* ed. T. T. Serafini, 123–7. Noyes, Park Ridge, NJ.

132. Albee, N. 1989. Adhesives for structural applications. *Adv. Compos.* November/December: 42–50.

133. Parker, B. M. 1994. The durability of adhesive bonded epoxy carbon fibre composite. In *Composites Bonding,* ed. D. J. Damico, T. L. Wilkinson, Jr., and S. L. F. Niks. ASTM, Philadelphia, PA. ASTM STP 1227.

134. Gilmore, R. B., and S. J. Shaw. 1994. The effect of temperature and humidity on the fatigue behaviour of composite bonded joints. In *Composite Bonding,* ed. D. J. Damico, T. L. Wilkinson, Jr., and S. L. F. Niks. ASTM, Philadelphia, PA. ASTM STP 1227.

135. Chin, J. W., and J. P. Wightman. 1994. Surface pretreatment and adhesive bonding of carbon fiber-reinforced epoxy composites. In *Composite Bonding,* ed. D. J. Damico, T. L. Wilkinson, Jr., and S. L. F. Niks. ASTM, Philadelphia, PA. ASTM STP 1227.

136. Kohli, D. K. 1992. Development of polyimide adhesives for 371°C (700°F) structural performance for aerospace bonding applications: FM® 680 system. *Proc. 37th Int. SAMPE Symp. Exhibit.,* ed. G. C. Grimes, R. Turbin, and G. Forsberg, pp. 430–9, March 1992, Anaheim, CA.

137. Bradshaw, D. R. 1989. Structural adhesive Q&A: Performance, cost and use. *Adhes. Age* September:44–8.

138. Bowtell, M. 1989. Adhesive activities in United Kingdom include seminars. *Adhes. Age* September:67–9.

139. Buckley, J. D., R. L. Fox, and J. R. Tyeryar. 1987. Seam bonding of graphite reinforced composite panel using induction heating. *Ind. Heat* March:32–4.

140. Joining composite structures in space. *Aerosp. Eng.* July 1989:9–11.

141. Brown, A. S. 1990. Thermoplastic composites—Material of the '90s? *Aerosp. Am.* January:28–33.

142. Zelenak, S., D. W. Radford, and M. W. Dean. 1992. The performance of carbon fiber reinforced PEEK subassemblies joined using a dual resin bonding approach. *Proc. 37th Int. SAMPE Symp.,* ed. G. C. Grimes, R. Turpin, and G. Forsberg, pp. 1346–56, March 1992, Anaheim, CA.

143. Swartz, H. D., and J. L. Swartz. 1989. Focused infrared melt fusion: Another option for welding thermoplastic composites. *Joining Compos. Conf.,* March 1989, Garden Grove, CA. EM89-175.

144. Swartz, H. D. Focused infrared bonds, shapes TPs. *Plast. Tech.* January 1989:32–41.

145. Robotic infrared heating technique used for joining PEEK composites. *Reinf. Plast.* 33(2):48, 1989.

146. Girardi, M. 1992. Plastics at TWI—An update. *TWI Bull. 5* September/October:100–3.

147. Gumbleton, H. 1989. Hot gas welding of thermoplastics: An introduction. *Joining Mater.* 5:215.

148. Watson, M. N. 1989. Hot gas welding of thermoplastics. *TWI Bull.* May/June:77–80.

149. Hot-gas welding system consolidates complex thermoplastics composites. *Adv. Mater.* 11 (16):4, 1989.

150. Watson, M. 1990. Plastics joining at TWI. *TWI Bull.* January/February.

151. Grim, R. A. 1990. Fusion welding techniques for plastics. *Weld. J.* March:23–28.

152. Hauber, D., and M. Pasanen. 1989. The process characterization of graphite/PEEK and graphite/J2 using the thermoplastic welding head. *Proc. 34th Int. SAMPE Symp. Exhibit.* pp. 360–69. May 1989, Reno, NV.

153. Powers, J., and W. Trzaskos. 1989. Recent developments in adhesives for bonding advanced thermoplastic composites. *Proc. 34th Int. SAMPE Symp. Exhibit.,* May 1989, Reno, NV.

154. Maguire, D. M. 1989. Joining thermoplastic composites. *SAMPE J.* 25(1):11–4.

155. Taylor, N. S. 1988. Ultrasonic welding of plastics. *TWI Bull* July/August:159–62.

156. Kempe, G., H. Krauss, and G. Korger-Roth. 1990. Adhesion and welding of continuous carbon-fiber reinforced polyetheretherketone (C_f-PEEK/APC2). *Proc. 4th Eur. Conf. Compos. Mater.,* pp. 105–12, September 1990, Stuttgart, Germany.

157. Stokes, K. 1988. Vibration welding of thermoplastics. Parts I–IV. *Polym. Eng. Sci.* 28(11).

158. Benatar, A., and T. G. Gutowski. 1988. Ultrasonic welding of advanced thermoplastic composites. Industrial Liaison Progress Report, Massachusetts Institute of Technology, Cambridge, MA. MIT-6-41-88.

159. Schwartz, W. H. 1990. Ultrasonic sound off. *Assem. Eng.* January:24–7.

160. Wolcott, J. 1989. Recent advances in ultrasonic technology. *Proc. 47th Ann. Tech. Conf. SPE,* pp. 502–5, May 1989.

161. Holmes, S. T., R. C. Don, J. W. Gillespie, Jr., et al. 1991. Sequential resistance welding of large scale IM7/PEEK double lap joints. CCM-91-55.

162. Wise, R. J. 1992. Approaching total joining technology for carbon fibre reinforced thermoplastics. *Bull. 1, TWI J.* January/February:4–6.

163. Leatherman, A. 1981. Induction bonding finds a niche in an evolving plastics industry. *Plast. Eng.* 4:27–9.

164. Smiley, A.J. 1989. A new concept for fusion bonding thermoplastic composites. *Joining Compos. Conf., SME,* March 1989, Garden Grove, CA.

165. Day, M. J., and M. F. Gittos. 1989. The application of microscopy to welded joints in thermoplastics. *TWI J.* 390/January.

166. Don, R. C., et al. 1990. Fusion bonding of thermoplastic composites by resistance heating. *SAMPE J.* January/February:59–66.

167. Eveno, E. C., and J. W. Gillespie, Jr. 1988. Resistance welding of graphite polyetherether ketone composites: An experimental investigation. University of Delaware, Newark. CCM88-35.

168. Eveno, E. C. 1988. Experimental investigation of resistance and ultrasonic welding of graphite reinforced poly ether ether ketone composites. University of Delaware, Newark. CCM88-30.

169. Don, R. C., J. W. Gillespie, Jr., and C. L. T. Lambing. 1992. Experimental characterization of processing—Performance relationships of resistance welded graphite/polyetheretherketone composite joints. *Polym. Eng. Sci.* 32(9):620.

170. Bastien, L., I. Howie, R. C. Don, et al. 1990. Manufacture and performance of resistance welded graphite reinforced thermoplastic composite structural elements. University of Delaware, Newark. CCM 90-33.

171. Xiao, X. R., S. V. Hoa, and K. N. Street. 1992. Processing and resistance welding of APC-2 composite. *J. Compos. Mater.* 26(7):1031–49.

172. Cantwell, W. J., P. Davies, P. E. Bourban, et al. 1990. Thermal joining of carbon fibre reinforced PEEK laminates. *Compos. Struct.* 16:305–21.

173. Connect. *TWI J.* May 1992:3.

174. Wise, R. J. 1992. Polymer coated material (PCM) welding: A novel technique for joining dissimilar materials—Preliminary study. *Bull. 1, TWI J.* 454 July.

175. Cook, L. C. 1989. The repair of aircraft structures using adhesive bonding—A survey of current UK practices, *TWI J.* 392.

176. Hall, R., M. D. Raizenne, and D. L. Simpson. 1989. Proposed composite repair methodology for primary structure. *Composites* 20(5):479–83.

177. Ramsey, C. 1989. Battle damage repair of composite structures. *Proc. 34th Int. SAMPE Symp. Exhibit.,* May 1989, Reno, NV.

178. Reinhart, T. 1989. Composites supportability in the U.S. Air Force. *Proc. 34th Int. SAMPE Symp. Exhibit.,* May 1989, Reno, NV.

179. Chichon, M. 1989. Repair adhesives: Development criteria for field level conditions. *Proc. 34th Int. SAMPE Symp. Exhibit.,* May 1989, Reno, NV.

180. Steelman, T. 1989. Repair technology for thermoplastic aircraft structures (REPTAS). *Proc. 34th Int. SAMPE Symp. Exhibit.,* May 1989, Reno, NV.

181. Purcell, B. 1989. Battlefield repair of bonded honeycomb panels. *Proc. 34th Int. SAMPE Symp. Exhibit.,* May 1989, Reno, NV.

182. Hartup, C. 1989. Battle damage repair of thermoplastic structure. *Proc. 34th Int. SAMPE Symp. Exhibit.,* May 1989, Reno, NV.

183. Sivy, G., and P. Briggs. 1989. Rapid low-temperature repair system for field repair. *Proc. 34th Int. SAMPE Symp. Exhibit.,* May 1989, Reno, NV.

184. Connolly, J., and D. Vannice. 1989. Development of PMR-15 repair concepts. *Proc. 34th Int. SAMPE Symp. Exhibit.,* May 1989, Reno, NV.

185. Livesay, M., N. Smith, and E. Castenada. 1989. Fast simple system for field repair of composite armor and structures. *Proc. 34th Int. SAMPE Symp. Exhibit.,* May 1989, Reno, NV.

186. Diberardino, M., J. Dominquez, and R. Cochran. 1989. Bonded field repair concepts using ambient temperature storable materials. *Proc. 34th Int. SAMPE Symp. Exhibit.,* May 1989, Reno, NV.

187. Meade, R. 1989. Repair of thermoplastic composites. *Proc. 34th Int. SAMPE Symp. Exhibit.,* May 1989, Reno, NV.

188. Bittaker, D. 1989. Manufacturing technology for bonded repair procedures. *Proc. 34th Int. SAMPE Symp. Exhibit.,* May 1989, Reno, NV.

189. Mahon, J. 1989. Induction bonded repair of aircraft and structure. *Proc. 34th Int. SAMPE Symp. Exhibit.,* May 1989, Reno, NV.

190. Sennett, M. S. 1989. Field repair of composite materials in Army service: Planning for the future. Materials Test Laboratory, Watertown, MA. TR-89-45, FR5/89.

191. Mohan, R. 1990. *Plast. Eng.* February.

192. Allen, R. C., J. Bird, and J. D. Clarke. 1988. *Mater. Sci. Technol.* 4.

193. Lees, W. A. 1988. *Mater. Sci. Technol.* 4.

194. Cheng, H. N. 1988. *Adhes. Age* December.

195. Taylor, N. S. 1990. Welds in thermoplastic composite materials. *Proc. 4th Eur. Conf. Compos. Mater.* pp. 325–32, September 1990, Stuttgart, Germany. Elsevier, London, U.K.

196. Bamborough, D., and P. M. Dunckley. 1990. *Adhes. Age* November.

197. Arah, C. O., D. K. McNamara, H. M. Hand, et al. 1990. *SAMPE J.* 25(4).

198. English, L. K. 1988. Field repairs of composite structures. *Mater. Eng.* September:37–9.

199. Armstrong, K. A. 1989. *Conf. Bonding Repair Compos.* July 1989.

200. Lee, F., S. Brinkerhoff, and S. McKinney. 1988. *Proc. 20th Int. SAMPE Tech. Conf.* September 1988, Minneapolis, MN.

201. Raghava, R. 1988. *J. Polym. Sci. B.* 26.

202. Assessment of fatigue damage in MMC and their joints: CRP project at EWI, Columbus, OH, 7(1):8, 1993.

203. McConnell, V. P. 1989. In need of repair. *Adv. Compos.* May/June:60–70.

204. Reinhart, T. J. 1991. *Proc. 3rd DOD/NASA Compos. Repair Technol. Workshop,* January 1991. WL TR-91-4054.

205. Guidance materials for the design, maintenance, inspection and repair of thermosetting epoxy matrix composite aircraft structures, 1991. International Air Transport Association.

206. Renbert, J. A., and J. C. Seferis. 1992. Fundamentals of composite repair. *SME Conf. Compos. Manuf.,* January 1992, Anaheim, CA. EM92-100.

207. Thorbeck, J. 1991. *Advanced Materials: Cost Effectiveness, Quality Control, Health and Environment.* SAMPE/Elsevier, San Diego, CA.

208. Xiao, X. R., and S. V. Hoa. 1990. Repair of thermoplastic composite structures by fusion bonding. *Proc. 35th Int. SAMPE Symp.,* pp. 37–43, Anaheim, CA.

209. Xiao, X., S. V. Hoa, K. N. Street. 1994. Repair of thermoplastic resin composite by fusion bonding. In *Composite Bonding,* ed. D. J. Damico, T. L. Wilkinson, Jr., and S. L. F. Niks, pp. 30–44. ASTM, Philadelphia. ASTM STP 1227.

210. Cogswell, F. N., P. J. Meakin, A. J. Smiley, et al. 1989. Thermoplastic interlayer bonding for aromatic polymer composites. *Proc. 34th Int. SAMPE Symp.,* pp. 2315–25, May 1989, Reno, NV.

211. Border, J., R. Salas, and M. Black. 1990. Induction heating development for aircraft repair. *Proc. 35th Int. SAMPE Symp.,* pp. 1411–8; *Proc. 3rd DOD/NASA Compos. Repair Technol. Workshop,* pp. 82–90, May 1991, Anaheim, CA.

212. Davies, P., W. J. Cantwell, P. -Y. Jar, et al. 1991. Joining and repair of a carbon fibre-reinforced thermoplastic. *Composites* 22(6):425–31.

213. Okine, R. K. 1988 and 1989. Analysis of forming parts from advanced thermoplastic composite sheet materials. *Int. SAMPE Tech. Conf.* 20:149, 1988; *SAMPE J.* 25(3):10, 1989, Minneapolis, MN.

214. Harper, R. C. 1992. Thermoforming of thermoplastic matrix composites. *SAMPE J.* 28(2): 9–17.

215. Chilva, T. E., and F. S. Deans. 1986. *Proceedings: Advanced Composites—The Latest Developments.* ASM International, Materials Park, OH.

216. Galli, E. 1988. *Plast. Mach. Equip.* 17(3):27.

217. Harper, R. C., and J. H. Pugh. 1991. Thermoforming of thermoplastic matrix composites. In *International Encyclopedia of Composites,* vol. 5, ed. S. M. Lee, pp. 496–530. VCH, New York.

218. Mallon, P. J., C. O'Bradaigh, and R. B. Pipes. 1988. Thermoforming of fiber reinforced thermoplastic matrix composites. University of Delaware, Newark. CCM 87-57.

219. Gunyuzlu, M. A. 1989. Processing of thermoplastic composites using drapeable preforms. *Proc. 34th Int. SAMPE Symp. Exhibit.,* pp. 341–52, May 1989, Reno, NV.

220. Silverman, E. M., W. C. Forbes, and K. K. Ueda. 1989. Property and processing performance of P75 and P100 graphite/PEEK thermoplastic prepreg tapes and fabrics. *Proc. 34th Int. SAMPE Symp. Exhibit.,* May 1989, Reno, NV.

221. Okine, R. K., D. H. Edison, and N. K. Little. 1987. *Proc. 32nd Int. SAMPE Symp.,* p. 1418, Anaheim, CA.

222. Smiley, A. J., and K. B. Pipes. 1988. Diaphragm forming of carbon fiber reinforced thermoplastic composites materials. University of Delaware, Newark. CCM 88-11.

223. O'Bradaigh, C., and P. J. Mallon. 1988. Polymeric diaphragm forming of continuous fiber reinforced thermoplastics. University of Delaware, Newark. CCM 88-08.

224. Mallon, P. J., C. O'Bradaigh, and R. B. Pipes. 1989. *Composites* 20(1):48; and CCM Report 87-57, November 1987, 34 pages. Thermoforming of fiber reinforced thermoplastic matrix composites.

225. Mallon, P. J., and C. O'Bradaigh. 1988. *Composites* 19(1):37.

226. Witzler, S. 1988. *Adv. Compos.* September/October:50.

227. Jar, P. -Y., P. Davies, W. Cantwell, et al. 1990. Manufacturing engineering components with carbon fibre reinforced PEEK. *Proc. 4th Eur. Conf. Compos. Mater.,* pp. 819–21, September 1990, Stuttgart, Germany.

228. Pantelakis, S. G., D. T. Tsahalis, T. B. Kermanidis, et al. Experimental investigation of the superplastic forming technique using continuous carbon fiber reinforced PEEK. *Proc. 4th Eur. Conf. Compos. Mater.,* pp. 79–85, September 1990, Stuttgart, Germany.

229. Baker, E. T., and T. L. Gesell. 1990. Autoclave molding of Avimid K® composites. *Proc. 35th Int. SAMPE Symp. Exhibit.,* ed. G. Janicki, V. Bailey, and H. Schjelderup, p. 979, Anaheim, CA.

230. Forsberg, G., and S. Koch. 1989. Design and manufacture of thermoplastic F-16 main landing gear doors. *Proc. 34th Int. SAMPE Symp. Exhibit.,* pp. 9–19, May 1989, Reno, NV.

231. Monks, R., and M. Naitove. *Plast. Technol.* May 1992:13; March 1991:48–55.

232. Stone, R. H., et al. 1991. Design and manufacture of advanced thermoplastic structures. WL-TR-91-8058.

233. Ramkumar, R. L. 1990. Design and manufacture of advanced thermoplastic structures (DMATS). PR 11, F33615-87-C-5242.

234. Bonazza, B. R. 1989. Ryton® PPS conductive composites for EMI shielding applications. *Proc. 34th Int. SAMPE Symp. Exhibit.,* pp. 20–34, May 1989, Reno, NV.

235. Monaghan, M. R., C. M. O'Bradaigh, P. J. Mallon, et al. 1990. The effect of diaphragm stiffness on the quality of diaphragm formed thermoplastic composite components. *Proc. 35th Int. SAMPE Symp. Exhibit.,* pp. 810–24, April 1990, Anaheim, CA.

236. Process produces practically perfect preforms. *Mach. Des.* March 1993:18.

237. Reinhard, D. 1990. Prototype forming processes for thermoplastic composites. *Proc. 22nd Int. SAMPE Tech. Conf.,* pp. 334–47, November 1990, Boston, MA.

238. Rogers, J. K. 1989. Fighter jet development spurs TP composite process innovations. *Plast. Technol.* July:15–19.

239. Kromrey, R. 1988. Therm-X (SM) process—A revolutionary new curing system for advanced composites, United Technologies Chemical Systems, Sunnyvale, CA. CSD-V45058.

240. Kromrey, R. B. 1988. *Proc. SME Tool. Compos. Conf.,* pp. 1–11, May 1988. EM 88-219.

241. Kromrey, R. B. 1988. *Proc. SME Tool. Compos. Conf.,* pp. 3–4, May 1988.

242. Niedermeier, M., S. Delaloye, and G. Ziegmann. Forming studies of continuous fibre reinforced thermoplastic composite parts with the diaphragm technology. *Proc. 16th Int. SAMPE Eur. Conf. Exhibit.* Salzburg, Austria, May/June 1995.

243. Kueterman, T. P. 1985. Advanced composites. *Proc. Conf. ASM Eng. Soc. Detroit, SAMPE, SPE, SPI,* pp. 147–53.

244. Robroek, L. M. J. 1992. Introduction to the rubber forming of thermoplastic composites. Technische University Delft, Netherlands, p. 57; *NTIS Alert,* 93(9), May 1993. LR-683, ETN-92-92877.

245. Gaspari, J. D. 1993. New equipment and processes for advanced composites. *Plast. Technol.* August:29–32.

246. Cattanach, J. J., and F. N. Cogswell. 1986. *Development of Reinforced Plastics 5, Process and Fabrication.* 20–5. Elsevier, London.

247. Strong, A. B., and P. Hauwiller. 1989. *Proc. 34th Int. SAMPE Symp.* p. 43, Reno, NV.

248. Krone, J. R., and J. H. Walker. 1986. Thermoforming woven fabric reinforced polyphenylene sulfide composites. *CoGSME Compos. Manuf. 5th Conf.,* pp. 112–23, January 1986, Los Angeles, CA.

249. Olson, S. H. 1990. Manufacturing with commingled yarn, fabrics and powder prepreg thermoplastic composite materials. *Proc. 35th Int. SAMPE Symp. Exhibit.,* pp. 1306–20, Anaheim, CA.

250. Yamaguchi, Y., Y. Sakatani, and M. Yoshida. 1988. Fabrication studies on carbon/thermoplastic matrix composites. *Proc. 4th Adv. Compos. Conf. ASM/ESD,* pp. 415–24, September 1988, Dearborn, MI.

251. Technology for producing carbon fiber-reinforced composite pipes. New Technology Japan, *JETRO* 19(10):34, 1992. 92-01-005-07.

252. Fibre-reinforced composite material for forming complicated shapes. New Technology Japan, *JETRO* 19(6):18, 1991. 91-09-001-01.

253. Offringa, A. 1992. From thermosets to thermoplastics: A logical evolution. *SME Copmpos. Manuf. Conf.,* January 1992, Anaheim, CA. SME-EM-92-113.

254. Offringa, A. R. 1992. Thermoplastics—Moving into series production. *Proc. 37th Int. SAMPE Symp.,* pp. 1028–39, March 1992, Anaheim, CA.

255. Butkus, L. M., A. K. Behme, Jr., and G. D. Meuer. 1990. Plastic media blast (PMB) paint removal from composites. *Proc. 35th Int. SAMPE Symp.,* ed. G. Janicki, V. Bailey, and H. Schjeldrup, pp. 1385–97, April 1990, Anaheim, CA.

256. Dwenger, B. 1992. Prototype machining of MMCs. *Mach. Requir. Met. Matrix Compos. Conf.,* September 1992, Dearborn, MI. SME-EM-92-251.

257. Schmenk, M. J., and W. J. Zdeblick. 1992. An economic study for machining MMC parts. *Mach. Requir. Met. Matrix Compos. Conf.,* September 1992, Dearborn, MI. SME-EM-92-253.

258. Stashko, D. A study of sintered diamond capabilities in machining 390 aluminum. In *Cutting Tool Engineering Diamond/Superabrasive Directory,* vol. 30, 59–61.

259. Slatin, N., ed. 1980. *Machining Data Handbook,* 3rd ed. Institute of Advanced Manufacturing Sciences, Cincinnati, OH.

260. Sacha, J. A. 1982. Machining 390 aluminum alloy. *ASM Met. Congr.,* October 1982, St. Louis, MO. 8201-050.

261. Miller, J. C., J. W. Sutherland, and R. E. DeVor. 1992. Surface roughness characteristics for turning 380 and 390 aluminum casting alloys. *Proc. 10th North Am. Manuf. Res. Conf., NAMRI/SME,* pp. 282–88, Dearborn, MI.

262. Berlin, B. R. 1992. Machining cost comparison of silicon carbide discontinuously reinforced aluminum, unreinforced aluminum, and titanium. *Mach. Requir. Met. Matrix. Compos. Conf.,* September 1992, Dearborn, MI. SME-EM-92-252.

263. Brun, M. K., F. Gorsler, and M. Lee. 1985. Wear characteristics of various hard materials for machining SiC reinforced aluminum alloy. *ASME Int. Conf.,* April 1985, and G. E. Corporate Research & Development, Report #84CRD327 (1984), Schenectady, NY.

264. Chadwick, G. A., and P. J. Heath. 1990. Machining metal matrix composites. *Met. Mater.* February:73–6.

265. Ginburg, D. M. 1987. Cost effective secondary fabrication processes for metal matrix composites. SME-MF-87-570.

266. Ricci, W. S. 1987. Machining metal matrix composites. SME-MR-87-827.

267. *Guide to Machining SXA® Engineered Materials.* Advanced Composite Materials Corporation, 1988, Greer, SC.

268. Allison, J. E., and G. S. Cole. 1993. Metal-matrix composites in the automotive industry: Opportunities and challenges. *J. Met.* 45(1):19–24.

269. Folgar, F., J. E. Widrig, J. W. Hunt. 1987. Design, fabrication, and performance of Fiber FP/metal matrix composite connecting rods. SAE, Warrendale, PA. 870406.

270. Urquhart, A. M. 1991. *Adv. Mater. Process.* July:25–9.

271. Lane, C. T. 1992. *Int. Sem. Mach. Tools Adv. Mater.,* October 1992, Milan, Italy.

272. Looney, L. A., J. M. Monaghan, P. O'Reilly, et al. 1992. The turning of an Al/SiC metal-matrix composite. *J. Mater. Process. Technol.* 33(4):453–68.

273. Schneider, G. 1990. High-speed machining: Solutions for productivity. *Proc. 1989 Conf. Soc. Carbide Tool Eng. ASM.* Dearborn, MI.

274. McConnell, V. P. 1990. Metal-matrix composites: Materials in transition. Part II. *Adv. Compos.* July/August:63–67.

275. Whiskered ceramic cutting grades. *Matl. Eng.* June 1989.

276. Weinert, K., and D. Biermann. 1993. Turning of fiber and particle reinforced aluminum. *Proc. Int. Conf. Mach. Adv. Mater.,* ed. S. Jahanmir, pp. 437–53, July 1993, Gaithersburg, MD. NIST SP 847.

277. Clark, I. E. 1994. A guide to machining metal matrix composites with Syndite PCD. *Ind. Diamond Rev.* 54(3):135–8.

278. Hung, N. P., F. Y. C. Boey, K. A. Khor, et al. 1995. Machinability of cast and powder formed aluminum alloys reinforced with SiC particles. *J. Mater. Proc. Technol.* 48(1–4):291–7.

279. Lane, C. 1992. Machinability of aluminum composites as a function of matrix alloy and heat treatment. *Proc. Mach. Compos. Mater. Symp., ASM,* ed. T. S. Srivatsan and D. M. Bowden, pp. 3–15, November 1992, Chicago, IL.

280. Coelho, R. T. 1994. MMC collaborative project. University of Birmingham, U.K. Internal Report.

281. Lin, J. T., D. Bhattacharyya, and C. Lane. 1995. Machinability of a silicon carbide reinforced aluminum metal matrix composite. *Wear* 181/183:2, 883–8.

282. Chambers, A. R., and S. E. Stephens. 1991. Machining of Al-5 Mg reinforced with 5 vol% Saffil and 15 vol% SiC. *Mater. Sci. Eng.* 135:287–90.

283. Ramulu, M., and M. Taya. 1988. An investigation of machinability of high temperature composites. *Metal Matrix Carbon Ceram. Matrix Compos.* ed. J. D. Buckley, pp. 423–31, January 1988, Cocoa Beach, FL.

284. Saito, N. 1984. Recent electrical discharge machining (EDM) techniques in Japan. *Bull. Japn. Soc. Precis. Eng.,* 18(2).

285. Hung, N. P., L. J. Yang, and K. W. Leong. 1994. Electrical discharge machining of cast metal matrix composites. *J. Mater. Process. Technol.* 44:229–36.

286. Hoover, W. 1991. Metal matrix composite-processing, microstructures and properties. *Proc. 12th Risø Int. Symp.,* ed. N. Hansen et al., pp. 387–92, Roskilde, Denmark. Risø.

287. Schwartz, M. M. 1992. *Handbook of Structural Ceramics.* McGraw-Hill, New York.

288. O'Reilly, P., J. Monaghan, and D. M. R. Taplin. 1989. The machining of metal-matrix composites—An overview. *Proc. Irish Fract. Durability Conf.,* ed. D. Taylor, pp. 81–93, September 1989, Dublin, Ireland.

289. Niskanen, P., and W. R. Mohn. 1988. Versatile metal-matrix composites. *Adv. Mater. Process.* March:39–41.

290. Yue, T. M., W. S. Lau, and C. Y. Jiang. 1993. Technology in pulsed ND:YAG laser drilling of an Al-Li based/SiC metal matrix composite. *Proc. Int. Conf. Mach. Adv. Mater.* ed. S. Jahanmir, pp. 549–53, July 1993, Gaithersburg, MD. NIST SP 847.

291. Feest, E. A. 1988. *Met. Mater.* May:273–78.

292. Chandrasekaran, H., and J. O. Johansson. 1993. The role of material and grinding parameters in grinding of alumina fiber-reinforced aluminum alloys. *Proc. Int. Conf. Mach. Adv. Mater.* ed. S. Jahanmir, pp. 55–70, July 1993, Gaithersburg, MD. NIST SP 847.

293. Lane, C. T. 1990. Machining characteristics of particle reinforced aluminium. *Proc. Conf. Fabric. Particulate Reinf. Met. Compos.,* pp. 204–10, September 1990, Montreal, Canada.

294. Matsubara, H., Y. Nishida, M. Yamada, et al. 1987; J. B. Pond. Rough cut. *Cutting Tool Eng.* 42:20–31.

295. Si_3N_4 whisker reinforced Al alloy composite. *J. Mater. Sci. Lett.* 6:1313–5.

296. Brun, M. K., M. Lee, and F. Gorsler. 1985. Wear characteristics of various hard materials for machining SiC-reinforced Al-alloy. *Wear* 104:21–9.

297. Andersson, C. H., J. E. Stahl, and M. Andersson. 1988. Plastic deformation and machining properties of some low volume fraction short fibre reinforced metal matrix composite material. *Proc. 9th Risø Int. Symp. Behav. Ceram. Metal. Compos.,* ed. S. I. Andersen, H. Lilholt, and O. B. Pedersen, Roskilde, Denmark.

298. Lau, W. S., T. M. Yue, T. C. Lee, et al. 1995. Un-conventional machining of composite materials. *J. Mater. Process. Technol.* 48(1–4):199–205.

299. Upadhya, K., ed. 1992. *Proc. Symp. TMS Ann. Meet.* March 1992, San Diego, CA.

300. Savrun, E., and M. Taya. 1988. Surface characterisation of SiC_w/2124 Al and Al_2O_3 composites machined by abrasive water jet. *J. Mater. Sci.* 23:1453–8.

301. Hashish, M. 1989. A model for abrasive-waterjet (AWJ) machining. *J. Eng. Mater. Technol.* 111(2):154–62.

302. Lau, W. S., T. M. Yue, C. Y. Jiang, et al. 1992. Pulsed Nd:YAG laser cutting of SiC/Al-Li metal matrix composite, *Proc. Mach. Compos. Mater. Symp., ASM,* ed. T. S. Srivatsan, and D. M. Bowden, pp. 29–34, November 1992, Chicago, IL.

303. Meinert, K. C., Jr., R. P. Martukanitz, and R. B. Bhagat. 1992. Laser processing of discontinuously reinforced aluminum composites. *Proc. Am. Soc. Compos. 7th Tech. Conf.,* pp. 168–77, October 1992, University Park, PA.

304. Ramos, T. 1995. Laser cuts beryllium fabrication costs. *High-Tech. Mater. Alert* August 11:2–3.

305. Kraus, S. A. 1993. Improved efficiency of composite preform cutting thru utilization of CAD/CAM based nesting software and laser cutting technology. *Fabtech Int. '93,* October 1993, Rosemont, IL. SME-MS-93-221.

306. Hoult, A. P., I. R. Pashby, and K. Chan. 1995. Fine plasma cutting of advanced aerospace materials. *J. Mater. Process. Technol.* 48(1–4):825–31.

307. Lavender, C. A., and M. T. Smith. 1986. Evaluation of waterjet-machined metal matrix composite tensile specimens. Pacific Northwest Laboratory, Richland, WA. PNL-5958, UC-25, DE-AC06-76 RLO 1830.

308. Neusens, K. F., P. K. Rohatgi, C. Vaidyanathan, et al. 1987. Abrasive waterjet cutting of metal matrix composites. *Proc. 4th U.S. Water Jet Conf.,* pp. 175–82, August 1987, Berkeley, CA.

309. Hamatani, G., and M. Ramulu. 1990. Machinability of high temperature composites by abrasive waterjet. *J. Eng. Mater. Technol.* 12(4):381–6.

310. Wise, R. J. 1992. Developments in techniques and research into bonding new combinations. *TWI Bull.* 33(1):4.

311. Devletian, J. H. 1987. SiC/Al metal matrix composite welding by a capacitance discharge process. *Weld. J.* 66(1):33–9.

312. Gittos, M., and P. Threadgill. 1992. Joining aluminum alloy MMCs. *Bull. 5, TWI J.*, September/October:104–5.

313. Cola, M. J., G. Martin, and C. E. Albright. 1991. Inertia-friction welding of a 6061-T6/Al$_2$O$_3$/15p metal matrix composite. EWI Research Report MR 9108, Columbus, OH.

314. Midling, O. T. 1991. Friction welding of SiC particle reinforced aluminum alloy. Selskapet for Industriell Og Teknisk Forskning, Trondheim, Norway. STF-34-A-91030. *NTIS Alert* 92(7):1.

315. Brusethaug, S. 1990. Mechanical properties and composition in A-357 (Al-7Si-Mg) PMMC. Hydro Aluminium N-Sunndalsra.

316. Ellis, M. B. D., M. F. Gittos, and P. L. Threadgill. 1994. Joining of aluminium based metal matrix composites: Initial studies. Technology Briefing. TWI, 501/1994.

317. Rosenwasser, S. N., A. J. Auvil, and R. D. Stevenson. 1989. Development of low-temperature, solid-state bonding approach for metal matrix composite joints. NSWC TR 89-302, LJ 89-040-TR, N60921-86-C-0279.

318. Min, B. K., M. Fisk, R. T. Hartle, et al. 1989. Spaceflight qualification level evaluation of graphite-aluminum tubes. *Proc. 34th Int. SAMPE Symp. Exhibit.* pp. 161–73, Reno, NV.

319. Min, B. K., and M. Fisk. 1989. LMSC-F176405-9.

320. Dunford, D. V., S. M. Flitcroft, D. S. McDarmaid, et al. 1992. Forming and diffusion bonding of Al-Li alloy containing 20 wt% particulate SiC. In *Aluminum-Lithium,* ed. M. Peters and P. -J. Winkler, vol. 2, 1087–92, Aachen. DGM.

321. Partridge, P. G., and D. V. Dunford. 1992. Interlayers and interfaces in diffusion bonded joints in metal matrix composites. In *Aluminum-Lithium,* ed. M. Peters and P. -J. Winkler, vol. 2, pp. 1145–50, Aachen. DGM.

322. Partridge, P. G., M. Shepherd, and D. V. Dunford. 1991. Statistical analysis of particulate interface lengths in diffusion bonded joints in a metal-matrix composite. *J. Mater. Sci.* 26 (4):953.

323. Partridge, P. G., and D. V. Dunford. 1991. The role of interlayers in diffusion bonded joints in metal-matrix composites. *J. Mater. Sci.* 26(2):255.

324. Sorensen, J. P. 1990. Titanium matrix composites. NASP Materials and Structures Augmentation Program. AIAA 90-5207.

325. Niemann, J. T. 1993. Titanium matrix composites. NASP Materials and Structures Augmentation Program. McDonnell Douglas Corporation, St. Louis, MO. Executive Summary, vol. 1. NASP Contractor Report 1145. Contract F33657-86-C-2126, 1993.

326. Fukumoto, S., A. Hirose, and K. F. Kobayashi. 1993. Diffusion bonding of SiC/Ti-6Al-4V alloy and fracture behaviour of joint. *Mater. Sci. Technol.* 9(6):520–7.

327. Altshuller, B., W. Christy, and B. Wiskel. 1990. GMA welding of Al-Al$_2$O$_3$ metal matrix composite. *Proc. Mater. Weldability Symp., ASM,* p. 305–9, Materials Park, OH.

328. Bhattacharyya, D., M. E. Bowie, and J. T. Gregory. 1992. The influence of alumina microsphere reinforcement on the mechanical behavior and weldability of a 6061 aluminum metal matrix composite. *Proc. Mach. Compos. Mater. Symp., ASM,* pp. 49–56, November 1992, Chicago, IL.

329. Irving, B. 1991. What's being done to weld metal-matrix composites? *Weld. J.* 70(6):65.

330. The weldability of particle reinforced aluminum. University of Wisconsin, Milwaukee. Internal Report, 1990.

331. Reynolds, G. H. 1990. Fusion welding of Al/Al$_2$O$_3$ composites—Phase I. MSNW, San Marcos, CA. MTL TR 90-38.

332. Gould, J. E. 1987. An examination of nugget developments during resistance spot welding using experimental and analytical techniques. *Weld. J.* 68(1):1s–10s.

333. Hirose, A., Y. Matsuhiro, M. Kotoh, et al. 1993. Laser-beam welding of SiC fibre-reinforced Ti-6Al-4V composite. *J. Mater. Sci.* 28:349.

334. Dahotre, N. B., S. Gopinathan, and M. H. McCay, et al. 1992. Laser joining of metal matrix composite. *Proc. Mach. Compos. Mater. Symp., ASM,* pp. 167–73, November 1992, Chicago, IL; *Lightweight Alloys Aerosp. 2nd Appl. Meet, TMS,* pp. 313–25, 1991, Warrendale, PA.

335. Kawall, S., and G. L. Viegelahn. 1991. Attempts at laser welding of an alumina reinforced 6061 aluminum alloy composite. *Proc. 7th Adv. Compos. Conf.* September/October 1991, pp. 283–90, Detroit, MI.

336. Dahotre, N. B., T. D. McCay, and M. H. McCay. 1994. Laser induced liquid phase reaction synthesis assisted joining of metal matrix composites. *Mater. Manuf. Process.* 9(3):447–66.

337. Dahotre, N. 1993. Joining metal matrix composites using laser improves joints. *Mater. Tech.* 8:190–2.

338. Fukumoto, S., A. Hirose, and K. F. Kobayashi. 1993. Application of laser beam welding to joining of continuous fibre reinforced composite to metal. *Mater. Sci. Technol.* 9(3):264–71.

339. Lienert, T. J., E. D. Brandon, and J. C. Lippold. 1993. Laser and electron beam welding of SiC$_p$ reinforced aluminum A356 metal matrix composite. *Scr. Metall. Mater.* 28:1341–6.

340. Lienert, T. J., and J. C. Lippold. 1994. Laser and electron beam welding of an SiC-reinforced aluminum metal matrix composite. EWI Research Report MR 9403, Columbus, OH.

341. Gedeon, S. A., and I. Tangerini. 1991. A new method for bonding metal matrix composite inserts during casting. *Mater. Sci. Eng.* A144:237.

342. Gedeon, S. A., R. Guerriero, and I. Tangerini. 1989. Process for obtaining a metallurgical bond between a metal material, or a composite material having a metal matrix and a metal-casting or a metal-alloy casting. European Patent 89202324.3.

343. Cox, A., W. A. Baeslack, III, S. Zorko, et al. 1993. Capacitor-discharge resistance-spot-welding of SiC fiber-reinforced Ti-6Al-4V. *Weld. J.* 72(10):479s–91s.

344. Suganuma, K., T. Okamoto, and N. Suzuki. 1987. Joining of alumina short-fiber reinforced AA6061 alloy to AA6061 alloy and to itself. *J. Mater. Sci.* 1580–4.

345. Blue, C. A., R. Y. Lin, J. -F. Lei, et al. 1993. Infrared joining of titanium matrix composites. NASA Lewis Research Center. NASP Contractor Report 1153, NGT-50982, WBS 4.1.03.

346. Khorunov, V. F., V. S. Kutchuk-Yatsenko, I. S. Dykhno, et al. 1990. Brazing of sheet composite materials with aluminum matrix. *Proc. Int. Conf. IIW,* pp. 143–4, July 1990, Montreal. Pergamon Press, New York.

347. Sudhakar, K. 1990. Joining of aluminum based particulate-reinforced metal-matrix composites. Ph.D. Thesis, Ohio State University, Columbus, OH.

348. Klehn, R. 1991. Joining of 6061 aluminum matrix ceramic particle reinforced composites. M.S. Thesis, Massachusetts Institute of Technology, Cambridge, MA.

349. Bergquist, P. R., S. P. Petne, and S. E. Wentworth. 1990. An assessment of the adhesive bondability of silicon carbide reinforced aluminum. *Proc. 22nd Int. SAMPE Tech. Conf.,* ed. L. D. Michelove, R. P. Caruso, and P. Adams, pp. 815–22, November 1990, Boston, MA.

350. Davies, R. J., and M. D. Ritchie. 1992. Future design concepts for the development of new pretreatments of aluminum alloy/metal matrix composites for adhesive bonding. *J. Adhes.* 38(3A):243–54.

351. Huang, X. X., Q. Lui, C. K. Yao, et al. 1991. Superplasticity in a SiC_w-6061 Al composite. *J. Mater. Sci. Lett.* 10(16):964–66.

352. Mahoney, M. W., and A. K. Ghosh. 1987. *Metall. Trans.* 18A:653.

353. Imai, T., et al. 1990. In *Metal and Ceramic Matrix Composites: Processing, Modeling and Mechanical Behavior,* ed. R. B. Bhagat et al., 235–42. TMS, Warrendale, PA.

354. Imai, T., M. Mabuchi, Y. Tozawa, et al. 1990. *J. Mater. Sci. Lett.* 9:255.

355. Mabuchi, M., and T. Imai. 1990. *J. Mater. Sci. Lett.* 9:763.

356. Higaschi, K., et al. 1991. *Scri. Metall. Mater.* 26(2):185.

357. Pickard, S. M., and B. Derby. 1992. *Mater. Sci. Eng.* A135:213.

358. Nieh, T. G., and J. Wadsworth. 1992. Superplasticity and super-plastic forming of aluminum metal-matrix composites. *J. Met.* 144(11):46.

359. Lasday, S. B. 1990. Production and properties of continuous silicon carbide reinforced titanium for high temperature applications. *Ind. Heat.* December:19–23.

360. Ravenhall, J. R., and W. E. Koop. 1990. Metal matrix composite fan blade development. AIAA 90-2178.

361. Miller, S., et al. 1989. Metallurgical design of novel metal matrix composites for aerospace applications. British Petroleum Company, p. 39.

362. Aylor, D. M., and P. J. Moran. 1985. *J. Electrochem. Soc.* 132:1277.

363. Trzaskoma, P. P. 1986. *Corrosion* 42:609.

364. Czyrklis, W. F. 1985. *Corrosion/85, NACE.* Paper 196, Houston, TX.

365. Aylor, D. M., R. J. Ferrara, and R. M. Kain. 1984. *Mod. Plast.* 23:32.

366. Keller, F., M. S. Hunter, and O. L. Robinson. 1983. *J. Electrochem. Soc.* 100:411.

367. Lin, S., H. Greene, H. Shih, et al. 1992. Corrosion protection of Al/SiC metal matrix composites by anodizing. *Corrosion* 48(1):61.

368. Trzaskoma, P. P., and E. McCafferty. 1983. *J. Electrochem. Soc.* 130:1804.

369. Ackerman, H., et al. 1987. *Metals Handbook,* 9th ed., vol. 13, 396.

370. Ridealgh, J. A., R. D. Rawlings, and D. R. F. West. 1990. Laser cutting of glass ceramic matrix composite. *Mater. Sci. Technol.* 6:395–8.

371. Guha, J. K. 1990. High-pressure waterjet cutting: An introduction. *Ceram. Bull.* 69(6):1027–9.

372. Petrofes, N. F., and A. M. Gadalla. 1988. Electrical discharge machining of advanced ceramics. *Ceram. Bull.* 67(6):1048–52.

373. Moreland, M. A., and D. O. Moore. 1988. Versatile performance of ultrasonic machining. *Ceram. Bull.* 67(6):1045–7.

374. Gadalla, A. M. 1992. Thermal spalling during electro-discharge machining of advanced ceramics and ceramic-ceramic composites. *Proc. Mach. Compos. Mater. Symp., ASM,* pp. 151–7, November 1992, Chicago, IL.

375. Gadalla, A. M., and W. Tsai. 1989. *J. Am. Ceram. Soc.* 72(8):1396.

376. Ramulu, M. 1988. EDM sinker cutting of ceramic particulate composite, TiB_2/SiC. *Advanced Ceramic Materials.*

377. Ramulu, M., and M. Taya. 1988. EDM machinability of SiC_w/Al composites. *J. Mater. Sci.* November.

378. Savrun, E., M. Taya, and M. Ramulu. 1987. A preliminary investigation of machinability of high temperature composites. University of Washington, Department of Mechanical Engineering, Pullman, WA, Technical Report.

379. Chryssolouris, G., J. Bredt, and S. Kordas. 1985. Machining of ceramic materials and composites. *Proc. Am. Soc. Mech. Eng.*, pp. 9–17, November 1985, Miami Beach, FL.

380. Tuersley, I. P., A. Jawaid, and I. R. Pashby. 1994. Review: Various methods of machining advanced ceramic materials. *J. Mater. Process. Technol.* 42(4):377–90.

381. Bifano, T., T. A. Dow, and R. O. Scattergood. 1988. *Proc. 3rd Int. Symp. Mach. Adv. Ceram. Mater. Components,* pp. 113–20, November/December 1988, Chicago, IL.

382. Sheppard, L. M. 1987. *Adv. Mater. Process. Inc. Met. Progr.* 12:40–8.

383. Lee, D. W., and G. Feick. 1972. The science of ceramic machining and surface finishing. National Bureau of Standards, Special Publication 348, pp. 197–211.

384. Rothman, E. P., G. B. Kenney, and H. K. Bowen. Potential of ceramic materials to replace cobalt, chromium, manganese, and platinum in critical applications. Industrial Liaison Report 9-11-85, Massachusetts Institute of Technology, Materials Processing Center, Cambridge, MA. Contract 333-6530.2.

385. Sarfaraz, M. A., Y. -W. Yau, and N. S. Sandhu. 1993. Electron beam machining of ceramic green-sheets for multilayer ceramic electronic packaging applications. *Nucl. Instrum. Meth. Phys. Res. B* 82(1):116–20.

386. Louh, R., and R. C. Buchanan. 1988. *Proc. Mater. Res. Soc. Symp.,* 100:659.

387. Abraham, T. 1990. *Ceramic Matrix Composites,* Business Communications Co., Norwalk, CT. GB-110R.

388. Zeng, J., and J. Munoz. 1994. Optimization of abrasive waterjet cutting—The abrasive issues. *Waterjet Mach. Technol. SME,* September 1994, Chicago, IL. MR94-247.

389. Kahlman, L., S. Karlsson, R. Carlsson, et al. 1993. Wear and machining of engineering ceramics by abrasive waterjets. *Am. Ceram. Soc. Bull.* 72(8):93–8.

390. Gochnour, S., J. D. Bright, D. K. Shetty, et al. 1990. Solid particle erosion of SiC-Al_2O_3 ceramics. *J. Mater. Sci.* 25:3229–35.

391. Adlerborn, J. E., and H. T. Larker. 1991. Encapsulated HIP of nonoxide composites. *Proc. 2nd Conf. Eur. Ceram. Soc.* September 1991, Augsburg, Germany.

392. Schwetz, K. A., L. S. Sigl, J. Grein, et al. 1995. Wear of boron carbide ceramics by abrasive waterjets. *Wear* 181/183:148–55.

393. Trubelja, M. F., S. Ramanathan, M. F. Modest, et al. 1994. Carbon dioxide laser cutting of a carbon-fiber-silicon carbide-matrix composite. *J. Am. Ceram. Soc.* 71(1):89–96.

394. Technological and economic assessment of advanced ceramic materials, vol. 6: A case study of ceramic cutting tools. Charles River Associates., 1984. PB85-113132, NBS GCR84-470-6.

395. Whitney, E. D., ed. 1994. *Ceramic Cutting Tools—Materials, Development, and Performance.* Noyes, Park Ridge, NJ.

396. Barrett, R., and T. F. Page. 1989. The interactions of an Al_2O_3-SiC-whisker-reinforced composite ceramic with liquid metals. *Wear* 138:225–37.

397. Thangaraj, A. R., and K. J. Weinmann. 1992. On the wear mechanisms: A cutting performance of silicon carbide whisker-reinforced aluminum. *ASME Trans.* 114(3):301.

398. Xiao, H., X. Ai, and H. S. Yang. 1993. Effect of whisker orientation on toughening behaviour and cutting performance of SiC_w-Al_2O_3 composite. *Mater. Sci. Technol.* 9:21.

399. Shintani, K., T. Oiji, Y. Fujimura, et al. 1989. Cutting performance of toughened ceramic tools. *Int. Conf. Evol. Adv. Mater.,* Associazione Italiana di Metallurgia Italy, pp. 267–72, Milan.

400. Smith, K. H. 1989. The application of whisker reinforced and phase transformation toughened materials in machining of hardened steel and nickel-based alloys, *ASM Int. Meet. SCTE Meet.,* ed. G. Schneider, pp. 81–8, November 1989, San Diego, CA.

401. Maloney, L. D. 1989. Make way for "engineered ceramics." *Des. News* March:64–74.

402. Withers, J. C. 1989. Ceramic composite cutting tool insert for increased productivity, ultra highspeed machining. REPT 0717891, NSF/ISI-89014; *NTIS Alert* 92(29):89, 1989.

403. Clarke, D. A. 1990. Industrial applications and markets for ceramic matrix composites. Rolls Royce reprint. PNR 90753.

404. Drozda, T. J. 1985. Ceramic tools find new applications. *Manuf. Eng.* 110–15.

405. Troczynski, T. B., D. Ghosh, S. Das Gupta, et al. 1988. Advanced ceramic materials for metal cutting. In *Advanced Structural Materials,* ed. D. S. Wilkinson, 157–68. Pergamon Press, New York.

406. Ashley, T. L. 1992. The application of sialon and silicon nitrides in metal cutting. *Manufacturing 1992,* September 1992, Chicago, IL. MR-92-355-8.

407. Ekstrom, T., and M. Nygren. 1992. SiAlON ceramics. *J. Am. Ceram. Soc.* 75(2):259.

408. Gruss, W. W. 1988. Ceramic tools improve cutting performance. *Ceram. Bull.* 67:993.

409. Whitney, E. D., and P. N. Vaidyanathan. 1987. Engineered ceramics for high-speed machining. *Proc. ASM SCTE Conf. Adv. Tool Mater. Use High Speed Mach.* ed. J. A. Swartley-Loush, pp. 77–82, February 1987, Scottsdale, AZ.

410. Agranov, D. 1986. *CIRP Conf.,* August 1986, Tel Aviv, Israel.

411. Klein, A. J. 1986. *Adv. Mater. Process.* 9:26.

412. Christopher, J. D., and N. Zlatin. 1974 and 1988. New cutting tool materials. SME TP MR 74-101, 1974; in D. W. Richerson, ed., *Ceramics Applications in Manufacturing,* pp. 105–9, SME, 1988, Dearborn, MI.

413. Billman, E. R., et al. 1988. Machining with Al_2O_3-SiC-whisker cutting tools. *Ceram. Bull.* 67(6):1016.

414. Schuldies, J. J., and J. A. Branch. 1991. Ceramic composites: Emerging manufacturing processes and applications. *SME Using Adv. Ceram. Manuf. Appl. Conf.,* June 1991, Dearborn, MI. SME EM 91-250.

415. Belitskus, D. 1993. *Fiber and Whisker Reinforced Ceramics for Structural Applications.* Marcel Dekker, New York.

416. Cawley, J. D. 1989. Joining of ceramic-matrix composites. *Ceram. Bull.* 68(9):1619–23.

417. Fernie, J. A., P. L. Threadgill, and M. N. Watson. 1991. Progress in joining of advanced materials. *Weld. Metal Fabr.* 59(4):179–84, 1991.

418. Rabin, B. H. 1990. Joining of fiber-reinforced SiC composites by in situ reaction methods. *Mater. Sci. Eng.* A130:L1–L5.

419. Rabin, B. H. 1992. Joining of silicon carbide/silicon carbide composites and dense silicon carbide using combustion reactions in the titanium-carbon-nickel system. *J. Am. Ceram. Soc.* 75(1):131–5.

420. DuPont CVI ceramic matrix composites. DuPont Composites, Newark, DE. Preliminary Engineering Data.

421. Sawyer, J. W., M. L. Blosser, and R. R. McWithey. 1984. Derivation and test of elevated temperature thermal-stress-free fastener concept. In *Welding Bonding and Fastening,* ed. T. Boles, B. Stein, and P. Royster, pp. 101–17, 1984 NASA Conf. Proc. 2387.

422. Thomson, B., and J. -F. LeCostaouec. 1991. Recent developments in SiC monofilament reinforced Si_3N_4 composites. *SAMPE Q,* 22(3):46–51.

423. Walls, P. A., and M. Ueki. 1992. Joining SiAlON ceramics using composite β-SiAlON-glass adhesives. *J. Am. Ceram. Soc.* 75(9):2491–7.

424. Lannutti, J. J., E. Park, and J. D. Cawley. 1992. Active metal brazing of ceramic matrix composites. Edison Welding Institute, Columbus, OH. Research Brief B9201.

425. Mizuhara, H., E. Huebel, and T. Oyama. 1989. High-reliability joining of ceramic to metal. *J. Am. Ceram. Soc.* 68(9):1591–1600.

426. Mizuhara, H., and K. Mally. 1985. Ceramic-to-metal joining using an active filler metal. *Weld. J.* 64(10):27–32.

427. Kingery, W. D. 1950. Fundamental study of phosphate bonding in refractories. I. Literature review. II. Cold setting properties. III. Phosphate adsorption by clay and bond migration. *J. Am. Ceram. Soc.* 33(8):239–50.

428. Gonzalez, F., and J. Halloran. 1981. Strength and microstructure of phosphate-bonded alumina refractories. *Am. Ceram. Soc. Bull.* 60(7):700–2.

429. Kuz'menkov, M., S. Plyshevskii, and I. Buraya. 1974. Interaction of an alumino-phosphate binder with some fillers. *Obshch. Prikl. Khim.* 6:53–7.

430. Karpinos, D., et al. 1982. Physicochemical processes occurring in nitride and oxide-nitride compositions with phosphate binders during heating. *Poroshk. Metall.* 5:50–4.

431. Gonzalez, F., and J. Halloran. 1980. Reaction of orthophosphoric acid with several forms of aluminum oxide. *Am. Ceram. Soc. Bull.,* 59(7):727–38.

432. Hummel, F. 1949. Properties of some substances isostructural with silica. *J. Am. Ceram. Soc.* 32(10):320–6.

433. Talmy, I. G., and C. A. Martin. 1990. Phosphate bonded non-oxide ceramics. *Proc. 14th Conf. Met. Matrix Carbon Ceram. Compos.,* Part 1, ed. J. D. Buckley, pp. 309–16, Cocoa Beach, FL. NASA CP-3097.

434. Martin, C. A., I. G. Talmy, A. H. Le, et al. 1990. Phosphate bonding in processing of whisker- and fiber reinforced ceramic composites. *Proc. 15th Conf. Met. Matrix Carbon Ceram. Matrix Compos.,* Part I, ed. J. D. Buckley, 281–87, Cocoa Beach, FL. NASA CP-3133.

435. Talmy, I. G., C. A. Martin, A. H. Lee, et al. 1990. Phosphate bonding in processing of non-oxide ceramics. *Proc. 15th Conf. Met. Matrix Carbon Ceram. Matrix Compos.,* Part I, ed. J. D. Buckley, 281–87, Cocoa Beach, FL. NASA CP-3133.

436. Kuecker, B., T. Bejarano, and T. Hackett, eds. 1993. Refractory composites, vol. 2B. NASP Materials and Structures Augmentation Program. General Dynamics Corporation, Ft. Worth, TX. NASP-CR-1148-V-2-B.

BIBLIOGRAPHY

Am. Mach. Automat. Manuf. September:37, 47, 51; August:23, 25, 26, 1986.

ANAMATEROS, E., M. MARCHETTI, D. PALMIERI, ET AL. Failure load and failure mode of single- and multiple-bolted joints in composite material for aeronautical applications. *Proc. 16th Int. SAMPE Eur. Conf. Exhibit.* Salzburg, Austria, May/June 1995.

BHATTACHARYYA, D. A. Drilling of Kevlar composites. *Mach. Compos. Mater. II, ASM,* Pittsburgh, PA, May/June 1993.

BONAZZA, B. R. Adhesive bonding of improved polyphenylene sulfide thermoplastic composites. *Proc. 35th Int. SAMPE Symp.,* ed. G. Janicki, V. Bailey, and H. Schjelderup, pp. 859–65, Anaheim, CA, April 1990.

BOWDEN, M. Machining of titanium matrix composites. *Mach. Compos. Mater. II, ASM,* Pittsburgh, PA, October 1993.

BURKES, J. M., AND M. R. LESHER. Advanced tooling and technology for drilling metal matrix composite materials. *Mach. Compos. Mater. II, ASM,* Pittsburgh, PA, October 1993.

BURKLE, N., AND G. KEMPE. Economic welding techniques for continuous fibre-reinforced thermoplastics. *Proc. 16th Int. SAMPE Eur. Conf. Exhibit.* Salzburg, Austria, May/June 1995.

CAMPONESCHI, E. T., JR., R. E. BOHLMANN, AND J. H. FOGARTY. Composite to metal joints for the ARPA man-rated demonstrated article. *J. Thermoplast. Compos. Mater.* 8(1):56–79, 1995.

CARMAK, M., AND A. DUTTA. Instrumented thermoforming of advanced thermoplastic composites. II: Dynamics of double curvature part formation and structure development from PEEK/carbon fiber prepreg tapes. *Polym. Compos.* 12(5):338–54, 1991.

Ceramics Technology: Automotive Gas Turbine Engine Component Applications. Nerac, Tolland, CT, 1993; *NTIS Alert,* p. 5, Feb. 15, 1994.

CHAMIS, C., AND P. L. N. MURTHY. Preliminary design of adhesively bonded composite joints. NASA Technical Brief, October 1992, pp. 78–80.

CHRYSSOLOURIS, G., P. SCHENG, AND W. C. CHOI. Three-dimensional laser machining of composite materials. *J. Eng. Mater. Technol.* 112(4):387–92, October 1990.

Composites improve aircraft engines. *Am. Mach.* 139(5):18–27, 1995.

COOK, J. Machining of SXA alloys. ARCO Metals Company, Silage Operation. SXA Alloy Data Sheet, 1990.

CORNELIA, R. H. Stripping without damage. *Aerosp. Compos. Mater.* 42–44, 1991.

COSTALAS, E., AND A. F. JOHNSON. Overinjected thermoplastic composites: Optimized bonding in the interface between short and continuous fibre reinforced materials. *Proc. 16th Int. SAMPE Eur. Conf. Exhibit.* Salzburg, Austria, May/June 1995.

CRUZ, C., AND K. P. RAJURKAR. Drilling of partially stabilized zirconia with ultrasonic machining. *Mach. Compos. Mater. II, ASM,* Pittsburgh, PA, May/June 1993.

CUPINI, N. L., AND J. R. FERREIRA. Turning of carbon and glass fiber hybrid cloth composite material. *Proc. Int. Conf. Mach. Adv. Mater.,* ed. S. Jahanmir, pp. 447–53, Gaithersburg, MD, July 1993. NIST SP 847.

Design and Manufacture of Advanced Thermoplastic Structures, PR3 3-6/88, Lockheed Aerospace Systems Company, Wright-Patterson Air Force Base, Dayton, OH. LR 31446, F33615-87-C-5333, June 8, 1988.

FENOUGHTY, K. A., A. JAWAID, AND I. R. PASHBY. Machining of advanced engineering materials using traditional and laser techniques. *J. Mater. Process. Technol.* 42(4):391–400, 1994.

Flexible assembly subsystem in cost-effective production. USAF Manufacturing Technology Program. Status Report, 1992, pp. 4–5.

FOX, W. J., AND M. J. LOUDERBACK. Rapid thermoforming of thermoplastic composite materials. *Proc. 25th Int. SAMPE Tech. Conf.,* Philadelphia, October 1993.

GADALLA, A. M. Machining of zirconium-boron composites. *Mach. Compos. Mater. II, ASM,* Pittsburgh, PA, October 1993.

GILMORE, R. Ultrasonic machining of ceramic matrix and metal matrix composites. *Mach. Compos. Mater. II, ASM,* Pittsburgh, PA, October 1993.

GOULTER, S. Laser processing of composite materials. *Mach. Compos. Mater. II, ASM,* Pittsburgh, PA, October 1993.

HANLEY, F., AND J. H. DORAN. Manufacturing methods for machining processes for high modulus composite materials. In *Composite Machining Handbook,* vol II. Convair Aerospace Division, General Dynamics, Ft. Worth, TX, 1988.

HOCHENG, H., H. Y, PUW, AND K. C. YAO. Experimental aspects of drilling of some fiber-reinforced plastics. *Mach. Compos. Mater. Symp., ASM/TMS,* ed. T. S. Srivatsan and D. M. Bowden, pp. 127–38, November 1992, Chicago, IL.

HOLMES, S. T., AND J. W. GILLESPIE, JR. Thermal analysis and experimental investigation of large-scale resistance welded thermoplastic composite joints. *Proc. 25th Int. SAMPE Tech. Conf.* Philadelphia, October 1993.

HUANG, X. X., AND Q. LIU, C. K. YAO, ET AL. Superplasticity in a SiC_w-6061Al composite. *J. Mater. Sci. Lett.* 10(16):964–6, 1991.

HUMPHREY, W. D., AND S. T. PETERS. Joining methods for filament wound structures. *Proc. 38th Int. SAMPE Symp. Exhibit.,* Anaheim, CA, May 1993.

JEONG, G. Laser drilling enhancements projects. IMIP-UTRC, Phase 2, Contract F33657-86-C-0011 P 00029, August 1993.

JERNIGAN, T. Environmental and safety attributes of waterjet cutting and cleaning. *Proc. 38th Int. SAMPE Symp. Exhibit.,* Anaheim, CA, May 1993.

JONES, S. L. Textron studies water-jet cutting techniques. *Metalworking News,* January 15:10, 1990.

KHAN, S., J. F. PRATTE, I. Y. CHANG, ET AL. Composites for aerospace application from Kevlar® aramid reinforced PEKK thermoplastic. *Proc. 35th Int. SAMPE Symp.,* ed. G. Janicki, V. Bailey, and H. Schjelderup, pp. 1579–93, Anaheim, CA, April 1990.

KLEHN, R., AND T. W. EAGAR. Joining of 6061 aluminum matrix-ceramic particle reinforced composites. Bulletin no. 385, Welding Research Council, New York, 1993.

KRISHNAMURTHY, R., G. SANTHANAKRISHNAN, AND S. K. MALHOTRA. Machining of polymeric composites. *Proc. Machin. Compos. Mater. Symp., ASM,* ed. T. S. Srivatsan, and D. M. Bowden, pp. 139–48, Chicago, IL, November 1992.

LACY, B. Comparative wear testing of competitive PCD products on two composite material systems. *Mach. Compos. Mater. II, ASM,* Pittsburgh, PA, May/June 1993.

LAMBING, C. L. Advancements in welding technology for polymer to metallic joints. *Proc. 38th Int. SAMPE Symp. Exhibit.,* Anaheim, CA, May 1993.

LANE, C. Drilling and tapping silicon carbide particle-reinforced aluminum alloy metal matrix composites. *Mach. Compos. Mater. II, ASM,* Pittsburgh, PA, May/June 1993.

LIN, H. J., AND C. S. TANG. Fatigue strength of woven fabric composites with drilled and moulded-in holes. *Compos. Sci. Technol.* 52(4):571–6, 1994.

LOPEZ, F., J. C. SUAREZ, F. MOLLEDA, ET AL. Spot weld/adhesive bonded joints' behaviour: Circular flaw propagation after failure of the weld. *Proc. 16th Int. SAMPE Eur. Conf. Exhibit.* Salzburg, Austria, May/June 1995.

Machine for molding advanced composite materials. *JETRO* 29(2):17. 92-05-001-02.

MAGUIRE, R. G. Finish cracking of composite surfaces. *Aerosp. Eng.* July:20–5, 1995.

Manufacturing Technology, International Institute for Production Engineering Research, France. Annuals of CIRP vol. 38, no. 1, 1989.

MAY, C. A review of machining studies for castables and preform infiltrated reinforced aluminum alloy metal matrix composites. *Mach. Compos. Mater. II, ASM,* Pittsburgh, PA, October 1993.

MEHTA, M., A. H. SONI, AND T. J. REINHART. Hole quality drilling and assessment issues in graphite fiber reinforced composite laminates. *Mach. Compos. Mater. II, ASM,* Pittsburgh, PA, May/June 1993.

MONKS, R. Two trends in composites. *Plast. Technol.* March:40–5, 1992.

NIEDERMEIER, M., S. DELALOYE, AND G. ZIEGMANN. Forming studies of continuous fibre reinforced thermoplastic composite parts with the diaphragm technology. *Proc. 16th Int. SAMPE Eur. Conf. Exhibit.* Salzburg, Austria, May/June 1995.

NIEH, T. G., AND J. WADSWORTH. High-strain-rate superplasticity in aluminum matrix composites. *Mater. Sci. Eng.* A147:129–42, 1991.

NIEMANN, J. T., S. M. SULLIVAN, AND D. J. THIES. Titanium matrix composites. NASP Materials and Structures Augmentation, vol. 4. NASP-CR-1145-V-4, 1993; MDC Report 91B0487.

NUMERICAL CONTROL COMPUTER SCIENCES (NCCS) STAFF. Cutting edge technology machines composite materials. *Modern Applications News,* pp. 14–7, 1995.

PHILLIPS, L. N. *Design with Advanced Composite Materials,* Design Council. Springer-Verlag, London, 1989.

PITCHUMANI, R., P. A. SCHWENK, AND V. M. KARBHARI. Knowledge-based expert systems for composites manufacturing. *Ceram. Ind.* March:53–7, 1994.

POON, S. K., AND T. C. LEE. Electrical discharge machining of particulate metal matrix composite. *Mach. Compos. Mater. II, ASM,* Pittsburgh, PA, May/June 1993.

Proc. Int. Symp. Adv. Mater. Lightweight Struct., European Space Agency, France, 1992. ESA-Sp-336.

Proc. 7th Adv. Compos. Conf., pp. 283–90, Detroit, MI, September/October 1991.

Proc. 6th Adv. Compos. Conf., ASM/ESD; ASM International, Materials Park, OH, 1990.

QUIGLEY, O., J. MONAGHAN, AND P. O'REILLY. Factors affecting the machinability of an Al/SiC metal-matrix composites. *J. Mater. Proc. Technol.* 439(19):21–36, 1994.

RAMARAJ, T. C., AND R. JANAKIRAM. Analytical investigation of the ultrasonic machining process applied to metal matrix composites, Greer, SC, 1989. SME EM89-822.

RAMKUMAR, R. L., J. OGONOWSKI, AND D. STOBBE. Design and manufacture of advanced thermoplastic structures (DMATS). Progress Report, 1991. F33615-87-C-5242.

SANTHANAKRISHNAN, G., R. KRISHNAMURTHY, S. K. MALHOTRA. Flank wear and spalling of ceramics during machining of polymeric composites. *Mach. Compos. Mater. II, ASM,* Pittsburgh, PA, May/June 1993.

SCHENG, P., AND G. CHRYSSOLOURIS. Investigation of acoustic sensing for laser machining processes. Part 1: Laser drilling. *J. Mater. Process. Technol.* 43:125–44, 1994.

SCHENG, P., AND G. CHRYSSOLOURIS. Investigation of acoustic sensing for laser machining processes. Part 2: Laser grooving and cutting. *J. Mater. Process. Technol.* 43:145–63, 1994.

SCHWARTZ, M. Water cuts composite aircraft parts. *Am. Mach.* March:103–5, 1983.

SCHWARTZ, M., AND S. KOSTURAK. 196 holes per shot in boron-epoxy. *Am. Mach.* October:130–31, 1982.

SHAW, D., AND C. -N. TSEUNG. Analysis of delamination in a laminate drilled by waterjet. *Proc. Mach. Compos. Mater. Symp., ASM,* ed. T. S. Srivatsan, pp. 89–95, Chicago, IL, November 1992.

SIORES, E., AND D. DO REGO. Microwave applications in materials joining. *J. Mater. Process. Technol.* 48(1–4):619–25, 1995.

SRIVATSAN, T. S., AND D. M. BOWDEN, EDS. *Proc. 8th Adv. Compos. Conf., ASM/ESD,* Chicago, IL, November 1992.

STANDARD, C. M. Shear forming techniques for metal matrix composite shell structures. *Proc. 34th Int. SAMPE Symp. Exhibit.* pp. 151–60, Reno, NV, 1989.

SULLIVAN, S., AND D. M. BOWDEN. Machining of titanium matrix composites for NASP components. *Mach. Compos. Mater. II, ASM,* Pittsburgh, PA, May/June 1993.

TEICHMANN, M., J. VAN LINDERT, AND G. ZIEGMANN. Rubber match die forming of continuous fibre reinforced thermoplastic composite parts. *Proc. 16th Int. SAMPE Eur. Conf. Exhibit.* Salzburg, Austria, May/June 1995.

TREGO, L. Composite machining improvements. *Aerosp. Eng.* July:7, 1995.

VIEGELAHN, G. L., S. KAWALL, R. J. SCHEUERMAN, ET AL. LASER cutting of fiberglass/polyester resin composites. *J. Mater. Process. Technol.* 42(4):143–7, 1994.

WALSH, R., M. VEDULA, AND M. J. KOCZAK. A comparative assessment of bolted joints in a graphite reinforced thermoset vs. thermoplastic. *SAMPE Q,* July:15–9, 1989.

WEST, P. European composite programs push ahead on primary structure applications. *Adv. Mater.* 14(9):1992.

WHITE, J. Adhesive bonding in the 1990s, 67–71. Sterling, London, 1994.

WITULSKI, T., J. M. M. HEUSSEN, A. WINKELMANN, ET AL. Near net shape forming of particulate reinforced Al-alloys by isothermal forming compared to semi solid forming. *J. Mater. Process. Technol.* 45(1–4):415–20, 1994.

WU, S. -I. Y. Adhesive bonding of thermoplastic composites. *Proc. 35th Int. SAMPE Symp.,* ed. G. Janicki, V. Bailey, and H. Schjelderup, pp. 866–72, Anaheim, CA, April 1990.

WU, S. -I. Y. Surface preparation of graphite/PEEK composite for adhesive bonding. *Proc. 6th Adv. Compos. Conf., ASM/ESD,* pp. 255–64, Detroit, MI, October 1990.

YAMAMOTO, T., AND T. HIROKAWA. Advanced joint of 3-D composite materials for space structure. *Proc. 35th Int. SAMPE Symp.,* ed. G. Janicki, V. Bailey, and H. Schjelderup, pp. 1069–76, Anaheim, CA, April 1990.

ZAWISTOWSKI, F. Investigation into operation of new special drills for drilling holes in thin components made of composite materials. Technische Hogeschool, Delft, Netherlands, 1987. LR 519.

ZHANG, Y. H., C. Y. JIANG, W. S. LAU, ET AL. An investigation into drilling of glass reinforced liquid crystal polymer composites. *Mach. Compos. Mater. II, ASM,* Pittsburgh, PA, October 1993.

ZWEBEN, C. Metal-matrix composites for electronic packaging. *J. Met.* 44(7):15–23, 1992.

3

Applications of Composite Materials

3.1 INTRODUCTION

Throughout the 1970 and 1980s, high-performance composites were on the leading edge of materials science. they offered the promise of new products with extraordinary strength, stiffness, and chemical and temperature resistance. However, the composites industry quickly learned that simply having the best material was no guarantee of success. Material performance had to be integrated with innovative design and affordable manufacturing processes to produce systems with real, measurable, and salable benefits for the marketplace.

In much of the advanced materials industry, the military was the primary customer. A well-defined need for faster and more survivable military aircraft and missiles sparked innovation and fueled the engine of materials development. In addition, the military and civil aerospace were often the only customers who could either afford the high prices of low-volume advanced composite materials or tolerate the craftsperson-like manufacturing processes and lead times that sometimes stretched into decades.

Such was the environment in which the advanced materials technology base evolved. Although the composites industry consisted of polymer and fiber plants, facilities for parts production, trained specialists, and a wealth of knowledge, it was naturally positioned to meet the needs of the military and not well suited to commercial requirements.

Dramatic world changes during the last 5 years, however, have had a major effect on the defense industry and, correspondingly, on composites. The U.S. Department of Defense engine is essentially stalled, and civil aerospace is in low gear. The "balance" in the composites industry that resulted from the preponderance of defense spending has been upset, leaving an unstable situation. Many companies have exited the field, and most others are struggling financially. Individually and collectively, industrial players are working to reposition the technology base in such a way as to sustain future growth.

Despite this outlook, the huge potential of composite materials still exists. However, the customer mix has changed, and the industry must change with it. Companies must apply what they have learned from their defense and aerospace experiences to make wise decisions regarding which markets to enter and how to penetrate them successfully.

3.2 PRINCIPLES FOR THE FUTURE

There are four recurring principles that will shape the future of advanced materials, systems solutions, economical manufacturing processing, diverse markets, and new technologies.

3.2.1 Systems Solutions

The industry must drive systems solutions for even the most down-to-earth markets. For maximum return, development of composite systems must be approached as an integrated process. Decisions regarding designs, processes, and materials must be made synergistically to ensure peak product performance.

One recent development, an aerodynamic composite bicycle wheel, is a good example. The materials involved—C and Kevlar aramid fiber, epoxy resin, Al, and urethane foam—are all well known. The manufacturing technology—RTM—is also familiar. Only the integration of these with a novel design based on both aerodynamic analysis and FEA has allowed the development of a composite system that gives cyclists a competitive edge.

Composites are everywhere: Olympic victories in the pole vault with a fiberglass pole; tennis tournament wins with graphite racket frames and handles; and use in automobiles, racing cars, guitars, windsurfing boards, bulletproof vests, and skis. Some composite fabrications have even been approved for implanting in the body to duplicate the strength and stiffness of bones.

3.2.2 Economical Manufacturing Processes

For composite material systems to grow successfully in the next century, manufacturing processes must be made more economical, productive, and efficient. Efforts in this area are already under way. For example, the Department of Defense has stimulated innovation in high-rate, high-flexibility manufacturing.

One approach draws on thermoplastics technology. Basic building blocks are bundles of reinforcing fiber (typically C or glass) impregnated with thermoplastic matrix resin and formed into a consolidated tow bundle. By briefly remelting the bundle, adding it to the growing surface of a composite, and then consolidating each new element of the structure as it forms, high-quality, thick-section cylinders have been produced. This avoids both the conventional composite curing process and some of the scale limitations of standard thermoset composites.

This technology has now matured to the point where larger and more complex structures can be produced with predictable, reliable mechanical properties. One example is an underwater vehicle chassis for a U.S. Navy program.

The next logical step in the evolution of this technology is toward thin, complex sections (e.g., aircraft structures), with forming, joining, and inspection all being carried out simultaneously. This will demonstrate the potential of this technology to dramatically streamline and simplify the manufacture of complex composite structures.

3.2.3 Diverse Markets

To obtain the best return on technology investment, systems solutions and more economical processes must be applied to new and diverse markets. There are many applications where existing technology might be successfully applied to commercial markets. For example, the outstanding high-temperature resistance of PI matrix composites makes them an ideal lightweight material for many aerospace structures. One component, a missile radome, sustains temperatures of more than 535°C during its brief lifetime.

Another example is hot glass handling. Here, a polymer composite replaces a monolithic graphite part, offering an exceptional combination of durability and low thermal conductivity. These properties are precisely those needed by bottle manufacturers for a durable takeout insert which is used to transfer the still-hot glass bottle out of the mold. A longer life for the insert means reduced bottle breakage and less time lost for replacement.

It is foreseen that in the next two decades airframe and engine materials will change from monolithic, metal-based alloys to ceramics, both monolithic and composite, and Gr/Ep composites. Candidate materials for airframes at high temperatures include C_f-reinforced thermoplastics and thermosets, such as cyanate esters and toughened BMIs. For engine materials, CMCs and TiAl provide future promise.[1] Leading aerospace contractors estimate approximately 35% of a future fighter's structural weight will include a combination of thermoset and thermoplastic composite materials.

In design studies and validation of new manufacturing processes and technologies for the production of thermoplastic structures for the next generation of fighter aircraft, engineers and technologists have built and validated full-scale hardware for current aircraft fighters.

Results for the F-16 main landing gear door have shown the following.

- An 11% reduction in parts count
- A 20% savings in weight
- A 22% projected cost decline.

For the F-15 forward engine access door in Figure 3.1, a 40% weight savings was demonstrated as compared to the current SPF/DB Ti door, with a projected cost savings of 25%.

3.2.4 New Technologies

The final key to success for advanced materials will involve taking experience from both military and commercial applications and seeking out new technologies that are appropriate to the future of the industry. The challenge is to identify and focus on the right problems and opportunities that will facilitate a successful shift from basic science to functioning technologies.

The procedures for materials selection and bounding of composite properties using

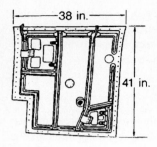

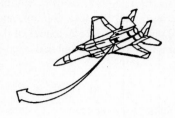

Figure 3.1 F-16 Main landing gear (MLG) door. 25.4 mm = 1 in.

selection charts and merit indices provide powerful general tools for matching materials to the needs of a design. Implementation of the approach in software greatly facilitates these procedures. The methodology gives

1. Quick, visually straightforward methods of exploring the potential of a new material for a given application
2. Vectors for the development of new materials to meet a specific design need.

Clearly, there are tremendous opportunities to apply composites to national needs, including transportation and infrastructure.

3.3 COOPERATION

To successfully drive all four principles—systems solutions, economical manufacturing processes, diverse markets, and new technologies—active cooperation is needed among industry, government, and academia. It is a formidable challenge, but the stakes are enormous and well worth the effort.

The future provides the opportunity for growth to a new and healthier balance, with a vibrant commercial sector delivering an improved quality of life and a stronger technology base possessing the agility and responsiveness necessary to support both commercial and national defense needs.

3.4 COMPOSITE AGE

When the history of humanity is looked back upon, it is seen that each age is named for called the type of material used in that age, starting from the Stone Age, the Bronze Age, and the Iron Age. The present age is called the Steel Age. However, now that composite materials are used in many fields and are completely integrated into daily life from cradle to grave, future generations may call the present age the "Composite Age" or the "Age of Engineered Materials."

Polymer composite materials, for example, still may offer many more and new potential applications even though half a century has passed since they were first developed. Their development is expected to continue in the future. On the other hand, the issue of how to preserve the earth's environments is becoming an extremely important one on which the survival of human life depends. Accordingly, composite materials with high performance properties must become ecological materials as well, to be further developed for the realization of their potential use. Fortunately, in the present period composite material concepts do not take much time to come true and change to components.[1]

3.5 APPLICATIONS

3.5.1 Aircraft Components, Especially Airframes

3.5.1.1 Military.   Starting in 1940 with glass fiber-reinforced polyester radomes and drop tanks, aircraft structural applications have grown steadily in both variety and quantity.

Glass-phenolic firewalls were introduced in 1944. It was in 1952 that Gl/Ep became available, and it has since been used extensively in radomes, rotor blades, floors, doors, control surfaces, empennages, wings, and fuselages. Gl/PI radomes, firewalls, and high-temperature ducts were introduced in 1960. By 1969, aircraft were using B/Ep composites in stabilizers, rudders, reinforcements, and patches for the repair of other structures. That same year, C_f/Ep composites were introduced in stabilizers, wings, fuselages, air brakes, control surfaces, spoilers, doors, floors, inlet ducts, cowlings, and thrust reversers. Kv/Ep was introduced in 1971 in radomes, fairings, and leading and trailing edges, and C_f/Ep engine inlet ducts and flaps were produced in 1974. C_f/BMI composites in wing flaps, strakes, wing and fuselage skins, stabilizers, engine casings, and thrust reversers were introduced in 1978. C_f-reinforced thermoplastics were first introduced in stabilizers and wing ribs in 1988.

Concomitantly, composite structural weight in U.S. aircraft has experienced a dramatic rate of growth. In 1960 the use of composites in the A-6A was just 90.7 kg, but by 1970 the F-14A was introduced with 236.3 kg of composites. The F-18A had 598.6 kg in 1978, the AV-8B used 732.6 kg in 1981, and now the AV-8B airframe alone uses 26% Gl/BMI and Gr/BMI composite materials, including the entire wing of the aircraft.

C-17.   This aircraft has a 1961.2-cm composite horizontal stabilizer which has been designed, constructed, and subsequently installed, evaluated, and tested. Additionally C_f/Ep is used in the large tail cone, winglet skins, fairings including the landing gear pod fairing, covers, and wing control surfaces. The total aircraft uses over 6% C_f/Ep material (Table 3.1).

AV-8B.   BMI composites are used on the aircraft's underside and wing trailing edges to withstand the high temperatures caused by vertical takeoffs. This aircraft with its composite wing components features the single largest composite structure on a fighter aircraft—utilizing 684.8 kg of C_f/Ep in wing skins, trailing-wing flaps, auxiliary flaps, ailerons, outrigger landing gear support, and ribs and spars in a torque box structure. C_f/BMI is used in the lower surface of the trailing-wing flap.

F/A-18 E/F.   Composites have helped to reduce the number of parts by 30%, although this new model is 25% larger than the C or D version. About 19% of the E/F structural weight consists of C_f/Ep composites, compared with the 10% used in the C and D models.

TABLE 3.1 Several Primary Military Aircraft Programs[5]

Program Designation	Composite Parts	Major Material Suppliers
F/A-18 E/F Hornet fighter jet	Carbon fiber-epoxy wing skins and wing structure, carbon fiber BMI fuselage skins	Hercules IM-7 fiber, Fiberite 9773 toughened epoxy
F-22 advanced tactical fighter jet	Carbon fiber/BMI skins, ducts, frames, bulkheads; carbon fiber-polyimide wing components; carbon fiber-epoxy primary structure	Narmco-BASF, Hexcel, Hercules, Fiberite, Quadrax, Amoco
A/FX carrier-based fighter jet	All-composite fuselage, probably carbon fiber-epoxy	Not yet decided
V-22 Osprey tilt-rotor	Carbon fiber-epoxy primary structure, S-2 glass-Kevlar rotor	Hercules, Fiberite, U.S. Polymeric, Owens Corning, Hexcel
RAH-66 Comanche helicopter	Carbon-fiber/epoxy primary structure, main rotor	Hercules IM-7/8552; second source, Fiberite
C-17A transport	Large tail cone, winglet skins, fairings and covers, control surfaces	Hercules Fiberite, Narmco-BASF, Hexcel, American Cyanamid, Ciba-Geigy

Contributing to the increase in use of composites is the E/F's longer fuselage and larger wing. In addition, the control surface redesign includes more cocured composites. Fuselage skins, principally those in the center and aft sections, have been converted to composite, as have the inlet ducts.

The use of C_f composites in the wing structure now includes the leading-edge flaps. Both are cocured structures, as are the composite trailing-edge flaps. The ailerons consist of C-skinned honeycomb, and the wing skins and tail skins are made of C_f for improved stiffness, strength, and toughness. The E/F's wingspan now measures 1368.6 cm, with wing tip missiles. That's about 25% more than the 46.45 cm² area of earlier models. The scaled-up tail has 36% larger horizontal and 15% larger vertical stabilizers.

F-18A. Gr/Ep composites are used in the wing skins as the primary structure of this production aircraft. Additionally the horizontal tails, speed brake, dorsal covers, and various access doors and covers are composites.

F-22 (ATF). Currently about 35% by weight of this craft's airframe materials consist of composites, with thermosets comprising 31% of the total and thermoplastics 4%. This compares with an 11% thermoplastic–10% thermoset split in the original YF-22.

Thermoplastic composites have been used in access panels and are also being employed mainly in places requiring high damage tolerance like main landing-gear doors, weapons bay doors, and upper forward-fuselage doors. A SiC/BMI composite also is used in the aircraft for "classified" purposes.

The predominant material being used for the F-22's composite wings, fuselage, and empennage structures is 5250-toughened BMI prepregs with intermediate-modulus carbon, ceramic, and glass fibers.

Other composite components used in the F-22 include the forward-fuselage frames for one of the fuel tanks and the fiber-placed composite pivot shafts that hold the fighter's

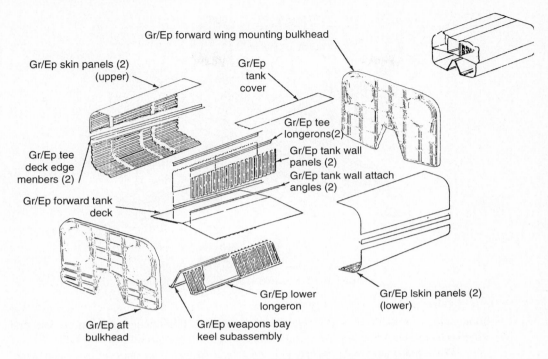

Figure 3.2   F-22 (ATF) forward-fuselage module.

horizontal tails in place. The tails are made from Al honeycomb cores with bonded BMI skins. A composite of C_f and toughened epoxy forms the aircraft's inlet duct skins, fuel floors, and shear webs. Wing boxes include both composite and Ti spars. Others include the forward-fuselage modules (Figure 3.2).

FS-X.   A next-generation close air support aircraft has been designed in Japan, and composites have played a major role in the construction and development of the aircraft. A left main wing has been built consisting of spars, major ribs, and underwing surface skins made as a single cocured piece of C_f-reinforced plastic. An upper surface made of the same material is joined with fasteners. A cocured composite wing box for the right wing has also been fabricated, as well as the aft-fuselage section.

F-4.   A one-piece windscreen kit comprised of a laminate windshield and composite frame to replace the existing three-piece windshield and metal components has been introduced in the A-4 aircraft. The windshield laminate is acrylic-polycarbonate, and the frame is made of Kevlar and S-2 glass with an Al ring in the aft arch for sealing the forward pilot canopy. This advanced composite frame on the forward windscreen repells bird strikes, offers improved visibility, and is available in kit form for easy assessibility.

F-111 and A-6E.   Modernization has permitted these two older aircraft to survive, especially with the substitution of composites for metallic structures. The F-111 two outboard spoilers have been retrofitted with composite materials, and improvements have transformed the A-6E into an even more formidable attack plane with the addition of a

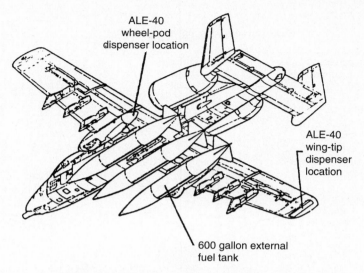

ALE-40
wheel-pod
dispenser location

ALE-40
wing-tip
dispenser
location

600 gallon external
fuel tank

Figure 3.3   EFA material distribution in an A-10 aircraft.

composite wing that makes it a revolutionary aircraft. The rewinging is one of the first large-scale uses of structural composite materials in military aircraft.

The replacement wing is stronger and longer-lasting than the original metal wing, and the composite material also offers the extra advantage of increased resistance to corrosion, which is an important benefit in a salt-air environment. The replacement wing is composed primarily of Gr/Ep composite materials. Control surfaces are made of Al, and Ti is used in other high-stress areas. A Ni-coated fabric is used on the wing surface to help neutralize the effects of possible lightning strikes. The composite wing will add 8800 flight hours, or about 15 years, to the aircraft's operational life.

A-10 Tactical Aircraft.   Designers have taken advantage of advanced thermoplastic materials and introduced them in three components of the A-10 shown in Figure 3.3.[2]

F-15 Fighter.   Landing gear systems account for 4–5% of an aircraft's gross take-off weight. Proponents of TMCs say they can deliver individual gear components that weigh half as much as steel. Besides providing a weight advantage, TMCs outperform steel in terms of resistance to corrosion, particularly stress corrosion. TMCs do not need to be coated for ambient-temperature operation, and although their initial cost is higher than that of steel or Al, they can last indefinitely.

Technicians and engineers expect to come up with a TMC landing gear system that can be made economically. However, TMC fabrication challenges are so great that some experts do not think the new material will ever play a significant role in landing gear production. Engineering stress and design analysts insist that the engineering complexity of landing gear systems and the manufacturing complexity engendered by a gear's numerous joints and welds make the metal matrix material unworkable. However, other experts disagree. A TMC project has been ongoing to evaluate heavy arresting hooks used by fighters when they land on aircraft carriers.

TABLE 3.2   Composite Structure Experience with Airbus Models

Matrix	Supplier	Process Temperature °C (°F)	Parts Fabricated
Cypac 7005	American Cyanamid	315–345 (600–650)	Feedhorns, access door, blade-stiffened panels
Cypac 7156	American Cyanamid	330–358 (625–675)	Test panels
Radel C	Amoco	338–360 (640–680)	Access door, F-16 MLG door
Radel 8320	Amoco	315–338 (600–640)	Access door, T-stiffeners, F-16 MLG door
Avimid K	DuPont	330–360 (625–680)	Hat-stiffened panels
APC-2	ICI	370–400 (700–750)	Access door, waffle panel, Z- and C-channels, T-stiffeners, cruciform stiffeners
HTA	ICI	338–370 (640–700)	Z-chanel, test panels
HTX	ICI	416–435 (780–815)	Access doors
PAS-2	Phillips	330–358 (625–675)	Forward-fuselage lower skin, T-stiffeners
PPS	Phillips	302–330 (575–625)	Z- and C-channels, T-stiffeners

The landing gear on the U.S. Navy fighter plane is expected to weigh as much as the fuselage, and calculations indicate that using TMCs in selected components could cut the 19,954-kg gear's weight by about 30%. Such savings could result in significant gains in payload and range for high speed civil transport (HSCT) aircraft, which is discussed further later in this chapter.

The importance of TMC is that the new lightweight landing gear represents a significant technology leap, and from a specific strength standpoint TMC is about as good as it gets.[3,4]

F-16.   The strake door was initially manufactured from Al and required numerous mechanical fasteners. The thermoplastic composite component is a fusion-bonded assembly with intermediate-modulus C_f and is 45% lighter.[5,6]

The F-16 main landing gear door has demonstrated the potential for coconsolidation of an inner and an outer skin using large bulk graphite tooling, and Table 3.2 reflects the use of thermoplastic composites in the aircraft.

Eurofighter Aircraft (EFA).   The monocoque of the center fuselage is a one-piece C_f composite skin with simultaneously cured frames and longerons. According to designers, the carbon fiber-reinforced plastic (CFRP) offers a substantial weight savings potential in the construction of a future fighter aircraft.[7,8] About 40% of the structural weight and more than 70% of the wetted area of the EFA is expected to be built from CFRP.

Advanced Jet Fighter.   Several MMC vertical tail stabilizers have been made from 2009 Al (A1-3.6% Cu-1.3% Mg)/SiC$_{p \text{ or } w}$ composites which have a much higher specific stiffness than steel, which means that they can greatly reduce the moment of inertia and metal fatigue. Measuring 320 cm long, 152.4 cm at the root cord, 76.2 cm at the tip cord, and 15.4 cm thick, these stabilizers are one of the largest MMC primary structures built to date and are undergoing testing.[9]

Miscellaneous Current and Future Fighters, Advanced Transports, and Antisubmarine Warfare Aircraft.   Various fighters need materials with high specific strength. Potential MMC applications as illustrated in Figure 3-4 include highly loaded parts such

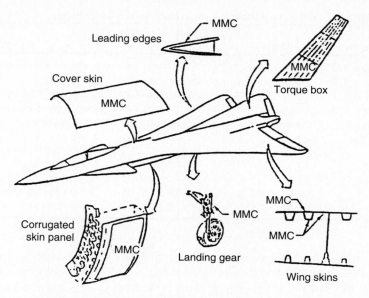

Figure 3.4   Potential MMC applications for high-speed aircraft.

as landing gear and arresting hooks. Larger applications such as stabilators and vertical tails utilize the high specific modulus of MMCs to perform more efficiently than conventional materials. Other potential applications identified to date are the following.

Pivot shafts	Fuselage stringers
Stiffener beams	Longerons
Stiffener hats	Beams
Fuselage frames	Fittings
Reinforced bulkheads	Spars

C-130 Transport Aircraft.   A series of tests in the C-130 belly skin program demonstrated the increased impact resistance of C_f-reinforced thermoplastic belly skins over that of the normal Al aircraft skins now in use. The belly skin program marked the first time thermoplastics were used in primary aircraft structures. In the program, a 100.58×304.8-cm C_f-reinforced thermoplastic panel was placed on the underside of a C-130, aft of the nose wheel. During normal use, this panel location is subject to damage from rocks and gravel tossed up during takeoff and landing. The area was chosen because it would give a true indication of how much abuse the test panel would take and whether thermoplastics could perform as primary structures.

The tests involved normal operations of a C-130 and included 55 takeoffs and landings over a year on unimproved airstrips. Severe weather conditions such as ice and slush were included in the testing of the panel.

Newer versions of the aircraft are flying with some new wing trailing-edge panels made of Gr/Ep, which will replace the currently used thin-gauge metal bonded Al panels.

C-141 Transport.   Engineers and technicians have used discontinuously rein-forced aluminum, to strengthen and improve the no. 2 emergency escape hatch on the U.S. Air Force's C-141 Starlifter aircraft. DRA uses whisker or particulate SiC_f that make the material stiffer, stronger, lighter, and more temperature-resistant and dimensionally stable than current Al alloys and other composites.

Additionally the U.S. Air Force Military Airlift Command and Lockheed Aeronau-tical Systems Company has developed a series of B/Ep and Gr/Ep repairs to strengthen the Starlifter. The company has already applied composite patches to operational C-141s to validate repair technology. The work has yielded composite repair kits for the Starlifter and other weapons systems, and it has obvious potential for aging aircraft in military and commercial service.

The Starlifter repairs are patches intended to stop cracks and doublers intended to redistribute stresses in Al structures. Composites promised a way to stop cracks, modify stress, and dramatically reduce inspection and repair time. Stiff B filaments lend them-selves to flat patches, whereas pliable Gr_f are more easily molded to curved or complex shapes.

A common patch was designed to stop cracks around wing riser rib clip holes and riser weep holes. The B/Ep patch used 4-mm fibers and 5505 resin from TSM. It was cured off the aircraft at 177°C and bonded in place under a vacuum bag. The apparently simple hole patch had 14 B plies with a Gl_f overlap and tapered in thickness to feed stress into the layup gradually.

Composite doublers can modify stresses by modifying metal structures. A B/Ep rear beam splice applied at C-141 wing station 405 reduced stresses in the lower wing surfaces by more than 12% and is expected to multiply fatigue life by 2.4.

The most complicated of the composite repairs and the most difficult to accomplish was a Gr/Ep patch to be cured in place at two points on a mainframe at fuselage station 998. The patch was intended to fix mainframe cracks without time-consuming disassem-bly. Other repair installations are shown in Figure 3.5.

C/KC-135 Transport and Tanker.   Repair and reinforcement of metallic struc-tures on this aircraft has been developed to fix floor beam cracks and corrosion with the use of composite doublers.

3.5.1.2 Commercial.

In the early 1970s reinforcing materials were not specifi-cally identified as the key materials for use in advanced composite materials (ACMs) made from B, C, and Kv_f since no one fully understood composite properties or their cost. Applications ranged from B_f F-4 rudders, 707 flaps, and F-14 horizontal tails to C_f 747 floor panels, 737 spoilers, and L-1011 floor support struts and Kv_f L-1011 wing fair-ing.

About 50% of the takeoff weight of large scale passenger aircraft is the empty weight, and the structural weight excluding engines, equipment, and interior furnishings is about 30%. The reduction in structural weight directly influences the reduction in fuel consumption and allows the downsizing of engines, wings, and so on, which produces a snowball effect resulting in further reduction. It is said that a weight reduction of 1 kg of airframe reduces the takeoff weight by 2–5 kg.[10,11]

If the design were made more suitable for ACMs in commercial aircraft, the struc-

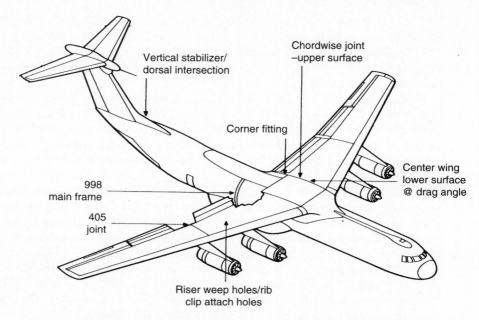

Figure 3.5   The Lockheed C-141B Starlifter with fast repairs using B- and C-reinforced patches that keep them in the air longer.

tural weight would be reduced by 36% and the takeoff weight would be reduced by 26%, which accounts for the high expectations for ACMs.[12] The A310 uses 2000 kg of composites, which has resulted in a reduction in the structural weight by 397 kg (1%). This is said to be equivalent to a fuel saving of more than 600 tons in a short-range service assuming 300 h/year flight time over a period of 20 years.[13]

ACMs were first used in small-scale aircraft. Business jets such as the Learfan 2100 and the Avtek 400 used composites for most of the structures and an innovative idea of using composites for joints was introduced in order to save fuel, but unfortunately the project was discontinued because of technical and economical reasons.[14] Thereafter, the Voyager accomplished the great feat of flying around the world without landing by using ACM for most of the aircraft body parts. The Beach Starship, made entirely of composites, has been flying since 1989. Business jets are discussed further later in the chapter.

At the beginning of the development and commercialization of ACMs for large-scale commercial transport airliners in the late 1970s and early 1980s, there were strong proposals that composites could be used for the main wings and fuselage, besides empannage and control surfaces, and that their use would be extended from 26% to 40–50% and maybe up to 65%. However, in the actual implementation, the major commercial aircraft companies and consortia took a conservative approach. In the case of McDonnell Douglas, from 1970 to 1983 several C_f-reinforced plastic components were designed for the DC10 and put into a flight service evaluation program, such as rudders (1976), ailerons (1981), and vertical stabilizers (1987). About 1000 kg of composites has been used for the MD 80 and more than 4100 kg for the MD 11.[10,15]

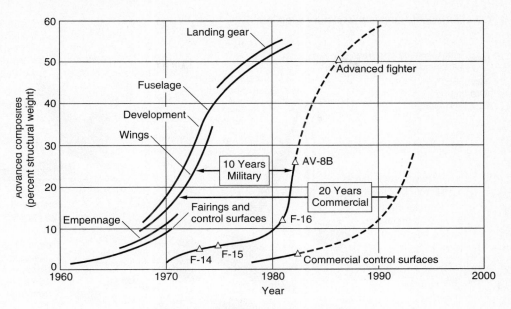

Figure 3.6   Implementation of advanced CCCs.

The growth of applications has continued, the Boeing 757 now having over 184 kg and the Boeing 777 using about 13,605 kg per vehicle. Since 1983, Airbus usage of composites has grown from 4490 kg in the A310 to 11,307 kg in the new A330 and A340 in 1991.

Based on Figure 3.6 it appears that it took 10 years from the start of composite development activities sponsored by the U.S. Air Force for equivalent structures to be applied to production military fighter aircraft. The same 10-year period from development by the U.S. Air Force to wing structure in the late 1970s to early 1980s. Therefore it is projected that the next fighter aircraft will have a primary structure of over 50% composite material.

It took almost 20 years from initial development by the U.S. Air Force for voluntary commercial production of C_f composite control surfaces to get started in several commercial transport divisions. If this 20-year trend holds and NASA activity and support continue, it appears that a major jump to primary commercial transport wing application can be expected from the mid-1990s to the year 2000.

The problem with commercial applications for composites was cost. Al, for instance, which makes up 75–80% of a passenger aircraft, is $5/0.4536 kg (1 lb) on average, compared with $35 for the competing composite material.

But the big companies may simply have been too early. Slowly, the cost is decreasing. Ten years ago Al cost $3 per 0.4536 kg (1 lb) and composites $70. By the end of the decade, the price differentials are expected to narrow further, giving a greater boost to composites. New uses are coming along almost every day.[16–22]

Boeing 777.   A 1087-cm composite rudder is currently being fabricated for this commercial airliner in production.

Fuselage side panels with a newly developed tow placement involving Korex

777 Lightweight composite structure suppliers

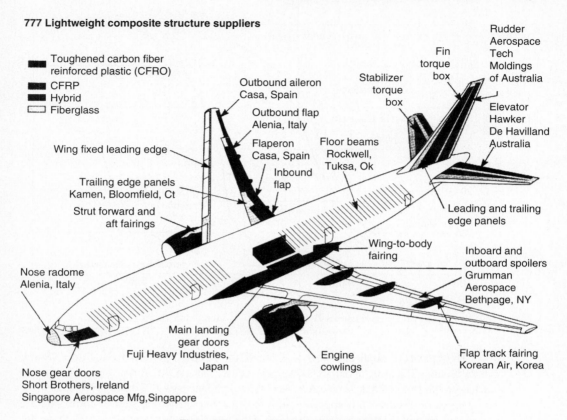

Figure 3.7 Boeing 777 lightweight composite structure.[16]

aramid honeycomb sandwiched between laminate sheets of AS4 C_f and 8552 epoxy are also in production.

Other composite components in production for the 777 include Gr/Ep horizontal and 1434-cm-tall vertical stabilizers. These stabilizers save 725.8 kg in weight and add 60 mi to the jetliner's range. In fact, nearly 9% of the plane's structural weight consists of composites. That's 10 times the amount used on the 757 or the 767. Each horizontal stabilizer (right and left) is approximately the size of a 737 wing. The 777 tail is one of the first major uses of composite primary structure in a Boeing commercial transport and saves about 454 kg compared with a comparable metal tail.

Shown in Figure 3.7 are the toughened C/Ep resin systems used in the design of the vertical fin torque box and the horizontal stabilizer boxes as well as the floor beams of the passenger cabin. The switch to a C/Ep system was made largely because of the composite's corrosion resistance qualities.

Other structural members of the 777 manufactured from composites include control surfaces, engine cowlings, landing gear doors, wing-to-body fairing, wing fixed leading and trailing edges, and the nose radome. Nonstructural composite parts can be found in the airplane's interior and in system ducts.

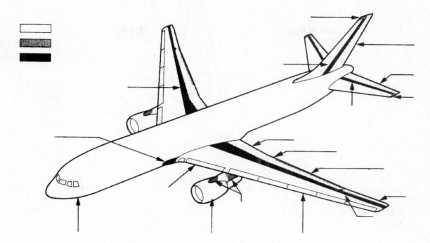

Figure 3.8   Boeing 767 composite applications.[17,18]

With the exception of the floor beams, all composite parts are cored with DuPont's Nomex aramid honeycomb. In addition, the same core material makes up the floors, ceilings, sidewalls, partitions, bulkheads, and baggage bins. Also, DuPont's Kevlar serves as an ingredient in the epoxy system for the (ECS) ducting system.[16,17]

Boeing 767 and 757.   The 767 uses composites in roughly 3–4% of the structural weight (Figure 3.8). The empennage is under consideration as a primary structural component made with composites. Additionally about 3% of the 767's exterior surfaces, for example, are made of composites, and approximately 30% of the wetted surface area is composite material. This represents about 1550 kg of the 767 aircraft weight. The largest production Gr/Ep panel on the Boeing aircraft is the composite rudder, which has dimensions approximately 11×2.5 m. Gr/Ep is also used to fabricate the 767 horizontal stabilizer.

Thrust reverser and blocker doors are made of composite materials in the 757 aircraft. The blocker doors are fitted in the bypass ducts of an aircraft engine to assist braking on landing. RTM has been found to produce composite doors that are cheaper and lighter than metal doors and also provide noise reduction. The design of the door incorporates front and back panels made from C_f cloth (T300 fiber), epoxy resin, and a Nomex honeycomb core. Continuous-fiber reinforcement has permitted integration of the hinge lugs into the back panel. In the hinge region unidirectional fibers run the length of the rib, around the hinge lug, and back along the rib to provide the necessary strength under tensile loading.

The 757 uses C composites for major components such as the primary control surfaces and landing gear doors (Figure 3.9). Typical Gl applications have been replaced by the lighter-weight C/Kv/Gl hybrid composites, although the latter components still retain

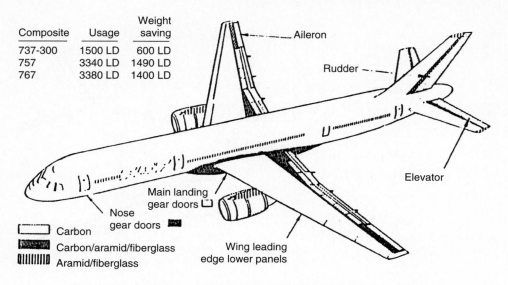

Composite	Usage	Weight saving
737-300	1500 LD	600 LD
757	3340 LD	1490 LD
767	3380 LD	1400 LD

Carbon

Carbon/aramid/fiberglass

Aramid/fiberglass

Figure 3.9   Advanced composite applications in a Boeing 757.[19] 0.4536 kg = 1 lb.

the supporting Al substructures. In the 757 airplane, approximately 1500 kg of composite material is used. As in other airplanes, C_f-reinforced composite material is utilized along with aramid, Gl_f and hybrid combinations.[16–19]

Boeing 737.   On all models of the Boeing advanced 737, the aileron, elevator, and rudder control surfaces are C_f composite. Approximately 700 kg of the structure is composite material on the 737-300 airplane.[20]

Boeing 747.   This aircraft has composite fairings, nacelle components, and winglets. B/Ep composite material utilized in bonded structural doublers has been successfully tested and is used in the 747 aircraft.

McDonnell-Douglas MD 80.   The original MD 80 series aircraft was an all-metal primary-structure design. Once an aircraft is in production, it is very difficult to change to a more expensive, even though lower-weight, component made from composite materials. Many features of the aircraft, representing 3% of the overall structural weight, were eventually converted to composite materials, including spoilers, ailerons, rudder, engine nacelles, wing trailing edges, and tail cone (Figure 3.10).

McDonnell-Douglas MD 11.   One of the most newly designed commercial transporters shown in Figure 3.11 reflects the extensive usage of C/Ep as engineers evaluate the effective life and flight hours of currently flying composite component structures and systems.

The MD 11 is a successor to the DC10 series of aircraft. Shown in Figure 3.11 is the center engine inlet duct which had a weight target of 227 kg compared with 364 kg for the DC10 equivalent duct. Toughened Gr/Ep with embedded Al mesh for lightning strike protection comprised the duct material system, allowing it to meet the above weight targets.

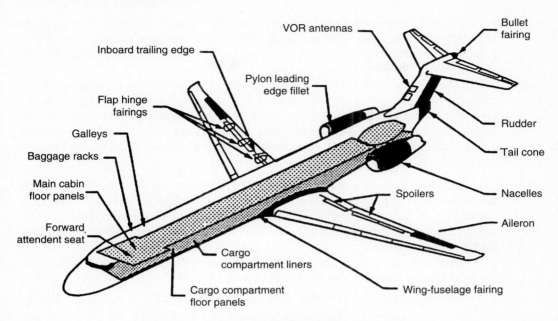

Figure 3.10   MD-80 advanced composites.[21]

Carbon/epoxy except as noted

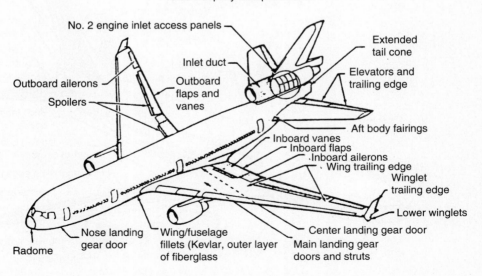

Figure 3.11   MD-11 composite construction.[21]

The MD 11 has composite applications in 20 separate components, however, none are initially in a primary structure. Additional conversions are expected in derivative models; likely candidates include the inboard flaps as well as floor and ceiling panels inside the plane.

Composite secondary structural components incorporated into the original MD 11 production design represent approximately 5% of the overall structural weight of the aircraft. Most control surfaces such as outboard ailerons, flaps and vanes, and spoilers and wing trailing-edge panels are C_f composites. Horizontal stabilizer elevators and trailing-edge panels are also made of C_f composites, as are winglet skins. Wing fuselage fairings and aft body fairings are Kv_f composites.

McDonnell-Douglas MD 90 Series (MD 91, 92, 94, etc.).   The MD 91 and 92 derivatives were to have featured more than the 4–5% composite makeup of the predecessor MD 80. Original plans called for a series to have C_f elevators in addition to the rudder, rudder tab, and extended tail cone from the MD 80, however, cost savings have delayed these changes.

There is a suggestion in the MD 94 design that probably a tail made from composites would be the initial primary composite structure.

McDonnell-Douglas MD 12X.   This aircraft (Figure 3.12), is designed for the future and reflects almost an 11% overall composite structural weight. All control surfaces are proposed to be C_f composite. As in the MD 11, fairings and various other components have also been proposed. In addition, a primary structure (C_f) is proposed for horizontal and vertical stabilizers. Internal fuselage cargo and passenger floor beams and support struts are also proposed for C_f.

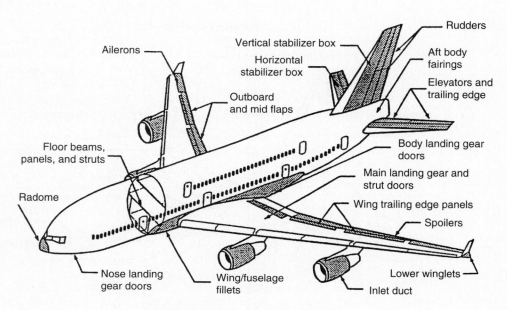

Figure 3.12   MD-12X composite structures.[21]

Airbus A300/310, and so on, Series.   Airbus composite experience started cautiously with the A300. Glass fiber-reinforced plastics (GFRPs) were used in the radome, pylon fairings, wing cover panels, fin leading edge and tip, apron, and wing and fuselage rear fairings. Cabin and cargo hold furnishings were also made of GFRP.

The A310-200 saw the introduction of aramid and C_f structures. These include the pylon fairings, fin leading edge and tip, fin trailing-edge panels, rudder, apron, wing and fuselage fairings, air brakes and spoilers, flap track fairings, main landing gear doors, thrust reversers, fan cowl, inlet cowl, and cooling air inlet fairings.

The next big step forward was the development of an entire vertical fin made of C_f, which entered airline service as a standard part on the A310-300. The fin was soon introduced in the A310-200 and the A300-600, and all Airbus aircraft now have weight-saving C_f fins.

The A300-600 has many of the composite structures found on earlier models with the addition of several new aramid (Kv_f) fiber-reinforced (AFRP) and C_f-reinforced plastic (CFRP) structures. These include the radome (AFRP), outer-wing trailing edge (CFRP), and nose landing gear doors (CFRP).

The A320 makes greater use of CFRP, particularly in the tail; both the horizontal and vertical stabilizers are mainly carbon. The fin has GFRP in the leading-edge and trailing-edge panels. Airbus has two other aircraft in production—the A330 and the A340. Similar uses for composites are found in the aircraft, including the flaps and empennage. The use of composites in floor beams and in the forward pressure bulkhead saves 12 kg in the latter case. Producing floor beams, of which there are many, will be quite a challenge with composites, however, the manufacturing cost will determine their use.[23,24]

The A340 represents a slight reduction in the percentage weight of composites used, although the total weight of composites is greater because the aircraft is larger. The A300-600 is about 4% composite (by weight), the A310-300 8%, the A320 15% (Figure 3.13), and the A340 12% (about 11,590 kg).[17]

Airbus does not employ Kv in loaded structures but makes intensive use of CFRP in these structures, sometimes along with layers of GFRP. This acts as a bridging layer between honeycomb and skin and also provides galvanic corrosion protection. It also forms a dielectric coating in areas likely to take lightning strikes.

Other thermoplastic composite applications in the A330/340 and A320 include control levers, cockpit furnishings, instrument panels, and electronic racks, with possible weight savings in the region of 200–300 kg.

The horizontal tailplane on the Airbus A320 represents the first use of a C_f/Ep advanced composite in a highly loaded primary structure in the production of this commercial jetliner. The use of C_f/Ep has resulted in a tailplane 15% lighter than its Al equivalent.[22] Both the horizontal and vertical tailplane structures are predominantly made from C_f-reinforced composites in both monolithic and honeycomb sandwich configurations.

The two outer flaps of the Airbus A330/340 wing mentioned above have a span of 60 m. Each flap is 12 m long and 1.2 m wide and weighs about 360 kg. The C/Ep composite flaps weigh about 20% less than Al flaps and offer higher resistance to fatigue and corrosion.

Fokker 50.   A C/PEI thermoplastic main undercarriage (MUC) door has been developed by Fokker Aircraft Company.[25] The MUC doors located on the bottom aft side of each nacelle are significantly cheaper and 53% lighter than the existing Al doors. Com-

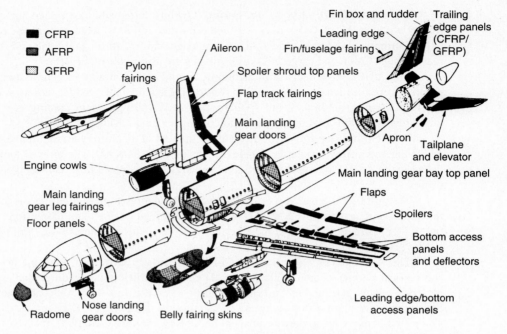

■ CFRP
▨ AFRP
▦ GFRP

Pylon fairings
Aileron
Spoiler shroud top panels
Flap track fairings
Fin box and rudder
Leading edge
Fin/fuselage fairing
Trailing edge panels (CFRP/GFRP)
Main landing gear doors
Apron
Tailplane and elevator
Engine cowls
Main landing gear leg fairings
Floor panels
Main landing gear bay top panel
Flaps
Spoilers
Bottom access panels and deflectors
Radome
Nose landing gear doors
Belly fairing skins
Leading edge/bottom access panels

Figure 3.13 The use of C_f-, Kv_f-, and Gl_f-reinforced plastics in the Airbus A-320 model.

pared to a new, optimized metal design, they are 30% lighter but more expensive. According to current Fokker standards for commercial aircraft, the weight savings justifies the extra cost, and thus the thermoplastic version is cost-effective compared to the new metal design. The above thermoplastic (PEI) composite nonstructural applications in the Fokker 50 and several other aircraft are listed in Table 3.3.

BAE 146. The floor beam strut and keel longeron bay tube in the BAE 146 aircraft were selected as components to demonstrate economical fabrication methods and the best material utilization of MMCs.[26] A cost and weight savings analysis for these components has estimated that up to 10% of the structural weight of a modern subsonic civil airliner could be made from discontinuously reinforced Al alloys (DRA) and exhibit cost-effective weight savings. The main areas of potential application are the upper wing and the internal structure of the fuselage.[26]

3.5.1.3 Business jets, gliders, and light planes

Beech Starship I. This first all-composite business jet aircraft has been certificated. It utilizes a sandwich construction (Gr/Ep laminate as face sheet and Nomex honeycomb as core material).

The maximum takeoff weight is 6773 kg; the empty weight is 4584 kg of which about 2364 kg is structure. And of this, over 1455 kg is composite honeycomb structure.

The all-composite sandwich airframe allows the molding of large, smooth seamless

TABLE 3.3   Applications of PEI Thermoplastic Composites in Aircraft[25]

Application	Model	Manufacturer or Fabricator
Brackets	Fokker	100 IPA-Belgium
Hat rack module	Boeing 737	Tremonti-Netherlands
Air diffuser	Airbus	MBB-Laupheim
Universal assy profiles	Fokker 100	IPA-Belgium
Window slide rail	Airbus	Fokker SP
X-ray covers	Medical	Tremonti-Netherlands
Radiotherapy panels	Medical	Fokker SP
Filter housing	Fokker 100	IPA-Belgium
Dado panel	Airbus	Fokker SP
Ice protection plates	Fokker 50	Westland Helicopters
Hinge cover	Fokker 100	Fokker SP
Duct support	Airbus	SEK-Germany
Air step	ATR	Fokker SP/Aeritalia
Switch cover	Fokker 100	IPA-Belgium
Kick panel	Fokker 50	Fokker AC
Cabin intercostals	Fokker 50	IPA-Belgium
Ice protection plates	DO 328	Fokker SP
Galleys	737/MD80	Various manufacturers
Heat shield	Fokker 50	Fokker AC
Speed brake fairing	Airbus	Aerospatiale
Landing flap ribs	DO 328	Dornier/Fokker
Cargo intercostals	Fokker 100	Tremonti-Netherlands
Relay box	Fokker 100	Tremonti-Netherlands
Radome	Harrier	Tremonti-Netherlands
Fire wall	Hovercraft	Hovertrans

aerodynamic surfaces. Major benefits are freedom from corrosion, ability to mold compound contours, relative insensitivity to fatigue, and lighter weight.[27]

The Beechcraft Starship is an example of what can be achieved in design and analysis of an all-composite sandwich airframe with the technology available today. And the key to making it all work successfully is proper application of composite sandwich material in the design engineer's mind as well as manufacturing simplicity in the manufacturing engineer's.

Dornier 328.   The HD-E6C-3 propeller assembly for Dornier Luftfahrt GmbH, a 30- to 33-seat regional airline, consists of six composite blades. The propeller is the first of a series of lightweight composite propeller systems that can be used in aircraft engines rated at more than 1800 hp.[28] Other significant applications and weight savings are listed in Table 3.3

Additionally Dornier's[24] new regional aircraft has C_f/Ep prepregs in the tailcone and vertical stabilizer primary structure. The rear bulkhead, the first all-composite pressure bulkhead, is made with Kv_f/Ep prepregs. The ATR engines use prepregs in structural parts such as wing spars (ATR 42 and 72), control surfaces, and the pioneering outer wing box for the ATR 72—the only complete C_f wing in a commercial aircraft.

Beginning in 1991, Kv/PEI ice protection plates for the Dornier 328 turboprop aircraft have been manufactured using thermofolding.

Stratos I and II (Grob Aircraft).   The Stratos I (Egrett) is a 10-ton, single-seat turboprop surveillance aircraft with the wingspan of a Boeing 727 made of fiberglass, carbon, epoxy, and foam construction.

Stratos II is a huge, all-composite twin piston engine craft with the wingspan of a Boeing 747. It is used for very high-altitude atmospheric research, carrying a crew of four in pressurized comfort. The refueling range of the Stratos II will be nearly halfway around the globe. Grob Aircraft has built over 3000 all-composite private aircraft, mostly gliders. A new Grob G-200 composite four-seater, a sleek, single-engine pusher-type sport and business design for the 1990s is currently being produced.

Ruschmeyer R-90.   The R-90 employs vinyl ester resin with Gl_f, claiming up to 70% parts reduction and superior properties compared to conventional aerospace resins in this new model for the 1990s.

Pezetal Gliders.   A range of Polish composite gliders (Pezetal), plus a C composite canopy frame, have been built and meet all FAR 23 requirements for the Orlik trainer. The 3-m-long structure was recently switched from woven glass to C/Ep material, obtaining yet another weight savings for this stiffness critical component.

Raytheon Premier I.   The fuselage shell of the new Premier I business turboprop jet shows more than a 20% weight savings over a comparable metal fuselage and related performance gains. The C/Ep honeycomb composite fuselage, produced by computer-controlled automated machines, requires fewer fasteners and only half the tooling costs to fabricate and mate to the wing. Composite fuselages provide greater safety in crashes, are easy to repair, and do not corrode.

Saab 340.   This is another regional aircraft that makes extensive use of a nonmetallic honeycomb core and Kv_f/Ep laminates for composite floors and ceiling and sidewall panels. The new Saab 2000 aircraft follows suit.

Cirrus VK30.   The Cirrus Design Corporation kit-built VK30 has high performance, handles well, and features composite airframe construction and cabin accommodations unavailable in factory-produced light aircraft. The airframe is constructed from vinyl ester resin and $E\text{-}Gl_f$ around polyurethane and PVC foam cores. The horizontal stabilizer, rudder, and elevator are made of Kv composite to reduce weight. The tapered wings use a single-piece C/Gr main spar and a 610-cm C/Gr front spar.

Europa.   This two-seat, light plane comes in kit form and uses advanced epoxy and phenolic fireproof resins. Gl, C_f, and high-density PVC foam are the sandwich materials. The composites make the airframe much lighter and help to reduce noise and vibrations from the four-stroke engine.

3.5.1.4 Future commercial space travel (HSCT, NASP, and others)

HSCT.   Development of the building blocks for tomorrow's HSCT has started with the more than 1600 components already made of composite materials. Based on this experience, the tougher material and simpler and more cost-effective design will allow

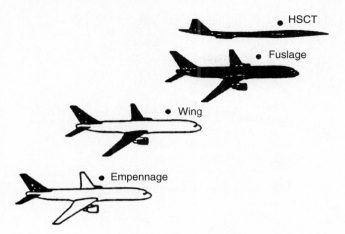

Figure 3.14   Phased implementation of composite structures.

for the expanded implementation of advanced composites in primary structures that will follow. Figure 3.14 shows the possible development steps for progressive implementation beginning with the empennage, followed by the wing structure and fuselage and the eventual construction of high-speed civil transport.

Empennage. Taking the first step, engineers have examined for example, the composite horizontal stabilizer and vertical fin for the Boeing 767.[29] New airplane design features include simple honeycomb ribs, laminate spars, and skins with cocured stringers. Cocured stringers on a laminate skin offer the reduction of fasteners from more than 12,000 in a comparable Al design to about 200.

Wing. Following the incorporation and successful service evaluation of the empennage, the design of a composite wing structure would draw on the empennage experience. The wing, with its unique technical issues and increased design loads, will utilize the material capability more fully.

Fuselage. Composite fuselage design focuses on the corrosion resistance and durability benefits as well as lower cost and weight over the existing metal designs.

The supersonic environment for the HSCT will require elevated-temperature materials and unique structural design configurations compared to the subsonic commercial aircraft materials and structures of today. Development of the HSCT and other primary structures as applications of composites will incorporate the same principle of large-scale developmental tests used on the empennage. Current studies indicate that composites are essential as prime structural material to enable the vehicle to be economically competitive with subsonic aircraft.

The HSCT would yield some clear economic rewards. Passenger traffic to and from the Pacific Basin will rise from 19% to 40% of world traffic by 2025, experts predict, and HSCT would be economically viable at competitive passenger rates. The market in the year 2000 will be able to support such aircraft in regular scheduled service to many parts of the world. Travel times to the Pacific from U.S. business centers will be 4 h or less at Mach 4–5 speeds. Such speeds would be more than double that of the Concorde.

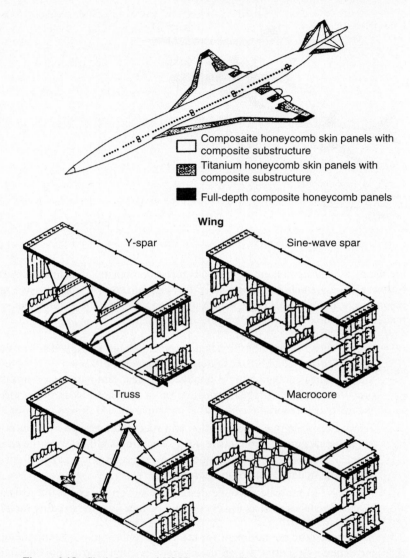

Figure 3.15 Shaded areas of HSCT represent likely use of Ti honeycomb and full-depth honeycomb composite materials. The remainder of the aircraft would be made of composite skin panels with composite substructures. Different HSCT wing structures include Y-spar, sine-wave spar, truss, and macrocore configurations.[30]

In general, a hypersonic transport with a 7000-mi range offers immediate (2 to 3 h) access between the markets of Asia, the United States, Western Europe, and Africa. A range of 10,500 mi opens a rapid link between the United States and Australia (Figure 3.15).[30,31] The aircraft is contemplated to have a top speed in the Mach 2.4–2.5 range. The speed is critical to the design in terms of materials selection because of frictional heating of the aircraft's skin. Above Mach 2.5, the heating increases rapidly with little over-all time gain for the passenger. Above this speed, the total trip time does not decrease that much because of time spent getting to and from the airport, checking baggage, and so on.

NASP.   The term *new space transportation systems* is meant to identify systems that no longer require expendable vertical launchers for takeoff as the shuttle, Buran, and Hermes do. New space transportation systems, rather, are fully reusable and are capable of taking off and landing horizontally, with a minimum of turnaround time, on any part of the globe. The main incentive for the development of such vehicles is the prospect of lowering the extremely high cost of placing payloads into orbit.

The national aerospace plane is a truly ambitious one-stage-to-orbit vehicle which will be propelled to Mach 25 by scramjets. Scramjets, that is, supersonic combustion ramjets, are a relatively new propulsion concept. In the case of the NASP, this concept will have to be advanced to its upper theoretical limits—at considerable cost and risk. Even so, the question is open whether the NASP can reach orbit with scramjets alone.

Apart from the intricacies of engine development, the design of the airframe is a challenge in its own right. This is largely because kinetic heating which, being more criti-cal in the ascent than in the descent phase, will lead to temperatures at stagnation points well above 2500°C with corresponding demands on the material selection.

Western European concepts are somewhat less demanding, partially because of fi-nancial constraints and partially because of a less well-developed technology base. Both Hotol and Sanger envision turbo- and ramjet propulsion to Mach 5–7, to be followed by rocket propulsion.[32]

The British Hotol project involves a one-stage-to-orbit craft with an expected pay-load capacity of 7 tons. It will be unmanned, its primary mission being the placement of satellites into low earth orbits.

The German Sanger project involves a two-stages-to-orbit system, and technologi-cally the lower stage of Sanger poses no major problems. A cluster of turbo- and ramjets will propel it to Mach 6.8 where the upper stage separates. The maximum temperatures over major areas of the airframe are in the 600°C range with local hot spots of up to 1100°C. From the materials viewpoint such temperatures can be accommodated by sili-con-coated Ti alloys. The aerodynamic loads are comparable to those of supersonic trans-ports with the exception of the stage separation phase.

NASA has designed a segment of a CCC elevon and plans to determine if the tech-nology can be applied to control surfaces of hypersonic vehicles such as the NASP. The first structural application of CCC mechanical fasteners was in a large CCC structure. At-tachment of the rib-to-torque tube fittings required development of CCC bolts and nuts. Specialized hardware was necessary because conventional metal fasteners could not with-stand the extremely high heat environment of the elevon.

Another part of the NASP program required the fabrication and testing of a full-size or near full-size article representative of the vehicle's structure. McDonnell Douglas chose to fabricate an integrated fuselage and tank structure, which is shown in Figure 3.16. The test section, shown at the upper left, was 122×244 cm overall. The TMC por-tion consisted of eight curved and four flat hat-stiffened panels. The other views in the figure show the test section with the load extension structure, made from Inco 909 panels and Al end plates, and the completed structure in the test fixture.

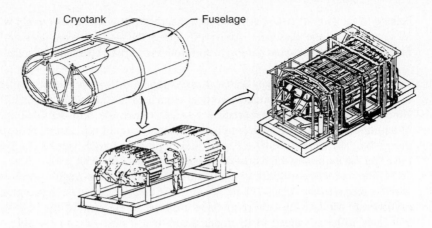

Figure 3.16   NASP vehicle integrated fuselage and LH$_2$ tank.

Waldman and Harsha[33] described other test components in design, fabrication and tests covering Gr/Ep LH$_2$ tanks, Gr/PI heat exchanger panels, CCCs for engine exhaust nozzles, and TMCs for a wing carry-through bulkhead and attach fittings as well as other highly loaded joints and stiffeners.[34,35]

3.5.2 Helicopters

Composite materials are not by any means new to helicopters. In 1953 Model 47H-1, a helicopter with much of its cabin built of Gl, was certified. For the next quarter-century, the use of composites was restricted to these kinds of secondary, or non-load-carrying, structures such as panels and access doors. However, in the last decade, airframe manufacturers have acquired certification for the primary structure of a commercial aircraft, and the S-76 has been certificated with a horizontal stabilizer made of composites. During the same period, Gl main rotor blades on the Model 214B were the first United States-made composite rotor blades to be approved by the Federal Aviation Administration (FAA).

Therefore, for almost 45 years fiber-reinforced materials have been successfully used in the helicopter industry. The most preferred applications of these materials can be seen in highly dynamically loaded main and tail rotor blades, but also in structural parts of the fuselage designed for a high performance-to-weight ratio and in some other special components. The excellent ratio between performance, reliability, and costs is one of the main benefits of using these materials. In addition, they provide the possibility of realizing new hingeless and bearingless rotor concepts with the advantage of high durability.

Helicopter manufacturers have established and maintained a position at the forefront of the development and application of composite materials. They are motivated by projections indicating that judicious use of composite materials can improve airworthiness and safety, increase operational performance capabilities, reduce the empty weight fraction of the aircraft, and make it more affordable.

There are two primary reasons for the influx of composites into the manufacture of helicopter structures. First, the advent of fibers and resins capable of being molded into lightweight, durable, complex shapes has made possible the manufacture of larger singular structures, thus eliminating many of the joining tasks and associated operational penalties. Second, because of the variety of properties displayed by the various materials used in creating an advanced composite, the engineer is given much latitude in matching composition to structural requirements.

3.5.2.1 Military helicopters

V-22 Osprey.   Extensive refinements in the design of the Bell Boeing V-22[36] are allowing the tilt rotor team to make significant progress toward achieving their goal of producing an economical, lightweight composite aircraft. Some examples are shown in Table 3.4.

Besides those shown in Table 3.4, another example is the aft fuselage for which the design has been changed from one requiring nine separate skin panels and more than 157 stiffeners to one featuring a single-piece component with 17 continuous cocured J stringers. Aft-fuselage sections made this way are produced at about 47% the cost of those employing hand layup techniques, and the scrap factor is reduced to 1.25. The aft-fuselage deck initial design consisted of more than 70 detail parts and required significant subassemblies. The new design eliminates 33 detail parts and 724 fasteners.[37]

A front landing gear door for the V-22 Osprey aircraft has been fabricated from APC-2 C_f reinforced PEEK. The door is produced from unidirectional prepreg tape, with the outer skin and corrugated inner skin being fabricated separately and then joined by fusion bonding.

Other composites utilizing the fiber placement manufacturing method include the sponsons, side skins, and drag angle—a nearly right-angle strip used to join the Osprey's fuselage skins and canted upper deck (roof). Also, a 366-cm V-22 wing box section using stringers of Graphlite was recently assembled. These hat-shaped stringers depend on the Graphlite C_f rods for strength. These rods have a 67% V_f and a Young's modulus of 549×10^4 MPa at zero strain.

The Graphlite retains its shape because the C_f are aligned before they are impregnated with epoxy and then cured. Because the fibers are locked in place, they do not move like a component made with Graphlite rods when it is cured. Preliminary results indicate that components made with Graphlite offer a tension strength on the lower wing surface and a compression strength on the upper surface that are almost equal. In tests tape with a waviness of 1% displayed a maximum compression strength of 1585 MPa, while Graphlite did not fail at 1931 MPa. Extrapolation of these results indicated that the C_f rods have a theoretical compression strength of 2657 MPa. In addition, a single layer of rods is as thick as 10 layers of tape. This means fewer passes and less effort are needed to lay down an equal amount of material.

TABLE 3.4   Several Composite Components and New Versus Old Process Improvements[36]

Component	Process
Wing skin	Manual layup, automated layup
Wing stringer	Manual stringer placement, mechanized stringer installation
Wing stringer	Manual ply-by-ply stringer fabrication, mechanized stringer fabrication
Grip	Filament-wound grip, fiber placement grip

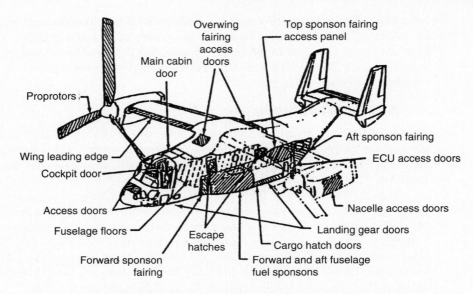

Figure 3.17 Honeycomb sandwich composite usage in the V-22 Osprey.[37]

Most experts feel that a tilt-rotor could have been made without composite materials, but that it would have been a much larger, heavier aircraft. In fact, the wing of the tilt-rotor is one of the best applications of composites, which make it easier to tailor stiffness and tune structures.[38]

Gr/Ep is 44% of the Osprey structure by weight, and Gl/Ep another 7%. Other materials are used where needed. But composites in the wing, fuselage, and proprotors are estimated to have made the V-22 25% lighter than a comparable metal airframe. Honeycomb sandwich structure is emphasized very much in the V-22 aircraft. The areas of usage of sandwich composites on the V-22 aircraft are shown in Figure 3.17.

EH101. Floor panels made from APC-2 thermoplastic reinforced with AS4 C_f have been introduced in the newly developed EH101 helicopter. The composite material is used in the form of skins over Al honeycomb.

AH-64A. Thermoplastic composites are now used for lower leading-edge fairings on this military helicopter in lieu of thermoset material. The thermoplastic composites initially offered slightly better strength-to-weight ratios than thermosets with the same reinforcing fibers. A Gl/PPS fairing and a Gr/PEEK structural skin for the vertical fin of the Apache demonstrated weight and cost savings. Riveted to the tail of a crashed AH-64 and shot with 23-mm cannon rounds, the thermoplastic parts also demonstrated slower crack propagation than thermosets for an added measure of ballistic tolerance in combat aircraft.

A horizontal stabilator for the AH-64 was built and tested and weighed 18% less than a comparable metal stabilator. Despite its apparently conventional internal structure, the thermoplastic (PEEK) stabilator used adhesive bonds which replaced individual fasteners—the 3300 fasteners in the original structure were reduced to just 148 in the test article. The metal stabilator had 36 pieces, while the thermoplastic version had just 25.

Despite the major effort expended the project was put on hold and canceled, however, there may be other applications in future helicopters.[39]

NH-90.   This new helicopter features a composite fuselage and blades, and another "first" is the NH-90's windshield, made up of two bonded half-shells places in a frame (composite like most of the airframe) which is also made of only two parts. In fact, 95% of the NH-90's airframe is of composite material, which may be the world's "first" for a helicopter of its size. This allows a 20% savings in weight, increased crash resistance, and a very substantial reduction in the number of airframe parts.

The Eurocopter (NH-90) is currently under test, and composite materials have clearly been used extensively in its structure. Completely modular, the structure is made from a variety of composites including C_f-reinforced plastic sandwich, hybrid Gl/C, and monolithic C.

SA365F and 366G1.   These Gazelle and Dauphin helicopters utilize a tail rotor, a *fenestron,* which is an original concept intended to overcome the major drawbacks of conventional tail rotors for light and medium-sized helicopters.

The fenestron rotor is housed in a shroud which protects it against most aggressions, reduces the radiated noise, and offers several significant advantages in operation. The fenestron has blades, a shroud, and a fin and in these models is made of composite materials (Figure 3.18).[40] Other components utilizing composite materials are shown in Figure 3.19.[41]

RAH 66 Comanche.   This newest military helicopter has an airframe comprised of more than 70% C_f/Ep or BMI, with blades of C/Gl$_f$/Ep.[42] Some of the composite components include a 772-cm-long Gr/Ep keel beam with cocured skins, (Figure 3.20).[43]

Figure 3.21 reflects the external usage of advanced composite material in the first developmental and test helicopter. The flex beam shown in Figure 3.21 is a critical element in the bearingless main rotor (BMR) designed for the Comanche helicopter. The BMR provides enhanced performance with lower maintenance requirements than conventional main rotor systems (Figure 3.22).[44]

S-70 Blackhawk.   Schwartz[20] describes many of the S-70 Blackhawk helicopter composite components, and Figure 3.23 shows the rear fuselage where composites have reduced the number of detail parts from 856 to 104, assemblies from 75 to 10, and fasteners from 13,600 to 1700.

3.5.2.2   Civil and commercial helicopters

BO 108(EC135), BK 117, BO 105.   Having gained design, fabrication, and test experience with composites on various models, Eurocopter Deutschland GmbH (ECD) has embarked on a very ambitious composite program with these helicopter models. Two BO 108 prototype helicopters were fabricated with different sections of the primary fuselage structure made of fiber-reinforced composite material. Additionally two EC 135 prototypes (civil versions of the BO 108) were constructed with composite fuselages. An experimental entire composite fuselage for ECD's BK 117 helicopter was designed with the continued use of Gl and aramid$_f$-reinforced polymers for secondary structures. The BO

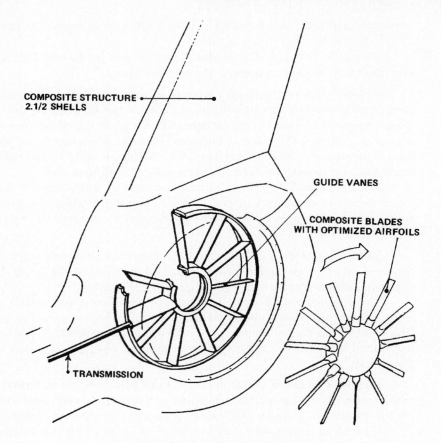

COMPOSITE STRUCTURE
2.1/2 SHELLS

GUIDE VANES

COMPOSITE BLADES
WITH OPTIMIZED AIRFOILS

TRANSMISSION

Figure 3.18   Light helicopter composite fan in fin.[40]

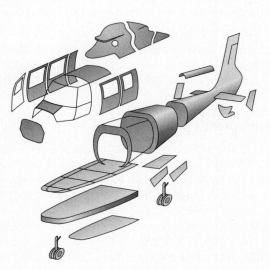

Figure 3.19   The use of composite materials (Kv/Ep and Gr/Ep) by all the above components.[41]

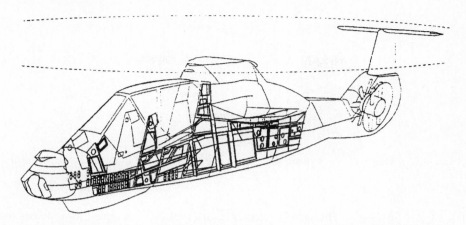

Figure 3.20   Seven 72-cm-long Gr/Ep keel beam used in the RAH 66 helicopter.[43]

105 and BK 117 have utilized composite materials in, for example, doors, engine cowlings, subfloor shell, stabilizer, and fairings.

Messerschmitt-Bölkow-Blohm GMBH has built a prototype bearingless rotor and installed it in an advanced technology BO 108 helicopter rotor blade featuring an integral composite flex beam that transmits loads to the blades. The composite beam eliminates roller bearings, bearing sleeves, and centrifugal elements in the rotor head. Redesign and parts integration have reduced the number of elements by approximately 50% and provided a weight savings of 50 kg compared with the current standard BO 105 rotor.

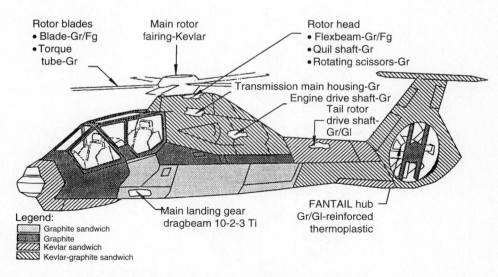

Figure 3.21   RAH 66 Comanche external composite usage.

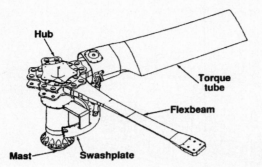

Hub

Torque
tube

Flexbeam

Mast   Swashplate

Figure 3.22   Comanche rotor system.

K-Max.   This new helicopter is an aerial truck designed for repeated load and un-
load cycles.

The aircraft's new composite blades include a Nomex honeycomb core, E- and
S-type Gl_f/Ep for skins and blade root reinforcement, and Gr composites for main structural members.

Servo flaps used to reduce rotor control forces at the pilot stick are also made of Gr
for stiffness, durability, and long life. The servo flaps act like the elevator on an airplane's tail and also helps to reduce helicopter weight by eliminating the need for expensive, complex hydraulic boost systems.

K-Max can land in areas inaccessible to ground equipment, eliminating destruction
to soil and landscapes. In a fire-fighting role, it provides a new and massive means of ap-

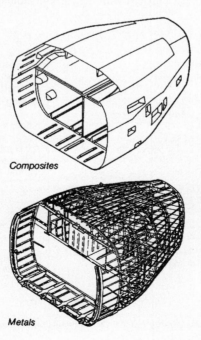

Composites

Metals

Figure 3.23   Blackhawk rear fuselage
comparison.

plying water and other retardants to forest fires. Other applications include power line in-stallations, construction, reforestation, mineral and oil field development and support op-erations, and seismic exploration.

MDX.   This McDonnell Douglas helicopter includes a composite tail boom, em-pannage, rotor system, and fuselage and incorporates the Notar (*no tail rotor*) system, the newest innovation in helicopter design.[45]

Bell 430/680.   The productionized 680 rotor is made of CFRP control cuffs[46] which transfer pitch forces to the composite flexures without the need for hinges or bear-ings (Figure 3.24).

Westland 30-300.   Engineers chose the tailplane (horizontal stabilizer and fin) of this model to explore and demonstrate thermoplastic composite manufacturing technol-ogy, as many elements incorporated in the assembly are directly applicable to other air-craft structures. The choice also exploited the opportunity to achieve a direct comparison between metallic and epoxy composite technology.[47]

The materials chosen for the project were C-reinforced PEEK and woven C-rein-forced PEI. The researchers found that thermoplastic composites offered considerable benefits in terms of weight savings and simplicity of manufacture. It was proved that these materials could be used in primary aircraft structures.[48] The cost-effectiveness of using these materials was found in automation of the processes involved (automated tape laying and filament winding).

Over the past 15 years, NASA has sponsored programs for building a database and establishing confidence in the long-term durability of advanced composite materials in transport aircraft structures.[49] Primary and secondary components have been installed in commercial aircraft to obtain worldwide flight service experience. Flight environments for transport aircraft and helicopters are quite different, and the behavior of composite components in the two environments may differ substantially.[50,51]

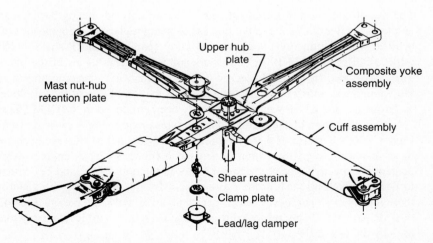

Figure 3.24   Bell-designed all-composite rotor system for model 680.[48]

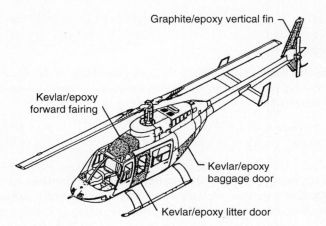

Graphite/epoxy vertical fin

Kevlar/epoxy
forward fairing

Kevlar/epoxy
baggage door

Kevlar/epoxy litter door

Figure 3.25   Composite components in flight service on Bell 206L helicopters.[52]

As a result, in 1978 NASA and the U.S. Army Research and Technology Activity (AVSCOM) initiated the first major program for evaluating helicopter composite components in flight service. Forty ship sets of composite litter doors, baggage doors, forward fairings, and vertical fins for Bell 206L helicopters were built and installed[52] (Figure 3.25). The specific objective of this program was to determine the long-term durability of composite airframe structures in the operational environment of light commercial helicopters.

The flight service performance of 4 horizontal stabilizers and 11 tail rotor paddles in Sikorsky S-76 helicopters was also to be evaluated and the residual strength of each composite component determined after removal from service. The composite components were production parts for the S-76.[53]

3.5.3 Engines and Engine Components

The use of composite materials in engines is not only advancing but increasing. For example, propeller blades are in the process of being replaced by composites of Al alloys to achieve weight reduction. Some use Al alloy spars and GFRP shells, while others use GFRP, CFRP, and AFRP for both spars and shells. The blades of the unducted fan engines developed by General Electric (GE) are thin, wide-sweeping ones made of CFRP. High toughness epoxy resin has been used for the fan blades of the GE 90 engine for the 777 airplane, and CFRP vanes have already been employed in the CFM 56 engine.[10,54,55]

An indication of the increased temperature requirements over the years is given in Figure 3.26.[56] Because of these requirements, CMCs and MMCs are expected to be used to an increasingly greater extent in aerospace engines.[57] In jet engines, for example, Ni content is expected to peak at about 45 wt% in the mid-1990s, and Ti content is expected to peak at about 30 wt% in the same period. By the year 2010, the quantity of each of these metals in jet engines is forecast to be about 10 wt%. On the other hand, percentages of CMC and MMCs are projected to increase from near zero levels today to over 25 wt% each by the year 2010.

In addition to the CMCs and MMCs being considered for some hot areas of en-

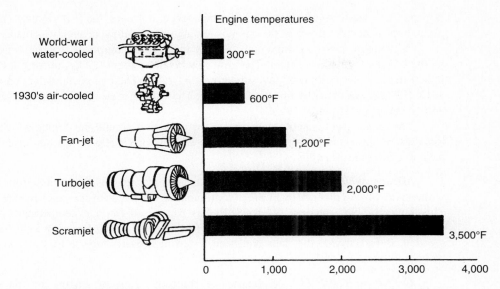

Figure 3.26 Increases in temperature requirements for engine components over the years. °F = 9/5°C + 32.

gines, organic composites that have initially brought weight reduction to nacelles and thrust reversers will, with BMIs and polymerization of monomer reactant polyimides (PMRs) in the future go into engine components themselves. Titanium metallic composites (TMCs) and intermetallic matrix composites (IMCs) with their capacity to operate at up to 800°C and beyond, are crucial in their development for their applications in jet engine components. CMCs are most suitable and promise operation at up to about 1100°C for uncooled turbine blades, vanes, and other hot path components.

R. Tape of Rolls-Royce, Inc.,[58] has painted a tantalizing vision of "tomorrow's civil engine," with PMC nacelles, doors, vanes and stators, and fan bearing supports. The fan may well be a PMC structure on a TMC ring. MMC bladed rings (blings) will be used in the compressor, IMs for the diffuser and combuster inner casings, and CMCs for the liner and nozzle guide vanes. CMC blades on a MMC disk are likely for the high-pressure (HP) turbine, and TiAl for the low-pressure (LP) turbine blades. IMCs could also form the LP disk and even the shaft. In this overview, tape forecast that engines would be 40 wt% and 50 vol% composite by the year 2010.

3.5.3.1 Military aircraft engines and components

F/A-18-GE F404 Engine.   C-reinforced PMR-15 PI has replaced Ti for the outer compressor ducts in this engine, and the payoff has been substantial. The composite ducts are at least 20% lighter, 30% less costly, and, unlike the chemically milled, fireproofing-lined Ti ducts, not prone to ignite as a result of foreign object damage (FOD).

According to the U.S. Navy, this last advantage was, surprisingly perhaps, the deciding factor in the decision to use composites, however, the other advantages and the performance capability of the composite ducts had been established previously. Weight savings alone was not sufficient to justify the change, however, the FOD-induced ignition problem provided the safety in the composite duct usage and the change was made.

A prototype environmental control valve made from composite material was tested on the F/A-18 U.S. Navy fighter. The new valve weighs about 0.9 kg, making it 0.45 kg lighter than the currently used investment-cast Al valve. The composite valve, composed of Gr_f/PEEK, was made by injection molding. The valve has sustained vibration levels of 26.6 g and temperatures of 177°C.

The engine inlet is another composite airframe part around the engine which is automatically fabricated by a robot that places ribbons of Gr_f/Ep in specific orientations prior to curing the assembly.

Mirage 2000-SNECMA M-53 Engine. This aircraft has flown numerous flights with engine nozzle inner flaps of SiC_f (continuous C_f are used for the outer nozzle flaps). This same SiC/SiC material (CERASEP) has been used for high-temperature leading edges on the European Hermes space plane (the nose of which heats up to 1900°C on reentry). Before flying, the flaps had been tested in acoustic fatigue to 4×106 cycles without failure.

C-17-P and WF-117. Exhaust gas charring from the engine of the stealth fighter's fuselage trailing edges has prompted the development of a new, processible, high-temperature composite resin. Potential commercial uses include lightweight thrust reverser components. The improved resin, designated AFR700B, increases the temperature capability of organic matrix composites by 66–371°C. The F117 engine additionally has composite fan exit guide vanes.

Future Engines like the Advanced Tactical Fighter. Figure 3.27 shows the potential application of CMCs and MMCs in advanced aircraft such as the ATF because engines could run much hotter and more efficiently with these types of materials.

Advanced Chinook and T 6000-Textron Lycoming T-55-L-712 Engine. A composite inlet housing has been tested at 4100 shp, exceeding the full takeoff power rating of 3750 shp. It is the first housing ever tested and was made from organic composite with braided fiber and produced by RTM. It was tested in a U.S. Army component improvement program to develop an interchangeable housing for the T55 series of engines. The composite housing weighed 13.5 kg less than an all-Al version (a 40% reduction), and the part count was reduced by over 30.[59]

GE-F l10 Engine. In this engine, if the low-pressure turbine shaft were made of MMC, over 67.5 kg could be saved. This includes a 9-kg reduction in the shaft weight and a 59-kg savings due to the elimination of a bearing compartment and a reduction in static hardware. As an additional benefit supercritcal shaft operations can be eliminated because of the vastly improved specific stiffness which designers can tailor to control vibratory margins. Finally, MMC shafts offer an increase in speed capability, reduced diameter, or an improved combination of both.[60]

Shaft materials used include SCS-6/Ti-6Al-4V, 35 fiber vol%, 22 ply, in a variety of orientations.[61,62] These same materials were used by GE to form a hollow truss core structure with 35 fiber vol%, 0° orientation, and 18 plies on both faces for MMC fan blades.[63]

Weight reduction is an obvious result of the hollow configuration. Lighter MMC

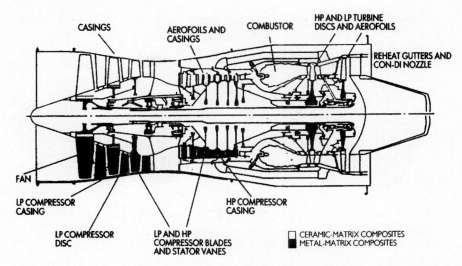

Figure 3.27   Many engine companies are investigating the use of CMCs and MMCs in many parts of an engine. Large performance increases are envisaged.

blades allow a reduction in static weight, reduced containment requirements, and potentially improved bearing life. Improvements in computational capability have resulted in advanced aerodynamic designs, and MMCs allow maximum use of the designs without limits imposed by monolithic materials.

Traditional compressors contain metallic disks which are heavy because of monolithic material limitations. Although MMCs offer lighter weight and other improved properties, simple material substitution in current monolithic metal designs does not take full advantage of these materials. Redesign based on MMC properties has permitted the development of ring structures capable of replacing the solid metal disks. MMC compressor rings have shown by far the most attractive weight savings of any MMC engine application—over 50% when compared to traditional designs. This savings pertains only to removal of the heavy disk region and does not include any synergistic savings that would result from lighter support structures, reduced gyroscopic moments, or lower bearing loads. MMC bladed rings are an enabling technology for advanced aerodynamics requiring higher rotational speeds and higher operating temperatures. Another significant payoff of this diskless rotor technology is the resulting cavity below the compressor rotor which could never exist in previous designs.

Other engine manufacturers have evaluated and tested similar components including MMC compressor rotors manufactured by Allison Gas Turbine Division[64] from SiC/Ti composite materials. The combination of compressor rotor and compressor bling weighs only 1.5 kg versus the 25 kg of the Ni-based ring.[65]

F/A-18E/F-GE F414 Engine. SiC_f/C composite secondary flaps and seals have been used and successfully tested in afterburner nozzles, and in service use of the lighter-weight flaps have the potential of saving 5.9 kg of weight per engine and even more airframe weight.

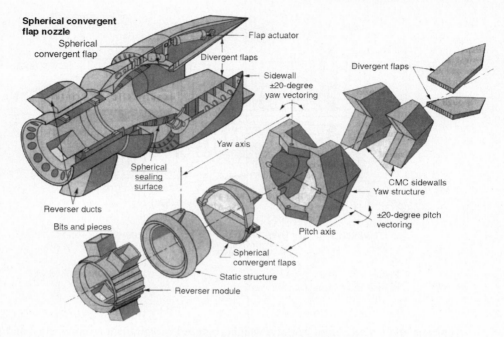

Spherical convergent flap nozzle

Spherical convergent flap

Flap actuator

Divergent flaps

Sidewall ±20-degree yaw vectoring

Divergent flaps

Yaw axis

Spherical sealing surface

CMC sidewalls Yaw structure

Reverser ducts

Bits and pieces

±20-degree pitch vectoring

Pitch axis

Spherical convergent flaps

Static structure

Reverser module

Figure 3.28   Spherical convergent flap nozzle.[67]

X-31-F404 Engine.   In a demonstration, CCC material has been used for vanes that encircle the engine exhaust region and serve as a primary vectoring nozzle.[66]

F-22-P and W-FII9-100 Engine.   Vectored thrust provides a key to vast improvements in jet fighter performance and maneuverability. Future aircraft, designed to use thrust vectoring, will require much smaller tail assemblies, or perhaps no tail at all. The latest in the series of multifunction jet engine nozzles is a spherical convergent flap nozzle (SCFN)[67] Figure 3.28 shows the SCFN where CMCs form the divergent section sidewalls and ceramic fiber liners insulate some of the flow paths within the reverser module. Organic matrix composites appear in other areas not directly exposed to exhaust gases. This engine may also be used in (IHPTET) testing of engine components.

Future Military Engine Materials and Applications.   Turbine rotors made of CCC materials have been successfully tested in a turbojet engine operating at above 1649°C. The significance of this demonstration is the use of CCC for rotating components in a turbojet engine. This is a new milestone in uncooled turbine engine operating temperatures, exceeding previous temperatures by more than 538°C.

Conventional engine technology uses internal cooling to maintain metal component temperatures of less than 1038°C. This cooling system, with its inherent performance penalties, is not needed with CCCs because of their increased material strength retention at very high temperatures.

Development and testing efforts aimed at breakthroughs in hot areas have been directed at the advanced high-temperature materials required for low-noise nozzles. Research and testing involve CMCs and monolithic metallic alloys and superalloys that can operate at temperatures up to 1038°C for 18,000 h for use in exhaust nozzles that weigh considerably less than those currently used. The importance of these nozzle weight savings becomes clear when one realizes that future HSCT propulsion systems will be large engines and that the noise reduction nozzle is larger than the core engine.

Other spinoffs from current developments include the design of a core engine's first three compressor stages using TMCs and eventually incorporating variable cycle capabilities as well as a five-stage, all MMC compressor.

Other development work using CMCs produced by microwave-assisted chemical vapor infiltration (MWCVI) appears to have a potential market in combustor cans, turbine bladed disks (blisks), heat exchanger tubes, and radiant burners. The potential for higher operation temperatures and improved efficiency utilizing components fabricated from CMC materials is expected to increase, specifically as economical processes are developed for production.

Several NASP spinoff applications[68] have produced structural CCC engine parts, CMC engine exhaust tubing, TMC impellers in turbine engines, and TiAl auto engine valves and Al MMC connecting rods (con rods). Future engines may have PMC intermediate cases joining the core to the fan section of the engine. The composite case is as strong as a conventional Ti case, but it weighs about half as much.

Aircraft turbine rotor failure is characterized by the release of high-energy fragments capable of penetrating the engine casing. The fragments could cause fires, loss of aircraft control, hull damage, and occupant injury. Lightweight materials commonly used as armor systems have the potential for application in turbine rotor containment systems at lower weight than in an equivalent steel system. Preliminary tests have shown that the most efficient system, on a weight basis, is one fabricated from Kv-phenolic. The most efficient system on a volume basis has used Ti, the containment capability requiring less volume than the Kv-phenolic system but at an increased weight.[69]

Another effective fiber material used in containment structures is polybenzbisoxazole (PBO). Tests conducted by Pepin and associates[70] showed that PBO was at least as effective as Kv at room temperature, and it is potentially one of the few choices for lightweight containment structures at elevated temperatures.

Other studies have indicated that low-cost, high-performance laminated metal composites will be available for aircraft engine containment and flight component applications. One composite material in development consists of 21 layers of Al metal and Al reinforced with SiC ceramic. The material resists fracture two to three times better than either of its components alone. In studies by researchers and others it has been shown that laminated metal composites offer excellent strength, toughness, and damping behavior.

Clyne and Withers[71] discussed MMC aero engine components (Ti-6Al-4V + SiC monofilaments) and their future use. They indicated that in a conventional engine, much of the weight consists of mechanical fixings and spacers, which are independently supported and direct flow through the engine. Figure 3.29 illustrates a series of modifications that could be made so as to reduce weight through exploitation of the high performance of Ti composites.[72] It should be noted that, as this is a rotating system, weight savings at the periphery are especially helpful because they reduce the necessary radial support. One-piece bladed disks reduce the parasitic weight, but much greater savings are possible

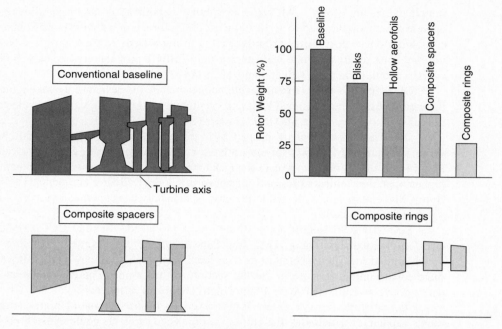

Figure 3.29   Upon incorporation of about 40% SiC$_f$, the stiffness of Ti can be doubled and its strength improved by 50%. In addition the density is markedly reduced (10%). These increases in performance offer exciting possibilities for engine design and weight reduction. This diagram suggests that if composite rings and spacers can be developed, weight savings of nearly 75% will be possible.[72]

if composite hoop-wound ring structures are employed. These eliminate the need for weighty support structures through their high circumferential specific strength and stiffness combined with acceptable radial properties (~50% lower).

Driver[73] has illustrated schematically in Figure 3.29 the potential advantages of MMCs in providing mechanically simple designs with considerable weight savings. Figure 3.29b shows a conventional monolithic metal compressor assembly comprising a large fan disk and blade with subsequent taper section disks and blades interspersed with independently supported spacers to direct the aerodynamic flow. This design provides the baseline. A significant proportion of a disk's weight results from the mechanical fixing required to allow blades to be inserted or removed from the disk. Integrally bladed disks offer one solution to reducing the parasitic weight, but this configuration precludes removal and replacement of blades and hence requires on-disk repair procedures. Hollow blading is a further weight reduction option (Figure 3.29a), while up to 75% weight savings can be achieved if both spacer and compressor disks are replaced by a composite ring structure (Figure 3.29d), which exploits the high specific tensile strength of continuously wound MMCs. Driver also opined on whether Al matrix systems with either particulates, short fiber or continuous reinforcement might offer more cost-effective possibilities than TMCs.

The Al$_p$/SiC$_w$ systems provide isotropic behavior with an attractive improvement in strength with increasing V_f of fiber. Unfortunately, although strength levels can be increased by 50% in Al alloys such as 6061 by the addition of up to 40% SiC_f, the elongation is correspondingly reduced from around 15% in the original alloy to little more than 1% in the composite; erosion resistance is also a problem, as the short fibers tend to be plucked from the component surface. Continuous-fiber Al composites have a restricted range of aero engine applications, but they are unattractive in comparison with Ti in terms of their temperature capabilities.

Japan's National Aerospace Laboratory and several industrial firms[74] have developed a new high-temperature composite that they hope to put into production for engines and space planes in 2–3 years. The material is a 3D weave of 9-μm-diameter Si/Ti fibers with an oxidized Si matrix. The material weighs a light 1.8 g/cm^3 and has a strength of 340 MPa at 1204°C. A surface coating is not necessary because of the material's inherent oxidation resistance.

Clarke[75] in his studies on CMCs estimates that engine components will represent 16% of CMC sales in the year 2000 compared to only 2.6% in 1988, a more rapid growth than that of many other applications.

In order to exploit the variety of PMCs, MMCs, CMCs, and IMCs in order to cut weight and boost engine performance many government agencies in the United States and Europe have funded and launched efforts[76] to move these materials from the laboratory and onto the factory floor to gain an edge in the global marketplace.

Applications include fan sections and compressor rings in fan sections whereby MMC stiffness reduces flow distortions, improving fuel burn. When MMCs are used in the initial rings of a compressor, weight reductions as great as 30% may be attained. Additional weight savings also could be realized if engine systems are redesigned to take advantage of the rotating rings' reduced mass.

Additionally fan blades of Gr/Ep, fan ducts made from AFR 700 (a material that can withstand temperatures of 316–343°C), fan stators, flaps, fan frames, self-lubricating variable stator bushings, aft-end fairings, and PMR-15 PMC outer ducts are also under study.

The researchers in Europe have considered PMCs primarily for engine fan blades and large static structures such as bypass ducts, fan frames, integrated fan case containment rings, combustors and exhaust guide vanes.

In the United States researchers[76] are evaluating SiC/SiC and SiC/Al_2O_3 for combustors, exhaust components including intermediate-pressure turbine guide vanes, flaps, and afterburner and exhaust nozzle components (Figures 3.30 and 3.31).

At a recent forum,[60] researchers, designers, and material technologists discussed the influence of new materials and manufacturing processes on the design of future aero engines. As an example, it was stated that fan blades would increasingly be manufactured with a wide chord in order to drive weight down, encouraging a hollow rather than a solid Ti or C/Ep composite. In the future is the prospect of blades that can be resin transfer-molded in a 3-D C composite. Ti aluminides and SiC/C_f (e.g., in nozzles) would enable future supersonic aircraft to be 25% lighter than those of the present generation (Figure 3.32).

3.5.3.2 Commercial aircraft engines and components.   Depicted in Figure 3.33 are tomorrow's civil engine and its materials according to Rolls Royce.[60]

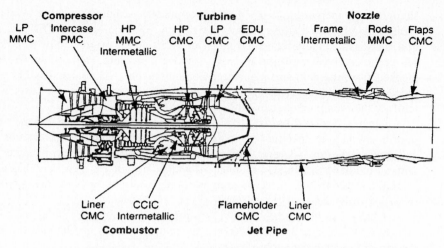

Figure 3.30   Tomorrow's military engine and its materials.[60]

McDonnell Douglas MD-11 Engine.   The air intake duct for the tail-mounted engine on the MD-11 is constructed from large C composite panels stiffened by preformed Rohacell foam ribs laid up as an integral part of the C_f laminations. The tail-mounted engine air intake assembly includes a forward duct about 3.6 m in diameter and 3.6 m in length and other panels that, when assembled, form the remainder of the 7.2-m-long duct. All of which are made of composite materials. The aforementioned air intake duct is 181.4 kg lighter than the previous metal design.

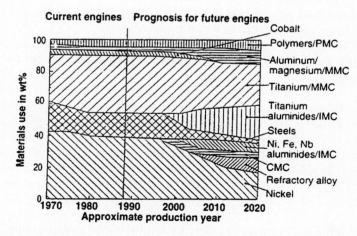

Figure 3.31   Material utilization in current engines versus tomorrow's engines.[60]

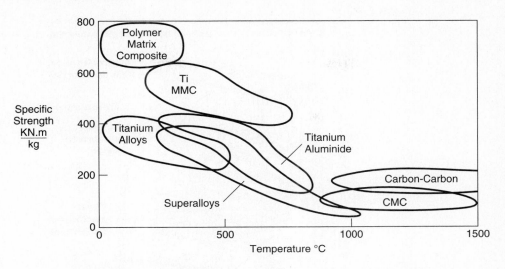

Figure 3.32   Strength and temperature envelope for various engine materials.[60]

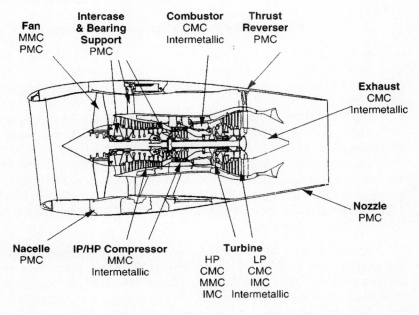

Figure 3.33   Tomorrow's civil engine and its materials.[60]

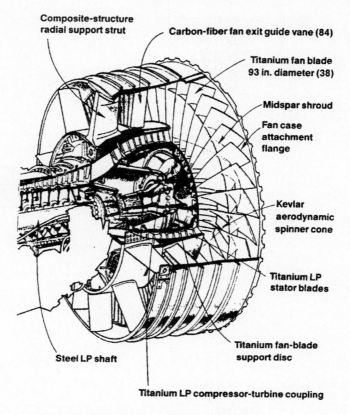

Figure 3.34 PW 4000 engine with C_f composites. 2.54 cm = 1 in.

Boeing 747/767 Aircraft and Engines. Four test cowls were used to demonstrate the feasibility of fiber placement of composite engine inlet cowls. These maximum-diameter 287-cm inlet cowls were built using a new compound-controlled fiber placement system.[77]

The P&W 4000 engine, a replacement for the JT8D, is 20.4 kg lighter as a result of the use of C_f parts in the fan exit guide vanes (Figure 3.34). C_f and single-direction tape placement allowed the C_f in the guide vanes to replace Al.

The vane shell consists of Celion types G30-500 (6K C_f) and G50-300 (12K C_f) for the vane core. The G30-500 has a minimum tensile strength of 3400 MPa and a minimum tensile modulus of 2.2×10^5 MPa. The G50-300 C_f has a minimum tensile modulus of 3.3×10^5 MPa and a minimum tensile strength of 2312 MPa. The prepreg material is 60% C_f and 40% Ep resin by volume.

V2500 Engine. The outer platform of the V2500 jet engine is made of an injection-moldable thermoplastic PI resin. The outer platform supports the outer vane, which rectifies the airstream coming from the engine. The outer platform must be corrosion-resistant and have high impact strength since jet engines often "inhale" birds while cruising and ingest sand or gravel during takeoff and landing. Historically, the outer platform has been constructed from a forged Al alloy.

The AURUM thermoplastic PI resin, reinforced with C_f, has half the weight of

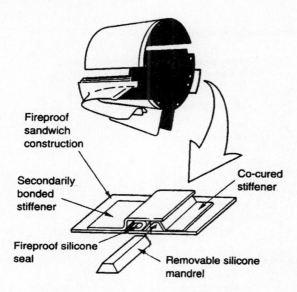

Fireproof
sandwich
construction

Secondarily
bonded
stiffener

Co-cured
stiffener

Fireproof silicone
seal

Removable silicone
mandrel

Figure 3.35 Reengined Boeing 727 with new composite components.

forged Al. It can be injection-molded and costs less to produce. During harsh environmental testing, including sand blasting and bird shooting, AURUM showed superior strength. In addition, it is recyclable.

Boeing 727 Aircraft and Engine.   The older 727 engine versions are being converted for third-tier airlines outside the United States. As a result, Figure 3.35 reflects the reengined 727 and the use of composites in the nacelle, struts, and acoustic tail pipe.

Boeing 757/737-300 Aircraft and Engines.   An air discharge duct, which is essentially a fluid barrier that attaches to the engine frames, is made of PEEK composite to protect the engine components in the event of fuel leaks in these models. When the engines for the 737-300 are started, the starter air discharge ducts serve a critical role in preventing air buildup in the engine compartment.

Boeing 777 and Engine.   This new model aircraft has a 312.9-cm fan, and to reduce weight in the engine GE has substituted Gr/Ep composites in place of their competition's hollow Ti blades. Composites have been used before in lower-speed parts, but GE is the first to try them for high-speed fan blades.

Garrett CFE-738 Engine.   A lightweight composite inner fan duct provides high-temperature service and engine fire containment for the CFE-738 turbo fan engine (Figure 3.36). The fairing weighs less than 3.6 kg, maintains structural integrity up to a maximum pressure differential of 0.051 MPa, and withstands operating temperatures of 121°C and heat soak to 232°C when the engine is shut down. This aerodynamic fairing is laminated as a cocured sandwich of BMI syntactic foam core and F-652 BMI/T-300, 3K, plain-weave Gr_f-fabric prepreg. Unlike other prepregs, this material maintains structural integrity and does not delaminate when exposed to flame contact.

Boeing Dash 8 Air Intake.[78]   The lower half of the Dash 8 nacelle contains the Kv/Ep air intake and debris dispenser as designed by Westland Aerospace, Ltd. The basic construction of the intake is honeycomb, with a Nomex aramid core and Kv

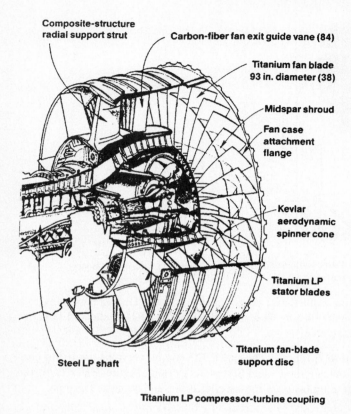

Composite-structure
radial support strut

Carbon-fiber fan exit guide vane (84)

Titanium fan blade
93 in. diameter (38)

Midspar shroud

Fan case
attachment
flange

Kevlar
aerodynamic
spinner cone

Titanium LP
stator blades

Titanium fan-blade
support disc

Steel LP shaft

Titanium LP compressor-turbine coupling

Figure 3.36   Fire-blocking composite fairing on a CFE-738 engine.

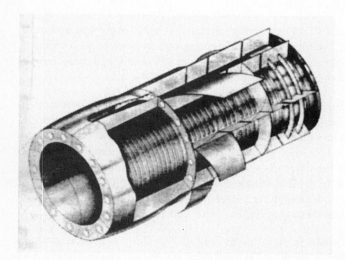

Figure 3.37   MD 11 inlet duct construction.[78] 2.54 cm = 1 in.

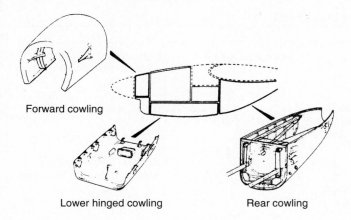

Forward cowling

Lower hinged cowling

Rear cowling

Figure 3.38   Dornier 328 intake duct components.[78]

face skins. This company has specialized in nacelle design and construction and has also been involved in the design and fabrication of the C/Ep inlet duct extension (MD11) (Figure 3.37) and the Dornier 328 inlet duct components composed of C_f/BMI (Figure 3.38).

Airbus A320 and A330 Engines.   A C_f/Ep sheet molding compound has been selected for the turbine blade in an emergency power system in the Airbus A320 and A330. The material provides an estimated weight savings of 25% over forged or machined Al blades and an estimated cost savings of 30–40%.

Initial feasibility studies indicated that C_f/Ep SMC could handle the static load and stiffness requirements for the blade. The larger composite blade, including an inte-gral Ti root fitting, has proven able to handle the critical loads and additional power. The compression-molded SMC blade has a metal plating of Cu/Ni for lightning protection.

The Rolls Royce RB211 Trent engine which currently powers the Airbus A330 aircraft (Figure 3.39), uses a C/BMI system because this material is capable of higher ser-

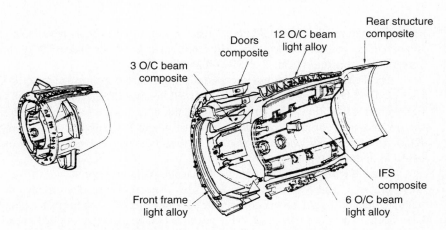

Doors composite

12 O/C beam light alloy

Rear structure composite

3 O/C beam composite

Front frame light alloy

6 O/C beam light alloy

IFS composite

Figure 3.39   Rolls Royce RB 211 Trent thrust reverser.

TABLE 3.5   Potential Automotive Applications of MMCs[88,89]

Model	Component	Materials	Comments
DC9-50	Inlet outer barrel	C/Ep tape	47 wt% savings versus metallic outer-barrel counterpart
DC-50			
B727	Lower doors	C/Ep tape	—
MD 80	Nacelle cowl system, inlet outer barrel, upper and lower cowl panels (engine access doors, apron)	Kv/Gr/Ep + Gr tape and Al H/C + Kv/Ep	152.4 cm in diameter, 426.7 cm long
P&W JT90-7R4	Engine nozzle—outer barrel, fan cowl panels, fan reverser translating sleeve	Kv/Gr/Ep	First filament-wound structure application in nacelle fabrication
GE CF6-80-C2	Nacelle—inlet outer-barrel, fan cowl panels	C/Ep + Cu screen and Ep adhesive film for lightning protection for outer barrel and the same for Al H/C for fan cowl doors	
V 2500	Translating sleeve	C/Ep	Sleeve comprises an upper and lower assembly which when latched together at the outboard edge form the external contour of the nacelle between the fan cowl and exhaust nozzle
Rolls Royce RB 211-535 E4	Thrust Reverser—fixed structure (inner C-duct)	C/PI (PMR-15)	Integration of upper and lower bifurcation ducts, upper track fairing, and precooler duct into a complex "coke bottle" sandwich structure with an acoustically treated (perforated) face sheet. Weight saving approximately 36 kg
GE unducted fan	Rings and access panels (8) for forward-stage fan and aft fan	C/PMR-15	

vice temperatures than standard epoxy resin systems and in most cases also provides an improved mechanical performance. The entire inner fixed structure (IFS) of this thrust reverser is manufactured from C_f/BMI composites.

The Fokker BVF-100 has successfully flight-tested a thrust reverser door that utilizes Gr/BMI tape skins and fabric ribs, a Gr/PI honeycomb core, and special erosion and lightweight strike films. The structure forms the nozzle for a high-bypass engine and, in the reverse mode, rotates into the jet exhaust where it is subjected to a temperature of 233°C. The assembly weight is less than 80% and the cost is less than 60% of that of the baseline Ti door it replaces.

Since the first advanced composite nacelle parts were produced in the mid-1970s, consisting of C and Kv_f in an Ep matrix, this concept has resulted in nacelle structures with significant weight savings over their previous metallic counterparts. Many advances have been made since then, and today advanced composites are applied even in acoustic structures utilizing PMR-15 PI matrix systems as well as Ep systems with Kv, C, and Gl_f (Table 3.5).

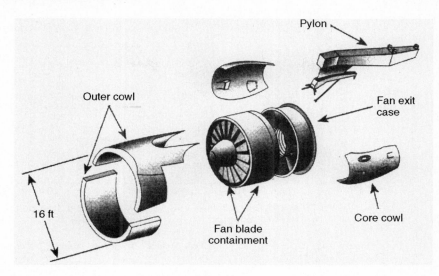

Figure 3.40   Components of a nacelle in a ducted prop engine.[79]

Finally, government technology investment programs[79] have now been funded to develop affordable composite materials for advanced aircraft engines (advanced ducted prop engines). The program covers polymer composites for use in making nacelles and fan cases.

Composite material is needed for advanced ducted prop engines because the 18,144- to 45,360-kg thrust power plants with 15:1 bypass ratios will have nacelles about 375 cm in diameter, a cross section similar to that of a narrow-body transport fuselage. The only way to keep the weight of the nacelle within reason and to make it thin enough to avoid excessive drag is to use composite material. The demand for composite materials may be as much as 1306.8 kg per engine because composite nacelles and other components could account for 30% of an advanced ducted prop engine's weight (Figure 3.40).

Composites that are being evaluated include Gr/Ep, Gr/BMI, and Avimid, and these materials must withstand temperatures of 177°C. The main reason for the limited number of materials is the interest of airlines in minimizing the variety of composite materials they have to deal with, and so reliability and maintainability are big issues (Figure 3.41).

3.5.4 Automotive Engines and Components

3.5.4.1 Automotive engines (all types).   ORNL and 3M have collaborated to develop a CMC filter for coal-fired power generation. The ceramic composite filter is made from a combination of continuous and chopped ceramic fibers reinforced with a SiC matrix. Because of the design, the filter is said to be stronger, lighter, and less brittle than its predecessors. The continuous fibers not only strengthen the filter but remain intact if the filter breaks, keeping impurities from damaging the turbine. The thin wall of the filter makes it less susceptible to breakage caused by extreme changes in temperature and pressure. These characteristics reportedly create many other uses for the technology, such as filters for industrial waste incinerators, metal smelters, and advanced diesel engines.

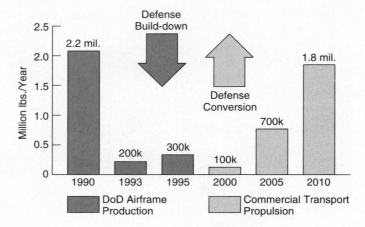

Figure 3.41 U.S. composite production. Courtesy of Pratt & Whitney.[79] Million lb/year = million kg/year; 0.4536 kg = 1 lb.

Advanced Refractory Technologies (ART) has developed ceramic reinforcements for MMCs in which SiC_w and a special Clarkson University coating technology are used to reinforce Al and Mg for use in automobile engine and other components. This new development, when successfully used in Al for pistons, could minimize unburned hydrocarbons and significantly reduce auto emissions.

MMCs appear to have potential as pistons where high-temperature performance and heat resistance are at a premium and con rods, which need high-temperature strength and stiffness. By combining MMC pistons, piston pins, and con rods, engine designers can reduce the rotating mass in an engine dramatically enough to do away with the weighted balanced shafts required to smoothe out the essentially unbalanced four-cylinder engine. This could amount to a savings of 9.07 kg in weight and $75 in costs. Engine efficiency increases as well, with engineers estimating a 5% improvement to be gained by reducing the rotating mass by one-third. Besides using MMCs in pistons and con rods, Honda has applied these composites in an engine block and cylinder liner for a new model automobile by employing a 12% Al_2O_3 Saffil$_f$ and a 9% Gr$_f$-reinforced cylinder liner in the engine block.

The NASP materials technology program for flight vehicles is being applied to current automobile engines which, according to materials scientists, could reduce the weight of their moving parts by 50%, increasing fuel efficiency and durability.

Candidate TMC components being considered include valves, con rods, crank shafts, and cam shafts. TMCs reinforced with Ti silicides have reportedly been developed for applications requiring high-temperature strength and corrosion resistance such as automotive engines.[80] Designated Ticad, one material is available in the form of granules. Parts may be fabricated by several conventional production methods including casting, forging, P/M, and SHS. Applications include turbocharger turbines, piston heads, con rods, and cylinder liners.

Toyota's MMC diesel engine piston is one of the most successful commercial applications of MMCs worldwide. Toyota started manufacturing them in 1983 and now produces more than 300,000 pistons/month. The total vehicle weight savings is insignificant, but in this application (and in the case of several other engine parts also) a tiny weight savings allows a higher r/min and thus a considerably higher power output.[81,82] The fiber content of the part is 3.5–8.0 vol%.

Koczak, Khatri, Allison, et al.[82] have described the extensive use of DRAs as well as many other fiber-, whisker-, and particulate-reinforced MMCs in automotive and nonautomotive applications in great detail.

Honda Motor Company has developed valve spring retainers made of Al_2O_3, ZrO_2, and SiO_{2p} dispersed in an Al alloy. The MMC valve spring retainers, which were made by a rapid solidification (RS) P/M method, weighed 60% less than conventional retainers while maintaining the same strength, stiffness, and wear resistance. The MMC valve spring retainers reduce the mass and inertia load of the valve system, which permits the use of weaker valve springs and also lessens the amount of resulting valve friction. Because less power is necessary to move valves with MMC retainers, engine rotation is faster and energy loss decreases.[81]

Yamaha Motor Company has employed SiC grain-reinforced Al alloys (instead of Fe-based materials) for shock absorber cylinders on its model TZ250 motorcycle to reduce the weight of the cylinder.

Toyota Motor has used an Al_2O_3-SiO_2 short fiber-reinforced Al alloy part for a crankshaft damper pulley in a 3-L in-line six-cylinder engine for its Crown luxury model car. The fiber content is less than 10% of the total weght of the pulley, and the mass of the pulley has been decreased by 20%. Toyota has thus decreased the mass of the crank damper and pulley while maintaining the necessary strength for the pulley.

Mitsubishi Motor Corporation has developed a version of its 2-ton small truck equipped with an experimental methanol engine that employs SiC_w-reinforced Al alloy pistons. The methanol engine featuring low emissions is, in general, likely to replace diesel engines as one way to solve the current diesel engine's nitrogen oxide and particulate emission problem. A conventional Al alloy piston is often cracked by severe thermal shock cycles because of alternating combustion and fuel injection and vaporization. To reduce the CTE, Mitsubishi employs Al matrix composites for the piston head exposed to the cycling. The SiC_w preform content is 20% of the total weight of the piston head.

Reducing the weight of automobile engine components obviously contributes to greater fuel efficiency. An innovative plastic composite manifold has been successfully fabricated for a Ford engine with 35% Gl_f-reinforced nylon 6/6 and Zytel 70G35 resin.[83]

Another thermoplastic intake manifold has been successfully designed and fabricated for a Cadillac V-8 engine. The manifold has more than 80 fewer components than the previous part, is environmentally friendly and consists of Gl_f-reinforced Ultramid nylon. The nylon 6/6 has been especially formulated to resist engine temperatures and attacks from oil, fuel, and underhood fluids. Additionally, the nylon 6/6 material can be recycled.

NASA Langley Center recently announced the development of a C_f-reinforced C piston that reportedly has excellent resistance to thermal shock and maintains high strength and stiffness at operating temperatures above 1400°C. Compared with Al pistons, the CCC piston is said to operate at higher temperatures, have greater reliability and lower mass, and use leaner fuel-air mixtures that result in better mileage and less pollution. It also has a lower CTE and higher thermal conductivity.

Potential commercial uses include high-performance internal combustion or diesel

TABLE 3.6   Selected Automotive Demonstration Components in Al Matrix Composites[89]

Reinforcement	Component	Property	Benefits	Manufacturer[a]
SiC(p)	Piston	Wear resistance, high strength	Reduced weight	Dural, Martin Marietta, Lanxide
$Al_2O_3(f)^b$	Piston ring groove	Wear resistance	Higher running temperature	Toyota
$Al_2O_3(f)^b$	Piston crown (combustion bowl)	Fatigue resistance, creep	Opportunity to use Al, reduced reciprocating mass	T&N, JPL, Mahle, and others
SiC(p)	Brake rotor, caliper, liner	Wear resistance	Reduced weight	Dural, Lanxide
Fiberfrax	Piston	Wear resistance, high strength	Reduced weight	Zollner
SiC(p)	Driveshaft	Specific stiffness	Reduction of parts and weight	GKN, Dural
SiC(w)	Connecting rod	Specific stiffness and strength, thermal expansion	Reduced reciprocating mass	Nissan
$Al_2O_3(f)^c$	Connecting rod	Specific stiffness and strength, thermal expansion	Reduced reciprocating mass	Dupont, Chrysler
Al_2O_3-SiO_2-C	Cylinder liner	Wear resistance, expansion	Increased life, reduced size	Honda
Gr(p)	Cylinder liner, pistons, bearings	Gall resistance, reduced wear	Increased power output	Associated Engineering, CSIR, IISc
TiC(p)	Piston, connecting rod	Wear, fatigue	Reduced weight and wear	Martin Marietta
Al_2O_3	Valve spring, retainer cam, lifter body	Wear, strength	Reduced weight, increased life	Lanxide

[a]CSIR, Council for Scientific and Industrial Research; IISc, Indian Institute of Science, Bangalore.
[b] Short fibers.
[c] Long fibers.

engines, oilless piston engines, ultralight aircraft engines, and other weight-critical applications.

Reference 84, McConnell,[85] and Kurihara[86] detail the latest developments involving polymer composites in automotive applications, and the latest innovations, and the move to drive cleaner, "greener" electric passenger vehicles, especially the use of C_f/Ep flywheels and trend toward modularity. Kurihara[86] discusses the application of PMCs in automobiles that has been gradually increasing for the last 20 years in Japan. However, his detailed assessment of PMCs and the component materials of mid-sized and compact passenger cars in Japan reveal that considerable work must be done with new materials and new manufacturing methods to overcome the problem of high costs and to produce more attractive automobiles.

Cole and Sherman[87] recently presented a review of the use of lightweight materials in automotive applications in which Al and Mg MMCs were discussed as possible engine block cylinder liners, pistons, and drivetrains and the future of these materials. Mroz[88] and Rohatgi[89] have shown that the potential automotive applications of MMCs, particularly Al matrix composites, are numerous (Table 3.6).

Particulate-reinforced CMCs are being given serious consideration for potential automotive applications. The toughened ZrO_2 in automotive components is very significant, particularly in diesel engine components. In this sector there is a requirement for heat insulation to reduce heat loss from the combustion chamber and to increase engine efficiency. The low thermal conductivity of ZrO_2 could be used to advantage in components such as piston crowns, piston liners, and cylinder liners. The additional wear resistance of the ceramic component should provide extended engine life. Where wear is a major limitation of current materials, such as in the valve train (cam, cam followers, tappet faces, and exhaust valves), ZrO_2 components are well suited to meet the stringent requirements of engine manufacturers.

3.5.4.2 Automotive and truck nonengine components.

Traditionally brake drums have been made of cast iron, but efforts to decrease the weight of cars and particularly to reduce the rotating unsprung mass have tilted the balance in favor of Al-based materials, especially Al matrix composites.

SiC_p MMCs are already used in a number of applications in which wear resistance is a crucial requirement. Duralcan has fabricated brake disks as well as MMC disks which have already been used extensively in U.S. racing cars.

The F3s Al alloy, which has a higher Si content at 10% and is easier to cast than A356 (F3A), has the best high-temperature mechanical properties (brake disks can operate at up to 250°C). The properties of the F3s alloy can be optimized by appropriate heat treatment.

Brake rotors of SiC_p-reinforced MMCs show wear as well as or better than those of cast iron. In fact, Al_2O_3- and SiC-reinforced Al rotors show such high durability that automakers may be able to warrantee the rotor for the car's life. Equally important, MMC rotors weigh only 33–60% as much as conventional parts. All this weight reduction is below the automobile's suspension springs. Cutting unsprung weight reduces vibration, improves handling, and allows a weight reduction in other support structures. Automakers appear willing to pay up to $9 for every kilogram of unsprung weight reduction (Table 3.7).

Prototypes are being prepared, and in Great Britain there are plans to market MMC brake rotors in 1997. The rotor is cast from a 20% SiC/Al composite.[90,91]

A MMC driveshaft has been tested for light vans; it is made of extruded 20% V_f

TABLE 3.7   Comparison Between MMCs and Other Materials

Property	Al 6061	Particulate SiC/6061	Fiber SiC/6061[a]	Polymer Composite[a]
Strength (MPa)	290	290–480	620–120	820–1680
Stiffness (GPa)	70	80–140	130–450	61–224
Specific strength	100	100–170	250–390	630–670
Transverse strength (MPa)	290	290–480	30–170	11–56
Transverse stiffness (GPa)	70	80–140	34–173	3–12
Maximum use (°C)	180	300	300	260
Fracture toughness (MPa $m^{1/2}$	18–35	12–35	N/A	N/A

[a] N/A, Not applicable.

Al_2O_3-reinforced alloy 6061 and has a 30% improvement in specific stiffness over steel and over unreinforced Al. Specific stiffness is the crucial design property for a propshaft, and MMCs offer a significant improvement in stiffness. Ordinary welding techniques can be used to join the shaft yoke caps to the extruded tube.

Kim and his associates[92] have successfully produced screw rotors for compressors, automobile turbo chargers, and pumps. The screw rotors were manufactured with the chopped C_f/Ep composite materials by RTM.

A C composite is the key to an automobile antenna that won't be bent out of shape at the car wash. Stainless steel antennas, which bend relatively easily, are costly to replace. However, like a Gr fishing rod, the composite antenna developed by Polygon bends readily but returns to its original shape when the bending force is released. The material has sufficient flexural strength to withstand repeated bending into a U-shape. When not bent, the 4-mm-diameter antenna stands straight, without any bowing or twisting. A smooth surface finish not only makes the antenna attractive but also improves reception. The C in the composite also does double duty, both providing the necessary conductivity and coloring the antenna black. A retractable Gr antenna is also under development.

Polygon's all-composite strut was designed as a replacement for an Al alloy forging in high-performance production automobiles, where it functions as a connecting link for the upper control arm in the front suspension system. Its performance characteristics exceed those of competitive alloy materials, including 30% weight savings, 20% greater tensile strength, 50% greater compressive strength, and a better than twofold increase in fatigue life (Figure 3.42).

A new concept has been developed and fostered by Rocky Mountain Institute (RMI).[85,93] The *hypercar* concept combines a very light aerodynamically slippery body with a hybrid electric drive. The dramatically lighter, yet crashworthy, body is made possible by changing its basic material from steel to an advanced polymer composite.

Hypercars will weigh three- to fourfold less than today's steel production cars and will shift essentially all structural mass from steel and other metals to polymer composites. Hypercar composite body manufacturing processes are of course fundamentally different for composites than for steel. Hypercars would cut fuel use by 5- to 10-fold, be able to burn almost any liquid or gaseous fuel, and cut conventional air pollution by at least one, and more likely two to three or more orders of magnitude. Hypercars would

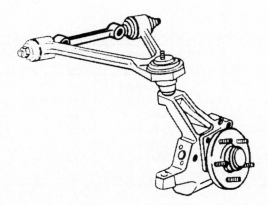

Figure 3.42   All-composite strut.

also produce many indirect environmental benefits. Collectively, their direct and indirect environmental benefits should far outweigh their tendency to encourage car ownership and driving.

The RMI report[94] cites the 1994 "avcar" (average car) as using 8% plastics and composites and weighing a total of 439 kg, whereas the 1998 hypercar uses 43% plastics and composites (21% C_f composites, 5% Gl_f composites, 4% $aramid_f$ composites, and 13% other polymers) and weighs a total of 482 kg. This represents an 86% increase in the use of composites and other plastics. Converting the entire U.S. auto industry to hypercars would increase total U.S. use of polymers by only 3% less than 1 year's normal growth, although it would mean an order-of-magnitude expansion in the composites industry ($10 billion in 1993) and a several hundredfold expansion in the production of C_f.[95]

Two types of driveshafts are under development—one made of MMC and another made of Al stiffened by a thin C_f-reinforced wrapping. Particulate-reinforced MMCs make sense because they can be extruded into long, stiff tubes that require little machining relatively inexpensively. The composite not only reduces vehicle weight but also improves power transfer and dampens vibration.

Other driveshaft tubes have been fabricated of $6061/Al_2O_3/20_p$ with a nominal 100-mm diameter and a 2-mm wall thickness. This material has the following properties: YS of 315 MPa, UTS of 50 MPa, tensile elongation of 2.1%, elastic modulus of 100 GPa, and specific stiffness of 35.3×10^8 mm. It should be noted that the stiffness, which is the primary reason for using this material, is not influenced by heat treatment and is a function of only the volume of Al_2O_3 in the composite. Hence a wide range of strengths, ductilities, and stiffnesses can be produced by varying the mechanical processing, the heat treatment, and the V_f of reinforcement.

To fabricate the Al composite tubes into assemblies, the forged 6061 Al yokes were welded to the ends of the tubes. The gas metallic arc welding (GMAW) process with 5356 Al filler wire was used in standard driveshaft assembly fixtures to produce components for evaluation.

A second generation of composite driveshafts for cars and trucks has now been produced. It features molded thermoplastic end fittings instead of cast iron parts. The improved shaft weighs 2.7 kg versus 5 kg for the first-generation product.

Price and Moulds reported their exhaustive work and findings, claiming that it can be said with some confidence that the necessary adhesives are currently available and that modeling, computational, and practical techniques are now sufficiently developed to allow production line assembly of bonded composite tubes.[96]

C_f/Ep composite driveshafts have been shown to offer substantial weight savings when replacing conventional steel shafts. One-piece Gr composite shafts can replace two-piece steel driveshafts and eliminate the requirement for a central bearing. The Dana Corporation is producing one-piece shafts by pultrusion of Gr_f over a 100-mm Al tube. A layer of Gl fabric between the two prevents galvanic corrosion, and Gl roving is spirally wound over the Gr to hold it in place. The composite shaft is so stiff that it does not whip at normal speeds and can be designed for performances up to at least 10,000 r/min.

Leaf Springs. A number of reports[97–99] have demonstrated the suitability of PMCs for use in major structural components such as leaf springs, driveshafts, and torsion bars. Composite leaf springs, based on continuous Gl_f/Ep composites, have been fitted as standard components on light commercial vehicles. Companies such as General Motors, Renault, Ford, Nissan, and DAF have adopted composite leaf springs of a conventional design. A weight savings in the region of 50% is achieved since a single-leaf composite spring offers the same energy storage function as a multileaf steel spring. The multileaf design configuration eliminated interleaf friction and provided improved ride comfort. Excellent fatigue durability has also been demonstrated.

Mitsubishi Motor Corporation[97] has developed a leaf spring made of fiber-reinforced plastic and plans to use it in small-truck suspensions. The composite springs are assembled with steel springs to form hybrid springs and are fitted to the front and rear suspension of the Mitsubishi 7257.6-kg fighter truck, and Mitsubishi also plans to use them in its small Canter trucks. The leaf spring is made of Gl_f/Ep, is 1 m long, weighs 2.8 kg, and has a spring rate of 59 N/mm. Reportedly this is the first time FRP has been used for a truck leaf spring. Finally, the composite spring has a weight savings advantage of 40 kg per vehicle over the existing multileaf steel springs and has improved ride and handling and a reduced interior noise level.[97]

The Delco Chassis Division of General Motors has been using springs made of composite materials—approximately 70% unidirectional E-Gl and 30% Ep resin—since 1981. The Delco Liteflex tractor spring, for example, is a monoleaf unit weighing only 18.1 kg (including 10.4 kg of composite material). This spring directly replaces the 45.4-kg traditional steel leaf spring, resulting in a 60% reduction in weight.[100]

Daimler Benz has reported the development of a novel design composite spring link. A rigid component that joins the axle support and the wheel carrier, the spring link is the most heavily loaded axle component. The thermoplastic spring link was fabricated in satin-weave C_f with PEI and has successfully withstood dynamic simulation testing equivalent to 2000 km on a poorly surfaced road. The company plans to explore the design and manufacture of a leaf spring link, combines the functions of the helical spring and steel spring link in one component, as illustrated in Figure 3.43.

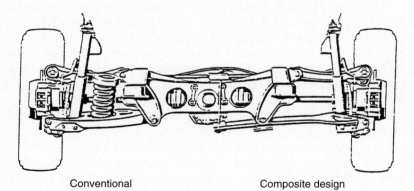

Conventional   Composite design

The composite leaf spring link replaces both the metallic spring link and the helical spring of a conventional rear suspension.

Figure 3.43   Prototype Daimler Benz design for composite leaf spring leak. The composite leaf spring link replaces both the metallic spring and the helical spring of a conventional rear suspension.

Bumper Beams, Radiator Support, Crankshaft Damper Pulley, Structural Beam, and Others.   High-Gl-content bumper beams in several thermoset and thermoplastic polymers are gaining penetration where major weight savings are required.

The radiator support on the new Ford Taurus and Sable models has been selected as the next component to be switched to composite because it allows more economical modular assembly and will proliferate to other high-volume vehicles.[98]

A significant measure of success has been achieved in semistructural and secondary structural parts, such as bumper beams, load floors, seating, and instrument panels (IPs), with such materials as polyester and vinyl ester (SMCs), Gl mat thermoplastics, and urethanes, and with processing techniques like compression molding, RTM, and SRIM.[99]

SMCs with a 40% Gl content are growing in use in structural parts. Most notable is Cambridge's structural beam, which debuted on Ford's 1995 Explorer and Ranger. It has been widely touted as a key advancement in structural design because of its multifunctionality. The system essentially supports the IP, incorporating the steering column support, air distribution system, mounting provision for electronic modules, and routing provisions for IP wiring.

Engineers at Toyota replaced the cast iron hub of a crankshaft damper pulley with an Al matrix composite to reduce weight and engine vibration. The pulley drives the alternator and power steering pump and acts to reduce torsional and bending vibration of the crankshaft significantly. When Al is reinforced with short Al_2O_{3f}, the composite resists deformation and expands less when heated. To fabricate a composite hub and crankshaft pulley, fiber preforms were first made by vacuum forming and then infiltrated with molten Al-Si-Mg alloy, JIS AC4C, or JIS AC8A which also contains Cu and Ni. The hub weight was reduced by 40%, and the crankshaft pulley weight was reduced by about 20%.

Body Panels.   A futuristic El Camino truck (XT-2) features body panels made from an interesting combination of a hybrid polyester-polyurethane resin (Xycon) and Thornel C_f. The composite material adds strength and toughness to the panels while reducing their weight. C_f were chosen for this high-performance car because they are four to five times lighter than steel and 50–100% stronger than fiberglass, and as a result the car can achieve higher speeds on a racetrack while burning less fuel.

Automotive Frames.   Pepin Associates, Inc.,[101] have developed a new polymer matrix composite for highly loaded structures such as automobile frames. The precursor of the composite is a commingled fiber tow, called CD tow, that mixes continuous thermoplastic or other meltable filaments with discontinuous structural filaments such as Gl_f. Recycled materials, including PET from postconsumer soft drink bottles, have been used to produce the thermoplastic filaments.

Compressed Gas Tanks.   The GM Ultralight, a 635-kg four-seat sedan built with a C_f/Ep body achieves 100 mi/gal in highway driving. HDPE liners reinforced with C_f wraps will replace Gl_f-reinforced Al in compressed natural gas (CNG) tanks for powering fleet vehicles. As a fuel, compressed gas has reduced emissions compared to gasoline, and C_f/HDPE (high density polyethylene) liner tanks are lighter, use less fiber, require thinner walls, and hold 20% more gas.[102]

Other engineers[103] are trying to develop a natural gas-fueled vehicle (NGV) that

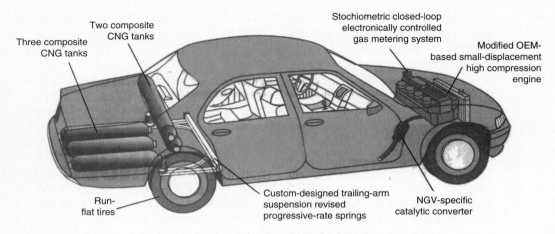

Three composite
CNG tanks

Two composite
CNG tanks

Stochiometric closed-loop
electronically controlled
gas metering system

Modified OEM-
based small-displacement
high compression
engine

Run-
flat tires

Custom-designed trailing-arm
suspension revised
progressive-rate springs

NGV-specific
catalytic converter

Figure 3.44   Advanced natural gas vehicle.[103]

matches or exceeds gasoline-powered cars in range, driveability, performance, and safety to meet California's strict standards for ultralow-emission vehicles. Researchers will focus on improving the range of a Geo Prism specially built for ultralow emissions that achieved a milestone 315-mi driving range between fueling, beating most of the 30,000 NGVs now on the road by a factor of 2.

To allow the Prism to carry more fuel, engineers are working to design and build an integrated fuel storage system that costs less and is lighter and safer than current designs. The system will feature thin-walled high-density polyethylene cylinders overwrapped with C_f and encapsulated in energy-absorbing foam. It will combine all valves, relief devices, and interconnected lines in an integrated package (Figure 3.44).[102]

The goal is to improve a storage system design that doubles the 125- to 150-mile NGV driving range and narrows by 50% the price gap between NGVs and gasoline-powered vehicles. These are not mutually exclusive goals. Gaining broader NGV market acceptance requires technology advances, cost reductions, and an increased driving range as part of a total package of improvements.

Brake Calipers. The 2618 Al alloy composite has a high modulus which is retained at elevated temperatures. This MMC costs less than higher-strength Al/Li and 7000 series composites, and so it serves cost-sensitive applications such as high-performance automotive brake calipers where high stiffness at temperature reduces distortion and improves braking performance. Other uses include con rods and pistons because the reduced CTE and temperature resistance improve engine efficiency.[104]

The Future. While the driving force behind the growth of SMCs in the automotive industry has been composite applications in exterior body panels, there has been an increasing use of SMCs for structural parts, too. A good example is the 1995 new-generation Chrysler Caravan and Voyager, which has a one-piece structural module called a plenum or windshield wiper motor housing. This is a light, strong replacement for several parts made with a variety of materials. This parts consolidation, coupled with low tooling costs, is a good example of the inherent advantages of composites as the material of choice.[105]

Krolewski and Busch[106] have developed cost models to evaluate five applications representative of different niches in the auto industry and the competitive position of several composite technologies relative to steel stamping.

They found that RTM and steel stamping dominated low- and high-production volumes, respectively, because of the low fixed costs for RTM and the low variable costs for steel stamping. Stampable thermoplastic composites (STCs) offer a lightweight low-cost option for structural parts at moderate production volumes. SMCs are the low-cost alternative for nonstructural parts at moderate volumes, particularly where good surface quality is required. SRIM competes most closely with STC but suffers from high preform costs. For the future, the liquid molding processes, RTM and SRIM, have the greatest potential for improvement by reducing the cost of structural preforms.

Chavka and Johnson[107] examined the critical preforming issues for large automotive structures and found that there were several technological barriers that must be overcome if structural composite materials are going to be utilized to a greater extent in the automotive industry. First, as with any engineering material, a valid knowledge base is required. Composite materials in particular must be further characterized with respect to their response at extremes, that is, high temperatures, low temperatures, and hostile environments. In addition, a knowledge base with respect to manufacturing data must be established. Such relationships as performance and physical properties and their variation with respect to changes in the manufacturing process are critical. More cost-effective manufacturing techniques should be developed. The overall cost of manufacturing a composite structure must be competitive with the overall cost of alternative materials. Improved design tools allow the simultaneous design of part geometry, part properties, and the manufacturing process. An optimized structure can be achieved by the simultaneous design of the part, its performance, and its manufacture, thereby producing a cost-effective composite structure. Process information must be improved and techniques developed not only to control the composite manufacturing process but also to control the quality of composites during and after manufacture without resorting to destructive testing processes.

3.5.5 Rail and Trains

3.5.5.1 All-composite boxcars.
An insulated composite boxcar made from Gl_f-reinforced composites has been designed and built using the SCRIMP process, a resin infusion method said to be especially effective in producing large structural parts.

This boxcar has fewer joints compared to steel boxcars, thus thermal leak paths that could jeopardize its ability to provide optimal energy-saving insulation are eliminted. In fact, the insulating capabilities of composite boxcars allow products to travel for longer periods and still retain proper temperature upon delivery, a dramatic improvement in insulating performance. The measured heat loss in terms of the heat transfer rate has been reduced by 60–70% by using composite materials.

The composite boxcar allows a railroad to enter new markets. Food and drink products that used to go by refrigerated boxcar or by truck can now be transported in an insulated composite boxcar because of its ability to maintain lower temperatures longer.[108]

3.5.5.2 Third-rail bracket.   A cast iron bracket supports the third rail of several electrified rail lines. Bolted to an extended railroad tie, it supports an insulator which in turn supports the contact rail.

Engineers traced the cause of failure of some cast iron brackets to corrosion caused by salty water from the street above. They also suspected the cast iron brackets of undesirable and possibly dangerous current leakage even though theoretically they are not in the electromagnetic circuit.

As an alternative to cast iron, engineers adapted composite molding compound technology to produce support brackets. The material was a Gl-reinforced vinyl ester thermoset composite. In addition to the necessary structural strength and corrosion resistance, it supplied extra insulation not provided by the cast iron.

Composite brackets cost almost 12% more than cast iron brackets. Self-extinguishing and impervious to the effects of arcing and chemicals, the material is UV-stable. The brackets weigh about one-third less than cast iron brackets, and their minimum life expectancy is 25 years.

3.5.5.3 Shipping containers.   A joint program of the Union Pacific Railroad and Goldsworthy Engineering Division[109] has engineered techniques for building large cargo containers using engineered pultrusions. The developers adapted a system long used by freight haulers, containerization, to transport vehicles. At the assembly plant, the vehicles are loaded and sealed in the containers or modules. They need not be handled again until they reach their final destination—the modules are simply lifted from the railcar to a truck trailer at the transfer point.

A special 3352.8-cm, articulated "spine" car has been developed to carry the modules—six per car, arranged in two three-high stacks. The fiberglass reinforcement, which averages 55% by weight, varies with the type of profile and the load it must sustain. Uni-axial roving and mat, sometimes supplemented by stitched fabrics, are used in various combinations to cope with a complex assortment of axial and off-axis loads.

Currently one spine car and a set of modules have been delivered for use in market-testing the concept to the automotive industry and refining the design if necessary.

3.5.5.4 Train brake disks.   Several research institutes and railway companies in Japan have been developing brake disks of Al_2O_3- and/or SiC_p dispersed Al alloy matrix composites (AMCs) for bullet trains. Reducing train weight is an effective way to increase train speed and save energy. The MMC hardens the disks, prevents heat seizure, and improves their wear resistance. The weight of the AMC brake disk decreases to half that of conventional disks made of Ni-Cr-Mo cast iron. The total weight of a 16-car bullet train could be decreased by approximately 7257.6 kg by using these Al brake disks. Laboratory experiments and field tests with an actual bullet train are being conducted to validate their reliability and durability.[81]

3.5.5.5 High-speed trains (TGV).   Extensive research on double-deck models for TGV is now under way in France. Sandwich panels of CFRP skins and honeycomb cores have been adopted, with which the projected speed of 350 km/h is being approached, resembling an airplane without wings. A German program for a magnetic levitation (maglev) high-speed train from Berlin to Hamburg has been initiated, and it is considered a possible application for lightweight composites.[10,110]

Burg and Loud[111] reported that high-speed rail lines represent a sizable but long-range potential new market for suppliers of advanced composites. There are two basic approaches—high-speed steel wheel trains similar to those operating in Europe and Japan and maglev systems that promise even higher speeds and may require greater use of advanced composites to be proven technically feasible.

The problems facing U.S. engineers and proponents of high-speed trains and/or magnetic levitation involve how advanced composites might be used on future high-speed trains that ride on steel rails. All agree that the rails will have to remain a high grade of steel and be much more precisely aligned, but it is possible that the rail ties may eventually be made of some new reinforced concrete that is toughened and strengthened with C or aramid. Likewise, railway bridges and their support structures offer many opportunities for using either fiber-reinforced concrete or high-strength composite structural components (to be discussed further later in the chapter).

A train with CCC brakes (not unlike those now in widespread use in all newer transport aircraft) to replace the steel brakes baselined on the TGV trains is currently being tested in France. Test results have been encouraging, and the prospect of not only new railcars, being equipped with CCC brakes but also supplying a huge potential retrofit market.

Other undercarriage components are also good candidates for the use of advanced composites, including the bogeys, possibly the wheels, various structural members such as supports for the flooring, doors, and even the outer panels of the cars. Reduced weight will pay dividends in fuel and energy savings, greater acceleration, and reduced braking requirements. Displays of composite usage have shown cross sections of a railcar to illustrate the variety of composites now being used to make wall panels, overhead bins, window reveals, and so on, just as in the main cabin of an Airbus or 767.

One idea[111] for maglev is to employ composites for mounting the over 20 superconducting magnets planned for each car, and because of the high loading combined with a minimum-weight requirement, these mounting assemblies can use C_f composites. Their design also calls for a guideway that is nonmagnetic and nonelectrically conductive, and so S-2 Gl-reinforced composites are under consideration. Total maglev system costs will be dominated by the guideway construction cost, involving pylon spacing, type and amount of structural materials, and linear synchronous motor stator windings and levitation strips. A maglev fully loaded car carrying 100 passengers will weigh 28,930 kg if made using composites wherever practical. (Other approaches employing fewer composites may result in a loaded car weight of more than 18,144 kg.)

3.5.6 Missiles and Space

3.5.6.1 Missiles

Missile Fin.   The 2014 Al alloy spray-formed SiC reinforcement in spray-formed Al composites gives greater stiffness. Increased Zn and Li levels increase the strength and lower the density. By comparison, conventional MMC alloys have high stiffness, but with a density penalty.[112]

Missile Nose Cone.   Researchers at Stevens Institute of Technology have designed a plastic composite nose cone for a shoulder-fixed missile for the U.S. Marines that should cost $40 each piece to manufacture versus $400 for the currently used Al cone.

Tactical Guided Missile Launcher (GML).   Boyer, Talmy, Powers, et al.[113] were tasked by the U.S. Navy to develop new ablators with enhanced erosion resistance for use on GMLs. They reported several general conclusions about the best constituents for use in an ablator exposed to solid rocket exhausts heavily laden with Al_2O_{3p}.

Their data indicated that the composition (melting point) of the reinforcement plays an important role in the magnitude of the ablation; silica- and quartz-reinforced ablators outperformed those containing Gl_f. The reinforcement geometry is also important; for example, ablators containing chopped rovings and chopped squares performed the best, those with continuous-filament mat performed well, and those with two-dimensional weaves tended to delaminate.

3.5.6.2 Space.

Space vehicles and satellites operate in a difficult environment, being subjected to wide variations in temperature, exposure to intensive UV radiation, atomic oxygen erosion, and outgassing. Traditionally, to satisfy the performance demands of space, Al and Ti alloys have been used. However, C_f/Ep composites have gained acceptance in a number of areas because of the weight savings to be gained, the high specific stiffness, and the ability to achieve a near-zero CTE. Major space application of composites are in satellites and launchers. They are being used in the fabrication of satellite dish reflectors, deployment arms, struts, and solar array panels. C_f metering trusses with a near-zero CTE have been designed by Boeing Company to maintain the separation and alignment of the primary and secondary mirrors on the Hubble telescope. Aerospatiale, (France) and Fokker (Germany) are using C_f composites in rigid arrays for various European satellite projects. Aerospatiale has also developed a novel unfurlable truss structure, based on a C_f composite, which may be used in the future for large-diameter antennas and for space stations.

Hercules Aerospace and Orbital Sciences Corporation have developed an all-composite air-launched space booster, Pegasus. C_f-reinforced structures are used in the fabrication of the wing, which has a 6.7-m span, the tail fin, and the fuselage, which is 15 m long and 1.3 m in diameter. The fuselage incorporates three filament-wound motor cases.

Lockheed Missiles and Space Corporation has evaluated a C_f/Ep composite that exhibits both high strength and high stiffness. The composite is based on Toray's M40J C_f and Hexcel's F584 toughened Ep resin. A laminated structure used for a spacecraft is reported to have a tensile strength of 2558 MPa and a tensile modulus of 236 GPa.

Several divisions of Courtaulds, Ltd., have collaborated on the development of a C_f/PEEK composite for the main structural material in a scientific space satellite. The company reports that the thermoplastic composite was selected in preference to Al because of the reduction in weight that is achieved: it is estimated that the launch cost for each kilogram put into space is on the order of $120,000. High stability of the satellite system has been achieved by designing a composite with a very high stiffness and a very low CTE. A thermoplastic composite was selected in preference to a thermoset composite because of the greater ease of fabrication and the reduced risk of outgassing in space.

According to Eaton and Slachmuylders of European Space Agency (ESA),[114] it may take about 5–10 years to build up sufficient confidence in a material for aerospace application. The space industry requires only a small amount of these materials, which are likely to be costly to produce, however, one cannot afford overly conservative requirements. For many applications the space environment is still more benign than those of the airframe world. This should allow space to be among the early demonstrators of new materials. Full advantage should be taken of the fewer load cycles experienced by space items when compared with those in aircraft applications. Opportunities for between-flight inspections should be exploited. Many of the emerging materials are likely to lead to components whose properties will be dependent on the design and the manufacturing process. This all adds emphasis to the need for correlated judgments involving the design, manufacture, test, and inspection stages.

With cost in mind engineers at LANL and Composite Optics, Inc.,[115] have developed a process for making strong, lightweight composite structures that snap together like a child's model aircraft. The researchers say that the technique is two-thirds faster and 60% less expensive than other fabrication design methods. The first all-composite spacecraft structure, the fast on-orbit recording of transient events (FORTE) satellite, was built from flat laminates of Gr/Ep composite which were precut using a high-pressure water jet. The parts, many of which are modular and interchangeable, were easily assembled by a built-in, self-fixturing technique not unlike the mortise-and-tenon joining used by woodworkers. Composite structures have zero thermal growth, in contrast to Al structures which expand and shrink as temperatures fluctuate.

The end result is a 40.8-kg framework that will provide space validation of technologies that will contribute directly to future nonproliferation satellite systems. This lightweight satellite will explore new techniques for detecting covert nuclear weapons tests while demonstrating the feasibility of using advanced, low-cost composite structures in small spacecraft.[116]

Electromagnetic pulse (EMP) detection experiments to be carried out on board the FORTE spacecraft will rely on advanced radiofrequency receivers and optical sensors that have broader applications beyond their original nonproliferation objectives. If these experiments are successful, advanced EMP detectors could be installed on global positioning system (GPS) satellites. It is also possible to deliver to orbit 22.7 more kilograms of payload than a similar-size satellite built with conventional materials.

Another example is the use of a honeycomb combination of materials in the international space station's payload racks. The station will have 55 of these racks to hold experiments and store equipment. By engineering and manufacturing the racks from composites rather than traditional Al, the weight of each rack was reduced by 27.2 kg. It will therefore require less energy to carry the racks into space.

Other space shuttle flights have carried a variety of experiments into orbit. For example, on the second international microgravity laboratory (IML-2) flight, scientists solidified five samples to study ceramic and metallic composites, semiconductor alloys, and liquid-phase sintering, a process for combining dissimilar metals using heat and pressure to join them without reaching the melting point of one or both metals. IML-2 scientists have also studied how the shape and composition of containers affect the solidification of liquids by performing containerless processing of liquids and other investigations.[117]

Numerous simulation computer programs and data analyses have been performed on space structure from launch vehicles, rocket motor cases, and nozzles to orbiting laboratories.[118–129]

Table 3.8 outlines many proposed applications of composites in space, while several other noteworthy composite applications are as follows.

TABLE 3.8   Proposed Composites in Spacecraft[130]

Spacecraft	Mission, Orbit, and Flight Life	Composite Structure	Materials
UARS	Research satellite, focus on stratosphere and mesosphere; LEO, 3 years	Truss with titanium end fittings	Carbon-epoxy using Ferro HMS/CE 339 prepreg
Mars observer	Planetary observer satellite studying Mars surface, interior, atmosphere and magnetic field; interplanetary, 5 years	High-gain antenna boom and struts, antenna reflector	ICI Fiberite using P100/7714-A pitch-based material, Amoco P75/1962 prepreg
EOS	Six massive space platforms with up to 20 instruments monitoring Earth; polar LEO, 5–10 years	Optical bench truss	ICI Fiberite using UHMS/7714-A with titanium fittings
AXAF	Third of great observatories, will study x-ray radiation; LEO, 15 years	Optical bench truss	Not specified
SIRTF	Fourth great observatory, will study infrared, probe young galaxies and search for brown dwarfs; HEO, 5 years	Suspension system for inner helium tank	Not specified
Cassini	Saturn probe satellite; interplanetary, 12 years	Struts on science booms	Carbon-epoxy with aluminum end fittings
CRAF	Asteroid rendezvous satellite; interplanetary, 12 years		
SDI	Space-based satellite defense system; LEO, mission-dependent	Buses, platforms, sunshade, truss for NPBSE	Carbon fiber-PEEK
Tech Stars	Small satellite, multimission configurations; LEO, mission-dependent	Trusses, buses, benches, solar array panels	Not specified
ALEXIS	Small satellite to map and monitor low-energy x-ray sources; LEO, 3 years	Solar paddles	Carbon-epoxy using Hercules Magnamite 3502 prepreg
ASSTD	Small satellite multimission configuration; LEO, MEO, HEO, GEO, 3–5 years	Primary bus, tanks, solar array, launch vehicle adaptor, spectronic bays	Carbon-epoxy sheets over aluminum, ARALL

- In the 3D C_f-reinforced C component for the space shuttle's new advanced solid rocket motor (ASRM), exhaust gases inside the nozzle are at temperatures in excess of 2760°C when they exit the rocket. The CCC component measures approximately 243.84 cm in diameter, being the largest CCC component of its kind ever made. CCC composite material reduces component degradation caused by erosion during the critical burn stage, thereby improving the rocket's reliability and performance.

- A 40% weight reduction and improved performance are the benefits of a CMC nozzle developed for the third stage of the European satellite-launching Ariane rocket. Able to withstand a calculated nozzle wall temperature of 1538°C, the CMC nozzle requires no cooling, unlike the hydrogen-cooled metal tube assembly currently in place. With all hydrogen burned as fuel, combustion efficiency and rocket power are increased.

     The nozzle is fabricated by SEP (France) using the Novoltex/SiC 3D preform infiltration system. To make the nozzle, a wooden mandrel is wound with 7.62-cm strips of woven C_f tape and needle-punched to entangle the layers. In a proprietary process, the preform is exposed to methyltrichlorosilane ($MeSiCl_3$), which decomposes into SiC and HCl. The SiC deposits on the C_f preform, forming the ceramic matrix.

- Development efforts at the University of Illinois have disclosed a C/BN composite material with properties superior to these of the CCC now used in aircraft brakes and spacecraft heat shields. The material, made from C_f in a BN matrix, can be created in a single baking that takes less than 2 days. The resulting composite can withstand temperatures roughly twice as high as CCC before oxidizing.

- The high-gain antenna (HGA) on the Galileo spacecraft en route to Jupiter has 18 Gr/Ep ribs that open like an umbrella, stretching a gold-plated Mo wire mesh between them. The ribs are covered with thermal blankets attached with double-sided adhesive tape.

- Jensen and Blair[124] demonstrated that innovative new material forms combining hydrophobic thermoplastic resins with ultrahigh-modulus (UHM) C_f could be successfully manufactured and subsequently consolidated into quasi-isotropic and unidirectional laminates using newly developed manufacturing methods that allow layup methods analogous to Ep composite processing and autoclave consolidation. The success of these techniques was verified by determination of the physical, mechanical, thermal, and moisture properties of several material systems. UHM C_f-reinforced thermoplastics can be produced in a variety of forms and can be formed and consolidated to meet the demanding needs of newer spacecraft design.

- Advanced composite materials traditionally have not been used for major structures on expendable launch vehicles. Weight reduction is important on these vehicles, but the superior performance of composites has not been attractive enough to justify the projected development and qualification costs necessary to replace current metallic designs. Recent mission models, however, are pushing existing launch vehicles near the limits of their performance capabilities, emphasizing the need for development of enhanced and/or larger vehicles to provide ensured access to space at affordable costs.[130]

     Robinson, Charette, and Leonard[131] reported studies describing development of lightweight, low-cost composite structures for the advanced launch system (ALS) that also provide viable technology options for existing expendable launch vehicles such as the Delta, Titan, and Atlas. They have indicated that because of their lower manufacturing costs and system cost benefits, advanced composite fairings and intertanks will help provide the affordable, reliable access to space that will be required in the coming decades.

- Hercules, Inc., fabricated a Gr/Ep truss structure that houses the corrective optics for the Hubble space telescope (HST).[132] The truss structure, called an *optical bench,* provides the stable platform required to maintain precise alignment for the corrective optics and holds this alignment throughout the wide day and night temperature ranges experienced in space.

According to Hercules, "The relatively small CTE of Gr material makes it ideal for this type of space structure application."

Finally a number of launch vehicles, motors, and interstage adaptors have been successfully fabricated of composite materials and launched into space. Among them are the following.

- *Pegasus XL vehicle.* Filament-wound IM7 C_f/Ep rocket motors and all-composite interstage adaptors between the motors hav been fabricated. The delta-shaped wing, three external fins, and aerodynamic wing and body fillet are made of C_f/Ep over a foam core, and the payload fairing on standard and XL Pegasus vehicles is C_f/Ep over honeycomb.
- Another family of Lockheed launch vehicles (LLVs) is expected to launch a 136.1-kg communications satellite, and the launcher uses 1360.8 kg of C_f/Ep composites in the solid rocket motor (SRM) case. Toray T1000$_f$ is used in the filament-wound case. The second stage is a filament-wound aramid-Ep SRM using 204.1 kg of aramid material.
- The first of NASA's science missions[133] to be launched into a low earth orbit in 1998 will be the far-ultraviolet spectroscopic explorer (FUSE) satellite. FUSE will make spectral observations at far-ultraviolet wavelengths (between 90 and 120 nm) to detect certain chemical elements, which in turn can help determine the evolution of galaxies and solar systems.

    Central to the FUSE instrument structure is a composite panel box surrounded by a truss of composite tubes (Figure 3.45). This 170-kg structure provides primary load-bearing paths, metering capability for the optical components, and space for instrument avionics. Square composite tubes are made from T50 C_f and 954-2A cyanate ester resin. The panels, which act as system optical benches and contain metallic inserts to provide attachment points, utilize the same resin system with Toray's M50J C_f in face sheets over Al honeycomb.

3.5.7 Marine

Many characteristics of composite materials make them well suited for shipboard applications. They can provide affordable design alternatives to conventional shipboard materials, such as steel and Al, and improve upon critical areas of ship design—weight, radar and sonar performance, stealth, and durability. Table 3.9 depicts how various composite properties contribute to specific marine needs. The importance of the advantages they offer varies with each application and type of ship.

Cost-effectiveness is highlighted in Table 3.9 because it is a primary driver of any marine design. Competitiveness with foreign markets and changing priorities in the defense budget have dictated this. Until recently, a focus on the raw materials cost of composites has limited their use. Today there is increased recognition of potential cost savings in manufacturing, installation, maintenance, and fuel that can far outweigh the initial material cost.[134]

Weight is important in ship design because it affects ship stability, speed, depth, and payload (cargo and weapons) capabilities. The low density of composite materials typically yields component designs that are 25–50% lighter than steel designs and up to 25% lighter than Al designs. Reduced weight produces savings in construction, operation, and maintenance.

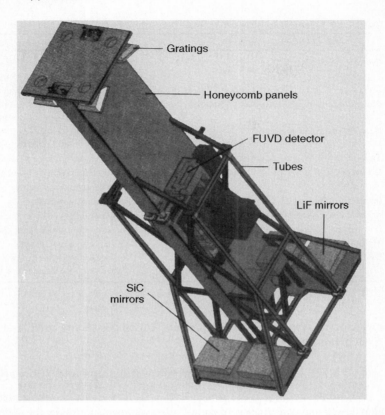

Figure 3.45 Dimensional stability and low CTE of composite panels and tubes in the FUSE spacecraft help decrease the risk of resonance at launch and provide precision observation accuracy.

Stealth and radar and sonar performance are factors that are especially critical in naval applications. The survivability of naval ships and, in particular, submarines depends upon these factors. Acoustic flow paths, magnetic signature, radar reflection, and radiated noise are some factors that determine stealth or the signature of a ship. Composites can easily provide affordable hydrodynamic shapes to streamline flow, and their nonmagnetic characteristics can reduce the detectable magnetic fields produced. Both their inherent damping property and the ability to tailor the stiffness of a component help in damping vibrations, thereby reducing the amount of noise generated. The nonreflective property reduces the radar cross section, while the ship's ability to detect other ships, submarines, or navigational obstacles, is improved. Composites used over radar and sonar equipment can be tailored so that they will not interfere with the equipment's performance. These properties are important in commercial shipping and recreational boating because they enhance comfort, reduce noise and vibration, and improve performance.

Durability has long been a concern of the marine industry. Steel corrodes, and many hours of labor and a significant amount of time in dock are spent on painting and

TABLE 3.9 Composite Properties and Their Advantages in Marine Use[134]

Composite Property	Advantage to Marine Use
Corrosion resistance	Longer life of component,[a] Reduced maintenance[a]
Lightweight	Greater payload capability,[a] increased depth, higher speeds, and easier handling and installation[a]
Monolithic seamless construction of complex shapes	Easier manufacturing of complex shapes,[a] consolidation of parts,[a] signature reduction
Near-net shape and good finish	Reduced need for secondary machining,[a] reduced material waste,[a] reduced painting needed[a]
Tailorability of design properties	Improved performance of component
Nonmagnetic	Signature reduction, reduced galvanic corrosion
Nonreflective	Reduced radar cross section
Inherently damping	Radiated noise reduction
Radar and acoustically transparent	Improved radar and sonar performance
Low thermal conductivity	Improved fire containment, good insulator[a]
Multiple domestic sources	Availability of raw materials
Design cascading effect	Improved performance of one component can reduce the size of or eliminate other system components

[a] Cost-effective.

replacing or repairing steel components. If composites are used, a ship will be able to spend more time at sea.

3.5.7.1 Commercial and recreational ships and submarines

Composite Ship Hulls and Decks. At first glance, it seems practical or useful for ship design and construction of composite components to follow the developments made by the aerospace community. However, advancements in aerospace composites resulted from focusing directly on the specific design requirements for aerospace applications. Ship design and shipbuilding have a different set of criteria and priorities than aircraft and need to examine different types of material systems, testing, and manufacturing methods.

The composite material most often used on marine vessels is glass-reinforced plastic (GRP) with a base resin system of polyester or Ep. Applications have included composite hulls and decks for small craft less than 45 m in length and large structures such as deckhouses, masts, fairwaters, and bow domes for larger ships and submarines. Gibbs and Cox[135] conducted a feasibility study of a GRP cargo ship with a length of 143 m. The conclusions of this study indicated that the design and fabrication of this ship were technically feasible with current technology but not practical because of unresolved questions concerning long-term durability, stiffness requirements, and fire safety. Significant incentives are required to justify a large GRP hull. Ships that perform unique jobs, such as minehunters and corrosive or cryogenic cargo ships, and ships 46–76 m in length appear to be better candidates for GRP hull construction. In some cases, major structural elements such as deckhouses and bow modules look promising for composite construction. Uses to date have generally followed the results of the Gibbs and Cox study because the unresolved questions posed in that study still exist. Until these problems are addressed and cost-effective manufacturing of large structures is developed, large GRP ship hulls may not be built.[134]

Composite material technology developments in the area of recreational boats have come the closest to matching the advances made for aircraft. Composite use has soared in the recreational marine industry because of economic and operational factors different from those involved in commercial and naval shipbuilding. Boat manufacturers began using composites with designs such as the 8.5 m Trilon, the 12.2-m Block Island, and the 4.3 m Sunfish.[135] These early designs were modifications of wood construction designs, providing cost advantages due to mass production and reduced maintenance over their service lives.

Since the early 1970s boat builders and designers have used more advanced construction techniques and materials for racing boats primarily to improve performance. An example is the new class of vessels created for the America's Cup competition which are 30% lighter, carry 40% more sail area, are 3 m longer, are 1.2 m wider, and are significantly faster than the Al vessels they replace. *Stars and Stripes,* is one of these boats and was used in the 1992 and 1995 America's Cup races. It has a prepreg C/Ep hull, deck, and mast for improved thickness uniformity and reduced weight.[136] From the experience gained, design and manufacturing techniques for lower-performance craft can be developed.

Advanced composites have found applications in specialist marine craft where both reduced weight and the highest performance are critical requirements, such as racing yachts, catamarans, canoes, kayaks, and hydrofoils. Many applications involve complex stress requirements to be met and demand combinations of high axial and transverse strength with high torsional and buckling stiffness. For these applications hybrid composites are often found to offer the best solution. In the last 5 years most of the yachts in the Whitbread Round the World Race were of composite construction. One of the boats used aramid-C hybrid composites in the outer-skin layups because they provide impact properties as good as those of 100% Kv laminates while retaining the good stiffness and compressive mechanical properties of 100% C composites.

A novel hydrofoil system for providing lift at the rear end of a boat has been developed in Italy in collaboration with Dornier of Germany. The system incorporates stabilizer wings fabricated in C_f-reinforced plastics. A composite construction was selected on the basis of its low weight, good dynamic behavior, and corrosion resistance. The wings, which are sickle-shaped, are multispar structures with multidirectional C_f-reinforced skins having a maximum thickness of 10 mm in the root area. It is reported that the stabilizers have demonstrated improved stability and controllability in rough seas and at high speeds.

Other applications utilize C_f composites in the manufacture of special features such as keels, rudders, and masts, especially composite masts for yachts. The market for such masts is reported to be racing multihulls and ultralight displacement boats.

Composite hulls for fishing trawlers have been in use for 35 years, and approximately 50% of the boats in the international commercial fishing fleet currently have GRP hulls, mostly small hulls less than 25 m[135]; in Japan 60% of trawlers up to 45 m in length have composite hulls. Composite construction enables fishing trawlers to carry more payload, have cleaner fishing holds, and spend less time at the dock for maintenance and repairs.[137]

Many foreign countries, especially Norway, have utilized GRP construction for passenger ferries to save on fuel, increase passenger capacity, and improve top speeds. In 1986 Norway built a 27-m catamaran for use as a 184-passenger commuter ferry. It is made of PVC foam-cored glass-reinforced polyester built on a timber skeleton mold. Italy built its Italcraft with an aramid-GRP hull of variable geometry. This ferry is 21 m long, and the use of a composite hull has resulted in 40% fuel savings and an improved top speed of 50 knots.[138]

Applications of composite materials on cargo ships and tankers have not included hull construction but have focused on large structural components such as deckhouses, false stacks, and masts, which are cost-effective. Composites are lighter and can be cheaper and more fire-resistant than conventional material. Also, the stiffness of the composite material can be tailored to reduce harmful hull-superstructure interaction. Composites improve the durability of topside structures in the marine environment as well as ship stability. Cargo ship lines have realized these advantages and incorporated GRP false stacks on their container ships.[139]

Additionally they have also used GRP piping in the ballast tanks of their container ships.[135] The good corrosion resistance and availability of standard-size pipe shapes make composites cost-effective in the corrosive environment of the ballast tanks. Also, the good corrosion and chemical resistance of composites make them candidate materials for fuel and water storage tanks. A variety of other applications, such as propulsion components, bow modules, and stern structures, are well suited for future composite use.

Although Gl-reinforced materials dominate boat hull construction, lighter, stronger, and stiffer C_f-reinforced plastics have also been evaluated for use in such applications. The emerging view is that C_f composites are too susceptible to impact damage to be used in sailboat hulls. Evaluations of hybrid composites containing both Gl and C_f, however, have indicated improved strength, stiffness, and impact resistance when used in Gl-C-Gl sandwich configurations.[140]

Submarines.   Composites have been used extensively in small submersibles for deep-sea research, investigations, and inspection. Unmanned submersibles making research observations have operated at depths of 458 m, and the French developed a manned submersible that could perform at depths of up to 1600 m for inspection of the *Titanic* wreckage site on the ocean floor. Many unmanned submersibles have also been developed for offshore industries. Complex geometries, easily manufactured with composites, are needed to perform inspection, maintenance, and repairs.

3.5.7.2 Military ships and submarines

Minehunters.   Composites are used extensively in minehunters. The need for a low magnetic signature and a cost that is competitive with that of the traditional material, wood, make composites an obvious choice. Sixty-meter minehunters have been built and have successfully passed underwater explosion tests to verify their design configuration, materials, and structural details.

Surface Ships, Submersibles, and Submarines.   Warships, carriers, and other large naval surface ships have only a limited composite content. Examples of applications include radomes, reflectors, masts, and deckhouses.

Deep submergence rescue vehicles (DSRVs), which perform deep submergence studies and mapping of oceanic parameters, have composites in their hydrodynamic fairings and control surfaces and in a bow with integral thruster tubes.

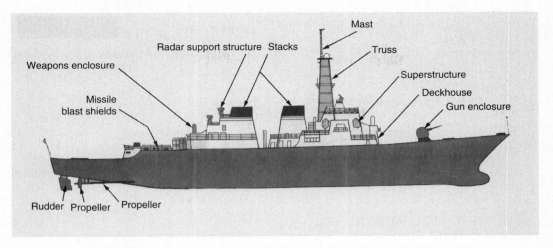

Figure 3.46 Major potential structural applications of composites on a large U.S. Navy surface ship. Others include boat davits, ventilation intakes and ducting, radomes, life rails, hull and deck machinery, deck drains, machinery foundations, ladders, hatches, lifeboats, and nonstructural partitions and false decks. (*Source:* NRC conference report. Courtesy of J. A. Corrado, David Taylor Research Center.)[139]

The use of composite materials in submarines has been limited to external, nonstructural, nonpressure hull applications such as fairings and acoustic windows. Fairings have been used either to cover an opening through the hull or to mold around a protruding object to provide favorable hydrodynamic flow. They require a surface that either matches the curvature of the hull or has a desirable hydrodynamic shape. Fairing contours are more economically produced using composites than traditional metals. Composite acoustic windows can significantly improve the performance of sonar equipment by providing acoustic transparency. Some typical GRP fairings include mast fairings, array fairings, and access covers.

Future naval applications are foreseen for masts, deckhouses, foundations, resilient mounts, ventilation ducts, ship stacks, propellers, propulsion shafting, gratings, decks, and piping (Figures 3.46 and 3. 47). For example, composite propulsion shafts are being developed to replace steel shafts that in some cases can make up 2% of a ship's total weight. A composite propulsion shaft intersperses Gr/Ep and Gl/Ep in a 10-cm-thick section, resulting in up to an 80% weight reduction for the steel shaft. The composite shaft meets all design requirements while improving corrosion resistance, bearing loads, magnetic signature, and fatigue resistance.[135–137, 139–140]

3.5.7.3 Driveshafts, pumps, and cable

Driveshafts.   A hybrid composite consisting of C, Gl, and aramid$_f$ is being produced in Sweden for the Swedish Royal Navy. The composite, it is claimed, can reduce the weight of driveshafts, rudder shafts, and rods up to 70% and lower installation and maintenance costs. The material is noncorrosive, nonmagnetic, and nonconductive and has higher damage tolerance and reduced vibration. The material is also lightweight, re-

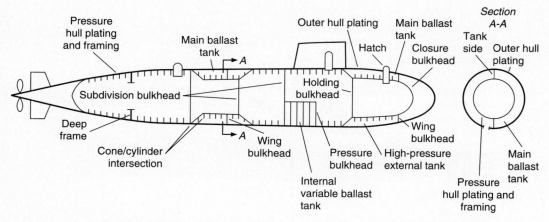

Figure 3.47 Schematic of a submarine hull showing potential advanced composite applications. "It is feasible to build a submarine entirely out of composites," says Rear Admiral T. W. Evans, director, Advanced Submarine R&D Office, Naval Sea Systems Comand. (*Source:* NRC conference briefing, "Composites R&D for Naval Ship Applications," by I. L. Caplan and J. A. Corrado, David Taylor Research Center.)[139]

ducing a traditional metal shaft from 6350.4 kg to 399.2 kg. The new composite can be found in an experimental *stealth* vessel, and after 4 years of extensive testing, it shows no sign of fatigue.

Pumps. PMC pumps are being used by large aquariums, such as the National Aquarium in Baltimore, Maryland, so that dissolved metals cannot affect marine life. Since its opening in 1981, the National Aquarium has used 27 composite pumps to circulate simulated seawater. Since their installation, these pumps have operated continuously, with their only downtime being the short periods required to replace seals, bearings, and lubricating oil. Debris is removed by screening systems. After more than 5 years of continuous operation, these pumps were inspected, and no corrosion, erosion, or cavitation of any composite pump castings, impellers, or other Gl-reinforced plastic components was reported.[140]

Cable. Traditional mooring materials such as steel chain can be highly susceptible to corrosion fatigue failures in the marine environment. As a result, failure of mooring systems for oceanographic instruments results in millions of dollars of replacement costs each year. More advanced materials such as Gr-reinforced plastics offer much higher resistance to corrosion and fatigue. To take advantage of these improved properties, a prototype Gr-reinforced thermoplastic cable has been manufactured using pultrusion and thermoforming techniques.

The basic building block of the cable is a small-diameter fiber-reinforced thermoplastic wire which was manufactured by pultruding several tow pregs together in a heated die. The finished wires were then thermoformed into a seven-wire strand by twisting them together at a temperature above the softening point of the thermoplastic. Several fiber-matrix combinations were evaluated in both the pultrusion and thermoforming operations. The best results were obtained from an AS-4/J-2 Gr-thermoplastic tow preg which produced a wire with a breaking strength of 1700 MPa and a weight of 1.6 g/cm.

Efficient utilization of composite cables depends on the development of effective terminations. Standard plug-and-sleeve wire rope terminations do not allow sufficient load sharing between individual wires in the composite strand because of the high stiffness of the Gr_f. A pretensioned thermoforming operation has been developed that substantially improves the load-carrying capacity of thermoplastic composite strands terminated with plug-and-sleeve fittings.[141]

3.5.7.4 Piers, pilings, and construction[142]

Pilings.   Graphlite, mentioned earlier in connection with helicopter usage, has also been considered for applications outside of aerospace. The material is being used by the U.S. Army Corps of Engineers in 2028.8-cm-long pilings and in a 676.3-cm section of concrete deck in a pier test bed. When Graphlite is used in the prestressed concrete, the components do not lose stength as they do when steel reinforcement corroded. This material is unique, and it should be another fiber form that can be used as well as fabric, roving, and tape.

Piers.   At its advanced water technology test site (AWTTS) the U.S. Navy has installed an all-composite catwalk on a test pier in order to test composites in marine structures.

The Navy is already testing two pier bays: a fiberglass composite deck and a concrete deck prestressed with C_f/Ep rods. Gl-vinyl ester pultruded rods posttension three pile caps—concrete sections that cap a row of pilings supporting the decks.

This "real-world laboratory" includes the world's first scaled-up, all-composite mobile pier. The 548.6×609.6-cm Gl_f-reinforced decking, made in both isophthalate polyester and vinyl ester versions, is modular and easy to assemble without the use of heavy equipment. Rectangular, 12.7×35.56-cm, pultruded box beams are used. Another project is to evaluate C_f/Ep-reinforced rebar in prestressed concrete decks.

In conjunction with this work the U.S. Navy is about to construct a 5410.2×676.3-cm structure which is the largest of its type ever built as a test pier. A 1352.5×676.3 cm-section is being used to test the effectiveness of composite tendons for prestressed concrete piling, pile caps, and decking. Also being studied is an all-composite superstructure mounted on composite-reinforced prestressed concrete piling. In addition, a full array of off-the-shelf composite fixtures, including grating, railing, ladders, lighting and utility poles, signs, and paneling, will be installed throughout the entire pier.

The nation's waterfronts—oceans, harbors, lakes, and rivers—are plagued by deteriorating wooden piers, docks, and other shoreside facilities. And since 60% of our waterfront structures are made of timber, this is no small problem. For generations, the traditional material used in these structures has been creosote-treated wood. The creosote treatment is used to help mitigate damages caused by wood rot and the ravages of marine borers and shipworms, water creatures that can devour the sturdiest wooden emplacement, even those made of exotic tropical hardwoods. But while creosote and other chemicals can deter the destructive effect of marine borers to some degree, this protection is not complete, and after relatively short periods of service, a wooden piling can deteriorate to a point where it can no longer function effectively.

In the past there was another deterrent to destructive marine borers. Many harbors and coastal waters were polluted to the extent that marine life, including marine borers, did not pose a threat to wooden waterfront structures. But now that we are making progress in cleaning up our lakes, rivers, and shorelines, wood-destroying marine organisms are making a comeback. Fortunately, though, today there are alternatives to traditional materials to combat the resurgence of the marine borer and to eliminate the use of hazardous chemicals in the marine environment.

Because of the availability of composites, modern techniques and materials can be used on newly installed submerged structures and to preserve existing ones. For example, polymer concrete provides an efficient casing impervious to marine borers and so can be used to extend the life and improve the appearance of underwater wooden piles. Also, coating metallic piles and sheeting with a combination of isophthalic polyester resin, fiberglass flakes, and corrosion-resistant fillers can deter the corrosive action of a marine environment.

Another new product that addresses these problems is a composite marine pile. The development of a continuous extrusion process has produced piling made of recycled high-density polyethylene and pultruded fiberglass reinforcing elements. Known as SEAPILE composite marine pile, this material is extruded in one piece, without joints or seams, thus providing increased structural strength. The plastic pile is derived from 100% recycled plastic products. The plastic matrix is enhanced by the addition of antioxidants and ultraviolet inhibitors. During the manufacturing process, the composite pile is reinforced with either steel or fiberglass reinforcing bars.

Given the nature of the composite pile's structure, it is a high-energy-absorbing product. When reinforced with fiberglass, it continues to bend beyond its elastic limit without breaking. While such an accidental overload may permanently deform the pile, it still absorbs energy and continues to perform its task, unlike timber piling which first bend and then breaks under such excessive load.

3.5.7.5 Marine fenders, tanks, and platforms

Marine Fenders. The SCRIMP process, a vacuum-assisted, closed-mold fabrication process, is capable of making large composite parts that are strong, lightweight, and resistant to corrosion and seismic shock. The technique is expected to impact the repair and upgrade of bridges, seawalls, walkways, and marine pilings. The key focus of research and development is to devise a process and an infrastructure that will allow SCRIMP to consistently yield quality composite parts at a repeat manufacturing facility. A 30% reduction in manufacturing costs and a 50% reduction in time to market are additional goals.

SCRIMP technology is used to manufacture a composite marine fender. The fender is designed to replace the timber and steel pile clusters traditionally used for pier fendering and vessel positioning. These clusters have been subject to breakage and abrasion from vessel impact, as well as rot and marine borer attack. The composite marine fender weighs nearly 2676.24 kg less than its steel counterpart, yet exhibits the strength and rigidity needed to absorb the impact of 2,721,600-kg ferries. It is resistant to corrosion and to the effects of salt water and ultraviolet light. Composite fenders also eliminate the problems caused by chemical leaching from creosote-soaked timbers and corroding steel faceplates.

Although high cost is a major factor, a number of technical issues also are holding back the broad introduction of composites into the large-structure marine market:

Military and commercial

Thick sections	Ultraviolet radiation
Compressive load behavior	Impact resistance
High-stress design	Scaling/modeling
Nondestructive evaluation	Reliability
Joints and joining	Residual-stress effects
Repair	Smoke and toxicity
Fire performance	Creep/stress rupture
Moisture absorption	

Primarily military

Shock performance	Acoustic behavior
Electromagnetic radiation	Ballistic performance

These issues can be refined into a set of five challenges to the marine industry and its suppliers:

- Develop new, advanced materials systems tailored for marine use at a cost competitive with other materials options
- Develop comprehensive tests of long-term performance
- Resolve fire, smoke, and toxicity issues
- Develop automated fabrication methods
- Develop appropriate design methods

Figure 3.48   Challenges and opportunities for the future of composites.[139]

Tanks.   Composite tanks are excellent devices for storing highly corrosive chemicals, including even concentrated hydrochloric acid. As such, filament-wound FRP tanks offer an economical alternative to metallic vessels.

A major producer of composite tanks recently began the construction of what may well be the world's largest filament-wound FRP tank, a 2768.7-cm-diameter vessel used for the containment of hot brine, a highly corrosive material. The tank is 338.1 cm high, with an open top. The total corrosion liner thickness will be 2.54 mm, typical of FRP tanks.

Platforms.   Oil platforms are another significant area for the use of composite materials, and Figure 3.48 reflects their future potential. There are significant cost and performance incentives to save weight in constructing the tension leg platform (TLP), a leading design for deep-water oil and gas production facilities. In general, reducing the weight of components on the platform by a kilogram leads to the removal of two additional kilograms of weight from the rest of the structure (hull, tethers, foundations, piles). The drilling structure, helicopter deck, process and utility equipment, tethers, risers, and drill are potential applications for advanced composites. Conventional composites currently are used for crew accommodations and core sample holders.[139]

3.5.8 Infrastructure

There is intense and concerted interest in the civil engineering construction industry (government and commercial) in developing markets for composite materials. The design approach combines aerospace knowledge of composite materials and structures with affordable manufacturing and assembly methods used in civil practice.

In this case, the market is the eventual need to replace the nation's highway bridges, which have deteriorated through age, lack of care, and normal wear and tear. There are approximately 580,000 bridges in the United States, and between 150 and 200 collapse partially or completely each year.

For example, a 4267.2-cm-long traffic-bearing Gl-polymer composite bridge was built in less than 5 weeks for less than $5/lb. This cost was four to five times that of a conventional short-span concrete bridge. However, that cost would be significantly reduced through the reduced cost of raw materials due to higher market demand for fiber and resin. Cost per kilogram would be further decreased through evaluation of the advantages of light weight, corrosion resistance, ease of transport, and rapid assembly with minimal traffic disruption. These reductions are expected to lower costs to the point of being competitive with conventional construction costs.

Infrastructure applications are the key to continued composite material usage and growth, and all indications are that steady growth will continue into the near future with the main potential in civil engineering applications including reinforcing bars, maritime structures such as piles and piers, wrapping for highway columns and utility poles, and bridge decking.

The state of the U.S. infrastructure is as follows.

- 23,000 of the 580,000 bridges are in a state of advanced decay
- 10% of highways require immediate repair.
- 40% of all roads fall below minimum standards.
- 50% of wastewater treatment plants are running at over capacity.
- 60% of all airports will be inadequate by the year 2000.[143]

The above statistics illustrate that the potential use of composites in structural infrastructure applications is great. Composite products are corrosion-resistant, a distinct advantage in applications such as bridge decks, railings, and lighting stands, which are exposed to the deteriorating effects of dampness, humidity, and motor exhaust. Composites also provide excellent strength-to-weight ratios, allowing civil engineers to reduce the weight of various components within a structure without sacrificing strength.

In addition, composite manufacturers can produce one-piece parts with intricate shapes, replacing assemblies of many different units. This parts consolidation reduces manufacturing costs and often provides a sturdier, trouble-free component. Another advantage of composites is their high dielectric strength. They exhibit outstanding electrical insulating properties and thus are ideal materials for current-carrying elements.

Unfortunately, however, several factors have restricted the use of new technologies by the construction sector. To begin with, the industry has an image that is anything but high technology. Second, the typical construction worker is regarded, at best, as semiskilled. And third, the construction industry is highly fragmented. Only a handful of large construction companies exist, and very few companies can afford any type of R&D activity.

Further, few firms in the construction sector can absorb the risk inherent in innovation. In addition to the typical risk associated with any new idea, construction is an area where minimizing risk to human safety is paramount. Bridges, roads, and structures must be safe, and "safe" often translates into using what is proven, even if technically dated.

The single greatest constraint facing the use of structural plastic in the construction industry is the lack of consensus standards for engineers to use in helping regulators approve the use of these new materials. Engineers need to convince the regulators that their innovative use of structural plastic and composite materials will not endanger the public.

3.5.8.1 Successful projects and the future.   Yet even with the difficulties in establishing composites as preferable materials for structural installations, numerous projects are in the works in such areas as offshore, building products, and bridges.

Offshore.   Phenolic grating for the Shell Mars oil and gas offshore platform is now being built for placement in the Gulf of Mexico off the Louisiana coast. The platform is held in place in water 91,440 cm deep by tendons anchored to the sea floor. The Duragrid phenolic grating was selected because it offers resistance to flame and high temperatures combined with low smoke and toxic fume emission. The nonflammable nature of phenolic is said to enable the grating to withstand direct flame contact for extended periods of time without major structural damage. Combined with low thermal conductivity, this provides fire protection not available with other materials and is a quality cited as a significant advantage at a site where oil and gas are being processed. The Duragrid grating is a pultrusion using phenolic resin and continuous Gl_f wrapped by a continuous-strand Gl mat.

Building Products.   For a one-piece Gl composite slider door sill that replaces wood, Al, and PVC sills, design requirements called for thin-walled, intricately shaped pultrusions providing the required strength plus UV stability. This product illustrates a basic advantage of composite fabrication: parts consolidation. Traditional sills are made of a combination of separate parts, screwed or bonded together, which tend to separate and weaken after a certain amount of use. The one-piece, integrated design is claimed to provide longer life.

Bridges.   Creative Pultrusions has produced a lightweight vehicular bridge for the Philadelphia Zoo. According to the company, the maintenance-free bridge has a load weight of 6804 kg and provides access for earth-moving equipment (backhoes) in the lion, tiger, and bear areas of the zoo. Designed in modular parts to span 304.8, 609.6, 914.4, and 1219 cm over moated sections, the bridge can also be reconstructed as a lifting frame for the animals. The pultruded components are made of fiberglass, isophthalic polyester resin, and vinyl ester, providing high strength and light weight. The heaviest individual component in the system weighs 40.8 kg, and the assembled 1219.2-cm unit totals 2721.6 kg.

The second example is the redecking and smart monitoring of bridges. Currently about 1000 bridges are lost every year from our national inventory of 580,000. At this time there are few mechanisms for determining the most likely candidates for failure because it is difficult to identify structural parameters such as creep, fatigue, and corrosion. High-performance composite materials need to be investigated in order to make future bridge structures more durable and more cost-effective over their lifetimes. Smart material systems that offer full-scale global monitoring for bridge structures would significantly reduce the cost of on-site inspection. Smart composite materials can be engineered to meet the stringent environmental and technical requirements of these sophisticated on-site measurement systems.

The previously mentioned Morrison Molded Fiber Glass Company plans to design new composite structural members using Gl and C_f in a resin matrix to optimize composite shapes and testing and hope this may go a long way toward repairing deteriorated bridges in the United States and reducing the cost of offshore oil rigs.

Tunnels. The current rehabilitation of Boston's Callahan Tunnel includes repair of its deteriorated sidewalls and modernization of its lighting and electrical systems.[144] The favorable performance of the Quazite polymer concrete panels installed in the Sumner Tunnel in 1982 led the tunnel administration to choose polymer panels for modernization of the Callahan Tunnel as well. The highly reflective, durable, white gel-coated panels perform like large pieces of ceramic tile but unlike tile are strong enough to resist frequent vehicular impacting.

The Callahan Tunnel employs polymer concrete panels that are used both as cladding panels and stay-in-place forms. Existing bench walls have been completely removed and reconstructed with new concrete. Quazite panels serve as the form work for the new concrete and will remain in place to become the interior tunnel finish.

The long-lasting, 40-mil gel-coated surface is an impervious coating that is fused to the polymer concrete during casting. The hard finish maintains its color, resists soiling from engine exhaust, and stands up to wear from detergent washings, brushes, high-pressure water jets, and deicing chemicals.

Oil Platforms. An offshore oil drilling platform for Shell Oil being built in the Gulf of Mexico illustrates the market's opportunities.[145] Approximately 696.75×10^5 cm^2 of phenolic grating has been manufactured (pultrusions). The material is about 30% lighter than steel and provides vital flame-retardant characteristics like low smoke and low toxicity as compared to vinyl ester.

The Shell Mars platform mentioned previously consists of 91.44-, 121.92-, and 152.4-cm-wide panels connected with I-bars, and it is the first major commercial application to use pultruded phenolic and a fiberglass roving compatible with the resin.

Housing and Building. To emphasize the opportunities awaiting the composites industry, here are two exciting examples. The first is housing. Currently the U.S. residential housing industry uses traditional materials in construction, with wood being the primary building material. But wood is in increasingly short supply and often of poor quality. Furthermore, it is the largest single contributor to the rapidly increasing cost of construction.

A major technological objective might be to fabricate large, lightweight, energy-efficient framing systems for residential construction using advanced materials technology. These composite building products could mimic the appearance of conventional construction such as brick, lap siding, and shingles. These new materials would offer tremendous advantages during construction, long-term low maintenance, and extraordinary energy savings in comparison to traditional construction materials.

The expanded use of composites in the housing industry has become a key issue being discussed by homebuilder groups conducting research on various composites as substitutes for traditional building products. The focus on the housing industry is part of the composite's industry's effort to encourage the use of structural composites in construction. Overall, the industry is seeking new markets and is diversifying as a way to overcome the limitations on growth and profitability imposed by the mature, cyclic businesses that make up most composite markets.

The use of structural composites in construction is just in its infancy. Construction accounted for 240,408 million kg of composite use in 1993, an increase of 9.7% from 1992. Market growth of 8.4% is forecast for 1995 and an average of 8.7% is forecast for 1996, 1997, and 1998.[142,143,145]

For construction and infrastructure repair, a new family of structural materials is being developed, including pultruded structural composites and traditional materials that incorporate composites. In one instance, structural beams made of reconstituted wood can be made more economical with the use of composite materials. A composite can be interleafed into the wood, enhancing its overall performance so that it can compete with steel and other materials in specific applications. As a result, a wood component of lower quality can be used.

The Chichibu Onoda Cement Corporation[146] has developed a lightweight Gl_f reinforced concrete (GRC) panel usable in constructing outside walls of high-rise buildings. The GRC curtain wall is used with the GRC plates reinforced with a framework of square steel tubes.

Compared with that of ordinary precast concrete panels, the weight per unit area is less than one-half, which translates into shorter high-rise building construction schedules. The panel is made into panel form by mixing low-alkaline cement, alkali-proof Gl_f and a lightweight aggregate with a mixer and then pouring the mixture into a mold. The concrete mixture, consisting of a low-water-ratio cement, features excellent fluidity, self-leveling attributes, and fiber dispersion.

The panel weighs 140 kg/m^2, which is less than one-half the weight of the precast concrete panel. The specific gravity is 1.5, the flexural strength is 90 kg_f/cm^2, and the compressive strength is 430 kg_f/cm^2, which are adequate for constructing high-rise buildings. The panel is much thinner than precast concrete panels which require a thickness of 15–20 cm to prevent corrosion of reinforcing steel. Contraction by drying is also one-half that of ordinary concrete products, and large panels can be produced since there is little warping or cracking.

The panel costs about 20% more than the precast concrete panel, but its lightness eases construction work and shortens project schedules, and so the construction cost is reduced. The new material, with its concrete characteristics, is being developed for ceiling and partition walls.

Grace, Bagchi, and Kennedy[147] conducted a comparative study of composite versus steel and concrete as a vibration control material, considering a single degree of freedom representation for beam and slab in an industrial building. The following conclusions were drawn from this study.

1. Both as a spring and floor material, composite (Gr/Ep) is the most suitable material because of its light weight, high modulus, and high damping value.
2. To achieve a desired stiffness or frequency, composite construction requires only one-third to one-sixth the weight of steel.

3. To meet the same vibration level, composite construction requires 0.29 times and 0.45 times the weight of steel for steady-state and impact vibration, respectively.

4. In all three types of vibration sources, and especially in the case of random vibration, the highly tuned composite structure can be utilized to take advantage of the low power spectral density function of ambient input.

5. Composite material is very attractive for the design of high-frequency platforms required for highly sensitive tools when the weight of the frame is an important design factor for the safety of the floor.

6. Finally, because of the limited availability of fabrication facilities for composites and the high expense of fabrication, their use in vibration control can become more widespread through mass production and standardization.

Numerous discussions and investigations have covered the strengthening of masonry structures with fiber composites. The application of C_f sheets (CFRP) opens up new possibilities for the strengthening of masonry walls in seismically endangered zones. Since the CFRP sheets are bonded to the wall surface, one is no longer restricted to orthogonal arrangement of the strengthening elements as is the case with conventional reinforcement placed within the material. The sheets can be arranged in the main stress directions to give optimum transfer of earthquake forces.

Since C_f is a strong material, strengthening of more than 300% can be achieved with only CFRP sheets only 1 mm thick. Also, compared with strengthening through steel screening, the use of polyester fabric confers the advantage of no corrosion. Since Ep resin bonding material is applied to the entire surface, there are no bonding problems between the masonry and the bonding material even at high earthquake loading.

More Bridges, Highways, and Tunnels. New opportunities are currently emerging for advanced composites in civil engineering markets. New markets inevitably require new materials or combinations of incumbent and new materials. These emerging markets are not only utilizing available composites but are also finding new composites like polymer matrix composites reinforced with new, ultrahigh-modulus, melt-spun Gl-ceramic$_f$ which could become major building blocks in these markets. These composites also offer higher strength than those reinforced with sol-gel-derived Gl-ceramic$_f$.

A prototype fiberglass polymer short-span highway bridge designed by Lockheed Martin has successfully held a 29,684-kg personnel carrier. Portable versions of composite bridges that are corrosion-resistant and lightweight also have potential military and disaster relief markets. Lockheed Martin's research laboratory has designed, analyzed, and built a 548.6 × 914.4-cm test span in less than 18 months, and material costs for the 10,432.8-kg-structure were less than $3/kg.

Existing composites technology (Table 3.10) is currently being explored to repair earthquake-damaged buildings, industrial chimneys, and defective bridge columns that show evidence of cracking. Over one-half of all the bridges in the United States have been declared defective for reasons ranging from surface damage to structural faults, and a major effort has been initiated at the federal and state levels to remedy the decay.

Much of the surface repair technology for defective bridge columns, smokestacks, and earthquake-damaged buildings originated in Japan and consists of either wrapping preformed composite sheets around structurally damaged column surfaces or jacketing

TABLE 3.10   Infrastructure Repair with Composite Jackets[148]

Defective Structures	Composite Jacket	Reinforcement
Bridge decks, columns	Preformed sheets	Glass, carbon
Smokestacks, columns	Filament winding	Glass, carbon
Earthquake-damaged buildings	Preformed sheets	Carbon

smokestacks by large-scale, on-location filament winding. The preferred reinforcing fibers are C and Gl. Structures repaired by a C_f sheet wrap were more damage-resistant after repair than the original structure.[148–156] Damaged surfaces repaired with Gl_f-reinforced sheets are believed to be adequate.

An economic cure for rotting support columns has been developed by New York City. A far too common sight these days is the deterioration of concrete support pillars. Two badly deteriorated columns supporting a bridge over a heavily traveled highway were found where a mixture of dampness and exhaust fumes in the environment had done heavy damage to the steel reinforcements in the two columns.

Replacement of the columns would have been a costly and complicated venture. However, this involved operation wasn't necessary because of a method of enveloping the columns in a composite wrap to prevent further deterioration. The wrap consisted of a high-strength fiber in an epoxy matrix.

This composite system, known as Tyfo S Fibrwrap, incorporated an Ep that had been tested to allow over 4% elongation of the fiber so that normal fiber elongation could be achieved with no matrix cracking. The principal advantage of Fibrwrap was that expansion of the reinforcing elements caused by corrosion could be accommodated without harm, and in fact the fiber was actually strengthened with elongation. The Ep had a 3-h working time, and there was no offensive odor.

Bridge projects involving Fibrwrap have now been completed in Pennsylvania, Wisconsin, New York, and Vermont. These were essentially repair installations. In California, wrap installations have been designed for seismic strengthening and increased ductility.

In November 1993, the Tyfo S system was selected to retrofit 42 short columns in a hotel close to the area affected by the January 1994 earthquake. Inspection after the quake showed that the Fibrwrap was not damaged, although signs of large displacements were evident in the adjacent walls. Throughout 1994, the material was used in 18 Fibrwrap projects involving over 700 retrofitted columns, including some in parking garages and buildings.

The cost of the Tyfo S Fibrwrap system is competitive with that of steel jackets, and the process has many additional advantages. It is corrosion-resistant, it conforms to the column shape (steel jackets on rectangular columns are elliptical), no grounding is required, it can be easily installed in tight spaces, materials are in stock (thus reducing installation time), and the wrap does not add stiffness to the column so that the whole structure retains its original stiffness characteristic.

Researchers at the State University of New York at Buffalo claim that adding small C_f to conventional concrete should make it easier to monitor internal flaws in bridges, roadways, and other structures. That's because the fibers are electrically conductive and

TABLE 3.11 New Hybrid Bridge Construction with Composites[148]

Structure	Composite	Reinforcement	Information
Pier column	Pultruded rods	Glass, carbon	U.S. Navy, CERL
Bridge deck	Woven sheets	Carbon	Mitsubishi-K
Bridge column	Woven sheets	Glass, carbon	CalTran, UCSD
Bridge base	Honeycomb	Glass	Mainland China

their resistance increases in the presence of a flaw. This change in resistance could be detected by electrical probes placed outside the concrete. In addition, C_f absorb more energy before bending, so researchers believe the substance could withstand stronger earthquakes.

In addition to bridge repair methods, a major effort is underway to develop new hybrid and all-composite construction techniques. Among the hybrid structures (Table 3.11) are concrete piers reinforced with rebars and Gl_f-reinforced pultruded rods. In addition, C_f-reinforced composite sheets are laid into the deck construction of new bridges, and new bridge columns are being wrapped with C_f-reinforced composite sheets.[154,157]

The world's largest all-composite foot bridge, which stretches across the River Tay near Aberfeldy, Scotland, was completed in 1993.[158] It was made of interlocking modules of fiberglass-reinforced plastic and aramid suspension cables (Table 3.12) and was assembled without heavy machinery at the remote site.

There were some problems to be overcome with composites. Some buildings stood very close to the bridge site, and this ruled out the use of a swing bridge, which requires a good deal of space during operation. Also, the bridge provided the only access to the busy town complex, and so the new bridge had to be strong enough to support heavy traffic, including heavily loaded tractor trailers.

After considering all the factors, a modular construction method known as the advanced composites construction system (ACCS) was chosen, which is based on interlocking structural units that can be joined together for a variety of uses. The components were pultruded, using an isopolyester resin reinforced with fiberglass. UV inhibitors were added to the resin to increase resistance to outdoor exposure. ACCS pieces were firmly linked together by mechanical interlocks and adhesively bonded without drilling or welding. Relatively unskilled workers quickly completed assembly without special tools or lifting equipment. And because of their light weight, the sectional pieces were easily handled and placed in position.

TABLE 3.12 New Bridge Construction with All-Composite Structures[148]

Bridge Structure	Composite	Fiber	Information or Source
Large footbridge	Deck modules	Glass	Vetrotex, U. Dundee
	Base modules	Glass	Vetrotex, U. Dundee
	Fiber cables	Kevlar	DuPont, U. Dundee
Full-service bridge	Hardcore deck	Glass	DuPont, UCSD
	Honeycomb deck	Kevlar	Fiberite, UCSD
	Base modules	TBD	UCSD, ARPA
	Fiber cables	Kevlar	DuPont, UCSD

TABLE 3.13   Incumbent Composite Reinforcing Fibers[148]

Reinforcing Fiber	E-Glass	Kevlar 49	Thornel 300
Fiber type	Glass	Aramid	Carbon
Processing method	Melt-spun	Gap-spun	From PAN
Fiber modulus (GPa)	72	117	227
Estimated fiber cost ($/kg)	2.20	35.00	18.00

A further advantage of the ACCS composite units was their resistance to corrosion. This was a big plus in the bridge application, where the parts were constantly exposed to a damp environment.

The lift bridge is 822.96 cm long and 457.2 cm wide. Its composite part weighs 4082.4 kg and can support heavy vehicles, including loaded trucks weighing up to 39,916.8 kg. The bridge is activated hydraulically. Lift bridges of this type, when built of metal, usually require a lifting mechanism as large as the bridge itself. Given the limited space available, this configuration would have been impractical, but because of the good weight-to-strength ratio of the composite components, the lightweight deck eliminated the need for a lifting tower and counterweight.

The first all-composite full-service bridge (Table 3.12) has recently been designed at the University of California at San Diego (UCSD) and will be built in 1996 and completed in 1997 at an estimated cost of $55 million to connect two parts of the campus.[148] The final deck will be made from modular sections of one of two prototype composite structures. Both prototype decks for this bridge were designed to meet the specification of the UCSD. The deck modules are less than 30.48 cm in thickness. One is a fiberglass-reinforced structure containing a minor amount of C_f. The other consists of a Nomex honeycomb core between woven Kv composite face sheets. Both prototypes are currently undergoing testing.[148–149,151–152]

Gl, C, and Kv_f-reinforced composites are being used in civil engineering markets for repair and/or new construction. Gl_f (E-Gl) have the lowest modulus and lowest selling price (Table 3.13). Kv_f have an intermediate modulus and a high selling price. C_f (Thornel 300) are the stiffest fibers and have an intermediate selling price.

Many combinations of reinforcing fibers (Gl and C) and reinforcing structures (sheet and honeycomb) are being explored in an attempt to utilize the strengths, and to minimize the weaknesses, of current reinforcing fibers while offering functional composites that may or may not be cost-effective in the end. In view of the fact that DuPont-Hardcore is designing bridge decks with Gl_f-reinforced composites and Fiberite with Kv-Nomex honeycomb composites,[148] it is evident that the search for goal functionality currently sidesteps cost considerations.

Thus there is no clear direction in which the field will eventually grow, but a new, low-cost, high-performance fiber may in the end determine all new product design and become the reinforcing fiber of choice for civil engineering composites.

For example, the moisture from the famous Niagara Falls mist can have a very corrosive effect on the installations put in place to enable tourists to view the falls safely.[159] The handrail and the light stands around the entire deck and the staircase leading from the observation point to the bottom of the falls were no longer serviceable and prior to replacement required constant scraping and repairing.

To avoid a repetition of this situation, the replacement for the metal components was to be composite materials to offset the corrosive effects of the constant spray from the falls. The decisive factor in selecting a composite material was its lower life cycle cost. The resin used in these components was isophthalic polyester, and the reinforcement agent was E-Gl rovings and mats. This formulation provided maximum corrosion resistance combined with the necessary strength and rigidity to ensure safety for Niagara tourists.

Throughout history, the quality and effectiveness of a nation's or empire's infrastructure has been defined by the development and innovative use of construction materials. The growth of civilization, whether in Europe, the Orient, the Americas, or elsewhere, bears testimony to this fact. Moreover, it is a relationship that appears destined to hold true as the United States, and indeed the world, prepares for the challenges of the twentieth-first century.

The quality of our infrastructure and our global competitiveness are inexorably linked. Simply put, a quality infrastructure enables a society to perform all functions more efficiently, whether in the private or the public sector. Infrastructure thus provides "value added" for all facets of the economy if it is in itself of high quality. In the United States, the hard fact is that much of our infrastructure is marginal or deficient in terms of its value added capability.

Because of their sheer impact on the infrastructure, concrete and steel have been accorded initial emphasis. However, composites are now on board in this national program. The future therefore looks promising. In fact, it may be said that infrastructure applications for composite construction materials will be limited, in the first instance, only by the vision and imagination of the industry. The ability to visualize the role of composites in existing infrastructure components and a vision of the currently unresolved infrastructure problems that composite materials can resolve are the keys to success. Beyond this, however, composites must also meet the pervasive challenges that face the introduction of any innovation in construction—challenges such as risk of failure; tort liability; a focus on initial instead of life cycle costs; technical education for designers, engineers, and the work force; development of standards; and so on. These challenges are primary targets of the proposed national high-performance construction materials and systems program.

In summary, the future for composites in the U.S. construction sector is promising, but it is not without challenges, challenges that the collective industry must address. The family of composites clearly promises a range of construction applications with the potential to dramatically influence the shape and utility of our twenty-first century infrastructure. Composites may indeed take their place among those construction materials that have, over millennia, fundamentally influenced humanity's social and economic well-being.[160]

3.5.9 Biomedical and Prosthetic Devices

It's getting harder for leg amputees to think of themselves as handicapped. Thanks to prosthetic devices, many can walk almost normally. Prostheses have been developed that provide almost natural foot movement. Unlike conventional artificial legs that come in two pieces with a leg and a foot, the Modular II Flex-Foot is a single unit with an inner metal bar that allows the ankle and foot to bend. At 0.2268 kg the C/Gr composite leg is

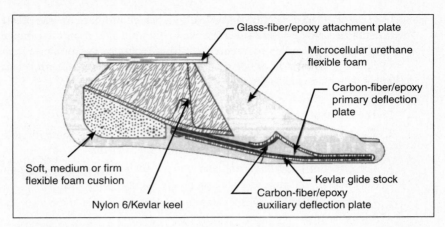

Figure 3.49 The Carbon Copy II, a prosthetic foot that comes very close to simulating natural foot action, relies on composites for critical parts—the deflection plates and the keel.[161]

durable enough to withstand running or uphill jogging. Other manufacturers[161] have designed and built prosthetic feet. The foot provides a natural toe-off action with two C_f/Ep springs attached to a central support of compression-molded nylon 6 reinforced with Kv fabric and enclosed in a special urethane foam body (Figure 3.49).

One industry particularly interested in the ability to re-form low-cost thermoplastic composites is the medical field.[162] Re-formability is appealing in this market because prosthetics and orthotics must be custom-formed for each individual patient. A variety of thermoplastic prepregs have been used by the industry, including C, Gl, and Kv_f and polypropylene, nylon, and polycarbonate resins. Companies looking at the applications of thermoplastics have considered the tradeoff between interfacial adhesion (nylon is better than polypropylene) and the forming temperature (polypropylene forms at a lower temperature). Many shops that customize orthotic and prosthetic devices have unsophisticated forming equipment, and so the lower forming temperatures may be preferred.

Also finding application in the medical field are high-performance thermoplastics. Orthopedic implants benefit from the inert nature of C/PEEK thermoplastics; that is, they do not react with body fluids or tissue.

Jet aircraft engines, nuclear pressure vessels, high-performance automobiles, and computer hardware have one thing in common—their performance does not depend upon a constantly changing, living environment. They can also be routinely inspected and serviced, whereas orthopedic implants cannot. Implant materials must be able to operate in the harsh and restrictive boundary conditions of the human skeletal system and, in the case of joint replacement, perform over the lifetime of the patient.[163] Because composites can be custom-built, so to speak, to optimize both material modulus and strength, there is an opportunity for them to meet the needs of future structural implant applications.

Composites were initially developed for reconstruction devices, specifically for hip joint replacement. Because only composites using high-strength fiber impregnated with polymer resins can most easily accommodate combined modulus and strength requirements, these types of composites are being pursued first. Significant research to date has focused on oriented C_f-reinforced polysulfone which provides an acceptable combination of strength and modulus. However, unlike metal implants, composite implants require an order of magnitude and more research to ensure long-term performance *in vivo*.[163]

Factors that must be understood in assessing the long-term performance of composite implants include the following.

1. The effect of bone cement on mechanical strength.
2. The effect of combined bending and torsion on fatigue strength and the tendency for matrix delamination, and other failure mechanisms.
3. The effect of body fluids and constituents such as lipids (including chlolesterol and fatty acids), oxidants, and glucose on mechanical properties.
4. The biocompatibility of fiber sizing, residual solvents in the matrix resin, and mold-release agents.

Other composite hip joint replacements have found that C/PEEK material offers better strength, stiffness, and biocompatibility. The APC-2 high-strength standard modulus C_f embedded in Victrex PEEK resin thermoplastic composite features continuous C_f that is evenly dispersed and thoroughly wetted in the PEEK matrix. Fiber content is 61% by volume and 68% by weight. A typical flexural modulus for this material is 119 GPa at 0°C and 8.8 GPa at 90°C. The short-beam shear strength is 105 MPa.

It has been found that biomedical implants of APC-2 will more closely emulate natural bone (Figure 3.50). With CAD and the anisotropic nature of the composite, high-modulus C_f can be placed in the appropriate orientation required to reproduce the stress and strain performance characteristics of the bone and joint they are replacing. Hip joints made of APC-2 will offer greater corrosion and chemical resistance and greater biocompatibility than currently used metals.[164–166]

Loss of bone around metal femoral stems implanted in humans was the impetus for studies made at Harvard Medical School and a Boston Veterans Administration Medical Center. More-flexible composite materials can reduce the stress between the prosthesis and surrounding tissue, which is considered a possible cause of bone loss. Gr/Ep composites and CCCs have been tried in animal experiments. But either cost in the case of the CCC implants or toxicity problems with some epoxies have discouraged researchers. Other materials investigated have been Gr/PS and Gr/PEEK. The PEEK-based material proved to be the best performer. Its compressive and flexural strengths were more than double those of the PS composite after immersion in a lactated solution meant to simulate implant environments. Predicting long-term performance from single-property testing, however, is a key issue. Fatigue, creep, and environmental factors interact in ways that are unpredictable. Animal testing with the implanted femoral stems can provide some of the answers.

If thermoplastic composites prove superior to metal versions, how to fabricate them is another question. Techniques proposed include a variety of innovative methods, including net shape molding, filament winding, pultrusion, and machining from flat or curved plates. To achieve a balance of strength, flexibility, biocompatibility, and fabrication ease will require considerably more research. Hybrid systems may evolve combining different fibers. Gr_f might be mixed with Kv, Gl, and quartz, for example. There is also a

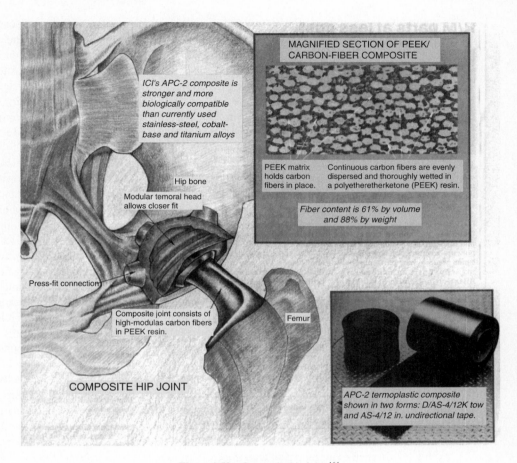

ICI's APC-2 composite is stronger and more biologically compatible than currently used stainless-steel, cobalt-base and titanium alloys

Hip bone

Modular temoral head allows closer fit

Press-fit connection

Composite joint consists of high-modulas carbon fibers in PEEK resin.

Femur

COMPOSITE HIP JOINT

MAGNIFIED SECTION OF PEEK/ CARBON-FIBER COMPOSITE

PEEK matrix holds carbon fibers in place.   Continuous carbon fibers are evenly dispersed and thoroughly wetted in a polyetheretherketone (PEEK) resin.

Fiber content is 61% by volume and 88% by weight

APC-2 termoplastic composite shown in two forms: D/AS-4/12K tow and AS-4/12 in. undirectional tape.

Figure 3.50   Composite hip joint.[164]

wide choice of material forms to choose from: prepreg sheets, collimated tapes, commingled and powder prepregs, felts, and various weaves and braids. In current and future developments of these materials and methods researchers will learn how valuable they can be in other implant applications. Currently scientists are using a material to make the tiles that protect the space shuttle from burning as it reenters the atmosphere that could someday solve medical problems on earth. To make this a reality, these researchers and scientists, at NASA's Ames Research Center, are working with physicians to determine whether ceramic surface insulation materials could be used as implants for skeletal reconstruction. They are still a long way from having a bone implant that's ready for human use, but they have data showing that the tile has excellent promise for this use. The tile is made of a SiO_2/Al_2O_{3f} and borosilicate Gl composite that mimics the structure of bone.

Composite materials and fiber-optic sensing may soon be at work in hospital delivery rooms. Doctors and scientists have collaborated to redesign the obstetric forceps used to position an infant in a mother's womb. Researchers hope to identify a composite for the forceps and design instrumentation to measure the forces being applied to an infant during delivery. Metal forceps currently used don't allow the attending physician to assess the force the instrument is placing on the infant. Currently, obstetricians must acquire a "feel" for their instruments during actual deliveries. The instrumented forceps will allow obstetrics students to learn to use forceps within safe limits before entering practice. Instrumentation will allow engineers to embed fiber-optic sensors in the composite material, which will help ensure a safer distribution of pressure and allow the doctor to monitor forces on the infant throughout delivery.

Finally, Merrill presented an interesting paper at a SAMPE meeting[167] in which he sought to establish a frame of reference in which to consider biomaterials according to application, with special reference to IN and MAN, that is, implants in the human body. The passive and active environments of the body are considered—*active* meaning biological response to the implant. Distribution of responsibility for successful functioning of the implanted device among the responsible parties—material producer ↔ material processor ↔ device manufacturer ↔ surgeon—is discussed along with comments on regulatory agency oversight and the possibility of litigation when an implant in a human being fails.

The greatest advances in implanted biomaterials in the twenty-first century will come from exploiting bioactive molecules (including biopolymers) grafted to or released from the surfaces of the implant to promote acceptance by the host or, stated the other way around, to prevent activation of an immune response, inflammation, thrombosis embolization, and unwanted cellular proliferation.

The second greatest advances will come as we transfer space age technology to medical devices—in the sense that we utilize advanced materials, especially composites, that have very long service lives in whatever function they perform [flexing, electric conduction, electrical insulation, molecular transport (as in membranes)], and in the further sense that these advanced materials are combined to form an implantable device with exquisitely refined manufacturing techniques.

Shoe inserts for athletic and medical podiatry applications form a significant commercial market for thermoplastics. More than 5 million orthotics such as shanks, arch supports, heel counters, toe protectors, and stiffeners have been delivered to users. Additionally, composites generally reduce the weight of a typical athletic shoe by 10–20% compared to traditional materials such as polyurethane (PU).

Taking a step from orthotics to prosthetics, companies are using C and Kv_f-nylon unitape for pylons in foot and leg-foot artificial limbs. These pylons offer energy damping, fatigue endurance, and tailorability to an amputee's specific gait and also feature C/Ep deflection plates.

Other firms have been successful with medical and sports shoe orthotics with multiply, multiaxis laminated sheet. The material is offered in Gl, Kv, and C_f with a variety of resins—from HDPE and PP through nylon and polycarbonate (PC) to PEI, PPS, and PEEK.[167]

Leg braces are an area where composites have shown their real advantage. A composite orthotic leg brace with one-third the weight, 40% higher stiffness, and twice the strength of its steel counterpart has been developed.[168] It is molded from a thermoplastic composite: nylon reinforced with long, discontinuous C_f. The thermoplastic allows a brace to be postformed under moderate heat and pressure, permitting the orthotist to adjust its shape to precisely fit the needs of the patient.

Current orthotic devices, especially long leg braces, are fabricated from monolithic aluminum and steel. Lightweight braces would significantly increase a person's mobility and reduce the fatigue associated with heavy braces. They may even allow some patients to walk who now are confined to wheelchairs.

This technology contributes to the field of orthotics by moving toward an ultralight orthotic system that can be applied to a wide range of external bracing devices.[168]

High-strength carbon was selected as fiber material because it provides the required strength margins (~2×) and the lowest cost As a long, discontinuous fiber, it offers high strength and stiffness and can be postformed without breaking or excessive buckling. Although weight could potentially be reduced by 30–40% by using a ultrahigh-strength fiber, material cost would go up by a factor of 2 or 3.

Nylon was selected as the optimal matrix material based on the following parameters.

- High strength and high modulus
- Good fiber wettability
- Moderate process temperature
- Low postform temperature
- Reasonable impact strength

Based on the data, therefore, the baseline material selected was T-300 C_f in a nylon matrix. It was obtained in a cowoven form, with C_f in the warp (lengthwise) direction and nylon fibers in the fill direction.

3.5.10 Sporting Goods and Recreational Equipment

Although advanced composites are being used increasingly in sporting goods, the industry has a long way to go before composite sports gear obtains a significant market share, say sporting goods manufacturers and composites industry analysts.

Studies in the early 1990s estimated the value of the U.S. recreational C_f market at $18.5 million with an annual growth averaging 15–20%. Golf shafts accounted for the largest share of the market (71%), followed by fishing rods (13%) and tennis rackets (6%). Other sporting goods such as archery and ski equipment made up the remaining 10%.

The sporting goods share of the advanced composites market clearly has increased since then because of increased demand in addition to a downturn in the aerospace industry. This trend is reflected in the sale of materials through the mid-1990s—in 1990 materials for sports equipment accounted for 49.6% of sales, and by 1995 that share had risen to 81.2%. Sales figures for military and commercial air applications dropped by approximately the same amount during that period.

Industry observers agree that golf shafts will remain the primary application for composite materials for some time, although rackets and bicycles are becoming more popular with consumers. Other segments of the sporting goods market, particularly equipment for winter sports, are likewise attracting more interest. Skates, hockey sticks, ski gear, and snowboards are at the forefront of these new growth areas. Decisions about which fiber will be used may be based on what marketing analysts think is the buzzword for this week or this season rather than on the material.

Much of the technology used to manufacture composite sporting goods has been copied from aerospace manufacturing processes. Familiar methods such as hand layup are not cost-effective. Reducing the cost of sports equipment to increase consumer demand for it will almost certainly require manufacturers to adopt new production technologies, industry observers believe.

Manufacturing skis, for example, depends on aerospace techniques including hand layup, vacuum bagging, core technology, bonding and adhesive approaches, and secondary bonding and mounting. If manufacturers are to see growth in skis and similar products, they must become innovative, look at alternative manufacturing processes where the technology has more recently advanced, and be willing to combine processes. In manufacturing skis, RTM offers several advantages, including molded-in-metal fittings, in-place core structures, and edge alignment and structural tie-in, which results in faster production, lower costs, and higher quality.

Manufacturers nust be able to use aerospace technology while simultaneously recognizing that commercial markets don't require that approach. Filament winding, pultrusion, and RTM, as low-cost manufacturing techniques, will influence market growth and product development concepts more rapidly than continued reliance on conventional aerospace technology. The move toward reduced aerospace manufacturing methods is already apparent used in the production of rackets, and bicycle manufacturers are looking for ways to do the same. For rackets, all you care about is low cost. You don't use aerospace technology because you don't need it.

Sports equipment has been generated at several companies out of a necessity to survive as aerospace demand dwindled for Delta and Titan rocket motor nozzles.

Manufacturers of rocket motor nozzles for defense industry found that business was drying up and therefore started their designers looking into the composite sporting equipment industry. Examples of the above include some cycling enthusiasts who developed a bike frame made from a thermoplastic reinforced with C_f.

Everyone has heard how Teflon migrated from the space program to the kitchen. Now, a wide array of new composite materials—tougher graphites, C_f and B filaments among them—are popping up on all kinds of playing fields. The properties for which B is prized—strength, lightness, and flexibility—are in demand by weekend golfers, tennis players, and mountain bikers. Composites are finding their way into ski poles, hockey sticks, and yacht masts.

3.5.10.1 Racing cars.

The 500-mi Indianapolis race and the complete Indy car series continue to be a unique proving ground that benefits drivers everywhere.

Within the strict rules for Indy racing, there's still room for experiments that racers hope will lead to a winning edge. Advanced materials that cut weight or boost performance provide a case in point. For example, Textron Specialty Materials new Hy-Bor composite should soon appear on the front and rear wings of Indy cars. Hy-Bor marries relatively large B_f with small diameter C_f in a resin matrix. The combination of large and small fibers boosts the fiber volume percentage in the laminate beyond the space-filling capabilities of either material alone. By selecting the size of the B and the C in a given laminate, it is possible to produce prepregs with a tensile modulus as high as 272×103 MPa.

This laminate combines high-modulus B_f with an intermediate-modulus C prepreg that gives increased stiffness, strength, and strain to failure compared to previous wing materials. The benefit to the race car comes from decreased droop in the wings under loads. A stiffer wing should lead to more predictable aerodynamic effects and thus a more consistent racing vehicle.

Engineers concede that these high-strength lightweight materials may not find applications in passenger cars soon. Nevertheless, government mandates for electric cars and the high-mileage so-called super car leave room for optimism.

Other components expanding the use of composite materials include wing supports made of SCS-6 SiC/Al. These 0.9-kg parts are 33% lighter than previously used Al elements. Another is a B/Gr hybrid composite for the front and rear wing material.

Thermoset composite valve covers first introduced in racing cars in 1990 are now on over 8 million GM vehicles. Weight and noise reduction are the primary benefits.

Rear-deck spoilers for the auto aftermarket use a polyurethane (PU) structural foam system that provides both strength and a good appearance. The spoilers, which can reportedly fit 90% of all cars sold in the United States, are produced by reaction injection molding (RIM). Eight models of the spoilers have been made weighing 2.3–2.9 kg.

In formula I (F1) racing, a team is using a chassis made from an ultrastiff Gr/Ep composite. The material makes the chassis stiffer, tougher, and lighter. Engineers fabricated the MP4/5 chassis from ultrahigh-modulus Gr_f and 8552 prepreg resin. Primarily an aircraft material, the UHM 8552 Gr composite produces a combination of high modulus and damage tolerance, both required for an F1 racing chassis.

UHM_f, made possible by precursor and fiber technology improvements, has a modulus of elasticity of 4.420×10^3 MPa. This is 30% higher than that of the $HMS-4_f$ used in the previous chassis. UHM_f has an elongation of about 0.9%, whereas the $pitch_f$ used in the past provided about 0.2% elongation. One cannot make very damage-tolerant structures with elongation that low. So UHM_f is more useful in a structure like a chassis.

The 8552 Ep portion of the composite is 50% tougher than the 3501-6 resin used before. It is tougher as measured by compression after impact. The high stiffness of the UHM 8552 Gr composite decreased the weight of the MP4/5 chassis by allowing a reduction in the number of layers. The walls of the new chassis are now a bit thinner but still have good flexural and tensile strength.

The use of composites in race cars is well documented, but safety dimensions also come into play for the drivers of these fast machines. There are three main reasons, other than weight reduction, for using composites in an Indy car's chassis. First, the cars are very sensitive to torsional flexing, and composites provide higher torsional stiffness in a smaller package.[169]

Second, high-performance materials enhance structural integrity. The composite chassis is designed to carry all the loads without using internal metal bulkheads, which gives drivers more leg room. For example, a driver could pull his or her legs back in the event of a head-on crash.

Third, composite materials enhance energy absorption. Championship auto rules require that the chassis structure be designed with driver safety as the main criterion. The material specifications dictate that high-performance materials be used, that these materials be processed to achieve the highest strength, and that a car's chassis be designed to absorb a certain amount of energy during two frontal impact tests.

3.5.10.2 Racing bicycles, mountain bicycles, and fun bicycles.   The lightweight Zipp 2001 bicycle is 18% more aerodynamic than a conventional bicycle. The hollow monocoque frame is Ep prepreg in a sandwich construction around a foam core; the rims are 100% C_f.[170]

A primary design focus when building bicycles is how best to use the beneficial properties of C_f when constructing frames. Beneficial properties include high strength-to-weight and stiffness-to-weight ratios, good shock damping, formability, pleasing aesthetics, and long-term structural integrity—all of which translate into durability and, just as importantly, safety for the end user.

One manufacturer of bicycle frames has created a distinctive design that uses molded-in reinforcing gussets at frame joints. Mitered C_f composite tubes are joined by laminated tubes that have intersectional C wraps at the joints. Extreme pressure and heat are then applied, resulting in the strengthening of gussets that meld the frame into one unit instead of creating the seams or parting lines found in some C frames.[169]

Conventional frame-building methods generally bond C_f tubes to either Al or C lugs. Because the fatigue life of fiber-reinforced joints is appreciably longer than that of Ep-adhesive joints, the possibility of catastrophic failure is virtually eliminated.

Additionally, attaching parts such as shifter and water bottle bosses and cable guides to C_f materials has always been a problem for manufacturers of C frames. As a result, many manufacturers drill holes in the structure, blind-rivet the part onto the C_f laminate, and reinforce the area or bond the pieces to the frame.

One company uses a Ti band design to affix add-on parts, which eliminates problems with frame integrity and creates a lighter frame overall. Ti was chosen because it resists corrosion and has a CTE similar to that of C_f composite.

Another manufacturer considers safety to be of paramount importance in designing and manufacturing its bicycle suspension systems, bicycles, and cycling-related products. All of its products have structural requirements that take safety factors into consideration.

When a new product is developed, a variety of materials and processes may be considered. After the proper material and process have been chosen, the design of the part takes material strengths, maximum anticipated loading characteristics, and a factor of safety into account. For example, this firm's bicycle rear suspension comprises two C_f support beams using a center laminate of elastomeric damping material as shown in Figure 3.51. The C_f beams function as a load support and spring mechanism for the rider's weight.

The upper and lower beams are formed around a lightweight PU foam core. The foam core is strongly reinforced with stringers of unidirectional E-Gl and C_f which are tightly woven over it. The foam core/Gl/C_f preform is then placed into a mold, and Ep resin is injected. RTM produces parts that have excellent resin-to-fiber ratios and a low volume.[170]

An English sports car manufacturer, Aston Martin, has developed a high-tech bicycle that uses a special aerodynamic steel and C_f frame and C_f wheels, with tires inflated to about 20.4×10^5 Pa.

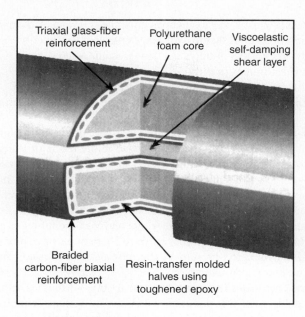

Triaxial glass-fiber reinforcement

Polyurethane foam core

Viscoelastic self-damping shear layer

Braided carbon-fiber biaxial reinforcement

Resin-transfer molded halves using toughened epoxy

Figure 3.51 Two C_f support beams sandwiching a center laminate of elastomeric damping material absorb the shocks of irregular road surfaces, which in turn cushions the rider from shocks and enhances comfort.

For the past decade the U.S. bicycle industry has been on a roll, becoming increasingly competitive as bike racing and recreational riding have grown in popularity.

A three-spoke, lost core injection-molded C_f-nylon bicycle wheel has been developed. The polyamide (nylon) is reinforced with 20–30% long C_f, a combination that offers a low density, a high modulus, and excellent impact resistance. The wheel weighs 1 kg and can withstand a 5.3-kN load compared with a standard 0.9-kg Al wheel which can withstand only 3.1 kN before failure. In addition, wind tunnel tests involving similar three-spoke wheels indicate that the design can save 5 min over a 100-km race. To complete the wheel, an Al rim is attached by adhesive bonding, and an Al hub is inserted.

Another joint venture of two firms produced a three-spoke prototype wheel. Made of C, Kv, and Gl_f/Ep resin and an Al rim, this specialized wheel is a futuristic hybrid of the traditional spoked wheel and the now-familiar disk wheel (which resembles a phonograph record and is commonly used by competitive cyclists and hip bicycle messengers). The wheel has some of the disk's advantages over 36-spoke wheels—fewer spokes translate to reduced drag created by air turbulence—and is free of some of the disk's shortcomings, such as instability in severe crosswinds.

In fact, it is claimed that the new types of spoke wheels can cover 1613 km in at least 4 min and as much as 10 min faster than a 36-spoke wheel. This obviously would be an enormous advantage in stage racing, but of considerably less or no value to the recreational rider.

A possible drawback of the wheel's strength is its weight: for road bikes, an average 36-spoke wheel weighs between 800 and 900 g; a disk wheel, anywhere from 600 to 2000 g; and these special wheels weigh 1.2 kg. But engineers believe that the aerodynamic advantage of the three-spoke wheel more than offsets the weight difference.

Many bicycle designers believe a stiff frame is the holy grail of racing bike design.

This is because even an Olympics-class cyclist generates only about 0.5 hp of energy and any flexibility along the bottom bracket of the bike dissipates some of this energy. A consensus among bicycle producers is that monocoque units manufactured from C_f/Ep prepregs have provided the stiffest frames thus far.

Both Duralcan and BP[171] have developed an Al alloy-10% Al_2O_3 or 20% SiC_p with high stiffness, low density, and good fatigue resistance for bicycle frames for commercial use.

The Duralcan 6061 Al-10 Al_2O_{3p} composite material is used[171] in the Stumpjumper M2 mountain bike, while BP 2124 Al/20 SiC_p material is used in the frame of Raleigh's racing bikes.[172] Both models have been successfully tested in extensive sports trials. In the former case, tubing of ~1.5-mm wall thickness is extruded via the porthole bridge tooling technique, which involves rejoining under pressure around the tube mandrel. The tube sections are then fusion-welded by GMAW using conventional techniques. In the case of the BP frame, the material is made by a powder route and then the tubing is adhesively joined. In addition to having improved specific stiffness, both bikes have proven to have exceptionally good fatigue endurance, presumably because of the enhanced value of ΔK_{th} compared with that for unreinforced material.

A bicycle wheel rim has been extruded from Duralcan Al_2O_{3p} MMC.[173] The extrusion was then bent and butt-welded into wheels for mountain bikes. Besides being lighter than conventional composites, metal matrix materials can be fabricated with metal-handling processes. The rims, consisting of 6061 Al/10 Al_2O_{3p}, are used on mountain bikes.

The main attraction of the MMC is a specific stiffness about 40% higher than that of C_f composites, which allows thickness and weight to be cut significantly without affecting strength and stiffness. The rims are about 10% lighter than rims made of C_f composite, and the company is also testing an off-road bike frame made of the MMC. The frame is 0.2268–0.4536 kg lighter than the lightest frame now on the market.[173]

The Kestrel 4000 all-composite bicycle has been produced and is 20% stiffer, 2.5 times stronger, and 30% lighter than bicycles made with steel frames. The Kestrel is constructed of C/Ep composite, and the tubes are formed from unidirectional fabric tapes. Although this method of construction is complex, it allows each ply angle to be optimized to provide the best load distribution in each tube.

Another advantage of the composite construction is improved vibration damping, reducing rider fatigue. In addition, tube shape can be modified to improve aerodynamics. For this reason, the Kestrel was recently banned from international competition because judges claimed that the designers used composites to give the bike an unfair aerodynamic advantage.

The use of composites also allows the frame to be molded in one piece, increasing the strength of critical load-bearing areas.

Sumitomo Heavy Industries, Ltd.,[174] has developed a forming machine for high-speed deep drawing of thermoplastic advanced composite materials. The new forming machine controls sheet fiber orientation during the process to permit wrinkle-free forming even when producing complicated shapes, and a rapid heating mechanism to allow optimum temperature control suited to the material characteristics.

So it is possible to form even an intermediate layer having resin impregnated in the fibers and consolidated in sheet form. The first product created using the new technology was a racing helmet for bicyclists. The ACM helmet is ultralight and impact-resistant.

It weighs 270 g, where a conventional product with the same strength weighs about 400 g.

The material used for producing the helmet is an ACM sheet having a Gl_f fabric consolidated with polyamide resin, and although the headgear thickness is only 0.75 mm, it contains an overall length of 300 km of Gl_f 0.014 mm in diameter. The cycle time for mass production is less than 1 min.

In the 1992 Olympics the U.S. cycling team featured a helmet with an inner expanded-polystyrene liner with an outer microshell, in which polyethylene terephthalate with graphite (PETG), polycarbonate (PC), and ABS have been tested. PC was the final choice for the microshell, as it can be molded in place with the liner rather than separately as a cap layer. Bicycle helmets are ventilated (in terms of heat and air intake or exhaust) through a series of holes. The use of a more dense expanded polystyrene (1.8–2.7 kg) allows the holes to be larger, or plastic rings can be foamed inside the helmet skins as crack retardants. The Olympic road helmet weighs approximately 0.2551 kg, while the track helmet weighs about 0.1984 kg.[175]

Another innovative application is a composite pedal shaped to reduce air resistance and to increase the bending angle on curves to 36°. The pedal is molded of LNP's Verton long-fiber-reinforced nylon, which provides high stiffness and impact strength and a good surface finish.[174]

3.5.10.3 Golf clubs: shafts and heads.

Typically when one thinks of filament winding, thoughts of rocket motor cases or automotive driveshafts come to mind. Recently Spaulding Composites–advanced composites technology (ACT) expanded their use of the filament winding manufacturing technique by creating an improved recreational product- Gr/Ep in golf club shafts.[176] Being a tubular structure, a golf shaft is well suited to filament winding. Although initially one might think of the design of a filament-wound golf club shaft as very simple task, it is actually quite complex. Combinations of shaft characteristics such as flexural rigidity, torsional rigidity, frequency of vibration, bend point, geometry, and weight all play a major role in influencing the "playability" of a golf club shaft.

Designers and engineers have found that filament-wound shafts offer greater consistency than their sheet-rolled counterparts. In the game of golf, as in most other sports, advanced technolgy is becoming more and more desirable.

Most manufacturers compete to produce top-quality metal or wood golf clubs, and the competition is every bit as fierce as that on the golf course. One company has developed a manufacturing process that combines a high-performance adhesive with a state-of-the-art Ep dispensing system, making the old threaded connection between the club and shaft a technique of the past.

A key factor in manufacturing golf clubs is the head-to-shaft attachment. Because of the stress at this point, the club needs a connection strong enough to withstand the golfer's swing and impact and durable enough to last for years. No manufacturing method has proved as effective as Ep adhesive bonding in attaching the composite shaft to the head.

Two new innovations in golf equipment[177] have been development of the new "Great Big Bertha" which is approximately 25% larger and yet 10% lighter than its older sister "Big Bertha." This new driver is made of Ti, which is extremely lightweight but strong enough to stand up under tremendous force. Great Big Bertha is not for everybody. With a Gr shaft weighing only 56 g and lofts of 10 and 12°, it is geared toward golfers with lower club head speeds up to about 90 mi/h.

The other major new technological advance in golf clubs comes from the introduction of new shaft technology that produces a grip 40% lighter than conventional ones. Burner Bubble shafts are manufactured by precise, high-pressure, inner molding process used to make parts for spacecraft and supersonic jets. Traditional Gr shafts are filament-wound or sheet-wrapped, where filament winding has the advantage of bringing consistency to shaft production but has limitations on shaft specifications, particularly weight distribution. Sheet wrapping adds flexibility in shaft design.

Inner molding blends the best of both methods. Gr material is placed around a bladder which is then inflated with pressurized air, expanding the Gr_f to a predefined shape as specified for the individual shaft's flex characteristics. The process also produces significantly denser compaction of the composite material, eliminating voids. This ensures durability and consistency from shaft to shaft.

3.5.10.4 Tennis rackets.

One of the major benefits of composite structures is their ability to absorb impact and vibration forces. This characteristic is used to advantage in tennis rackets, for instance, where the frame absorbs the vibration of ball contact before it travels through the grip and into the arm. This helps reduce the threat of tennis elbow.

One example is the Ultra FPK racket from Wilson Sporting Goods, which consists of Kv_f and ceramic Fiber FP along with Gr. The racket reportedly combines the play characteristics of a stiffer racket with the comfort of a soft racket.[173]

Fiber FP is a highly pure form of Al_2O_3 that has a higher modulus and better vibration-damping properties than the silica-based fibers typically used in ceramic rackets. The Gr contributes strength and stiffness, providing durability and reducing racket head deflection. The Kv helps absorb vibration, and its high-impact strength also enhances durability. The combination of fibers provides durability, power, stroke control, and a soft but solid feel (Figure 3.52).

Engineers are developing tennis rackets with increasingly larger and unusually shaped heads for greater accuracy, and with wider-bodied frames for more power, but are

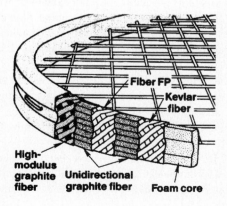

Figure 3.52 Tennis racket consisting of high-modulus Gr, unidirectional Gr, Fiber FP, and Kv, all wrapped around a foam core.[173]

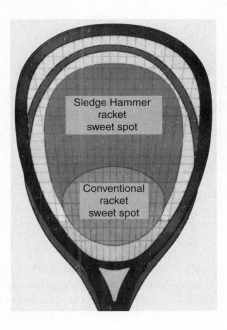

Figure 3.53   The Sledge Hammer sweet spot.[177]

light enough to swing with ease and velocity. To do this, they're using more and more exotic space-age materials that offer both strength and light weight.[177] For example, the "sweet spot" or optimal hitting area has been stretched to nearly fill the face of Wilson Sporting Goods' newest "Sledge Hammer" tennis racket (Figure 3.53). The racket frame is made of 90% Gr and 10% Gl.

3.5.10.5 Sailboats, catamarans, trimarans, racing shells, canoes, and paddles.   Water sports represent a growing market for suppliers of sporting goods. They are turning to composites to provide stronger, lighter equipment that enhances a user's performance. In addition, they are finding that composites can cut manufacturing costs.

As an example, a new racing shell features an advanced composite hull made of Kv 149. The material makes the shell lightweight, stiff, and impact-resistant. These structural characteristics translate into high speed and durabilty.[173] The design was driven by an effort to create a shell that would minimize lost oar stroking power. Long, thin shells tend to bend and twist from the oaring force generated by the crew, and this action detracts from the forward force vector. A proprietary knitting process has been developed to produce Kv 149 fabric with the fibers aligned in the ±45° direction, eliminating extra labor costs and material waste that would have resulted from using conventional ±90° fibers (Figure 3.54). The end result was a 1828.8-cm-long racing shell that weighs only 99.79 kg fully rigged. Preliminary races indicate that the composite construction provides a decided speed advantage over conventional shells.

In the area of recreational water sports, high-performance sailboards are now manufactured from Amoco Chemical's Xycon hybrid resins, PU foam core, E-Gl, and

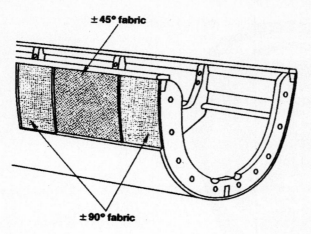

Figure 3.54   Hull skin of a racing shell.[173]

aerospace-grade C_f, which includes nylon scrim. This composite construction provides a lightweight, stiff, strong fin and board. Fully rigged with mast, boom, and sail, the board weighs only 21.77 kg. Use of the RTM method produces a seamless sailboard, eliminating a common point of failure. In addition, critical areas of the board have been stiffened to provide better durability in the face of strong ocean waves.

Besides increased strength and stiffness, composites also offer improved wear resistance and the ability to mold several parts into one. These capabilities have been incorporated in the redesign of fishing reels produced by the Zebco Division of Brunswick Corporation.[173] When a series of reinforced thermoplastic polyester composites are used, the redesigned reel has only 47 parts, several of which perform more than one function. The body is molded of a 25% Gl-reinforced polybenzothiazole (PBT). The roller pickup arm is a PBT composite with 25% Gl reinforcement and 15% polytetrafluoroethylene (PTFE) added for lubricity and increased wear resistance. The pickup arm wheel is a PBT reinforced with 30% pitch C_f for maximum compression strength.

Sails and sailboats are revolutionizing the composites industry. With yacht races often won by mere seconds, designers are using the latest CAD/CAM and FEA technology to create the perfect sail. Borrowing a technique from the aerospace industry, engineers have developed a method of building 3-D sails that combines weaving and sailmaking in one process. This technique allows the fabrication of unique sails, called 3D laminate (3DL) sails that more closely match what top-notch boat designers are developing on their computers.

Large molds of Al panels match the size and shape of the sail being built. The molds are precisely positioned according to settings generated by a computer model of the sail. A layer of Mylar is laid down on the mold, and then a six-axis machine with a knitting head lays yarns onto it. The head, mounted on a six-axis gantry follows the shape of the computer model, and so accurate placement of the mold is crucial. The yarns, made out of Kv, polyester, C_f or almost any other material, give the sail its strength. After curing under infrared lights, the sail is trimmed and attachments such as grommets are added.

The common method of making sailboats using thermoset FRP laminates may be tossed overboard for a new approach Advanced Composite Process (ACP) a patented manufacturing system to produce sailboats, developed by USA of East Haddam, Connecticut, in cooperation with JY Sailboats, Noank, Connecticut. The ACP process is a method for making a three-layer structure that has a thermoformed thermoplastic skin, a foam core, and a Gl_f reinforcing layer. The ACP process is used for making sailboat hull and deck structures. It is possible to now make a sailboat that is stronger and lighter than standard FRP boats but requires much less manufacturing time and less user maintenance.[178]

Engineers have also now created a 1310.64-cm yacht powered by the wind, diesel engines, or both. Using CAD software, engineers designed the monohull, which was vacuum-molded with Ep resin reinforced with knitted unidirectional Gl and C_f. A prototype has been built and used to complete a trans-Atlantic journey.

Canoeing enthusiasts looking for some relief for their backs and shoulders might find it in a new, lightweight paddle. The enhanced, high-performance oar weighs 0.2268 kg—approximately half the weight of the lightest wooden paddle. Contributing to the extremely low weight is the paddle's blade. It's constructed of Rohacell IG, a closed-cell, rigid polymethacrylimide foam, and the material is used in the RTM manufacturing process. The RTM technique has been designed to achieve high fiber volumes in paddle components and requires core materials that do not break down quickly. Rohacell IG provides a high strength-to-weight ratio yet withstands the extreme pressures and high temperatures of the RTM process. As a result, the process attains a 55–65% fiber volume range with Rohacell, as compared to a 30–50% fiber volume with other materials.

Originally the rigid foam functioned as a core material in aerospace applications that use fiber-reinforced cover skins. Other sporting equipment using C_f composites include rowing shell seats and short-track speed-skating boots.

3.5.10.6 Workout machines, baseball bats, and sailplanes.   C and Gl_f are structurally foamed with nitrogen gas in a polypropylene homopolymer matrix to form the black base of a Nordic Trak rowing machine. The composite offers inherent sound damping, the overall base does not require secondary painting, and the fiber and molding process create a smooth surface with a texturized appearance. It's reported that the composite base has allowed a reduction in parts by 20%, and assembly labor by 22%. An intricate ribbing system is used throughout for reinforcement and to transmit forces away from high-pressure points such as screw holes in the composite.

Even the national pastime has benefited from the use of composites. Braided Gr composite has been used in producing a baseball bat, the TRX-1 Boston Bat. The bat combines the light weight and high strength of Al with the feel and sound of wood. The braiding process permits yarns to be combined in any orientation over a mandrel of virtually any shape or size. The braided preform is then placed in a two-part resin transfer mold, and resin is injected under pressure and vacuum. The cured part comes out of the mold with an excellent surface finish and good dimensional tolerances.

The Boston Bat is molded around a steel core pin, and the molded shell is filled with cellular urethane. The sweet spot can be broadened and moved along the barrel by altering filler location. The ends are then capped and the bat painted black.

The higher strength and lighter weight of Gr make it possible to provide a range of balances within a player's acceptable weight range. In addition, the material is more durable than Al.

A sailplane, *Scimitar,* has been designed, built in the United States, and flown for the first time in 30 years in competition in New Zealand. It consists of a wing that is flexible in bending and stiff in torsion as a result of this unique combination of an S-Gl spar and Kv skins.[178]

3.5.10.7 Sport shoes, in-line skates, scooters, dune buggies, and arrows.
Whether they are designed for team or individual sports, all athletic equipment shares one design element: a reliance on cutting-edge materials—often originally developed for other applications—to help athletes push the performance envelope.

Foremost among examples is footwear. For almost every sport, there is specially designed footwear, and athletic shoes provide a rigorous testing ground for new materials. Witness the new Instapump Fury running shoe from Reebok. The arch and midsole of this shoe are constructed of Graphlite, the Gr/Gl composite originally developed for aerospace use.[175,179]

A honeycomb cushioning material called Hexalite owes its rigidity and lightweight characteristics to aircraft wing design and provides heel cushioning in the Fury, which has a low overall weight of 0.2268 kg.

Other shoes feature a composite rollbar that resists breakdown for long-term stability. Biomechanical models suggest that the rollbar can reduce excessive pronation, one of the major causes of injury among runners. The rollbar consists of two hybrid C_f plates and an Ellastollan thermoplastic urethane (TPU) core.

Rollerblade manufacturers incorporate a special composite shoe inside a skate frame. Users detach the skate frame upon arrival at their destination and fold it up until the skate blades are required.

The body of the E24B electric scooter from Mitsubishi Kasei[170] functions as both the structural frame and the exterior panel. The company has developed a SRIM system and a matrix resin to produce the unit. Thermoforming was used to reinforce an Ep-acrylic hybrid resin with randomly oriented Gl_f mats. The scooter is smaller and lighter and has more consolidated parts than a model with a traditional steel pipe frame.

The use of textile composites pared 68.04 kg from a dune buggy assembled by students at North Carolina State University at Raleigh. C_f fabric was used in the frame. The roll cage combined C and Gl fibers and part of the roll cage was made with a triaxial braider.

Gr composite arrows are lighter, slimmer, faster, and tougher than conventional Al arrows. The shafts are made by combining continuous unidirectional tows of C_f and thermosetting resin. Shaft diameter is smaller than that of Al arrows, reducing drag and increasing velocity. This results in less wind drift, flatter trajectories, and deeper penetration.

3.5.10.8 Ice hockey sticks.
The primary methods of manufacturing ice hockey sticks involve filament-wound prepregs, filament-wound RTM, and in-line braided pultrusions.

The anhydride Ep resin systems used are proprietary blends featuring high-T_g (177°C), high-impact resistance, and high-elongation properties. The toughened system also features high ductility so that under ultimate failure load the stick has a tendency to collapse and fold rather than to snap or explode as would a more brittle system. The shafts employ a hybrid of 24×10^3 MPa C and 6.9×10^3 MPa E-Gl in wind pattern variations of from 5 to 55° of inclination. Longitudinal flexural rigidity is obtained by varying the wind angle within a specific pattern. The sticks are processed using a filament-winding RTM method.

3.5.10.9 Sport leg braces and wheelchairs.

With increased participation in sports activities comes a greater risk of injury, with the knee being especially vulnerable. Typically, knee braces have been heavy and cumbersome, limiting the ability of players to return to their favorite pastime. Now, Analog Orthotics Unlimited, Inc., has replaced the metal in their braces with a thermoplastic composite of Ryton PPS from Phillips 66. Weight has been cut by one-third to only 0.6804 kg. In addition, the material can be molded to an exact fit, increasing patient comfort and ensuring that the brace stays in the correct position.[173]

The Quickie carbon performance wheelchair has a C_f/Ep frame that absorbs shock and resists corrosion. A molding process gives the frame high fiber density and large, strong joints while keeping total chair weight below 8.62 kg.[170]

3.5.11 Miscellaneous Applications

3.5.11.1 Military.

The U.S. Army plans to design composite parts that could lower the weight of an M-113 tank to 34,473.6 kg versus the designed and fabricated 63,504 kg. Processes to be used in the weight-reducing program include injection molding and RTM.

The air intake system of the M1A1 Abrams main battle tank has been redesigned and a prototype made using advanced composite materials. Problems of the current Al part such as air leaks, weld cracking, and reproducibility were eliminated, and required structural properties were maintained with the composite design and use of the RTM process. It allowed fabrication of the airbox-plenum as a one-piece structure that was 38% lighter than the current Al plenum and economically competitive. The composite airbox-plenum also enhanced airflow by integrating rounded corners and more gradual directional changes into the design.

PVC core E-Gl laminate turret structures appear to be feasible for naval turrets. They are adequate structurally for a 25-mm Gatling cannon at various rates of fire. Structures of this type are compatible with fabrication techniques that have been used in the boat industry for over a decade. They are competitive in price with metal turrets and offer advantages in corrosion resistance as well as ease and flexibilty of manufacturing techniques, and it is possible to design the laminate to yield a given stiffness for the mount.

The U.S. Air Force has developed a wind-corrected munitions dispenser (WCMD), a canister that can be filled with a variety of munitions and dropped at high altitude from a global positioning system-equipped aircraft to strike within 3048 cm of a target.

One payload option for the WCMD is a classified warhead of specially treated C filaments. After being dispersed by an explosion or a gas generator cartridge, the fibers cause arcing and shorting in electrical equipment. The cloud of C_f and/or C_w is very dense, and each filament is small enough—measured in micrometers–to float in the air for some time. Because of their small size, the fibers can enter air-conditioning or equipment cooling ducts of electronic vans and avionics black boxes.

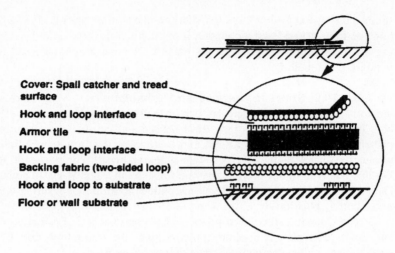

Cover: Spall catcher and tread surface

Hook and loop interface

Armor tile

Hook and loop interface

Backing fabric (two-sided loop)

Hook and loop to substrate

Floor or wall substrate

Figure 3.55   Portable armor concept.

The WCMD's C_f warhead can affect everything from the avionics of an aircraft sitting on a runway to the power grid of an electricity-generating plant. A C_f warhead can disable radar antennas and associated equipment without injuring the crews, making it a potential air defense weapon for the suppression of an enemy.

The new portable ceramic composite armor (Figure 3.55), made of a ceramic-metal matrix bonded to Kv, is being attached to C-141 aircraft walls, floors, and portable oxygen bottles with heavy-duty hook-and-loop material. The armor comprises several mats of 10.16-cm tiles and is designed to protect flight crews from small-arms fire such as a 7.62-mm bullet fired from an AK-47 rifle when an aircraft is flying at several hundred meters.

The material selected was a SiC/AlN ceramic matrix armor material. The ceramic-metal mixture was bonded to Kv, creating a composite-composite—actually a composite armor bonded to a composite polymer. By bonding the two materials into a seamless sandwich, the Kv becomes "a catcher's mitt for ceramic fragments" that are expelled when the tile is hit by a bullet.

The armor can be installed in approximately 2 h, for example, in the flight deck walls and floors of a C-141 aircraft. As currently configured, a bullet striking a tile will create a hole two to three times the round's diameter. The Lanxide ceramic matrix tile cracks, but remains intact inside the sandwichlike mat.

The lightweight version replaces approximately 1064.64 kg of steel armor with about 621.04 kg of ceramic composite material. Although roughly the same surface area is covered, the ceramic matrix armor weighs only 36.6 kg/cm^2 compared to steel's 58.6 kg/cm.^{2}

CCC battery cases will help make future spacecraft power systems up to 400% more efficient. The new containers are capable of housing new 75 to 200-W·h/kg sodium-sulfur batteries, and tests indicate that the CCC battery cases resist corrosion caused by contact with molten sulfur and sodium polysulfides. They also offer excellent electric and thermal conductivity. Other potential uses include civil and military remote power sources.

While composites continue to dominate in volume and value, the 1980s ended with increased interest in such ceramics as SiC and AlN. Research has focused on reducing manufacturing costs, producing near-net- and net-shaped parts and defending against kinetic energy and shaped charge projectiles. The beginning of the new decade has already seen the introduction of innovative CMCs.

The market for composite armor materials will experience strong competition among traditional, improved, and new materials. At the lightweight, high-performance end, aramid and polyethylene fabrics and composites will go head to head for use in vests, helmets, and blankets, as well as in spall and backing applications. High-strength Gl will compete both with these materials and traditional, heavier, inexpensive E-Gl.

Composites have been applied to a new rollbar design, resulting in a weight savings of more than 60% on the 105-mm M102 howitzer. The current steel rollbar weighs 106.6 kg, requires time-consuming manual installation, and makes the howitzer operations awkward. The composite rollbar was developed with an impressive weight savings (18.14 kg), ease of handling, and detachability and had a filament-wound Gl construction mounted on an Al substructure.

A standard hull composed of 23 Al parts for the Bradley fighting vehicle weighs 5443.2 kg. However, an experimental three-piece fiber composite for the composite infantry fighting vehicle hull weighs only 4082.4 kg. This hull is said to be the first nonmetallic type of an armored vehicle that meets the Army's severe structural and ballistic requirements.

A good structural composite requires a strong fiber-to-resin bond, while a good ballistic matrix needs a weak bond. The composite hull using an S-2 Gl_f system has reduced weight by 25% and is expected to cut manufacturing costs by 20% and increase fatigue strength by 60% with equal ballistic protection. Unlike a metal hull, the composite hull does not span or fragment from artillery hits. Also, heat loss is one-third that of a standard hull. In addition, reductions in noise and vibration alleviate crew fatigue and provide a lower signature against infrared, acoustic, and radar detection methods.

The composite hull is composed of two halves and a straight belly plate, the halves being joined with bolts and bonding. An Al subframe holds suspension members and diffuses forces transmitted to the hull. The hull surrounds an Al turret cage and ranges in thickness from 1.91 to 4.44 cm. It is thickest at the front and thinnest under the sponson, a horizontal portion of the hull just above the tracks.

In another armor application C-130 aircraft have been outfitted with Kv/Gl ceramic armor to protect aircrews from small-arms fire (Figure 3.56). The armor system has panels covering the entire floor, observation windows at floor level to the left and right of the pilot and copilot, and a crew bunk area. In addition, the liquid oxygen converter is fully enclosed in armor because it could explode if hit. The aircrew seats have also been partially enclosed in armor.

The Armourtek laminate is 14.5-mm thick and weighs 26.4 kg/m.[2] This material can be molded to fit curved surfaces such as aircraft seat backs because the armor system uses a formable Gl ceramic bonded to a Kv composite multiply laminate manufactured with a thermoplastic resin. The Gl ceramic offers a 44% weight savings over Al_2O_3, which is often used for ballistic protection, and one-tenth the cost of B_4C; it is also easy to machine and drill.

In another development a ceramic-metal composite is being used for lightweight ar-

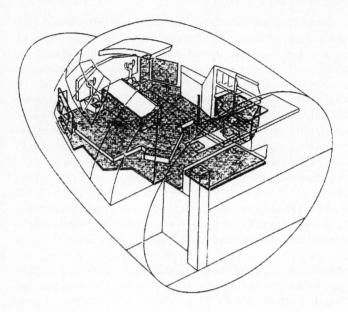

Figure 3.56 Cockpit armor in a C-130 aircraft.

mor in U.S. Air Force aircraft and has potential applications in golf club heads and computer disks. The process consists of binding B_4C_p in an Al matrix, which results in an extremely hard yet tough material. It measures about 90 on the Rockwell C hardness scale, is six times stiffer than Al, and at 2.6 g/cm^3 is lighter than Al. Replacing current armor in the previously discussed C-130s with this composite could cut aircraft weight by up to 1330.8 kg. Near-term use is limited by the high \$10/kg cost, but scientists believe this could drop to \$5/kg with volume production.

When in flight, rocket motors create temperatures in the 2023–3034°C range. Finding a material that can withstand this extreme heat for the tube housing the rocket's motor igniter created a problem which was solved with the use of Gl_f/PEEK. The igniter serves as the firing mechanism for rocket motors in military air-to-ground and ground-to-ground missiles.

A 30% Gl-reinforced Victrex PEEK retained the greatest amount of material after the immense heat generated by the firings. Additionally, the nonmetallic tube cut machining costs considerably compared to the metal model. In fact, the PEEK tube cost 80% less than the Al counterpart.

3.5.11.2 Commercial and industrial.

Almost every industrial process uses heat exchangers to transfer heat between process streams or to recover waste heat normally vented to the atmosphere. Until now, such *recuperators* used metal tubes that were limited to airstreams of 711°C or cooler. A new ceramic composite heat exchanger allows manufacturers to recover energy from waste streams such as flue gases at temperatures exceeding 1011°C. The Al_2O_3/ZrO_{2f} material resists corrosion, wears better than metal, and exhibits a high strength-to-weight ratio. At a prototype incinerator installation, researchers estimate that the ceramic heat exchanger reduced fuel consumption 35–38%. Other applications are in scrap metal remelting and advanced gas-fired turbines.

A new aircraft being planned will, according to its inventor, be able to take off and land vertically and fly efficiently at low speeds like a helicopter. It also will be able to match the high speeds, long range, high altitude, and efficiency of a regular plane. The new aircraft has a rotor blade on top to provide all the lift needed for vertical takeoff and slow-speed flight, and a compound turbocharged 300-hp GM Nascar racing piston engine and propeller for forward thrust. It differs from a helicopter in that the rotor is not powered by a driveshaft throughout flight.

The bearingless two-bladed rotor is a continuous tip-to-tip composite structure comprising a blade and a hub. The gearless drive system for the rotor and the twin-bladed C-Gr propeller is an efficient, lightweight poly chain toothed belt. The rotor drive is sized for short duration and low power because it is used only to prerotate the rotor to jump takeoff speed.

The streamlined, pressurized, round fuselage is also made of a composite material and the flapped, swept wing contains enough fuel for long-range flying. Twin booms support the vertical and horizontal tail surfaces and house the retractable main landing gear.

This design could potentially revive the general aviation market, especially because of its high-speed, over-the-weather, high-altitude flight and its long, economical cruising range.[180]

Manufacturers and users of compressed natural gas (CNG) fuel tanks find themselves at a critical juncture as federal and local governments sort out the future of alternative fuels for passenger vehicles. As a result of users' experience, Ep resin-based composites reinforced with E-Gl are favored in CNG applications over alternate resin systems.

Plastic composites are more actively replacing steel in heavy-wear applications such as bearings, bushings, and wear rings.

XC-2 composites, C_f/PEEK, are now serving in centrifugal pumps. These composite parts rival steel parts because they resist galling and seizure and can run dry—preventing catastrophic failure when pumps leak fluid. Besides handling fluid loss, XC-2 composites can endure impeller and shaft vibrations and extreme temperature excursions common to centrifugal pumps.

With a 2040-MPa tensile strength and an operating temperature of up to 303°C, XC-2 composites have excellent chemical and abrasion resistance and a thermal expansion close to that of metal. Their toughness and impact resistance exceed those of common plastics. Current applications for these composites include pump rings and bushings in industries such as oil refining, chemical processing, pulp and paper mills, utilities, and nuclear power generation.

TiB_2 and TiB_2/Al_2O_3 composites are said to offer excellent electric conductivity, high compressive and mechanical strength, resistance to chemical reactions and thermal shock, high thermal stability, and high operating temperatures. Their hardness is superior to that of WC, their thermal conductivity is better than that of cubic BN, their fracture toughness is greater than that of Si_3N_4, and their stiffness-to-weight ratio is excellent.

Originally developed to be used in armor for U.S. Army tanks, their commercial applications include cutting tools, dies, and electrodes. With up to five times the durability and one-fifth the weight, using filament-wound C_f/Ep composites for power transmission systems in driveshafts, rollers, and other wear- and corrosion-resistant parts can boost production line run time, decrease maintenance, and reduce back injuries.

There are over 4000 driveshafts in operation, including other applications in high-speed (up to 4500 hp) turbine-driven boiler feed pumps and vertical pump drives for sewage treatment plants. Future developments are expected in high-torque, high-horse-power power transmission applications—up to 20,000 r/min.

A fast-growing market for filament-wound, braided, and roll-wrapped C_f/Ep tubes is in rollers used in web-driven paper, printing, and plastic film conversion equipment. By replacing heavy metal rollers, the lightweight composites may help reduce the high rate of insurance claims for back injuries caused by repeated roller lifting. Lightweight rollers also need less power to turn and have a higher natural frequency than metal, and these benefits can speed up production lines.

A 70 to 73-vol% SiC_p/Al matrix composite is being vacuum-diecast to produce 3-D parts having excellent quality and performance. The composite is said to offer a beneficial combination of low CTE and very high thermal conductivity, which can improve both package reliability and heat dissipation in semiconductor packaging components.

The efficiency and capability of continuous-fiber ceramic composite (CFCC) processing (SCS-6) in the fabrication of industrial components were demonstrated in the manufacture of ceramic composite furnace tubes and gas turbine combustors.

A piezoelectric ceramic-Ep composite said to display superior ultrasonic performance compared with conventional monolithic materials has been developed. The piezocomposite ceramic reportedly produces transducers having high-gain, broad bandwidths and a high signal-to-noise ratio, a combination that is not possible with conventional materials.

Probes using this technology provide high gain because the piezocomposite has a relatively low acoustic impedance, which matches more efficiently low-impedance materials such as water and plastic. The material has a broadband response because of its inherent damping characteristics.

Liquid crystal polymers (LCP) are finding their way into more medical applications because of their superior chemical resistance. A new dental syringe gun is one example of LCP not only withstanding the harsh chemicals used in sterilization but also meeting new design requirements. The switch from polyarylsulfone (PAS) to LCP eliminated a 2% return rate due to environmental stress cracking from repeated thermal and chemical sterilization. Extended wear life is a key issue in the industry; at the same time cross-contamination and infection are issues, and so sterilization is crucial.

The nondisposable device requires high strength and durability as a dispensing unit, and it applies light-curable composite resins for both cosmetic restoration and as a replacement for traditional metal-based amalgam fillings. PAS is weakened after 500–1000 cycles of various sterilization methods, however, LCP is expected to achieve a minimum of 2000 cycles.

Engineers know that flywheels can shrink the package size of energy components for electric vehicles, as well as add efficiency and trim overall vehicle weight. The cost of materials for such a demanding application, however, can be prohibitive. Now, a less-expensive, high-performance composite flywheel rotor for electric vehicle motors and other applications has been developed. The composite promises to make flywheel technology more practical by creating strong, durable rotors for as little as 20% of the material costs of conventional designs.

The composite uses a Dow-UT process called *polar weaving* to add strength to the rotor without increasing weight. By aligning reinforcement fibers in the radial direction in addition to the conventional hoop direction along the rotor's circumference, engineers can replace Gr_f with E-Gl_f. In some applications, the composite replaces Gr_f entirely. Engineers are pairing the new composite rotor with a high-density motor to evaluate automotive hybrid propulsion systems for the future.

Fiberglass composite wind blades convert wind into electric energy. The blades are grouped and assembled into three-bladed propellers which, when mounted on towers, capture the wind's energy. These wind-energy converters will be clustered on wind farms in Texas to generate energy for southwestern utility companies.

Toray Industries, Inc., has established a molding technology for manufacturing rollers made of C_f-reinforced metal (CFRM) by using the reinforcing C_f in long lengths, allowing the manufacture of long, high-strength, highly elastic rollers. More specifically, uniformly filling molds with easily entangled long C_f and efficiently removing air bubbles generated inside allowed the manufacture of 1.8-m rollers for use in textile- and film-producing machines and printing machines.[181]

The use of composites (usually C_f-reinforced Ep) in winding shafts offers several benefits that stem from the weight savings (generally about 40%) they provide over steel and Al. The most notable is the potential for reduced insurance and workman's compensation claims resulting from back injuries in lifting shafts. Manufacturers of composite winding shafts claim that despite their higher price, processors and converters provide a tangible return on investment from reduced insurance costs and fewer workdays lost because of injury.

Moreover, because of their stiffness and light weight, composite shafts are also said to accept faster winding speeds with no sag or vibration than their steel and Al counterparts.

A lug has been developed that is fabricated of nylon 66 with 30% Gl_f reinforcement. It is one-fourth the weight of a steel lug, and when used with a composite shaft, it can reduce the total assembly weight by 5%. A processor or converter that uses a composite shaft with the reinforced nylon lug would thus be able to achieve a weight reduction of 45% over steel.

A composite journal, which is still in development, can be fabricated by RTM. The use of such a component with a composite shaft would reduce total shaft weight by 50–60% where the target weight reduction is 50%. A composite shaft and journal, with the reinforced nylon lug, would thus weigh up to 60% less than a comparable assembly in metal.

Researchers at the Department of Energy's (DOE's) ORNL have combined a ceramic and a metal to form a new ceramic-metal composite that can be used for rock- and coal-drilling equipment or dies. The new composite combines the strengths of both the ceramic and the metal and overcomes some of the weaknesses of the ceramic-metal alloy, forming a material that is very hard and tough.

The material is made from WC chemically bonded to a modified Ni_3Al alloy. It offers several advantages over currently used commercial methods. Tests show that the new material is harder and may last longer than WC bonded to a Co composite used commercially throughout the world for dies, in drilling equipment, and in other cutting tools. The new ceramic-metal composite also is less expensive and contains metals that are readily available.

Applications that encounter high temperatures can benefit from bearings made of polyimide resin-based composites and components made entirely of the C_f-reinforced polymeric composite. The composite can withstand hostile environments that produce continuous surface temperatures of 225–359°C and spot temperatures up to 455°C. The material is self-lubricating and corrosion-free. In addition, the composite parts weigh 60% less than comparable metal components and can be compression-molded to near-net shape. Aerospace applications are common, but composite components also have uses in the petroleum, dental, and textile industries.

The Patriot's unique design replaces the standard lead-acid battery with a C composite, 55.45-kg flywheel energy storage (FES) system. The flywheel converts latent electric energy to rotational energy and stores it by spinning at up to 58,000 r/min in a near-perfect vacuum on almost frictionless bearings. The flywheel's C_f have a tensile strength of 432 MPa and act as buffers that allow the turbine to operate at a constant speed. Its latent energy is converted back to electricity and used when needed for maximum acceleration out of corners and down straightaways. When the car is not running at top speeds, the flywheel spins faster, storing the excess generated energy. In a hybrid-powered passenger car, the recaptured braking energy potentially could produce a 10–15% increase in fuel mileage.

A vest made of Spectra, a lightweight, high-strength synthetic fiber helps protect professional rodeo riders from the hooves and horns of bucking broncos and bulls. The vest combines Spectra fiber and ballistic material with closed-cell foams to provide protection, comfort, and freedom of movement. The vest absorbs up to 72% of the energy from an impact. Spectra fiber was also used in making the sails, ropes, and cordage of the *America,*[3] winner of the 1992 America's Cup, and in building the *Earthwinds* balloon, a lighter-than-air vehicle designed to fly around the world.

Lightweight, low-cost, high-precision mirrors are needed to support a number of near- and far-term submillimeter space-based astronomical telescopes. The development of these highly specialized mirrors was a challenge associated with (1) the processing and manufacturing required to produce high-precision, lightweight mirrors, and (2) the determination of materials and structural mirror configurations that produce the thermal stability needed for specific classes of applications. The technical approach used in the design, manufacture, testing, and analytical simulation of lightweight Gr/Ep mirrors is described by Freeland and Johnston.[182]

SiC_w/Al_2O_3 cutting tools have been available and in use for the past dozen years. The whisker reinforcement produces a twofold increase in the fracture toughness of the composite relative to that of monolithic Al_2O_3. SiC_w/Al_2O_3 has received widespread acceptance in the aerospace industry where it is considered the state-of-the-art cutting tool material for rough machining of superalloys.[183,184]

SiC/Al_2O_3 tools can run up to 10 times faster than carbide tools because they can survive temperatures above 1000°C at the cutting tip versus 600°C for carbide tools. Fabricators are able to run machines longer without replacing inserts, which justifies the four to five times higher cost of SiC/Al_2O_3.

SiC/Al_2O_3 inserts can also be used to cut hardened (>45 Rockwell) steel but cannot be used with Ti and Al because of chemical interactions. SiC/Si_3N_4 tools have now appeared on the market, however, they are expensive.[183,185]

Researchers have developed a hot-gas candle filter that offers greater durability and longer life expectancy than existing technology. Used as a gas stream filter in power generation systems, the ceramic composite filter combines continuous and chopped ceramic fibers reinforced with a strong SiC matrix in a lightweight, thin-walled design that provides resistance to corrosion, thermal shock, and brittle fracture. The chopped fibers form the filter surface. A braid of continuous fibers creates a tube sheet covering the filter, which prevents mechanical abuse and further brittle fracture.

A composite external fixator was recently produced to repair serious fractures. C/Ep laminated pieces and C/Ep pultruded rods reduced weight by 20% compared to that of the previous stainless steel-Al design. The Ep resin pultrusion was reinforced with 206×10^3 MPa modulus C_f, and radiolucency was introduced so physicians could x-ray through the device to monitor bone growth and healing. Finally, the designed and fabricated part was able to withstand sterilization in an autoclave, which typically uses steam at 126°C at 0.21 MPa.

Tabs made from a composite of Gr and PMR-15 PI allow the high-temperature tensile testing of flat coupons of CMCs. Tabs made of the Gr/PMR-15 composite can also be used in high-temperature tensile testing of flat coupons of ceramics, metals, and MMCs.

Previously, high-temperature tensile testing of ceramic composite specimens involved the use of metal tabs. However, the significant difference between the CTE of the specimens and of the tabs resulted in delamination of the specimens and the tabs. The metal tabs are glued to the specimens with a commercial Ep-based structural film adhesive, and must be held at a temperature of 177°C to cure this adhesive. Upon cooling, the residual stresses caused by differential thermal shrinkage cause failure of the adhesive joints. In contrast, tabs made of Gr/PMR-15 are sufficiently compliant to prevent delamination.

Shikibo, Ltd., has developed a CCC bolt with excellent strength. Shikibo used its own 3-D woven fabric structure to produce the bolt, which showed higher strength and better shear characteristics than existing CCC bolts. The newly developed bolt also attained a tensile strength of 395 kg_f which was four times the prevailing strengths. The new composite is also so highly resistant that Shikibo hopes to develop new bolt applications for structural materials, which must be strong under high-temperature conditions, and in the field of high-temperature reactors.[186]

Two French companies have created a carbon brake systems company to produce Sepcarb, a carbon brake material. Aircraft C brakes usually have an operational life of 1500–2000 cycle compared with 1000 cycles for conventional steel brakes. An enhanced 2500-cycle C brake system is being evaluated on the Airbus A 330 transport.

Advanced composites are most often used in aerospace applications, but the materials are a viable option for fluid power jobs as well. Composite hydraulic and pneumatic cylinder tubing is a cost-saving alternative to metals for many low- and high-pressure applications. Cylinder bodies are typically constructed of fiber-reinforced thermoset Ep matrix materials, and they offer a number of performance benefits. Among them are the following.

Weight reduction. Gl/Ep composite cylinders weigh substantially less than their metal counterparts. At approximately 25% of the weight of steel or brass and 75% of that of Al, not only is the composite cylinder tubing a good alternative in applications where system weight is a concern but the composites are also easier to handle and less expensive to ship.

Corrosion resistance. Composite materials stand up to many corrosives that quickly destroy metals. They give a trouble-free performance even when exposed to moisture or to many acid, caustic, and salt solutions plus other adverse environments that would normally corrode or impair the operation of metal components.

Reduced maintenance. If the inside of a cylinder is impregnated with low-friction additives, the cylinder will run nonlubricated for millions of cycles without appreciable wear. The smooth, self-lubricating internal surface also prevents pistons from sticking, even after extended idle periods.

Elimination of honing. Composite cylinders may be manufactured with a 5 to 15 rms µin. internal diameter (ID) finish. This ID surface performs as well as a honed or chromed surface without the added costs of machining or plating.

Composite materials offer a number of other advantages as well:

- The material does not fail catastrophically if subject to overpressure conditions, a possibility with metal cylinders. Instead, the composite tends to develop microcracks and a first-ply failure mode which safely releases pressure.
- Unlike metals, the material does not dent. Typical impact strength is about 54.02 J (izod).
- The material bonds to most metals using a two-part Ep adhesive. Thus either tie rod or bonded end-cap cylinder configurations are possible.[187]

With an eye toward cellular phone users, researchers at the DOE's ORNL have developed a cloth that protects electronic equipment from stray electronic signals. This C cloth could protect people from the electromagnetic fields associated with cellular phones and electronic appliances.

Previously, Fe foil and C cloth in Ep were sandwiched to make lightweight shielding. The new method uses CVD to impregnate cloth woven from C_f with Fe. The C cloth is heated to about 200°C, and vapors of heated Fe pentacarbonyl liquid are carried by Ar gas into the chamber containing the heated C cloth. A layer of Fe is evenly deposited on the cloth to a thickness of about 0.1–2 µm.

A new fiber-reinforced ceramic composite material provides significant design opportunities for components used in thermal processing and furnace systems. This CMC consists of approximately 30% NEXTEL 312 ceramic fibers in a matrix of 70% chemical vapor-deposited SiC. This structure allows production of lightweight components in a variety of sizes and complex shapes. Typical are thin-walled components such as radiant burner tubes, burner inserts, heating element liners, and protective shielding for thermocouples.[188]

Typically, multilayered personal computer (PC) boards rely on metal cores sandwiched between the layers for heat conduction and control of cyclic thermal stress. Without the core, which constrains thermal expansion, there would be a high probability of circuit failure. But the core can cause problems. SiC_p/Al MMCs may offer a viable solution. In such systems, particles of SiC_p are embedded in an Al matrix. The system so created exhibits the best properties of both materials. With a density of 2989.43 kg/m³, SiC_p/Al MMC has the weight advantage of Al and a Young's modulus of 240×10^3 to 260×10^3 MPa.

With a thermal conductivity of 180 W/m·C, SiC_p/Al MMCs exhibit the thermal conduction advantages of Al. The thermal conductivity of the metal matrix determines how much heat the assembly can dissipate. An increase in the thermal conductivity of a board's core permits engineers to design modules with either higher power dissipations or lower operating temperatures for the same power dissipation. Whichever choice one makes, the bottom line is better reliability.

When comparing SiC_p/Al MMC materials with other materials, engineers see a 70% weight savings. This reduction can make a big difference in the designing of avionics subsystems.

Despite such advantages, MMC materials are strong, however, they are not ductile. Because of their high volume percentage of crystalline fill material, 60–75% of MMC materials tend to behave rather like ceramics. In this respect, they resemble much stronger versions of cofired multilayer ceramic PC boards. Potential applications include those requiring materials that must match the CTE of a ceramic, such as multichip module bases and chip carrier cooling fin assemblies.

A new MMC for electronics packaging has the electric and thermal conducting properties of a metal and the CTE and stiffness of a ceramic. The new SiC/Al material, MCX-622, has about one-third the weight of all-Kovar or one-sixth the weight of W/Cu packages. A pressureless metal infiltration fabrication process produces an isotropic compound, and a ceramic can be joined to the material with void-free bonds. Complex shapes can be produced with injection molding, but machining MCX-622 is not easy. Initial applications include chip carriers, SEM-E boards and heat sinks.

A composite Gl-ceramic is being developed for use in a sodium-based cell. Experiments show the MgO-Al_2O_3-BaO-CaO-B_2O_3 system has many promising compositions for use as ceramic-to-ANL (42 Na_2O-8 Al_2O_3-5 ZrO_2-45 SiO_2)-Gl seals. Hermetic ANL-Gl-to-MgO seals have been made and thermally cycled 10–14 times from ambient temperature to 450°C. Additional studies are now needed to optimize the MgO-Al_2O_3-BaO-CaO-B_2O_3 composition and to test the candidate ceramics for good chemical stability in Na/S and $Na/NiCl_2$ cell environments.[189]

Over the past three decades, power stations, refineries, and large process industries have become more and more dependent on cooling towers, which have over this period grown in size as a result of economics of scale. Increasingly, large fans inside these towers have required equally long driveshafts with support bearings to distribute weight. The problem was eventually overcome by combining a composite material with adhesive bonding technology. A C_f tube was manufactured by filament winding, with the fibers wound at an angle of 12° with the horizontal. This gave the tube the best blend of longitudinal stiffness and torque capability.

The composite driveshaft offered many benefits which include the following.

- Reduced weight—one-tenth that of a steel shaft
- Longer single spans—up to 8 m long without a central bearing
- Elimination of thermal bowing—zero CTE achieved
- Corrosion resistance—virtually impervious to corrosive atmospheres
- Lower bending loads—center section approximately one-tenth the weight of steel
- Higher fatigue load resistance—combined with lighter weight and better balance.

Similar systems that could benefit from this application are as follows.

- Driveshafts in deep well pumps
- Agitators in sewage ponds
- Stern tube driveshafts in ships.

REFERENCES

1. Shercliff, H. R., and M. F. Ashby. 1994. Design with metal matrix composites. *Mater. Sci. Technol.* 10 (6):443–51.

2. Tang, D., and C. L. Frank. 1989. Air Force applications of injection molding techniques. EM89-103. *Compos. Manuf. 8,* January 1989, Anaheim, CA.

3. Piellisch, R. 1993. Beyond steel: TMCs for lighter landing gear. *Aerosp Am.* 31 (7):42–3.

4. Macy, W. W., M.A. Shea, R. Perez, et. al., 1990. Advanced materials for landing gear. *Aerosp. Eng.* July:17-21.

5. McConnell, V. P. 1993. Next-decade defense: Less is more. *Adv. Compos.* January/February: 18–28.

6. Anderson, M. H. Processing of high temperature carbon fiber reinforced polymers.*High Temp. Compos. Clin.,* August 1992, Los Angeles, CA. EM92-215.

7. Lemmer, L., and G. Kagerbauer 1992. The design development of the monolithic CFRP centre fuselage skin of the European fighter aircraft, *37th Int. SAMPE Symp.,* pp. 747–51, March 1992, Anaheim, CA.

8. Klenner, J., F. Grier, H. Kriegelstein, et al., 1992. The production of a monolithic CFRP fuselage skin for the European fighter aircraft, *Proc. 37th Int. SAMPE Symp.,* pp. 1170-7, March 1992, Anaheim, CA.

9. Cassidy, V. M. 1990. New aluminum composite is stiffer, lighter. *Mod Met.* December:54–5.

10. Matsui, J. 1995. Polymer matrix composites (PMC) in aerospace. *Adv. Compos. Mater.* 4(3): 197–208.

11. Sakatani, Y., and Y. Yamaguchi 1980. *J. Japn. Soc. Compos. Mater.* 6:43–8.

12. Nakai, E., 1980. *J. Japn. Soc. Aeronaut. Space Sci.* 28:370; 28:417.

13. Mehdorn, E. 1992. *13th SAMPE Eur. Int. Symp.* p. 13.

14. Ojio, K. 1985. *Japn. Soc. Compos. Mater.* 11:114–8.

15. Ashizawa, M. 1991. Composite technology growth leading to application to the MD-11 and to civil transport aircraft of tomorrow. *2nd Proc. Int. SAMPE Japn. Symp..*

16. Tortolano F. W. 1994. Why composites are still soaring. *Des. News,* September 12:70–5.

17. Stover, D. 1989. The outlook for composites use in future commercial transports. *Adv. Compos* May/June:49–58.

18. Smith, B. D. 1990. The cautious approach. *Aerosp Compos. Mater.* November/December: 4–30.

19. Joynes, D. 1990. Inservice experience and maintenance of advanced composite structures in airline service. *Proc. 22nd Int. SAMPE Symp.,* pp. 1131–45, November 1990, Boston, MA.

20. Schwartz, M. M. 1983 and 1992. *Composite Materials Handbook,* lst and 2nd eds. McGraw-Hill, New York.

21. Palmer, R. 1992. Techo-economic requirements for composite aircraft components, NAS 1-18862. *Fiber-Tex 1992, 6th Conf. Adv. Eng. Fibers Text. Struct. Compos.,* Philadelphia, PA, pp 305–42, October 1992, NASA CP3211.

22. Cardaba, A. B., F. R. Lence, and J. S. Gómez. 1990. Design and fabrication of the carbon fiber/epoxy A320 horizontal tailplane. *SAMPE J.* January/February:9–13.

23. Dekok, R. E., 1993. Airbus floor beam: Towards a cost-effective composite design and manufacturing research project sponsored by Airbus Industries. *NTIS Alert* June 1:9–10.

24. Marsh, G. 1993. Weathering the storm. *Aerosp. Mater.* 5(1):13–7.

25. Schijve, W. 1994. Fokker 50 thermoplastic main undercarriage door: Design and cost-effectiveness of thermoplastic parts. *J. Adv. Mater.* 25(3):2–9.

26. Charles, D. 1991. Unlocking the potential of metal matrix composites for civil aircraft. *Mater. Sci. Eng.* A135:295–97.

27. Wong, R. 1992. Sandwich construction in the Starship. *Proc. 37th Int. SAMPE Symp.* pp. 186–97, March, 1992, Anaheim, CA.

28. Trego, L. E. 1993. Composite propeller system for Dornier 328. *Aerosp. Eng.* December: 7–11.

29. Walden, D. C. 1990. Applications of composites in commercial airplanes. *Proc. 22nd Int. SAMPE Symp.,* pp. 77–82, November 1990, Boston, MA.

30. Smith, B. A., P. Proctor, and P. Sparaco. 1994. Airframers pursue lower aircraft costs, *Aviat. Week Space Technol.* September 5:57–8.

31. Hatakeyama, S. J., 1995. Wing structures development for the HSCT. *Aerosp. Eng.* May: 25–30.

32. Bergmann, H. W. 1990. Materials and structural concepts for new space transportation systems. *SAMPE Q.* 22(1):51–61.

33. Waldman, B. J., and P. T. Harsha. 1992. NASP: Focus on technology. *AIAA 4th Int. Aerosp. Planes Conf.,* December 1992. Conference held in Orlando, FL. National Aerospace Plane National Program Office, Palmdale, CA. AIAA-92-5001, A-9322278.

34. Wilson, T. M. 1993. Applications of titanium matrix composites to large airframe structure, *Proc. 77th Meet. AGARD Struct. Mater. Panel,* September 1993, Bordeaux, France. AGARD Report 796.

35. National aero-space plane: Restructuring future research and development efforts. Report to Congressional requesters, 1992. General Accounting Office, Washington DC. GAO-NSIAD-93-71.

36. Harvey, D. S. 1994. Building V-22s smarter, *Rotor Wing.* June:26–30.

37. Schulze, E. J. and W. J. Kesack. 1990. Honeycomb sandwich composite structures used on the V-22 Osprey fuselage. *Proc. 22nd Int. SAMPE Tech. Conf.,* pp. 1059–69, November 1990.

38. Colucci, F. 1989. Graphite wonder. *Aerosp. Compos. Mater.* Fall:25–7,49.

39. Colucci, F. 1991. Trying thermoplastics. *Aerosp. Compos.* Mater. Summer:11–30.

40. Vuillet, A. 1989. Operational advantages and power efficiency of the fenestron as compared to a conventional tail rotor. *Vertiflite* 35 (5):24–8.

41. deBriganti, G. 1984. Aerospatiale regroups for the '90s. *Rotor Wing,* July:24–30.

42. McConnell, V. P. 1995. Aerospace applications: Affordability first. *High Performance Compos.* 3(2)27–33.

43. Kandebo, S. W. 1993. Boeing Sikorsky findings underscore RAH-66 stealth. *Aviat Week Space Technol.* July 19:22–3.

44. Colucci, F. 1993. Composite Comanche. *Def. Helicopter* August/September:28–30.

45. Jouin, P. H., J. C. Heigl, and T. L. Youtsey. 1992. Rapid fabrication of flight worthy composite parts. *Proc. 37th Int. SAMPE Symp.* pp. 878–91, March 1992, Anaheim, CA.

46. Donaldson, P. 1995. Belle of bells. *Helicopter World.* April:14–6.

47. Griffiths, G. R., W. D. Hillier, and J. A. S. Whiting. Thermoplastic composite manufacturing technology for a flight standard tailplane, *SAMPE J.* 25(3):29–33.

48. Brahney, J. H. 1989. Composites: Helicopters leading the way. *Aerosp. Eng.* May:19–26.

49. Dexter, H. B. 1987. Long-term environmental effects and flight service evaluation of composite materials. NASA TM-89067.

50. Baker, D. J. 1990. Evaluation of composite components on Bell 206L and Sikorsky S-76 helicopters. AVSCOM Technical Report 90-B-004. NASA TM 4195.

51. Composite materials and structures for rotorcraft. *Proc. 2nd Int. Workshop,* September 1989, Troy, NY. Compendium of Abstracts and Viewgraphs. Army Reserve Office, ARO 26588-1, AHS and RPI, AD-A217 189.

52. Zinberg, H. 1984. Flight service evaluation of composite components on the Bell helicopter model 206L—First annual flight service report. NASA CR-172296.

53. Mardoian, G. H., and M. B. Ezzo. 1986. Flight service evaluation of composite components—Third annual report. NASA CR-178149.

54. Masaki, S. 1988. *J. Japn. Soc. Compos. Mater.* 14:2–8.

55. Japan Society Aeronautical and Space Society, ed. 1992. *Koku-Uchu-Kogaku Binran,* 2nd ed., 547.

56. Steinberg, M. A. 1986. Materials for aerospace. *Sci. Am.* 225(4):67–72.

57. Highton, D. R., and W. J. Crispin. 1989. Future advanced aero-engines—The materials challenge. In *Application of Advanced Material for Turbomachinery and Rocket Propulsion,* pp. 4-1 to 4-4. NATO Advisory Group for Aerospace R&D. AGARD CP 449.

58. Marsh, G. 1993. Engine drivers. Part 2. *Aero. Mater.* 5(3):9–11.

59. T 55 runs with composite housing. *Aero. Mater.* 5(2):3, 1993.

60. Marsh, G. 1993. Engine drivers. *Aero. Mater.* 5(2):6–9.

61. Koop, W. E., and C. J. Cross. 1990. Metal matrix composites structural design experience. *Proc. 26th Jet Propul. Conf., AIAA/SAE/ASME/ASEE,* July 1990, Orlando, FL.

62. Johnson, S. R., and R. Ravenhall. 1987. Metal matrix composite shaft research program. AFWAL-TR-87-2007.

63. Salemme, C., and R. Ravenhall. 1982. Advanced reinforced titanium blade development. AFWAL-TR-82-4010.

64. Zordan, R., R. Fannin, and G. Doble. 1993. Metal matrix composite (MMC) reinforced compressor rotor. Res & Dev. October:57.

65. Kandebo, S. 1992. Allison tests variable cycle fighter/attack core engine. *Aviat. Week Space Technol.* February 24:130–1.

66. Francis, M. S. 1995. X-31: An international success story. *Aerosp. Am.* February:22–31.

67. Boggs, R. N. 1993. Nozzle combines vectored and reverse thrust. *Des. News.* April 19:99–100.

68. *Adv. Compos.* March/April:16–7, 1992.

69. DeLuca, J. J., and S. P. Petrie. 1990. Evaluation of lightweight material concepts for aircraft turbine engine rotor protection. *Proc. 22nd Int. SAMPE Tech. Conf.,* pp. 319–33, November 1990, Boston, MA.

70. Pepin, J. N. 1993. Fiber reinforced structures for turbine engine fragment containment. *Proc. 29th Jet Propul. Conf., AIAA/SAE/ASME/ASEE,* June 1993. AIAA-93-1816, A-9349704.

71. Clyne, T. W., and P. J. Withers. 1993. *An Introduction to Metal Matrix Composites.* Cambridge University Press, Cambridge.

72. Driver, D. 1989. Towards 2000—The composite engine. *Meet. Aust. Aeronaut. Soc.,* Sydney, Australia.

73. Driver, D. 1989. Metal matrix composites and powder processing for aero-engine applications. *BNF Int. Conf. Mater. Revol. through 90's,* July 1989, Oxford, U.K.

74. NAL develops hot composite. *Aviat. Week Space Technol.* April 12, 1993:17.

75. Clarke, D. A. 1990. Industrial applications and markets for ceramic matrix composites. Rolls-Royce, Ltd., Cambridge, Great Britain. PNR 90753.

76. Kandebo, S. 1994. New materials may boost engine performance. *Aviat. Week Space Technol.* September 5:63–7.

77. Trego, L. 1995. Fiber placement for composite engine cowls. *Aerosp. Eng.* January/February:10.

78. Curran, M. J. 1990. The application of advanced composite materials to nacelle structures. *Aerosp. Technol. Conf.,* October 1990, Long Beach, CA. SAE 901980.

79. Hughes, D. 1994. TRP award boosts nacelle composites. *Aviat. Week Space Technol.* March 28:61–3.

80. Titanium matrix composite developed for gas turbines. *Advanced Materials & Processes (AM&P)* October 1995:7–8.

81. Kume, Y. 1994. Metal matrix composites in Japan: Commercial applications and current research. SRI International Business Intelligence Program, D94–1821.

82. Koczak, M., S. C. Khatri, J. E. Allison, et al. Metal-matrix composites for ground vehicle, aerospace, and industrial applications. In *Fundamentals of Metal-Matrix Composites,* ed. S. Suresh, A. Mortensen, and A. Needleman, 297–324. Butterworth-Heinemann, Newton, MA.

83. Peterson, C. 1994. Composite manifold lightens the load. *Plast. Des. Forum* January/February:22–3.

84. Pater, R. H. 1989. *New Developments in Polymer Composites for Automotive Applications: Collection of Technical Papers Presented at 1989 Symp.,* Detroit, MI. SAE, Warrendale, PA, 1989. SAE SP 784.

85. McConnell, V. P. 1995. Electric Avenue. *High-Performance Compos.* July/August:16–21.

86. Kurihara, Y. 1995. Polymer matrix composite materials in automobile industries. *Adv. Compos. Mater.* 4(3):209–87.

87. Cole, G. S., and A. M. Sherman. 1995. Lighweight materials for automotive applications. *Mater. Charact.* 35(1):3–9.

88. Mroz, C. 1993. Silicon carbide-whisker preforms for metal-matrix composites by infiltration processing. *Am. Ceram. Soc. Bull.* 72(9):89–93.

89. Rohatgi, P. 1991. Cast aluminum-matrix composite for automotive applications. *J. Met.* April:10–6.

90. Grimm, W. 1992. Advances in materials technology. *Monitor,* Issue No. 26. U.N. Industrial Development Organization, Reinforced Plastics, Salzburg, Austria.

91. Bowman, T. J., B. S. Cumming, and P. R. Gilmore. 1992. Fibre-reinforced composite engine. Ford Motor Company Ltd., Essex, U.K. Contarct R11 B-0115-D(B), EUR 13249 EN.

92. Kim, Y. G., D. G. Lee, and P. K. Oh. 1995. Manufacturing of composite screw rotors by resin transfer molding. *J. Mater. Proc. Technol.* 48(1–4):641–47.

93. Hypercar materials *SAMPE J.* 31(4):18, 1995.

94. Hypercars: Materials and policy implications. RMI, 1994, Snowmass, CO.

95. Kretschmer, J. 1988. Composites in automotive applications—State of the art and prospects. *Mater. Sci. Technol.* 4:757–67.

96. Price, A., and R. J. Moulds. 1992. Bonded propshafts: Design and assembly of composite tubes and metal end fittings. *Adhes. '92 Conf.,* September 1992, St. Louis, MO. SME-AD-92-233.

97. Fiber-reinforced truck spring. *Automot. Eng.* 102(5):60, 1994.

98. Chamberlain, G. 1995. Metals vs plastics: And the winner is. *Des. News,* October 9:122–8.

99. Grande, J. A. 1995. Automotive composites target processing hurdles. *Mod. Plast.* October 53–7.

100. Chamberlain, G. 1995. Carmakers ponder future of materials, *Res. Dev.* November:45–8.

101. Auto frames from recycled pop bottles. *Adv. Mater. Process.* 145(5):4.

102. Stevens, T. 1992. Hit the road. *Mater. Eng.* May:9.

103. Next NGV will go farther, cost less. *Mach. Des.* November 23:27, 1995.

104. Cassidy, V. M. 1989. Composite metal auto parts only five years away. *Mod. Met.* December: 62–6.

105. Taggart, H. 1995. Growth is in automotive SMC. *Compos. Des. Appl.* Fall:20–2.

106. Krolewski, S., and J. Busch. 1990. The competitive position of selected composite fabrication technologies for automotive applications. *Proc. 35th Int. SAMPE Symp.,* pp. 117–27, April 1990, Anaheim, CA.

107. Chavka, N. G., and C. F. Johnson. 1991. Critical preforming issues for large automotive structures. *Proc. 7th Adv. Compos. Conf., ASM/ESD,* pp. 413–22, Detroit, MI.

108. Lindsay, K. F. 1995. All-composite boxcar rejuvenates rail fleet. *Compos. Des. Appl.* Fall:16–7.

109. Miller, B. 1991. Pultruded shipping boxes keep cars factory-fresh. *Plast. World,* April:24.

110. Suzuki, Y., and K. Satoh. 1995. High speed trains and composite material, *SAMPE J.* 31(5): 18–21.

111. Burg, M., and S. Loud. 1992. Prospects for advanced composites in the high-speed rail industry. *Proc. 37th Int. SAMPE Symp.,* pp. 41–49, March 1992, Anaheim, CA.

112. McGuire, P. F. 1992. Aluminum composites come in for a landing. *Mach. Des.* April 23:71–4.

113. Boyer, C. T., I. G. Talmy, J. W. Powers, et al. 1993. Development and evaluation of erosion-resistant polymer matrix composite ablators. *Proc. 31st Aerosp. Sci. Meet.,* January 1993, Reno, NV. AIAA 93-0842, A-9324909.

114. Eaton, D. C. G., and E. Slachmuylders. 1993. Potentials and problems when considering the use of advanced structural materials for space applications. *Space Technol.* 13(6):545–52.

115. Thompson, T. 1995. Lego-like composites could cut aircraft costs dramatically. *Des. News* September 11:13.

116. Scott, W. B. 1995. Composite satellite to detect covert nuclear tests. *Aviat. Week Space Technol.* November 27:48–9.

117. McKenna, J. T. 1994. Shuttle flight to find materials, life studies. *Aviat. Week Space Technol.* July 4:28–9.

118. Machlis, S. 1995. "What-if" analysis moves to the desktop. *Des. News* January 23:58–63.

119. Robinson, M. J., R. O. Charette, and B. G. Leonard. 1990. Advanced composite structures for launch vehicles. *Proc. 22nd Int. SAMPE Tech. Conf.,* pp. 957–71, November 1990, Boston, MA.

120. Darais, M. A. 1990. Effect of delaminations on 63-inch Kevlar/epoxy composite rocket motor case. AIAA 90-1976.

121. Asker, J. R. 1994. Lockheed details RLV plans. *Aviat. Week Space Technol.* December 5:20–1.

122. *Adv. Mater. Process.* October:16, 1995.

123. Sater, J. M., and M. Rigdon. 1993. Graphite-reinforced polycyanate composites for space and missile applications. Institute for Defense Analyses, Alexandria, VA. IDA-D-1454, IDA/HQ-93-44719.

124. Blair, C., and G. A. Jensen. 1992. Process development and characterization of ultra high modulus, drapable graphite/thermoplastic composites for space applications. *Proc. 37th Int. SAMPE Symp.,* pp. 115–27, March 1992, Anaheim, CA.

125. Becker, W., T. Haug, J. Heitzer, et al. 1992. Design concepts for new composite materials. Dornier, Deutsche Aerospace A.G. Friedrichshafen, Germany.

126. Hercules test fires graphite epoxy motor. *Aviat. Week Space Technol.* May 2, 1994:15.

127. U.K. Atomic Energy Authority (AEA). 1995. New materials to benefit space program. *Ceram. Ind.* October:24–6.

128. *Aviat. Week Space Technol.* June 27, 1994:23.

129. Kourtides, D. A., H. K. Tran, and S. A. Chiu. 1992. Composite flexible insulation for thermal protection of space vehicles. *Proc. 37th Int. SAMPE Symp.,* pp. 147–58, March 1992, Anaheim, CA.

130. McConnell, V. P. 1991. *Adv. Compos.* July/August 1991; *HTMIAC Newsl.* 3(1):2, 1991.

131. Robinson, M. J., R. O. Charette, and B. G. Leonard. 1991. Advanced composite structures for launch vehicles. *SAMPE Q.* 22(2):26–37.

132. Hercules Inc. supplies critical structure to house corrective optics for the Hubble space telescope. *Adv. Mater. Process.* 145(2):48–9, 1994.

133. McConnell, V. P. 1995. Aerospace applications: Affordability first. *High-Performance Compos.* 3(2):27–33.

134. Smith, L. E., Jr., and M. J. McCowin Smith. 1992. Shipbuilding and ship design perspectives on applications of composite materials: Capability drivers and technical issues. *Proc. 37th Int. SAMPE Symp.,* ed. G. C. Grimes, R. Turpin, and G. Forsberg, pp. 63–77, Anaheim, CA.

135. Greene, E., et al. 1990. Use of fiberglass reinforced plastics in the marine industry, Ship Structure Committee, U.K.

136. Ciba-Geigy composites. *SAMPE J.* 27(4):98, 1991.

137. Smith, C. S. 1990. *Design of Marine Structures in Composite Materials.* Elsevier, Essex, U.K.

138. *Engineered Materials Handbook,* vol. I: *Composites,* 27–39, 801–9, 837–44. ASM International, Materials Park, OH, 1987.

139. Staff Report. 1992. Alternate materials reduce weight in automobiles. *Adv. Mater. Process.* August:16–20.

140. Sedriks, A. J. 1994. Advanced materials in marine environment. *Mater. Performance* 33(2): 56–63.

141. Seymour, R. J., and F. E. Sloan. 1991. Pultrusion and thermoforming of a graphte-reinforced-thermoplastic marine cable. *Proc. 36th Int. SAMPE Symp.* pp. 105–15, April 1991, San Diego, CA.

142. Toensmeier, P. A. 1994. Composite industry eyes civil engineering as next big market. *Mod. Plast.* April:17–89.

143. Taggart, H. 1995. Infrastructure applications key to continued growth. *Compos. Des. Appl.* Winter:20–1.

144. Barnaby, D. 1994. Rehabilitating Boston's Callahan Tunnell. *CI Compos.* December/January:2–3.

145. Grande, J. A. 1995. Advanced composites sector heads for diversification. *Mod. Plast.* 72(7):36–43.

146. Lighweight glass fiber reinforced concrete panel. *JETRO* August 1995:29. 95-08-006-02.

147. Grace, N. F., D. K. Bagchi, and J. B. Kennedy. 1991. Fiber-reinforced composite versus steel and concrete for vibration control of industrial building. *Proc. 7th Adv. Compos. Conf., ASM/ESD,* pp. 597–603, Detroit, MI.

148. Wallenberger, F. T. 1995. Affordability of glass-ceramic fiber reinforced composites in civil engineering, automotive, transportation and metal matrix uses. *J. Adv. Mater.* 26(3):42–8.

149. Ma, G. C. L. 1994. Composite wraps stabilize earthquake-prone bridges. *Adv. Mater. Process.* April:9.

150. McConnell, V. P. 1994. Advanced composites—Practical uses lay foundation for the future. Plastics Design Forum, April:17–23.

151. Karbhari, V. M. 1993. Development of composite materials and technology for use in bridge structures—A preliminary assessment. University of Delaware, Center for Composite Materials, Newark. CCM 93-05.

152. Grande. J. A. 1995. Infrastructure repair: New composites market. *Mod. Plast.* March:25–9.

153. Arockiasamy, M., R. Sowrirajan, and R. Qu. 1995. Experimental studies on the feasibility of use of carbon fiber reinforced plastics in repair of concrete bridges. FL/DOT/RMC-641-7748; *NTIS Alert* 95(18):17–8.

154. Lindsay, K. 1995. FRP rebar: The next generation, *Compos. Des. Appl.* Winter:7.

155. Xie, M., S. V. Hoa, and X. R. Xiao. 1995. Bonding steel reinforced concrete with composites. *J. Reinf. Plast. Compos.* 14(9):949–64.

156. GangaRao, H. V. S., S. S. Faza, S. V. Kumar, et al. 1995. Experimental behavior of concrete bridge decks reinforced with reinforced plastic (RP) rebars. *J. Reinf. Plast. Compos.* 14(9): 910–22.

157. Rogers, J. K. 1993. Composites sector gets serious about conversion. *Mod. Plast.* June:72–3.

158. Advanced composite bridge spans English canal. *Des. News.* March 6, 1995:15.

159. Taggart, H. 1994. Composites in the mist. *CI Composites,* October/November:2–3.

160. McConnell, V. P., and K. Fisher. 1995. Advancing composites, *High-Performance Compos.* 1996 Sourcebook, 10–24.

161. White, L. 1986. Putting the best foot forward. *Adv. Compos.* 1(3):45–7.

162. Fisher, K. J. 1993. Low-cost thermoplastic composites: On the verge. Part I. *Adv. Compos.* 8(1):30–7.

163. Davidson, J. A., and F. S. Georgette. 1986. State of the art materials for orthopedic prosthetic devices. *SME Implant Manuf. Mater. Technol. Conf.,* ed. T. Drozda, pp. 397–422, Dearborn, MI.

164. Composite HIP promises better fit. *Adv. Compos.* May/June 1989; *Mach. Des.* January 11, 1990:46.

165. Kwarteng, K. B., and C. Stark. 1990. Carbon fiber reinforced PEEK (APC-2/AS-4) composites for orthopaedic implants. *SAMPE Q.* 22(1):10–4.

166. Spector, M., E. J. Cheal, R. D. Jamison, et al. 1990. Composite materials for hip replacement prostheses. *Proc. 22nd Int. SAMPE Symp.,* pp. 1119–27, November 1990, Boston, MA.

167. Merrill, E. W. 1990. Biomaterials in the bionic man of the 21st century. *Proc. 22nd Int. SAMPE Symp.,* pp. 1106–17, November 1990.

168. White, M. 1993. Composite leg brace wins 1993 Dupont/ASM award. *Adv. Mater. Process.* September 1993:47–8.

169. Manji, J. F. 1994. Sports safety spurs innovation by design. *Plast. Des. Forum* January/February:26–30.

170. Spurling, N. 1993. National conference to showcase innovative composite applications. *Adv. Compos.* September/October:34–5.

171. Dixon, W. 1991. Processing, properties and applications of Duralcan composite tubes. *ITA Sem.*

172. Tarrant, A. 1991. Particle reinforced aluminium alloys: Maximizing the benefits—Minimizing the risks. Metal Matrix Composites—Exploiting the Investment, London, Institute of Metals.

173. Beercheck, R. C. 1989. Composites at play. *Mach. Des.* June 8:80–6.

174. Machine for molding advanced composite materials. *JETRO* 20(2), 1992. 92-05-001-02.

175. *Plast. Des. Forum* May/June 1992:28–32.

176. Howell, D. D. 1992. The design of filament wound graphite/epoxy golf shafts. *Proc. 37th Int. SAMPE Symp.,* ed. G. C. Grimes, R. Turpin, and G. Forsberg, pp. 1392–405, Anaheim, CA.

177. Braham, J. 1995. Boom, boom, boom! And Bertha's even bigger! *Mach. Des.* September 14:36–40.

178. A whole new way to make sailboats. *Plast. Technol.* November 1995:92.

179. Baker, A. 1994. Sports steals some winning materials. *Des. News* April 25:21–2.

180. Braham, J. 1995. Revolutionary gyroplane to blend best of copters and planes. *Mach. Des.* December 14:20–2.

181. Toray Industries, Inc. Public Relations Section. 1994. Manufacture of long rollers made of carbon fiber reinforced metal. *JETRO* 22(8):28–8. New Technology JAPAN. 94-11-005-03.

182. Freeland, R. E., and R. D. Johnston. 1993. Development of structural composite mirror technology for submillimeter space telescopes. *Acta Astronaut.* 29(7):537–45.

183. *Advanced Ceramic Materials.* Charles River Associates, Noyes, 1985, Cambridge, MA.

184. Blinn, L. 1990. *New Materials Society: Challenges and Opportunities,* New Materials Science and Technology. Bureau of Mines, Washington, DC.

185. Schwartz, M. M. 1992. *Handbook of Structural Ceramics.* McGraw-Hill, New York.

186. Shikibo Ltd. develops world's highest strength C/C composite bolt. May 1993:30–1. JPRS-JST-93-031L.

187. Nuck, F., and J. Plout. 1993. Composite cylinder tubes cut weight, costs. *Mach. Des.* December 12:44–6.

188. Lasday, S. B. 1993. Ceramic/ceramic composite components advance furnace systems and processes. *Ind. Heat.* April 11:26–9.

189. Bloom, I., and M. C. Hash. 1990. High-MgO composite ceramics for use in sodium-based cell. *Mater. Lett.* 10(3):105–8.

BIBLIOGRAPHY

Advanced composites for aerospace applications. Nerac, CT, 1995; *NTIS Alert* 13(95):5, 1995.

ALLBEE, N. Tapping the market for industrial applications. *Adv. Compos.* July/August:49–52, 1989.

ASHLEY, S. High-tech rackets hold court. *Mech. Eng.* 115(8):50–5, 1993.

ASHLEY, S. Metal-matrix walls. *Mech. Eng.* 116(2):24, 1994.

BAAKLINI, G. Y., L. D. PERCIVAL, AND R. N. YANCEY. NDE of titanium alloy MMC rings for gas turbine engines. NASA Lewis Research Center, Cleveland, OH, 1993. N94-13138.

BALLINGER, C. A. Advanced composites in the construction industry. *Proc. 37th Int. SAMPE Symp.*, pp. 1–14, March 1992, Anaheim, CA.

BEARDMORE, P. Ford tests structural body composites. *Des. News* November 20:28, 1989.

BERRISFORD, R. S. Advanced composites in compressed gas applications. *Proc. 37th Int. SAMPE Symp.*, pp. 34–40, March 1992, Anaheim, CA.

BORSTELL, H. Manufacturing technology for complex composite fuselage shapes, Phase I. AFWAL-TR-88-4085, 1983.

BRADLEY, R. J., AND R. R. FENBERT. Manufacturing technology for large aircraft composite primary structure (fuselage). Boeing Co. Interim Technical Report IR 489-3(I). Contract F33615-83-C-5024.

BRAHAM, J. Designed to be different. *Mach. Des.* July 11:54, 1994.

BRAUN, W. R., J. C. GARNER, AND S. M. TURNER, ET AL. Graphite/epoxy heat sink/mounting for common pressure vessels. Department of the Navy, Washington, DC. AD-D017 491. *NTIS Alert* 95(22).

BROWN, A. S. Auto composites may lighten aircraft's load. *Aerosp. Am.* March:22–3, 1995.

BURD, B. A., I. HOWIE, J. M. LAMBERT, ET AL. A stealth composite frame for generators. *Mech. Eng.* 115(3):74–6, 1993.

CAHN, R. W., AND D. F. WILLIAMS. *Materials Science and Technology: Comprehensive Treatment—Medical and Dental Materials,* Material Science and Technology, vol. 14. VCH, Weinheim, Germany, 1992.

CARLSON, C. W. Polymer composites: Adjusting to the commercial marketplace. *J. Met.* 45(8): 56–7, 1993.

Ceramic matrix composites: Technology and industrial applications. Tech Trends, Innovation 128, Paris, France, 1990.

CHAMBERLAIN, G. Reinforcing rod solves steel rebar corrosion problems. *Des. News* November 6:14, 1995.

CHANG, I. Y. Thermoplastic matrix composites development update. *Proc. 37th Int. SAMPE Symp.*, ed. G. C. Grimes, R. Turpin, and G. Forsberg, pp. 1276–90, Anaheim, CA, March 1992.

CHARLES, D. Unlocking the potential of metal matrix composites for civil aircraft. *Mater. Sci. Eng.* 135:295–7, 1991.

CHEON, S. S., J. H. CHOI, AND D. G. LEE. Development of the composite bumper beam for passenger cars. *Compos. Struct.* 32(1–4):491–99, 1995.

CHING, F. Composite flight control actuators. *Aerosp. Eng.* September:17–9, 1990.

CLARKE, D. A. Fabrication aspects of glass matrix composites for gas turbine applications. Rolls Royce reprint PNR 90752, Cambridge, U.K.

Composite materials used in shipbuilding. Nerac, Tolland, CT, 1994; *NTIS Alert* 95(2):14.

Composite parts roll into action. *Mach. Des.* November 23:14–5, 1989.

Composites in transportation. *Gorham/Intech. Conf.*, Atlanta, GA, July 1994.

Composites '91—Best of new composite design. *Plast. World* April:54–5, 1991.

Composite wheel gets a whirl. *Des. News,* October:36–9, 1989.

Concept car helps ease composites into production. *Mach. Des.* November 23:14, 1989.

DELLACORTE, C., H. E. SLIVEY, AND M. S. BOGDANSKI. High-temperature, self-lubricating ceramic/ metal composites. NASA Technical Brief, January 1994. LEW-15678.

DELUCA, P. L. Composites in the defense arena. *SAMPE J.* 31(3):76–82, 1995.

DE LUCCIA, J. J., R. E. TRABOCCO, AND J. WALDMAN. Materials pace aerospace technology. *Adv. Mater. Process.* May:39–50, 1989.

Development of stitched/RTM primary structures for transport aircraft. McDonnell Douglas Aerospace, Long Beach, CA, 1993. NAS 1.26:191441, MDC-93K0265, NASA-CR-191441; *NTIS Alert* 95(22):24, 1995.

DICK, W. E. Advanced composites in the petroleum and natural gas fuel container industries. *Proc. 37th Int. SAMPE Symp.,* pp. 29–33, March 1992, Anaheim, CA.

DUDGEON, C. D. Overview of unsaturated polyester resins—Growth opportunities in automotive and marine applications. *Proc. 34th Int. SAMPE Symp.,* pp. 129–41, May 1989, Reno, NV.

EPSTEIN, G., AND S. RUTH. Composites: A key to space systems, *SAMPE J.* 32(1):24–9, 1996.

FEHRENBACHER, L. L., AND J. A. HANIGOFSKY. Ceramic composite combustor cans for expendable turbine engines, 1994. WL-TR-95-2001.

FITE, E. B. Fabrication of composite propfan blades for a Cruise missile wind tunnel model. NASA TM 105270, 1993.

GIRARDI, M. Not sticking to tradition. *Connect* January:6, 1994.

GLATZ, J., AND D. L. VRABLE. Applications of advanced composites for satellite packaging for improved electronic component thermal management. *Acta Astronaut.* 29(7):527–35, 1993.

GRAHAM, D. Composite pressure hulls for deep ocean submersibles. *Compos. Struct.* 32(1–4): 331–43, 1995.

Green machine. NASA Technical Brief, October 1995, p. 21.

GRUDKOWSKI, T. W., A. J. DENNIS, T. G. MEYER, ET AL. Flywheels for energy storage. *SAMPE J.* 32(1):65–9, 1996.

HOWARTH, G. F., G. GLINECKI, AND B. TAMBUSSI. High-temperature composite ducts. *SME Fabr. Conf.,* Hartford, CT, June 1985. MF85-501.

IMMARIGEON, J. -P., R. T. HOLT, A. K. KOUL, ET AL. Lightweight materials for aircraft applications. *Mater. Charact.* 35(1):41–67, 1995.

INNACE, J. Lower-cost materials, processing reposition composites sector. *Mod. Plast.* June:55–60, 1994.

IRVING, R. R. Advanced materials seen playing vital solar car role. *Metalworking News* July 23:10, 1990.

JANICKI, J., L. CLEMENTS, H. SCHJELDERUP, ET AL., eds. *Proc. 38th Int. SAMPE Symp. Exhibit.* Anaheim, CA, May 1993.

KAEMPEN, C. E. Building and transportation systems that provide a new growth market for structural composites. *Proc. 37th Int. SAMPE Symp.,* pp. 15–28, March 1992, Anaheim, CA.

KANDEBO, S. W. HSCT propulsion team mulls flight test issue. *Aviat. Week Space Technol.* August 31:64–7, 1992.

KASTEN, T., AND W. WEST. High technology comes down to earth. NASA Technical Brief, June 1993, pp. 16–20.

KATZMAN, H. A., J. J. MALLON, AND W. T. BARRY. Polyarylacetylene-matrix composites for solid rocket motor components. *J. Adv. Mater.* 26(3):21–7, 1995.

KING, R. H. Honda tries an Indy comeback. *Des. News* May 22:58–63, 1995.

LARRODE', E., A. MIRAVETE, AND F. J. FERNANDEZ. A new concept of a bus structure made of composite materials by using continuous transversal frames. *Compos. Struct.* 32(1–4):345–56, 1995.

LOUD, S. Three steps toward a composites revolution in construction, *SAMPE J.* 32(1):30–5, 1996.

LYNCH, T. Indy racing: An engineering crucible. *Des. News* May 23:42–6, 1994.

MALARIK, D. C., AND R. D. VANUCCI. High molecular weight first generation PMR polyimides for 343°C applications, 1991. NASA-TM-105364, E-6742.

Materials Edge, Metal matrix composites in automotive industry, U.K., 1993.

MCCONNELL, V. P. Update on thermoplastics. *High-Performance Compos.* September/October: 42–8, 1994.

MILLER, B. Composites are off to the races. *Plast. World* April:42–52, 1991.

Next Tulip from France will be electric two-seater. *Mach. Des.* September 28:20–2, 1995.

NOWAK, B., AND B. SCOULAR. Switching to composites: New thinking required. *Mater. Eng.* February:33–6, 1989.

OGASA, T., J. TAKAHASHI, AND K. KEMMOCHI. Polymer-based composite materials in general industrial fields. *Adv. Compos. Mater.* 4(3):221–35, 1995.

PARA, P. R., AND D. E. WEERTH. FRP military combat vehicle hull and turret design. *Proc. 4th Adv. Compos. Conf., ASM/ESD,* pp. 119–26, Dearborn, MI.

PASKIN, M. D. Composite matrix experimental combustor. NASA CR-194446, ARL-TR-334, 1994.

PATER, R. H. *New Developments in Polymer Composites for Automotive Applications: Collection of Technical Papers Presented at 1989 SAE* Int. Cong. Mtg. Detroit, MI. SAE, Warrendale, PA. SP 784.

PETERSEN, C. Growing pains. *Adv. Compos.* March/April:20–1, 1994.

PRICE, A., AND R. J. MOULDS. Bonded propshafts: Design and assembly of composite tubes and metal end fittings. *Mater. Sci. Technol.* 9(6):528–35, 1993.

Proc. 4th Ann. HITEMP Rev. Cleveland, OH, October 1991. NASA CP 10082.

RAZELLI, G. B. Design and engineering materials. *Proc. 2nd Int. Conf. Eng. Mater.,* Bologna, Italy, June 1988; *Mater. Eng.* 1(1):3–22.

ROCKWAY, J. W., J. C. LOGAN, J. H. SCHUKANTZ, ET AL. Intermodulation interference (IMI) testing of composite materials. Naval Command, Control and Ocean Surveillance Center, San Diego, CA, 1994, NRAD-TD-2705. *NTIS Alert* 95(22).

SCHINDLER, R. Polymer composites for helicopters: Applications and experience. Deutsche Aerospace AG (DASA), Munich, Germany, 1994; Whertechnisches Symp. bei der BAKWVT, Erding, March 1994.

SCHOFIELD, J. A. Technology tacks toward the cup. *Des. News,* April 10:54–62, 1995.

SEN, R., AND L. LIBY. Repair of steel composite bridge sections using CFRP laminates. CEM/ST/94/2, August 1994; REPT-0510616, University of South Florida, Tampa, FL; *NTIS Alert* 95(19): 27–8.

Solar-car rollout preview to GM Sunrayce USA. *Adv. Comp.* July/August:11–21, 1990.

Structural adhesives simplify manufacture of composite sailboats. *Adhes. Age* December:31, 1993.

Structural composites: A reality. *Plast. Technol.* September:109–17, 1989.

TAYLOR, A. H., AND P. O. RANSONE. Advanced composite pistons, NASA Technical Brief, September 1990, p. 90.

TEAGUE, P. E., D. COLUCCI, AND W. WINGO. A glimpse at the supercar. *Des. News* October 10:110–18, 1994.

THOMAS, G. E., AND D. E. WEERTH. FRP material considerations for military combat vehicle hull and turret applications. *Proc. 4th Adv. Compos. Conf., ASM/ESD,* pp. 127–33, September 1988, Dearborn, MI.

THOMPSON, B. S. Biomimetic materials: Was Leonardo mistaken? Part 1. *SAMPE J.* 32(1):38–43, 1996.

WILDER, R. V. Resin transfer molding finally gets some real attention from industry. *Mod. Plast.* July:44–50, 1989.

WITZLER, S. Helicopter programs: Composites for today and tomorrow. *Adv. Compos.* November/December:52–8, 1989.

Index